MONOGRAPHS ON NUMERICAL ANALYSIS

GENERAL EDITORS: L. Fox and J. Walsh

NUMERICAL FUNCTIONAL ANALYSIS

COLIN W. CRYER
University of Wisconsin-Madison

Clarendon Press · Oxford
Oxford University Press · New York
1982

Oxford University Press, Walton Street, Oxford OX2 6DP
London Glasgow New York Toronto
Delhi Bombay Calcutta Madras Karachi
Kuala Lumpur Singapore Hong Kong Tokyo
Nairobi Dar es Salaam Cape Town
Melbourne Wellington

and associate companies in
Beirut Berlin Ibadan Mexico City

Published in the United States of America
by Oxford University Press, New York

LIBRARY OF CONGRESS CATALOGING IN PUBLICATION DATA

Cryer, Colin W.
Numerical functional analysis.

(Monographs on numerical analysis)
Bibliography: p.
Includes index.
1. Numerical analysis. 2. Functional analysis.
I. Title. II. Series.
QA297.C79 1982 515.7 82-14077
ISBN 0-19-853410-8

BRITISH LIBRARY CATALOGING IN PUBLICATION DATA

Cryer, Colin W.
Numerical functional analysis.—(Monographs on numerical analysis)

1. Numerical analysis 2. Functional analysis
I. Title II. Series
511 QA297

ISBN 0-19-853410-8

Printed in the United States of America

Preface

The present book arose out of a year-long course in numerical functional analysis given from time to time since 1970 at the University of Wisconsin. This volume introduces the basic techniques of functional analysis and applies them to linear problems. It also lays the foundations for volume 2 which will consider boundary value problems for elliptic equations, and nonlinear problems.

Particular features include:

(1) The concepts of functional analysis are developed systematically, with frequent pauses to introduce applications. Detailed indexes are provided for the reader who wishes to use the material in a different order.

(2) Solutions are given for all the problems except the last few, which are very specialized. This should help students who have taken basic courses in analysis but have little experience of providing proofs.

(3) A large number of counterexamples are given, to show that various results are not true if the hypotheses are weakened. This introduces the reader to an important aspect of mathematics which is often neglected — one must not only *prove* conjectures but also *disprove* them.

Parts of the text were used as course material by John Halton, who made several useful suggestions. Ennio Stacchetti read the first eight chapters, solved many of the problems, and made an invaluable contribution.

The camera-ready copy was typed by Marilyn Wolff, who showed not only great skill but also limitless patience.

Throughout the period of writing the author received support from the National Science Foundation (most recently under grant No. MCS77-26732) and the U.S. Army (most recently under grant No. DAAG29-80-C-0041.

Colin Cryer
Cambridge

Contents

MONOGRAPHS ON NUMERICAL ANALYSIS

GENERAL EDITORS: L. Fox and J. Walsh

1
Introduction

In this book we are concerned with the applications of functional analysis to numerical analysis. It is assumed that the reader is familiar with calculus, matrix theory, and basic numerical methods. Some knowledge of Lebesgue integration would be helpful, but the essential facts are summarized in Chapter 4. No knowledge of functional analysis is presumed, and the various aspects of functional analysis are developed as required. However, the emphasis is on the applications, and we sometimes state and explain, but do not prove, theorems with lengthy or uninteresting proofs in order to leave time to show how the theorems can be applied.

Functional analysis is a subject which has developed during the twentieth century. There were at first a number of scattered, but important, papers, and the subject perhaps first gained the stature of a discipline with the publication in 1932 of the monograph by S. Banach, *Théorie des opérations linéaires*. There is no generally accepted definition of functional analysis. From our point of view we may define functional analysis as infinite-dimensional analysis: that is, functional analysis extends, so far as possible, the concepts of matrix theory and calculus for a finite number of dimensions to an infinite number of dimensions. For example, consider the *Fredholm integral equation of the second kind*,

$$x(s) + \int_0^1 k(s,t)x(t)dt = f(s), \quad 0 \le s \le 1 . \tag{1.1}$$

Formally, this equation is rather similar to the matrix equation

$$x + Kx = f , \tag{1.2}$$

where x and f are n-vectors, and K is an $n\times n$-matrix. Indeed, many numerical methods of approximating eqn (1.1) will lead to an equation of the form (1.2). The solution x of eqn (1.2) involves only the finite number of components x_i of x, but the solution $x(s)$ of eqn (1.1) involves the infinitely many values $x(s)$, $0 \le s \le 1$. However, much of the theory of eqn (1.2) can be extended to eqn (1.1). Indeed, beginning with the beautiful paper of Fredholm (1900), the analysis of integral equations has been the source of many ideas in functional analysis.

The first volume of this book covers the basic theory of linear spaces. The second volume will begin by considering weak topologies, which serve as an introduction to Sobolev spaces, and then turn to the theory of nonlinear problems. While the subjects covered are standard, the emphasis given to the various topics is different to that in most texts on functional analysis for two reasons:

(1) We judge a particular topic by its usefulness rather than its depth or beauty. For example, the principle of uniform boundedness is discussed at great length in Chapter 5 because of its many applications.

(2) A numerical analyst may be thought of as a mathematician with one hand tied behind his back, because while a mathematician can use either constructive or non-constructive methods, a numerical analyst must use constructive methods. Therefore, a constructive approach such as a contraction mapping is much more useful than a non-constructive approach such as the Schauder fixed point theorem because the constructive approach immediately leads to a numerical method.

Numerical functional analysis began with the publication in 1948 in the Russian journal *Uspekhi Matem. Nauk* of a long paper by L. V. Kantorovich entitled 'Functional analysis and applied mathematics'. In the introduction to this paper Kantorovich wrote:

> 'Explicitly, we want to show that the ideas and methods of functional analysis may be used for the construction and analysis of effective practical algorithms for the solution of mathematical problems with just the same success as has attended their use for the theoretical investigation of these problems.'

It took quite a long time before the ideas of Kantorovich became widely known. A helping factor was the publication in 1952 by the U.S. Bureau of Standards of a translation due to C. D. Benster and G. E. Forsythe, but even this translation was a rarity. Another early reference, which drew attention to the value of using functional analysis to analyse numerical methods for matrix problems, was the book of Faddeeva, the first chapter of which was translated into English in 1952.

The goals of numerical functional analysis are:

(1) To simplify and unify by treating whole classes of problems at once.

(2) To bring the power of the general results of functional analysis to bear upon the problems of numerical analysis.

Numerical functional analysis may be expected to play an important rôle in the following situations:

(1) When the mathematics of a problem depends heavily upon functional analysis. For example, functional analysis plays an important rôle in the modern theory of partial differential equations, and it is therefore to be expected (and is indeed the case) that the theory of numerical methods for partial differential equations will also depend heavily upon functional analysis. (See Section 5.7)

(2) When whole classes of numerical methods are being considered, as, for example, when one considers all convergent quadrature formulae. (Section 5.3)

(3) When proving the existence of a numerical method with certain properties (e.g. Example 7.4).

On the other hand, numerical functional analysis may be expected to play a less important rôle when a specific method for a specific problem is being considered, and numerical functional analysis has nothing to say about the practical implementation of a numerical method.

The goals of numerical functional analysis are best illustrated by a famous example, Newton's method. Newton's method for solving the equation in a single real variable

$$f(x) = 0 \,, \tag{1.3}$$

is given by

$$x_{n+1} = x_n - f(x_n)/\dot{f}(x_n) \,. \tag{1.4}$$

If f is twice continuously differentiable then it is possible to show that Newton's method is quadratically convergent, and to establish conditions which ensure its convergence. How can we generalize Newton's method to the case when there are two equations

$$\begin{aligned} f(x,y) &= 0 \,, \\ g(x,y) &= 0 \,, \end{aligned} \tag{1.5}$$

and two unknowns x and y? Well, we note that eqn (1.4) is equivalent to

$$f(x_n) + (x_{n+1}-x_n)\dot{f}(x_n) = 0 \,.$$

That is, we have expanded $f(x)$ in a Taylor series about x_n, and retained only the first two terms. Applying the same argument to eqns (1.5) we obtain

$$\begin{aligned} f(x_n,y_n) + (x_{n+1}-x_n)\frac{\partial f}{\partial x}(x_n,y_n) + (y_{n+1}-y_n)\frac{\partial f}{\partial y}(x_n,y_n) &= 0 \,, \\ g(x_n,y_n) + (x_{n+1}-x_n)\frac{\partial g}{\partial x}(x_n,y_n) + (y_{n+1}-y_n)\frac{\partial g}{\partial y}(x_n,y_n) &= 0 \,, \end{aligned} \tag{1.6}$$

which is Newton's method for eqns (1.5). In his 1948 paper previously mentioned, Kantorovich considered Newton's method from the viewpoint of functional analysis. He established quadratic convergence, as well as conditions which ensure convergence. Kantorovich's proof held not only for eqn (1.6) but also for any finite number of dimensions, and, indeed, for many infinite-dimensional problems. Hence, in this case the methods of numerical functional analysis were triumphantly vindicated.

At present, the methods of functional analysis have so permeated numerical analysis that an understanding of functional analysis is essential for an understanding of modern numerical analysis. However, two warnings should be sounded:

(i) Functional analysis seeks to generalize so that in casting a particular problem into functional analysis form some essential features may be lost. For example, if the kernel $k(s,t)$ in eqn (1.1) is non-negative then certain numerical methods for solving eqn (1.1) may exhibit desirable properties such as monotone convergence, but these properties will be 'invisible' in any functional analysis approach which does not make use of the non-negativity of $k(s,t)$.

(ii) Ideas are not always generated by logical processes. An engineer may have a 'feeling' for a problem which may lead him to a method of solution. A functional analyst may not be led to the method of solution thought of by the engineer because his mind is working along different paths. The reverse is of course also true.

In summary, numerical analysis is not just a branch of functional analysis but, rather, functional analysis is a powerful tool for use in numerical analysis.

A detailed example of the beneficial interaction between physical ideas, functional analysis, and classical numerical analysis, is given by Radon's integral equation which is considered at length in (Section 9.7, p. 332).

2
Topological vector spaces

In this chapter we discuss the concept of a topological vector space which, as the name suggests, is a space possessing a topology (open sets) and permitting linear operations (as for vectors). Functional analysis is concerned with the properties of, and mappings between, topological vector spaces.

2.1 *PRELIMINARY REMARKS*

In elementary real analysis one is concerned with one topological vector space namely the *real line* R^1,

$$R^1 = \{x : -\infty < x < +\infty\} .$$

R^1 is the prototypal topological vector space, and we develop the concept of a topological vector space by generalizing properties of R^1. By successively introducing more and more assumptions, we obtain a hierarchy of spaces, beginning with the lowly topological spaces and ending with the Hilbert spaces. As more assumptions are made about a space, more properties can be deduced but fewer concrete examples exist. If too many assumptions are made there will be only one concrete example, the real line. There is, therefore, a balance to be achieved: one wishes to make enough assumptions to be able to derive interesting properties, but not so many assumptions as to exclude interesting applications.

The examples in this chapter and the following chapter are intended to illustrate basic ideas, and often have no direct connection with numerical analysis. In Chapter 4 we list and discuss most of the spaces which are widely used in numerical analysis. Later chapters give applications.

2.2 *SET THEORETIC NOTATION*

The following notation from set theory will be used (Halmos [1960]). The symbol $\in$ denotes membership in a set so that $x \in A$ means that x is an *element* or *point* of the set or *collection* A , that x *belongs* to A , and that A *contains* x . Sets are often defined by enumeration: the set $A = \{a,b,c\}$ consists of the three elements a, b, and c . Sets whose elements are indexed will often be called *families*; thus we may speak of a family of sets $\{A_i\}$ for $i \in I$, where I is an *index set*. The *empty set*, the set which contains no elements, is denoted by $\emptyset$.

We will often be concerned with a given collection of elements, one or more sets containing these elements, and one or more collections of these sets; as an aid in comprehension we will when possible denote the original elements by lower case Roman letters, the sets containing these elements by upper case Roman letters, and the collections of sets by Greek letters; for example,

$$x \in C \in \tau .$$

If $P(x)$ is a proposition concerning the elements x of a set A then $\{x \in A: P(x)\}$ denotes the set of elements of A for which $P(x)$ is true. For example, the set $\{x \in R^1: x \geq 0\}$ is the set of non-negative real numbers. The statement 'if $x \in \emptyset$ then P' is true for every proposition P .

If A and B are sets and $\{A_i: i \in I\}$ is a family of sets, then the *union* and *intersection* of A and B are denoted by $A \cup B$ and $A \cap B$, respectively, while the union and intersection of the A_i are denoted by

$$\bigcup_{i \in I} A_i \quad \text{or} \quad \bigcup A_i , \quad \text{and} \quad \bigcap_{i \in I} A_i \quad \text{or} \quad \bigcap A_i ,$$

respectively.

If every element of a set B is an element of a set A we say that A *includes* B , or A *contains* B , or B is a *subset* of A , and we write $A \supset B$ or $B \subset A$; in particular, $A \supset A$ for every set A . By convention, the empty set is a subset of every set. If $A \supset B$ and $A \neq B$, B is a *proper subset* of A . If $A \neq B$, A and B are *distinct*.

$A \setminus B$ or $A - B$ denotes the elements of A which do not belong to B ; $A \setminus B$ is the *complement of B with respect to* A . If A is understood from the context, we may speak of the *complement of B* and denote it by B^c .

The set A is *finite* if A contains a finite number of elements. The set A is *denumerable* or *countable* if there exists a one-to-one correspondence between the elements of A and a subset of the positive integers. The real interval $[0,1]$ is not denumerable. (Natanson [1954, p. 13])

The symbol □ is used to denote the end of a logical entity, such as the proof of a theorem.

If P and Q are statements such that whenever P is true then Q is true, then P *implies* Q and we write $P \Rightarrow Q$. If, furtheremore, $Q \Rightarrow P$ then P and Q are *equivalent* and we write $P \Leftrightarrow Q$; another way of saying this is 'P if and only if Q' or, in abbreviated form, 'P *iff* Q'.

The logical operators AND, OR, and NOT will be used.

Logical notation helps to clarify the chain of arguments in the proof of theorems. To prove that $P \Rightarrow R$ it is often convenient to prove first that $P \Rightarrow Q$ and then that $Q \Rightarrow R$. If, furthermore, $R \Rightarrow P$, then $P \Leftrightarrow R$. We occasionally prove that $P \Rightarrow Q$ by proving the equivalent statement

$$\text{NOT } Q \Rightarrow \text{ NOT } P \ .$$

2.3 *TOPOLOGICAL SPACES*

The first property of R^1 that we generalize is the concept of open and closed sets, such as the *open interval in* R^1

$$(a,b) = \{x\colon a < x < b\} \ ,$$

and the *closed interval in* R^1

$$[a,b] = \{x\colon a \le x \le b\} \ .$$

DEFINITION 2.1. A set X is a *topological space* with *topology* τ if τ is a collection of subsets of X satisfying the following three axioms:

(1) The empty set $\emptyset$ and the whole set X belong to τ .

(2) If $G_\alpha \in \tau$ for $\alpha \in A$, where A is an index set, then $\bigcup_{\alpha \in A} G_\alpha \in \tau$.

(3) If $G_1 , G_2 \in \tau$, then $G_1 \cap G_2 \in \tau$. □

The elements of a topology τ are called *open sets*, and axioms (2) and (3) of Definition 2.1 may be stated as follows: the union of arbitrarily many open sets is open; the intersection of finitely many open sets is open.

Strictly speaking, a topological space should be described as a pair $\{X,\tau\}$, but it is usually clear from the context, or by convention, which topology τ is associated with X, and in such cases we speak of the topological space X.

Some examples of topological spaces are:

EXAMPLE 2.1. $X = \{a,b,c\}$, $\tau = \{\emptyset,X,\{a\},\{b,c\}\}$, where a,b, and c are three elements. □

EXAMPLE 2.2. $X = R^1$. $G \in \tau$ iff for every $x \in G$ there exists an interval $(a,b) \subset G$ such that $x \in (a,b)$. This is the usual topology on R^1, and this topology will be used unless explicitly stated otherwise. □

There may be more than one topology on a set. Let two topologies be introduced on a set X by means of τ_1 and τ_2. If $\tau_1 \supset \tau_2$, we say that τ_1 is *stronger* (or *finer*) than τ_2, and that τ_2 is *weaker* (or *coarser*) than τ_1.

EXAMPLE 2.3. $X = \{a,b,c\}$, $\tau_1 = \{\emptyset,X, \{a\}, \{b,c\}\}$,
$\tau_2 = \{\emptyset,X, \{a\}, \{b\}, \{c\}, \{a,b\}, \{a,c\}, \{b,c\}\}$,
$\tau_3 = \{\emptyset,X, \{b\}, \{a,c\}\}$.

The topology τ_2 is stronger than both τ_1 and τ_3. □

Many spaces have two special topologies, the 'weak' topology and the 'strong' topology, which play an important rôle in the theory of such spaces.

Given a set X with two topologies τ_1 and τ_2 it can happen that neither topology is stronger than the other: in Example 2.3, $\tau_1 \not\subset \tau_3$ and $\tau_3 \not\subset \tau_1$, so that neither τ_1 nor τ_3 is stronger.

If $H = X \setminus G$ where G is an open subset of the topological space X, we say that H is *closed*. A set can be both open and closed. In Example 2.1 every element of

τ is both open and closed. There is a connection between closed sets and limits of sequences of elements which is discussed in Chapter 3 (see Theorem 3.2).

In Example 2.2 the topology on R^1 was defined by means of special sets, namely sets of the form (a,b). We now generalize this approach.

DEFINITION 2.2. Let X be a topological space with topology τ. A set $N_x \subset X$ is a *neighbourhood* (in the topology τ) of a point $x \in X$ if there is a set $G \in \tau$ such that $x \in G \subset N_x$. A collection β_x of neighbourhoods of a point x is called a *base of neighbourhoods* (in the topology τ) *of* x if β_x is non-empty and if for every neighbourhood V_x of x there exists $N_x \in \beta_x$ such that $N_x \subset V_x$. A family $\beta = \{\beta_x : x \in X\}$ of bases of neighbourhoods is called a *base of neighbourhoods for the topology* τ. □

REMARK 2.1. The subscript x on N_x in Definition 2.2 is not necessary and may be dropped, but it is often a helpful reminder of the link with x. □

REMARK 2.2. It follows immediately from Definition 2.2 that a non-empty open set is a neighbourhood of each of the points that it contains. □

REMARK 2.3. There are several slightly different definitions of neighbourhoods and bases in the literature. Sometimes a neighbourhood of x is defined to be an open set containing x, and a base β for a topology τ is defined to be a subfamily of τ such that for each x and each neighbourhood U of x there exists $V \in \beta$ satisfying $x \in V \subset U$. □

There may be more than one base for a topology as is shown by:

EXAMPLE 2.4. For R^1 (with the usual topology) one base of neighbourhoods $\beta' = \{\beta'_x\}$ is given by

$$\beta'_x = \{[a,b) : a < x < b\} .$$

That is, the base β'_x of neighbourhoods of x consists of the half-open intervals $[a,b)$ which strictly contain x. Another base of neighbourhoods for R^1 is given

$$\beta''_x = \{(a,b) : a < x < b\} ,$$

and this is the base that is usually used. □

We can now turn the above process around and ask the following question. Given a set X and a family of (families of) subsets $\beta = \{\beta_x : x \in X\}$, is β a base of neighbourhoods for some topology, and, if so, what topology? This question is answered by the following two theorems, the proofs of which are given in detail because they illustrate the concepts discussed so far.

THEOREM 2.1. *Let* $\beta = \{\beta_x : x \in X\}$ *be a base of neighbourhoods for the topology* τ *on a topological space* X . *Let* $G \subset X$. *Then the following are equivalent:*

(1) G *is open.*

(2) *For each* $x \in G$ *there exists* $N_x \in \beta_x$ *such that* $N_x \subset G$.

Proof. *(2)⇒(1)*: Suppose that for each $x \in G$ there exists $N_x \in \beta_x$ such that $N_x \subset G$. If G is not empty then, since N_x is a neighbourhood of x, for every $x \in G$ there is an open G_x such that $x \in G_x \subset N_x \subset G$; thus $G = \cup G_x$ and hence G is open by (2) of Definition 2.1. If G is empty then G is open by (1) of Definition 2.1. *(1)⇒(2)*: Suppose that G is open. If G is not empty let $x \in G$; since G is open, G is a neighbourhood of x and, since $\beta = \{\beta_x\}$ is a base, there exists $N_x \in \beta_x$ such that $N_x \subset G$. If G is empty, the statement 'for each $x \in G$ then P' is true for any proposition P . □

THEOREM 2.2. *For* $\beta = \{\beta_x : x \in X\}$ *to be a base of neighbourhoods for a topology* τ *on* X , *it is necessary that:*

(a) *For each* $x \in X$, β_x *is not empty.*

(b) *If* $N \in \beta_x$ *then* $x \in N$.

(c) *If* $N_1, N_2 \in \beta_x$ *then there exists* $N \in \beta_x$ *such that* $N \subset N_1 \cap N_2$.

(d) *For every* $N_x \in \beta_x$ *there exists* $N'_x \in \beta_x$ *such that*

(d1) $N'_x \subset N_x$,

(d2) *If* $y \in N'_x$ *then there exists* $N_y \in \beta_y$ *satisfying* $N_y \subset N_x$.

Conversely, if $\{\beta_x\}$ *is such that conditions (a) to (d) are satisfied, then* $\beta = \{\beta_x\}$ *is a base of neighbourhoods for the topology* τ *defined by:* $G \in \tau$ *iff for each* $x \in G$ *there is an* $N_x \in \beta_x$ *satisfying* $N_x \subset G$.

Proof. Necessity: We assume that β is a base of neighbourhoods for a topology τ . Then (a) and (b) follow immediately from the definition of a base (Definition 2.2).

To prove (c) we note that if $N_1, N_2 \in \beta_x$ then there exist open sets G_1 and G_2 satisfying $x \in G_1 \subset N_1$ and $x \in G_2 \subset N_2$. The intersection of open sets is open by (3) of Definition 2.1 so $G = G_1 \cap G_2$ is open. Since $G \subset N_1 \cap N_2$ we conclude that $N_1 \cap N_2$ is a neighbourhood of x . But β_x is a base of neighbourhoods of x so there exists $N \in \beta_x$ with $N \subset N_1 \cap N_2$.

To prove (d) let $N_x \in \beta_x$. Then there exists an open G with $x \in G \subset N_x$. G is a neighbourhood of x so there exists $N'_x \in \beta_x$ such that $N'_x \subset G$. Clearly, (d1) holds. To prove (d2) we observe that if $y \in N'_x$ then $y \in G$; but G is a neighbourhood of y so there exists $N_y \in \beta_y$ with $y \in N_y \subset G \subset N_x$.

Sufficiency: Assume that $\beta = \{\beta_x\}$ satisfies conditions (a) to (d). We define the collection τ to consist of all sets $G \subset X$ such that for each $x \in G$ there exists $N_x \in \beta_x$ satisfying $N_x \subset G$. We must show that τ is a topology with respect to which β is a base of neighbourhoods.

To prove that τ is a topology on X conditions (1) to (3) of Definition 2.1 must be verified. Clearly $\emptyset \in \tau$ and $X \in \tau$ so that (1) holds.

To prove (2), let $G_\alpha \in \tau$ for $\alpha \in A$, and set $G = \cup G_\alpha$. If $x \in G$, then $x \in G_\alpha$ for some α so that there exists $N_x \in \beta_x$ for which $N_x \subset G_\alpha \subset G$. Thus $G \in \tau$.

To prove (3), suppose that $G_1, G_2 \in \tau$, and let $G = G_1 \cap G_2$. Let $x \in G$. Then there exist $N_1, N_2 \in \beta_x$ such that $N_1 \subset G_1$ and $N_2 \subset G_2$. By (c) there exists $N \in \beta_x$ with $N \subset N_1 \cap N_2 \subset G$. Thus $G \in \tau$, and hence (3) holds.

We have thus shown that τ is a topology and may call the elements of τ open sets.

To show that β is a base of neighbourhoods for τ, we must show that β satisfies the requirements of Definition 2.2. The first requirement is that β_x is non-empty for all $x \in X$, and this is guaranteed by (a).

The second requirement is that the elements of β_x be neighbourhoods of x (in the topology τ). To show this choose $N_x \in \beta_x$. Let U be the set of points $u \in N_x$ such that there exists $N_u \in \beta_u$ with $N_u \subset N_x$. Then $x \in U \subset N_x$, so that it suffices to show that $U \in \tau$.

Choose $u \in U$. From the construction of U there exists $N_u \in \beta_u$ such that $N_u \subset N_x$. By (d), there exists $N'_u \in \beta_u$ such that for each $y \in N'_u$ there exists $N_y \in \beta_y$ satisfying $N_y \subset N_u$; remembering that $N_u \subset N_x$, we see that this implies that $N'_u \subset U$. Since u was any point in U, it follows from the definition of τ that $U \in \tau$.

The third and last requirement which must be satisfied by β is that if N is any neighbourhood of x (with respect to τ) then there must exist $N_x \in \beta_x$ such that $N_x \subset N$. To see this we note that if N is a neighbourhood of x then there exists $G \in \tau$ such that $x \in G \subset N$. From the definition of τ we see that there exists $N_x \in \beta_x$ such that $N_x \subset G$, and hence $N_x \subset N$. □

In the following examples, Theorems 2.1 and 2.2 may be used to verify that the sets β_x are indeed bases of neighbourhoods. See Problems 2.1, 2.2, and 2.4-2.7.

EXAMPLE 2.5. (1) For any set X, the *discrete topology* is the topology in which every set is open; $\beta_x = \{\{x\}\}$ is a base of neighbourhoods of x, and every set containing x is a neighbourhood of x. (2) The *indiscrete* or *trivial*

topology is the topology in which the only open sets are $\emptyset$ and X ; $\beta_x = \{X\}$ is a base of neighbourhoods of x .
(3) For the real line, $\beta_x = \{[x,b) : b > x\}$ gives rise to a base of neighbourhoods for a topology which is stronger than the usual topology on R^1 . (See Problem 2.15.) □

EXAMPLE 2.6. X is the set of all real functions x defined on $[0,1]$. $\beta_x = \{N_x(T;\varepsilon)\}$, where ε is any strictly positive real number, T is any finite set of points in $[0,1]$, and

$$N_x(T;\varepsilon) = \{y \in X : |y(t) - x(t)| \le \varepsilon \text{ for all } t \in T\} .$$

For example, if $T_1 = \{\frac{1}{2}\}$, $\varepsilon_1 = 10^{-1}$, $T_2 = \{1/3, 2/3\}$, $\varepsilon_2 = 10^{-2}$, then

$$N_x(T_1;\varepsilon_1) = \{y \in X : |y(\tfrac{1}{2}) - x(\tfrac{1}{2})| \le 10^{-1}\} ,$$

$$N_x(T_2;\varepsilon_2) = \{y \in X : |y(1/3) - x(1/3)| \le 10^{-2} \text{ and } |y(2/3) - x(2/3)| \le 10^{-2}\} .$$

This is the topology of *pointwise convergence*. It is of some interest in numerical analysis, because one is comparing function values at the finite number of points T , which is all that one can do numerically. □

DEFINITION 2.3. A real-valued function $\rho(x,y)$ defined for all x and y in a set X is a *metric* or a *distance function* if, for all $x, y, z \in X$:

(1) $\rho(x,y) \ge 0$, and $\rho(x,y) = 0$ iff $x = y$, (*non-negativity*).
(2) $\rho(x,y) = \rho(y,x)$, (*symmetry*).
(3) $\rho(x,y) \le \rho(x,z) + \rho(z,y)$, (the *triangle inequality*). □

EXAMPLE 2.7. Let X be a non-empty set with a metric ρ . For $x \in X$ and $\varepsilon > 0$ let $B(x;\varepsilon)$ be the *open ball* with centre X and radius ε ,

$$B(x;\varepsilon) = \{y \in X : \rho(x,y) < \varepsilon\} ,$$

and let $B[x;\varepsilon]$ be the *closed ball* with centre x and radius ε,

$$B[x;\varepsilon] = \{y \in X : \rho(x,y) \le \varepsilon\} .$$

Finally, let

$$\beta_x = \{B(x;\tfrac{1}{n}) : n = 1,2,\ldots\} ,$$
$$\beta = \{\beta_x : x \in X\} .$$

It is readily verified that β satisfies conditions (a) to (d) of Theorem 2.2. For example, to verify (c) we note that if $N_1 = B(x;\frac{1}{n})$ and $N_2 = B(x;\frac{1}{m})$ then we may take $N = B(x;\ 1/(m+n))$. Hence β is a base of neighbourhoods for a topology on X. A topological space X whose topology can be defined in this way is called a *metric space* (*with metric* ρ). (For examples see Problems 2.3 and 2.12). □

EXAMPLE 2.8. Let s be the set of all real-valued sequences (not necessarily convergent), so that if $x \in s$ then $x = (x_1,x_2,\ldots)$. If $y = (y_1,y_2,\ldots) \in s$ let

$$\rho(x,y) = \sum_{k=1}^{\infty} \frac{1}{2^k} \frac{|x_k-y_k|}{1+|x_k-y_k|} .$$

The function ρ clearly satisfies conditions (1) and (2) of Definition 2.3. To prove condition (3) we note that $t/(1+t)$ is a monotonically increasing function of t so that

$$\begin{aligned}\frac{|x_k-y_k|}{1+|x_k-y_k|} &\le \frac{(|x_k-z_k|+|z_k-y_k|)}{1+(|x_k-z_k|+|z_k-y_k|)} ,\\ &= \frac{|x_k-z_k|}{1+(|x_k-z_k|+|z_k-y_k|)} + \frac{|z_k-y_k|}{1+(|x_k-z_k|+|z_k-y_k|)} ,\\ &\le \frac{|x_k-z_k|}{1+|x_k-z_k|} + \frac{|z_k-y_k|}{1+|z_k-y_k|} .\end{aligned}$$

Thus, s is a metric space with metric ρ, the topology being defined as in Example 2.7. (See also Problem 2.18 for an alternative definition of this topology) □

2.4 *MAPPINGS*

We now consider mappings from one set into another. If X and Y are sets, the notation $f: X \to Y$ means that with each element $x \in X$ there is associated a point $f(x) \in Y$; f is a *mapping* or *function* or *operator* or *transformation* with *domain* $D(f) = X$, and with *range* $R(f)$,

$$R(f) = \{y \in Y : y = f(x) \quad \text{for some} \quad x \in X\} .$$

$f(x)$ is called the *value* of the mapping f at the point x, and the process of determining $f(x)$ is called *evaluating* f *at* x. Often a mapping f is defined by giving the values of $f(x)$ for all x:

$$f : x \in X \to f(x) \in Y .$$

The mapping $f: x \in X \to x \in X$ which maps each point in X onto itself is called the *identity mapping on* X and is often denoted by I.

It is important to note that, in distinction to the usage in elementary analysis, the domain of a mapping is part of the definition of the mapping. For example, if $f: R^1 \to R^1$ and X_1 and X_2 are distinct subsets of R^1, then the mappings

$$f_1 : x \in X_1 \to f(x) \in R^1 , \quad \text{and} \quad f_2 : x \in X_2 \to f(x) \in R^1 ,$$

are different mappings.

The choice of the domain of a mapping f often plays a crucial role in the analysis of a problem. If the domain is too small the problem may not have a solution, while if the domain is too large then valuable information may be lost. We will encounter examples in later chapters, and for the moment the following trivial example must suffice. Let $X \subset R^1$ and let $Y = R^1_+$, the set of nonnegative real numbers. If $f(x) = x^2$ consider the problem of finding $x \in X$ such that $f(x) = y$ when $y \in Y$ is given. If $X = D(f)$ is the set of the positive rational numbers, then the problem cannot always be solved; if $X = R^1_+$, then the problem has the unique solution $x = \sqrt{y}$; while if $X = R^1$ then the problem has the solution(s) $\pm\sqrt{y}$, and the fact that one solution is nonnegative may be 'lost'.

If the mappings $f : X \to Y$ and $g : Z \to Y$ are such that $X \supset Z$ and $f(x) = g(x)$ for $x \in Z$, then f is an *extension* of g to X and g is a *restriction* of f to Z; the restriction of f to a subset Z of its domain is often denoted by $f|_Z$. There are many possible extensions of a mapping, and one is usually interested in those extensions which preserve properties such as differentiability or linearity.

There are a number of terms which are used in the literature to express properties of the domain and range of a mapping. If $f : X \to Y$ we say that f is a mapping *from* X *into* Y; if, in addition, $R(f) = Y$ we say that f *maps* X *onto* Y, that f is a *surjection* of X onto Y, or that f is *surjective*; if $f(x_1) = f(x_2)$ iff $x_1 = x_2$ we say that f is *one-to-one*, that f is an *injection* of X into Y, or that f is *injective*; if f is a one-to-one surjection, we say that f is a *bijection* or that f is *bijective*.

EXAMPLE 2.9. The mapping

$$f : x \in [0,1] \to \sin(x) \in R^1 ,$$

has domain the real interval $[0,1]$, and range the interval $[0, \sin(1)]$. f maps $[0,1]$ into R^1. f is an injection of $[0,1]$ into R^1, or f is one-to-one. f is not a surjection and f is not a bijection. The mapping

$$g : x \in [0,\tfrac{1}{2}] \to \sin(x) \in R^1 ,$$

is the restriction of f to the subset $Z = [0,\frac{1}{2}]$ and could be denoted by $f|_{[0,\frac{1}{2}]}$.

The mapping

$$h : x \in R^1 \to \begin{cases} x & \text{if } x \le 0 , \\ \sin(x), & \text{if } 0 < x < 1 , \\ x \sin(1), & \text{if } x \ge 1 , \end{cases}$$

is an extension of f to R^1. h maps R^1 onto R^1, and h is one-to-one, so that h is a surjection and a bijection.

□

Given a mapping $f : X \to Y$ we are often concerned with solving the equation $f(x) = y$ for given y. We have:

(1) The equation is solvable for all $y \in Y$ iff f maps X onto Y, that is, iff f is surjective.

(2) The solutions of the equation (if they exist) are unique iff f is one-to-one, that is, iff f is injective.

(3) The equation has a unique solution for all $y \in Y$ iff f is one-to-one and surjective, that is, iff f is bijective.

Given mappings,

$$f : X \to Y \quad \text{and} \quad g : Y \to Z ,$$

we can form the *composite mapping*

$$g \circ f : X \to Z ,$$

defined by

$$g \circ f : x \in X \to g \circ f(x) = g(f(x)) \in Z .$$

It is a matter of convenience whether one uses the notation $g \circ f(x)$ or the notation $g(f(x))$. One often also writes gf and $(gf)(x)$ instead of $g \circ f$ and $g \circ f(x)$.

When considering composite mappings the fact that the domain of a mapping is part of the definition of the mapping can provide a very useful check:

EXAMPLE 2.10. In the approximate solution of equations (see Chapter 9) we are often concerned with spaces and mappings with the interrelationship shown in Figure 2.1:

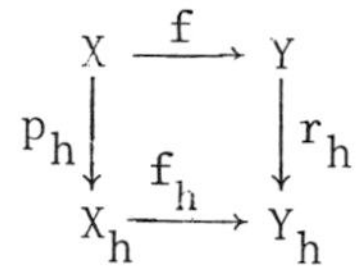

Figure 2.1: Approximation of an equation $f(x) = y$

The original equation is

$$f(x) = y ,$$

where we must find $x \in X$ given $y \in Y$. This is approximated by the equation

$$f_h(x_h) = y_h ,$$

where $x_h \in X_h$ and $y_h \in Y_h$. The spaces X_h and Y_h and the mapping $f_h : X_h \to Y_h$ are chosen so as to be approximations to X, Y, and f, respectively. The mappings

$$p_h : X \to X_h ,$$

$$r_h : Y \to Y_h ,$$

are also chosen. We can consider the composite mappings $r_h \circ f$ and $f_h \circ p_h$, but composite mappings such as $r_h \circ p_h$ and $f \circ r_h$ are not defined in general. □

If $f : X \to Y$ and U is a subset of X then f(U) denotes the set $\{f(u) : u \in U\}$. Notice that the mapping f preserves the ordering of inclusion: if $U \subset V$ then $f(U) \subset f(V)$.

Up to this point we have been considering *single-valued mappings* f, that is, mappings which associate with a point x a single point f(x). Unless otherwise stated, mappings will be assumed to be single-valued. A mapping f is a *set-valued mapping* (or *multivalued mapping* or *correspondence* or *multi-function*) from X to Y if with each $x \in X$ there is associated a set $f(x) \subset Y$, where f(x) may be empty. A single-valued mapping $f : X \to Y$ can be redefined as the set-valued mapping $F : x \to \{f(x)\}$, but in most instances this creates unnecessary complications in notation.

If $f : X \to Y$ and f is not injective, then the *inverse mapping*

$$f^{-1} : y \in Y \to \{x \in X : f(x) = y\} ,$$

is set-valued. Until about 1960 inverse mappings were the only set-valued mappings normally encountered in analysis. Recently, however, it has been found that set-valued mappings can be usefully employed in the theory of optimization (and elsewhere).

EXAMPLE 2.11

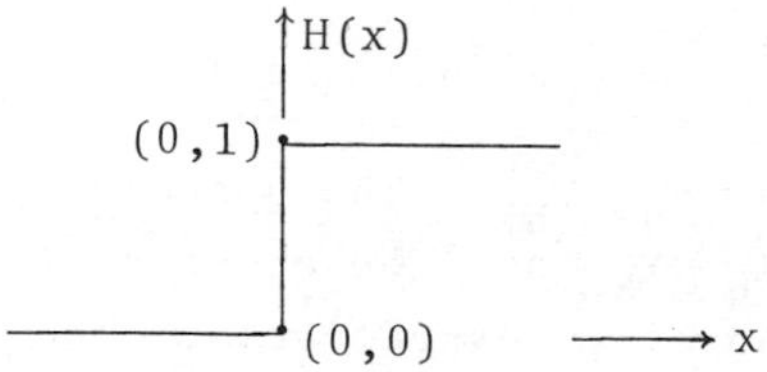

Figure 2.2: The Heaviside step function H(x)

The mapping

$$H(x) = \begin{cases} 0, & \text{if } x < 0 , \\ [0,1], & \text{if } x = 0 , \\ 1, & \text{if } x > 0 , \end{cases}$$

is a set-valued mapping (the *Heaviside step function*) from R^1 into R^1 whose graph is shown in Figure 2.2. □

If $f : X \to Y$ where X and Y are topological spaces, we can make assumptions about f which are related to the topologies on X and Y. f is *open* if f maps open sets onto open sets. f is *continuous at the point* x of X if to each neighbourhood V of f(x) there corresponds a neighbourhood U of x such that if $x \in U$ then $f(x) \in V$; that is, $f(U) \subset V$. If f is continuous at each point of X then f is said to be *continuous*. If f is a bijection and both f and f^{-1} are continuous, f is *bicontinuous*.

THEOREM 2.3. *If* $f : X \to Y$, *where* X *and* Y *are topological spaces, then the following three conditions are equivalent:*

(1) f *is continuous.*

(2) $f^{-1}(V)$ *is an open set in* X *for every open set* V *in* Y.

(3) $f^{-1}(W)$ *is a closed set in* X *for every closed set* W *in* Y.

Proof. We indicate the proof in one case: (1) ⇒ (2). Let $V \subset Y$ be open. Consider $f^{-1}(V)$. If $f^{-1}(V)$ is empty, it is open. If $f^{-1}(V)$ is not empty, let $x \in f^{-1}(V)$. Then $f(x) = y$ for some $y \in V$. Since V is a neighbourhood of y, and since f is continuous, there exists a neighbourhood N_x of x such that $f(N_x) \subset V$. Thus $x \in N_x \subset f^{-1}(V)$. Hence, by Theorem 2.1, $f^{-1}(V)$ is open. □

Condition (2) of Theorem 2.3 is the condition for continuity that is most frequently used. In words: f is continuous if and only if the inverse image of open sets is open. The definition of continuity in terms of neighbourhoods is a generalization of the classical epsilon-delta definition: if $f : R^1 \to R^1$ then f is continuous at x if, given $\varepsilon > 0$, there exists $\delta > 0$ such that $|f(x)-f(y)| \le \varepsilon$ provided that $|x-y| \le \delta$. This definition is less elegant than condition (2) of Theorem 2.3, but it is often useful and it also corresponds to the intuitive concept of continuity of many analysts. (If U is closed $f(U)$ may not be: Prob. 6.30).

We now consider product topologies. If X_1 and X_2 are sets then the *Cartesian product* $X_1 \times X_2$ is the set of ordered pairs (x_1,x_2):

$$X_1 \times X_2 = \{(x_1,x_2) : x_1 \in X_1 \text{ and } x_2 \in X_2\} .$$

If X_1 and X_2 are also topological spaces with bases of neighbourhoods $\beta^{(1)} = \{\beta^{(1)}_{x_1}\}$ and $\beta^{(2)} = \{\beta^{(2)}_{x_2}\}$, respectively, then the *product topology* on $X_1 \times X_2$ is defined by the base of neighbourhoods $\beta = \{\beta_{(x_1,x_2)}\}$ where each $\beta_{(x_1,x_2)}$ is formed by taking all possible Cartesian products of neighbourhoods in $\beta^{(1)}_{x_1}$ with neighbourhoods in $\beta^{(2)}_{x_2}$. That is,

$$\beta_{(x_1,x_2)} = \{N^{(1)}_{x_1} \times N^{(2)}_{x_2} : N^{(1)}_{x_1} \in \beta^{(1)}_{x_1} \text{ and } N^{(2)}_{x_2} \in \beta^{(2)}_{x_2}\} ,$$

$$N^{(1)}_{x_1} \times N^{(2)}_{x_2} = \{(u_1,u_2) : u_1 \in N^{(1)}_{x_1} \text{ and } u_2 \in N^{(2)}_{x_2}\} .$$

It is readily verified that β satisfies conditions (a) to (d) of Theorem 2.2 so that β is a base of neighbourhoods for a topology on $X_1 \times X_2$.

An example of a product space is the real two-dimensional space R^2 which may be defined as the product space

$$R^2 = R^1 \times R^1 ,$$

with the product topology.

Given a mapping $f : X_1 \times X_2 \to Z$, we can obtain mappings from X_1 or X_2 into Z by fixing the value of x_2 or x_1, respectively. In such a case the value which is not fixed is denoted by a dot. Thus $f(x_1,\cdot)$ maps X_2 into Z,

$$f(x_1,\cdot) : x_2 \in X_2 \to f(x_1,x_2) \in Z .$$

Equivalently, $f(x_1,\cdot)$ is the restriction of f to $\{x_1\} \times X_2$,

$$f(x_1,\cdot) = f\big|_{\{x_1\}\times X_2} .$$

If X_1, X_2, and Z are topological spaces and $f : X_1 \times X_2 \to Z$ then it follows from the definition of continuity that f is continuous in the product topology iff for each $z = f(x_1,x_2)$ and each neighbourhood N_z of z there exist neighbourhoods N_{x_1} and N_{x_2} of x_1 and x_2, respectively, such that $f(N_{x_1},N_{x_2}) \subset N_z$. In particular, $f(N_{x_1},x_2) \subset N_z$ and $f(x_1,N_{x_2}) \subset N_z$, so that the mappings

$$f(\cdot,x_2) : x_1 \in X_1 \to f(x_1,x_2) \in Z ,$$

and

$$f(x_1,\cdot) : x_2 \in X_2 \to f(x_1,x_2) \in Z ,$$

are continuous. In words: if the mapping $f : X_1 \times X_2 \to Z$ is *jointly continuous* in the variables then it is continuous in the variables separately. The converse is not true as is shown by the following example.

EXAMPLE 2.12. The mapping $f : R^1 \times R^1 \to R^1$ defined by

$$f(x_1,x_2) = \begin{cases} \dfrac{x_1x_2}{x_1^2 + x_2^2} , & \text{if } (x_1,x_2) \neq (0,0) , \\ 0 , & \text{if } (x_1,x_2) = (0,0) , \end{cases}$$

is clearly continuous in x_1 and x_2 separately. It is not jointly continuous at $(0,0)$ because $f(0,0) = 0$ while $f(\varepsilon,\varepsilon) = 1/2$ for all $\varepsilon \neq 0$. □

There is an elegant extension of the product of two spaces to the product of an arbitrary number of spaces. Let $\{X_\alpha\}$ be a family of spaces for $\alpha \in A$. The *Cartesian product* $\prod_{\alpha \in A} X_\alpha$ is the set of all functions f mapping A into $\bigcup_{\alpha \in A} X_\alpha$ such that $f(\alpha) \in X_\alpha$ for all $\alpha \in A$. If $A = \{1,2,\ldots,n\}$ we write

$$\prod_{\alpha \in A} X_\alpha = \prod_{\alpha=1}^{n} X_\alpha = X_1 \times X_2 \times \ldots \times X_n ,$$

and call $f(\alpha)$ the α-th *component* of f.

EXAMPLE 2.13. We define the real n-dimensional space R^n to be the Cartesian product

$$R^n = \prod_{i=1}^{n} R^1 .$$

An element f in R^n is a mapping from $\{1,2,\ldots,n\}$ into R^1, so that f corresponds to the ordered n-tuple $(f(1),f(2),\ldots,f(n))$ comprising the n components of f. □

EXAMPLE 2.14. Let $A = (a_{ij})$ be an $n \times n$ real matrix so that

$$A : x \in R^n \to Ax \in R^n .$$

It is readily verified that the problem of finding a real eigenvector $x = (x_i)$ and corresponding real eigenvalue λ of A is equivalent to solving the equation $f(u) = 0$ where

$$f : (x,\lambda) \in R^n \times R^1 \to (Ax - \lambda x, 1 - \sum_{i=1}^{n} x_i) \in R^n \times R^1 . \quad \square$$

The *product topology* on $\prod_{\alpha \in A} X_\alpha$ is defined by the base of neighbourhoods $\beta = \{\beta_f\}$ where

$$\beta_f = \{ \prod_{\alpha \in A} N_\alpha : N_\alpha \text{ is a neighbourhood of } f(\alpha) \text{ in } X_\alpha \text{, and } N_\alpha = X_\alpha \text{ except for at most a finite number of } \alpha \} .$$

Clearly $\beta = \{\beta_f\}$ satisfies the conditions of Theorem 2.2 and so β is a base for a topology.

EXAMPLE 2.15. If the topological space X is as defined in Example 2.6, then $X = \prod_{\alpha \in [0,1]} R^1$, and X is equipped with the product topology. □

It is often necessary to recover the components of an element in a Cartesian product of spaces, and this is done by using projection maps. If $X = \prod_{\alpha \in A} X_\alpha$ and $\gamma \in A$ then the *projection map* p_γ is the mapping of X into X_γ defined by

$$p_\gamma(f) = f(\gamma), \quad \text{for } f \in X .$$

If V is a neighbourhood of $f(\gamma)$ in X_γ, let

$$U = \prod_{\alpha \in A} N_\alpha ,$$

where $N_\gamma = V$, and $N_\alpha = X_\alpha$ if $\alpha \neq \gamma$, Then U is a neighbourhood of f in X and $p_\gamma(U) = V$; we conclude that the projection maps p_γ are continuous.

We assert that a mapping $g : Z \to \prod_{\alpha \in A} X_\alpha \equiv X$ is continuous iff $p_\alpha \circ g : Z \to X_\alpha$ is continuous for each $\alpha \in A$. *Sufficiency:* Suppose that $p_\alpha \circ g$ is continuous for each α. Each neighbourhood V of $g(z) \in X$, contains a neighbourhood W of the form

$$W = \prod_\alpha N_\alpha ,$$

where $N_\alpha = X_\alpha$ except for a finite number of $\alpha, \alpha = \alpha_1, \ldots, \alpha_k$, say, and where N_α is a neighbourhood of $p_\alpha \circ g(z) \in X_\alpha$. Since $p_\alpha \circ g$ is continuous there exist neighbourhoods $U_{\alpha_j} \subset Z$ of z such that

$$p_{\alpha_j} \circ g(U_{\alpha_j}) \subset N_{\alpha_j} , \quad 1 \le j \le k ,$$

so that

$$g(U_{\alpha_j}) \subset p_{\alpha_j}^{-1}(N_{\alpha_j}) \ , \quad 1 \le j \le k \ .$$

Hence, if $U = \bigcap_j U_{\alpha_j}$ then

$$g(U) \subset \bigcap_j p_{\alpha_j}^{-1}(N_{\alpha_j}) = \prod_\alpha N_\alpha = W \subset V \ .$$

That is, g is continuous.

Necessity: Since the composite of continuous maps is continuous (Problem 2.9), if g is continuous then $p_\alpha \circ g$ is continuous. □

Example 2.15 illustrates an important aspect of product topologies. In the product space

$$X = \prod_{\alpha \in [0,1]} R^1 \ ,$$

we can conceive of many types of neighbourhoods. On the other hand, the neighbourhoods for the base in the product topology are very special: they are 'rectangular', and of 'infinite dimension' except in a finite number of coordinates. In general, we may expect the product topology on a product space to be weaker than many of the other topologies on the same space. Indeed, it is easy to prove that the product topology is the weakest topology for which all the projection maps p_α are continuous.

2.5 *VECTOR SPACES*

So far, we have only considered topological spaces, but the space R^1 also possesses an algebraic structure: the elements of R^1 can be added and multiplied. We now consider generalizations of these properties.

In addition to considering multiplication by real numbers in R^1 we sometimes consider multiplication by complex numbers. The space of complex numbers $z = x + iy$ is denoted by $\mathbb{C}$. $\mathbb{C}$ is a metric space with metric

$$\rho(z_1,z_2) \equiv \rho(x_1+iy_1,x_2+iy_2) = |z_1-z_2| = \sqrt{(x_1-x_2)^2 + (y_1-y_2)^2} \ .$$

DEFINITION 2.4. A set X is called a *real (complex) vector space* or *linear space* if for all $x, y \in X$ and all $\lambda \in R^1$ ($\lambda \in ₵$), the *vector sum* $x+y \in X$ and the *scalar product* $\lambda x \in X$ are defined and satisfy:

(1) $(x+y) + z = x + (y+z)$, (*associative addition*).
(2) $x + y = y + x$, (*commutative addition*).
(3) An element $\underline{0} \in X$ exists such that, for all x, $\underline{0}+x = x$.
(4) For each $x \in X$ there is an element $-x$ such that $x + (-x) = \underline{0}$. We write $x - y$ for $x + (-y)$.
(5) $(\lambda\mu)x = \lambda(\mu x)$ (*associative multiplication*).
(6) $(\lambda+\mu)x = \lambda x + \mu x$
(7) $\lambda(x+y) = \lambda x + \lambda y$ (6)–(7) (*distributive laws*).
(8) $1x = x$.

These imply that

(9) The element $\underline{0}$ is unique. We call it the *origin*. When no confusion is possible, we write 0 instead of $\underline{0}$.
(10) The element $-x$ is unique and $(-1)x = -x$.
(11) $\lambda\,\underline{0} = \underline{0}$ for all λ .

The elements $x \in X$ are called *vectors* and the elements $\lambda \in R^1$ (or ₵) are called *scalars*. □

If U, V are subsets of a real (complex) vector space X and $\Lambda \subset R^1$ (₵) then we denote by $U\underline{+}V$ and ΛU the sets

$$U\underline{+}V = \{u\underline{+}v : u \in U \text{ and } v \in V\}\ , \text{ and } \Lambda U = \{\lambda u : \lambda \in \Lambda \text{ and } u \in U\}\ .$$

We write $u\underline{+}V$ and λU instead of $\{u\}\underline{+}V$ and $\{\lambda\}U$.

In most spaces of interest to us, it is possible to define a vector sum and a scalar product in such a way that we obtain a vector space, but that this is not always so is shown by the following example.

EXAMPLE 2.16. The *extended real line* $\overline{R}^1 = [-\infty, +\infty]$ is obtained by adjoining two new elements, $+\infty$ and $-\infty$, to R^1 . A base of neighbourhoods is defined by $\beta = \{\beta_x\}$, where $\beta_x = \{\beta_x(n) : n = 1,2,\ldots\}$ and

$$\beta_x(n) = \begin{cases} (x - \frac{1}{n},\ x + \frac{1}{n}), & \text{if } x \text{ is finite,} \\ (n, +\infty], & \text{if } x = +\infty , \\ [-\infty, n), & \text{if } x = -\infty . \end{cases}$$

It is easily seen that conditions (a)-(d) of Theorem 2.2 are satisfied, so that $\overline{R}^1$ is a topological space.

For $\overline{R}^1$ to be a vector space, we must define addition and multiplication for $+\infty$ and $-\infty$ so that the conditions of Definition 2.4 are satisfied. From Definition 2.4, part 4, we have that $\infty + (-\infty) = 0$. Now let x be any finite real number, and set

$$y = x + \infty . \tag{*}$$

If y is finite, then, using the associativity of addition, we conclude that $y - x = \infty$, which is impossible since $y - x$ is finite. If $y = +\infty$ then, adding $-\infty$ to both sides of eqn (*) we obtain $x = 0$, which is, in general, impossible. Finally, if $y = -\infty$ then we find that

$$x = -\infty - \infty = (-1)\infty + (-1)\infty = ((-1) + (-1))\infty = -2\infty ,$$

so that $\frac{1}{2}x = \frac{1}{2}(-2)\infty = (-1)\infty = -\infty$, which is yet another contradiction. In summary, there is no $y \in \overline{R}^1$ which satisfies eqn (*) and all the conditions of Definition 2.4. □

Further examples of spaces which are not vector spaces are given in Problems 2.12, 2.13 and 2.14.

Let $f : X \to Y$ where X and Y are real (complex) vector spaces. If

$$f(\lambda_1 x_1 + \lambda_2 x_2) = \lambda_1 f(x_1) + \lambda_2 f(x_2) ,$$

for all $x_1, x_2 \in X$ and all real (complex) λ_1, λ_2 then f is a *linear mapping* or *linear operator* or *linear function*. If $Y = R^1(\mathbb{C})$ then f is a *(real (complex)) linear functional*.

Let $f : X \times Y \to Z$ where X, Y, and Z are vector spaces. If

$$f(\cdot, y) : X \to Z ,$$

and

$$f(x, \cdot) : Y \to Z ,$$

are linear mappings then f is a *bilinear operator* or *bilinear mapping*; if the mappings $f(\cdot, y)$ and $f(x, \cdot)$ are linear functionals then f is a *bilinear functional*.

A vector space X is *finite-dimensional* if there exist a finite number of elements in X, $x_1, x_2, \ldots, x_n$, say, such that every $x \in X$ is a *finite linear combination* of the x_i; that is,

$$x = \sum_{i=1}^{n} \lambda_i x_i \, , \tag{*}$$

for some scalars λ_i (which depend upon x of course). X is *infinite-dimensional* if X is not finite-dimensional. If X is finite-dimensional, then the *dimension* of X, is the smallest number n for which a representation of the form (*) is possible, and the corresponding elements $x_1, \ldots, x_n$ form a *finite basis* for X.

A subset E of a vector space X is a *linear subspace* (or *linear manifold*) if E contains all finite linear combinations of its vectors. That is, if $x_1, \ldots, x_n \in E$ then the linear combination (*) belongs to E for all choices of the scalars λ_i. We usually write 'subspace' instead of 'linear subspace'.

Given a subset A of a vector space X we can form the the set E *spanned* by A,

$$E = \operatorname{span}(A) = \{x \in X : x = \sum_{i=1}^{n} \lambda_i a_i \, , \text{ for some } n \, , \text{ some } \lambda_i \, , \text{ and some } a_i \in A \, .\}$$

Clearly E is a (linear) subspace of X.

The vectors $x_1, \ldots, x_n \in X$ are *linearly dependent* if there exist $\lambda_1, \ldots, \lambda_n$, not all zero, such that

$$\underline{0} = \sum_{i=1}^{n} \lambda_i x_i \, ; \tag{**}$$

otherwise, $x_1, \ldots, x_n$ are *linearly independent*.

For example, the elements x_i in a finite basis are linearly independent since if eqn (**) holds with $\lambda_j \neq 0$ for some j, then x_j can be represented as a finite linear combination of the remaining x_i,

$$x_j = \sum_{\substack{i=1 \\ i\neq j}}^{n} -(\lambda_i/\lambda_j)x_i \ . \qquad (**)$$

Combining eqns (*) and (**) we see that every $x \in X$ is a finite linear combination of the (n-1) elements x_1, x_2, x_{j-1}, $x_{j+1},\ldots,x_n$, which contradicts the assumption that $x_1,\ldots,x_n$ form a basis.

We refrain from giving examples of vector spaces, since the reader can readily provide these from linear algebra.

The concept of a basis for a space will be discussed in greater detail in Chapter 8. For many problems in numerical analysis it is necessary to approximate an infinite-dimensional space X by a finite-dimensional subspace X_h, say, and the choice of a suitable X_h and a suitable basis for X_h is of crucial importance. (see Chapters 8 and 9).

2.6 *TOPOLOGICAL VECTOR SPACES*

We now combine the concepts of a topological space and a vector space:

DEFINITION 2.5. A *real (complex) topological vector space* (or *linear topological space*) X is a vector space with a topology such that vector addition and scalar multiplication are continuous (as mappings from $X \times X$ into X and $R^1(\text{or}\,\mathbb{C})\times X$ into X, respectively). That is,

(1) If $x+y=z$ and N_z is any neighbourhood of z, there exist neighbourhoods N_x and N_y of x and y such that $N_x + N_y \subset N_z$.

(2) If $\lambda x = z$ and N_z is any neighbourhood of z, there exist neighbourhoods N_x and N_λ of x and λ such that $N_\lambda N_x \subset N_z$. □

The requirement that vector addition and scalar multiplication be continuous in a topological vector space is non-trivial, and there are many examples of vector spaces which are topological spaces but not topological vector spaces.

EXAMPLE 2.17. Consider the real line with the discrete topology (Example 2.5), and the usual definitions of addition and multiplication. This is a real vector space and a topological space.

Vector addition is continuous since if $x + y = z$ and N_z is a neighbourhood of z we may choose $N_x = \{x\}$ and $N_y = \{y\}$ and then $N_x + N_y \subset N_z$.

However, scalar multiplication is not continuous. To see this, let $\lambda x = z \neq 0$. Then $N_z = \{z\}$ is a neighbourhood of z. We seek neighbourhoods N_x and N_λ satisfying $N_\lambda N_x \subset N_z$. Since $\beta_\lambda = \{(\lambda-1/n, \lambda+1/n) : n = 1,2,\ldots\}$ is a base of neighbourhoods of λ there must be an n such that

$$(\lambda-1/n, \lambda+1/n)\{x\} \subset N_\lambda N_x \subset N_z = \{z\} = \{\lambda x\},$$

which is impossible. We conclude that the real line with the discrete topology is not a topological vector space. □

In a topological vector space the concept of a base of neighbourhoods can be simplified.

THEOREM 2.4. $\beta_{\underline{0}} = \{N_{\underline{0}}\}$ *is a base of neighbourhoods at the origin in a topological vector space* X *iff* $\beta_x = \{x+N_{\underline{0}}\}$ *is a base of neighbourhoods of* x.

Proof. Let $x \in X$. We assert that $V \subset X$ is a neighbourhood of $\underline{0}$ iff $T = V + x$ is a neighbourhood of x.

To see this, we note that if V is a neighbourhood of $\underline{0}$ there exists an open G satisfying $\underline{0} \in G \subset V$. Let $U = G + x$ so that $U + (-x) = G$ or $U = \phi^{-1}(G)$ where $\phi : y \to y + (-x)$. Since addition is continuous, U is open. But, $x \in U \subset T$. Hence, T is a neighbourhood of x. Similarly, if T is a neighbourhood of x then V is a neighbourhood of $\underline{0}$. □

In a topological vector space a base of neighbourhoods of $\underline{0}$ is called a *local base*. It follows from Theorem 2.4 that in a topological vector space it suffices to consider local bases. Theorem 2.2 characterized bases of neighbourhoods in a topological space. The following theorem, the proof of which is left to the reader (Problem 2.16), characterizes local bases in topological vector spaces.

THEOREM 2.5. *Let* X *be a topological vector space and let* β *be a local base. Then:*

(a) *For each* $U, V \in \beta$ *there is a* $W \in \beta$ *such that* $W \subset U \cap V$.

(b) *For each* $U \in \beta$ *there is a* $V \in \beta$ *such that* $V + V \subset U$.

(c) *For each* $U \in \beta$ *there is a* $V \in \beta$ *such that* $\lambda V \subset U$ *for all scalars* λ *satisfying* $|\lambda| \le 1$.

(d) *For each* $x \in X$ *and* $U \in \beta$ *there is a scalar* λ *such that* $x \in \lambda U$.

(e) *For each* $U \in \beta$ *there is a* $V \in \beta$ *and a circled set* W *such that* $V \subset W \subset U$ (A set W is *circled* or *balanced* if $\lambda W \subset W$ for all λ satisfying $|\lambda| \le 1$).

(f) *If* X *is a Hausdorff space then* $\cap \{U : U \in \beta\} = \underline{0}$. (X is a *Hausdorff space* or a *separated space* if for every $x, y \in X$ with $x \neq y$ there exist disjoint neighbourhoods N_x and N_y.)

Conversely, let X *be a vector space and* β *a nonempty family of subsets which satisfy* (a) *to* (d), *and let* τ *be the collection of all sets* $W \subset X$ *such that for each* $x \in W$ *there exists* $U \in \beta$ *with* $x + U \subset W$. *Then* τ *is a topology for* X *and with this topology* X *is a topological vector space. Furthermore, if* (f) *holds, then* X *is a Hausdorff space.* □

There is a relationship between the properties of a topological vector space and the properties of its local base, as is illustrated by Theorem 2.7 below and by the properties of locally convex topological vector spaces (which we now define).

A subset K of a vector space is *convex* if $\lambda x + (1-\lambda)y \in K$ for all $x, y \in K$ and all real $\lambda \in [0,1]$. Let λ_i, $1 \le i \le m$, be m non-negative real numbers with sum 1; then $x = \sum_{i=1}^{m} \lambda_i x_i$ is a *convex linear combination* of the points x_i. It follows by induction that if K is convex, then every convex linear combination of points $x_i \in K$ also belongs to K.

A topological vector space is *locally convex* if there is a local base of convex neighbourhoods (of the origin). The class of locally convex topological vector spaces is very important because such spaces have a large number of linear functionals (see Chapter 7), and all the useful spaces with which we will be concerned will be of this type.

EXAMPLE 2.18. As is readily checked (Problem 2.17), the space ℓ^p, $0<p<1$, of real sequences $x=\{x_n\}$ satisfying

$$\sum_{n=1}^{\infty} |x_n|^p < \infty ,$$

is a topological vector space with metric $\rho(x,y)=\sum_{n=1}^{\infty}|x_n-y_n|^p$ and local base $\{B[\underline{0};\varepsilon] : \varepsilon>0\}$. ℓ^p is not locally convex since, if it were, the unit ball $B[\underline{0};1]$ would have to contain a convex neighbourhood U, which would in turn contain some ball $B[\underline{0};\varepsilon]=\{x : \rho(x,\underline{0})\le\varepsilon\}$ in the local base. Let $x^{(r)}=\{x_n^{(r)}\}\in\ell^p$ with r an integer and

$$x_n^{(r)} = \begin{cases} 1, & r=n, \\ 0, & r\ne n, \end{cases}$$

and let

$$y^{(s)} = s^{-1}\sum_{r=1}^{s} \varepsilon^{1/p}\, x^{(r)} .$$

Then $\varepsilon^{1/p}x^{(r)}\in B[\underline{0};\varepsilon]\subset U$. Since U is convex, and $y^{(s)}$ is a convex linear combination of the $x^{(r)}$, $y^{(s)}\in U\subset B[\underline{0};1]$. But, for large s, $\rho(y^{(s)},\underline{0})=\varepsilon s^{1-p}>1$ and we have a contradiction. □

There is a very useful way to construct locally convex topological vector spaces:

DEFINITION 2.6. A non-negative finite real-valued function $p(x)$ defined on a vector space X is a *seminorm* if

(1) $p(x)\ge 0$,
(2) $p(\lambda x)=|\lambda|p(x)$,
(3) $p(x+y)\le p(x)+p(y)$. □

THEOREM 2.6. *Given any family* $Q = \{p_\alpha ; \alpha \in A\}$ *of seminorms on a vector space* X *there is a coarsest topology on* X *in which every seminorm in* Q *is continuous. With this topology* X *is a locally convex topological vector space and a local base of open neighbourhoods is formed by*

$$\beta = \{N(J;\varepsilon) : \varepsilon>0 \text{ and } J \text{ a finite subset of } A\},$$

where

$$N(J;\varepsilon) = \{x : \max_{\alpha\in J} [p_\alpha(x)] < \varepsilon\} .$$

We speak of the topology induced on X *by* Q .

Proof. Using the properties of seminorms, conditions (a)-(d) of Theorem 2.5 can be verified. □

EXAMPLE 2.19. Let X be the space of real continuous functions on [0,1] . Then X is a real vector space. Let

$$p_\alpha(x) = |x(\alpha)| , \text{ for } \alpha \in [0,1] .$$

Then p_α is a seminorm. Thus, by Theorem 2.6, the local base of convex neighbourhoods of the form

$$N(\alpha_1,\ldots,\alpha_n;\varepsilon) = \{x : |x(\alpha_i)| < \varepsilon \text{ for } 1\le i\le n\} ,$$

where $\varepsilon > 0$ and $\alpha_i \in [0,1]$, defines a topology on X for which X is a locally convex topological vector space. This topology is the topology of pointwise convergence which has been defined in different, but equivalent, ways in Examples 2.6 and 2.15. □

EXAMPLE 2.20. Let X be the space of real infinitely differentiable functions on [0,1] . For $n \ge 1$ let

$$p_n(x) = \max_{0\le t\le 1} |x^{(n)}(t)| .$$

Then $p_n(x)$ is a seminorm, and the family of seminorms p_n induces a topology on X . These seminorms play a rôle in the theory of distributions. □

In the next two sections we consider two important types of locally convex topological vector spaces: normed spaces and inner product spaces.

2.7 *NORMED SPACES*

We now introduce the class of spaces which is most frequently used in numerical functional analysis.

DEFINITION 2.7. A *norm* $\|\cdot\|$ on a vector space X is a real-valued finite function such that

(1) $\|x\| \geq 0$, and $\|x\| = 0$ iff $x = 0$.
(2) $\|\lambda x\| = |\lambda| \, \|x\|$.
(3) $\|x+y\| \leq \|x\| + \|y\|$. (*triangle inequality*) □

That is, a norm is a seminorm with the additional property that $\|x\| = 0$ implies $x = 0$. A vector space with a norm is a *normed space* or *normed linear space*.

If X is a normed space then

$$\rho(x,y) = \|x-y\| ,$$

is a metric, so that X is also a metric space. As for a metric space,

$$B[x;1/n] = \{y \in X : \|x-y\| \leq 1/n\} ,$$

denotes the closed ball with centre x and radius $1/n$. It follows, as in Example 2.7, that if

$$\beta_x = \{B[x;1/n]: n = 1,2,..\} ,$$

and

$$\beta = \{\beta_x: x \in X\} ,$$

then β is a base of neighbourhoods for a topology on X, the *norm* or *strong* topology on X.

In a normed space vector addition and scalar multiplication are continuous in the norm topology so that such a space is a topological vector space with respect to the norm topology. This can be proved by checking that β_0 satisfies the conditions of Theorem 2.5, or by using Theorem 2.6. It is instructive to give a direct proof:

(a) *Vector addition in a normed space*

Let $x + y = z$ and let N_z be a neighbourhood of z. Then there exists a ball $B[z;\delta] \subset N_z$ for some $\delta > 0$. If $\tilde{x} \in B[x;\delta/2]$ and $\tilde{y} \in B[y;\delta/2]$ then,

$$\begin{aligned}||\tilde{x}+\tilde{y}-z|| &= ||\tilde{x}+\tilde{y}-(x+y)|| ,\\ &\le ||\tilde{x}-x|| + ||\tilde{y}-y|| ,\\ &\le \delta/2 + \delta/2 ,\end{aligned}$$

so that $B[x;\delta/2] + B[y;\delta/2] \subset B[z;\delta]$, and vector addition is continuous.

(b) *Vector multiplication in a normed space*

Let $\lambda x = z$ and let N_z be a neighbourhood of z . Then there exists a ball $B[z;\delta] \subset N_z$ with $\delta > 0$. Set

$$\delta_1 = \begin{cases} \delta/2, & \text{if } |\lambda| < 1,\\ \delta/(2|\lambda|), & \text{if } |\lambda| \ge 1,\end{cases}$$

$$\delta_2 = \delta/[2(||x|| + \delta)],$$

and let

$$\begin{aligned}\tilde{x} \in B[x;\delta_1] &= \{y\in X : ||x-y|| \le \delta_1\} ,\\ \tilde{\lambda} \in B[\lambda;\delta_2] &= \{\mu\in R^1 (\text{or } \mathbb{C}) : |\mu-\lambda| \le \delta_2\} .\end{aligned}$$

Then

$$\begin{aligned}||\tilde{\lambda}\tilde{x}-z|| &= ||\tilde{\lambda}\tilde{x}-\lambda x|| ,\\ &\le ||\tilde{\lambda}\tilde{x}-\lambda\tilde{x}|| + ||\lambda\tilde{x}-\lambda x|| ,\\ &= |\tilde{\lambda}-\lambda|\ ||\tilde{x}|| + |\lambda|\ ||\tilde{x}-x|| ,\\ &\le \delta_2||\tilde{x}|| + |\lambda|\delta_1 ,\\ &\le \delta_2(||x|| + \delta_1) + |\lambda|\delta_1 ,\\ &\le \delta_2(||x||+\delta) + |\lambda|\delta_1 ,\\ &\le \delta/2 + \delta/2 ,\end{aligned}$$

so that $B[\lambda;\delta_2]\ B[x;\delta_1] \subset B[z;\delta]$, and vector multiplication is continuous. □

In Example 2.17 we gave an example of a space which was a vector space and a topological space but not a topological vector space, and there are (rather esoteric) examples of metric spaces which are also vector spaces but not topological vector spaces. The following definition is therefore not redundant.

DEFINITION 2.8. A topological vector space with a topology defined by a metric ρ is called a *linear metric space. If* $\rho(x,y) = \rho(x-y,0)$ then ρ is *invariant.* □

It is often possible to introduce different norms on a vector space. To distinguish between norms on different spaces and different norms on the same space a variety of self-explanatory notation is used such as:

$$\|x\|_1,\ \|x\|_2,\ \|x\|_X,\ \|x\|_Y,\ \|x;R^1\| \ .$$

Let X be a vector space, and let $\|\cdot\|_1$ and $\|\cdot\|_2$ be two norms on X. These norms define topologies τ_1 and τ_2, respectively, with bases of neighbourhoods of the form

$$\beta_x^{(1)} = \{B_1[x;\delta] = \{y \in X : \|y-x\|_1 \le \delta\} : \delta > 0\} ,$$

and

$$\beta_x^{(2)} = \{B_2[x;\delta] = \{y \in X : \|y-x\|_2 \le \delta\} : \delta > 0\} ,$$

respectively. We now show that $\tau_1 = \tau_2$ iff $\|\cdot\|_1$ and $\|\cdot\|_2$ are *equivalent*, that is, iff there exist strictly positive real constants α_1, α_2 such that

$$\frac{1}{\alpha_1}\|x\|_1 \le \|x\|_2 \le \alpha_2 \|x\|_1, \quad \text{for all } x \in X . \tag{2.1}$$

Sufficiency. Suppose that $\|\cdot\|_1$ and $\|\cdot\|_2$ are equivalent. Let $G \in \tau_1$ and choose any $x \in G$. By Theorem 2.1 there exists a neighbourhood $B_1 = B_1[x;\delta]$ such that $B_1 \subset G$. By eqn (2.1), if $\|x\|_2 \le \delta/\alpha_1$ then $\|x\|_1 \le \delta$, and so $B_2[x;\delta/\alpha_1] \subset G$. That is, for each $x \in G$ there is a neighbourhood B_2 of x which is contained in G. By Theorem 2.1 we see that $G \in \tau_2$, and hence $\tau_2 \supset \tau_1$. Similar arguments show that $\tau_1 \supset \tau_2$, and hence $\tau_1 = \tau_2$.

Necessity. Suppose that $\tau_1 = \tau_2$. The unit ball $B_1 = B_1[\underline{0};1]$ is a neighbourhood of $\underline{0}$ in τ_1, so there exists $G \in \tau_1$ with $\underline{0} \in G \subset B_1$. But $G \in \tau_2$ so there exists a ball $B_2 = B_2[\underline{0};\delta_2]$ with $\underline{0} \in B_2 \in G$. Now let x be any nonzero point in X. Then $(\delta_2/\|x\|_2)x$ belongs to B_2 and hence also to B_1. That is

$$\|(\delta_2/\|x\|_2)x\|_1 = \delta_2\|x\|_1/\|x\|_2 \le 1 ,$$

so that

$$\delta_2\|x\|_1 \le \|x\|_2 ,$$

and this is, of course, also true for $x = \underline{0}$. Similarly, there exists a constant δ_1 such that

$$\delta_1\|x\|_2 \le \|x\|_1 .$$

We conclude that $\|\cdot\|_1$ and $\|\cdot\|_2$ are equivalent with constants $\alpha_1 = 1/\delta_2$ and $\alpha_2 = 1/\delta_1$. □

EXAMPLE 2.21. R^n is a normed linear space. If $x = (x_i) \in R^n$, three norms are:

$$\|x\|_1 = \sum_{i=1}^{n} |x_i| , \quad \text{(the } \ell_1 \text{ *norm*)};$$

$$\|x\|_2 = [\sum_{i=1}^{n} x_i^2]^{\frac{1}{2}} , \quad \text{(the *Euclidean norm*)};$$

$$\|x\|_\infty = \max_{1\le i\le n} |x_i| \quad \text{(the *maximum norm*)}.$$

These norms are equivalent. For example,

$$\|x\|_\infty = \max_i |x_i| \le \sum_{i=1}^{n} |x_i| = \|x\|_1 ,$$

and

$$\|x\|_1 = \sum_{i=1}^{n} |x_i| \le \sum_{i=1}^{n} \|x\|_\infty = n\|x\|_\infty ,$$

so that

$$\frac{1}{n}\|x\|_1 \le \|x\|_\infty \le \|x\|_1 . \quad □ \qquad (2.2)$$

The relationship between different norms plays a very important rôle in numerical analysis:

(1) Suppose that it is possible to obtain a bound for an element x in $\|\cdot\|_1$; $\|x\|_1 \le K_1$, say. If $\|\cdot\|_1$ is equivalent to $\|\cdot\|_2$, then we immediately obtain the bound $\|x\|_2 \le \alpha_2 K_1$. However, if $\|\cdot\|_1$ is not

equivalent to $\|\cdot\|_2$ then a different approach must be used to bound x in $\|\cdot\|_2$, and it is even possible that a bound does not exist.

(2) The size of the constants of equivalence α_1 and α_2 in eqn (2.1) is often crucial. For example, in order to apply the contraction mapping theorem (see Volume 2) to a mapping $f : X \to X$ in a normed space X we must have that $\|f(x)-f(y)\| \le L\|x-y\|$, for some constant $L < 1$. The value of L depends upon the choice of norm. If $\|\cdot\|_1$ and $\|\cdot\|_2$ are equivalent, with constants α_1 and α_2 then

$$\|f(x)-f(y)\|_1 \le L\|x-y\|_1 ,$$

implies only that

$$\|f(x)-f(y)\|_2 \le L\, \alpha_1\alpha_2\, \|x-y\|_2 .$$

Thus f may be a contraction mapping for $\|\cdot\|_1$ but not for $\|\cdot\|_2$.

(3) The constants α_1 and α_2 in eqn (2.1) can be very large — take $n = 50{,}000$ in eqn (2.2). If x represents an error in the solution of a problem, a respectable error bound $\|x\|_1 \le K_1$ may become a terrible bound $\|x\|_2 \le \alpha_2 K_1$. This is sometimes the case in numerical linear algebra where it is often possible to obtain good error bounds in the Euclidean norm while one would really like error bounds in the maximum norm.

We conclude this section by pointing out that the norms on a linear space sometimes have useful additional properties. A norm is *strictly convex* if $\|x+y\| = \|x\| + \|y\|$ implies that either $x = 0$ or $y = \lambda x$ for some λ. A norm is *uniformly convex* if for every constant $\mu \in (0,2]$ there exists $\delta = \delta(\mu) > 0$ such that if $\|x\| = \|y\| = 1$ and $\|x-y\| \ge \mu$ then $\|x+y\| \le 2(1-\delta)$. These properties are of importance in the theory of the approximation of functions, particularly as regards the uniqueness of approximations (see Theorem 6.22), and also manifest themselves in the geometric properties of the unit ball, $B[\underline{0};1]$ (see Figure 4.1). In R^n, $\|\cdot\|_1$ and $\|\cdot\|_\infty$ are neither strictly convex nor uniformly convex. (Problem 2.19).

2.8 *INNER - PRODUCT SPACES*

DEFINITION 2.9. A *real (complex) inner product* (x,y) on a real (complex) vector space X is a real-valued (complex-valued) function on $X \times X$ such that

(1) $(x,y) = \overline{(y,x)}$ [the bar denotes the complex conjugate].
(2) $(\lambda x_1 + \mu x_2, y) = \lambda(x_1,y) + \mu(x_2,y)$.
(3) $(x,x) \geq 0$, and $(x,x) = 0$ iff $x = 0$.

A vector space with an inner product is called a *real (complex) pre-Hilbert* space or *inner-product* space. □

Let X be an inner-product space with inner product $(\cdot,\cdot)$. Choose $x,y \in X$ and let $(x,y) = |(x,y)|e^{i\theta}$. For any real number r set $\lambda = re^{i\theta}$. Then, by the conditions of Definition 2.9,

$$\begin{aligned}(x+\lambda y, x+\lambda y) &= (x,x) + (\lambda y,x) + (x,\lambda y) + (\lambda y,\lambda y) ,\\ &= (x,x) + 2\,\mathrm{Re}[\lambda\overline{(x,y)}] + |\lambda|^2(y,y) ,\\ &= (x,x) + 2r\,|(x,y)| + r^2(y,y) , \qquad (*)\\ &= p(r), \text{ say},\\ &\geq 0 .\end{aligned}$$

Now, if the quadratic polynomial $p(r) = ar^2 + 2br + c$, say, is non-negative for all r then the roots of p must either coincide or be complex; that is, $b^2 - ac \leq 0$. Applying this to eqn (*) we obtain that

$$|(x,y)|^2 \leq (x,x)\,(y,y) ,$$

which is the *Schwarz inequality* for the inner product.

Given an inner-product space X define

$$||x|| = (x,x)^{\frac{1}{2}} .$$

Using the three conditions of Definition 2.9 together with the Schwarz inequality for the inner product, it is readily seen that $\|\cdot\|$ satisfies the conditions of Definition 2.7 and is a norm. Thus an inner-product space is a special type of normed space. The norm is strictly convex (Problem 2.28).

EXAMPLE 2.22. The space of real n-tuples $x = (x_1,\ldots,x_n)$ with inner product

$$(x,y) = \sum_{i=1}^{n} x_i y_i \ .$$

is a real inner-product space , E^n . □

Inner-product spaces have many beautiful properties. For example, the concept of two vectors in E^n being orthogonal can be extended to inner-product spaces, and this has many important consequences (see Section 8.3). For extensive treatments of inner-product spaces see Dunford and Schwartz [1966a] and Halmos [1967].

Inner-product spaces do arise frequently in numerical analysis, but there are many more problems which involve normed spaces but not inner-product spaces.

2.9 *CHARACTERIZATIONS OF SPACES*

It is of interest to determine whether a given topological vector space is *normable* or *metrisable*, that is whether the topology can be defined by a norm or a metric. This is particularly pertinent for us, because our justification for discussing general topological spaces is that while most problems in numerical analysis can be formulated using normed linear spaces this is not always possible.

We recall that a space X is a Hausdorff space if for each distinct pair $x, y \in X$ there exist neighbourhoods N_x and N_y with $N_x \cap N_y = \emptyset$. Clearly, every normed space or metric space is a Hausdorff space. Furthermore, normed spaces and metric spaces have countable local bases, that is, local bases of the form $\beta = \{N(n);\ n = 1,2,\ldots\}$. Finally, a set A in a topological vector space is *topologically bounded* iff for each neighbourhood U of $\underline{0}$ there exists a real number λ

such that $A \subset \lambda U$. (See Problem 2.25). It can be shown (Kelley and Namioka [1976, p. 44 and 49]) that

THEOREM 2.7. *A topological vector space is metrisable iff it is a Hausdorff space and has a countable local base.*

A topological vector space is normable iff it is a Hausdorff space and there is a topologically bounded convex neighbourhood of $\underline{0}$. □

Finally (Dunford and Schwartz [1966, p. 393]):

THEOREM 2.8. *A normed linear space with norm* $||\cdot||$ *has an inner-product satisfying* $(x,x) = ||x||^2$ *iff the parallelogram identity holds, namely that*

$$||x-y||^2 + ||x+y||^2 = 2[||x||^2 + ||y||^2] . \quad \square$$

In connection with these theorems see Problems 2.20, and 2.22-2.24.

2.10 *CONDITION NUMBERS*

The concepts of topological spaces can give both qualitative and quantitative insight into the solution of numerical problems. In a typical problem in numerical analysis we are given input data z in a topological space Z and are required to provide output x in a topological space X. We assume that the output is given by a single valued mapping $f : Z \to X$.

EXAMPLE 2.23. Consider the problem of solving the equation $A(x) = b$ where $b \in R^2$, and $A : R^2 \to R^2$ is defined by

$$A : x = \begin{pmatrix} x_1 \\ x_2 \end{pmatrix} \to \begin{pmatrix} a_{11}x_1 + a_{12}x_2 \\ a_{21}x_1 + a_{22}x_2 \end{pmatrix} .$$

Here, the input data consists of the four coefficients a_{ij} of the matrix A and the two coefficients of the vector b so that $Z = R^4 \times R^2$. The output data consists of the two components of the solution $x = A^{-1}b$ so that $X = R^2$. □

In some problems the input data z is obtained from experiments and is subject to error. In other cases, the input data may be known exactly but errors occur during the process of solution. Thus, rather than being given a single piece of data z , we are usually concerned with a neighbourhood N_z of z , which is mapped by f into a set $S_x \subset X$ (Figure 2.3).

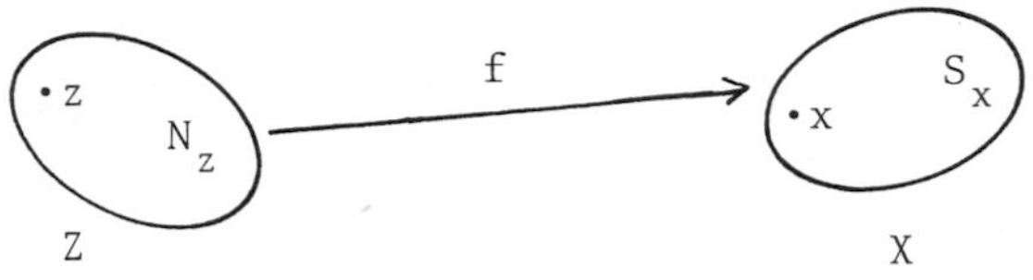

Figure 2.3: The solution mapping f

If small changes in the input data z give rise to large changes in the output x then the problem is *ill-conditioned*; otherwise it is *well-conditioned*. Naturally, ill-conditioned problems give rise to severe numerical difficulties. Examples of ill-conditioned problems are given in Problems 2.26, 5.31, Example 9.3 and Remark 9.6.

If X and Z are normed spaces we can quantify these concepts. If $\|z\| \neq 0$ and $\|f(z)\| \neq 0$ then the *relative condition number* $C_r(z)$ *of the problem at the point* z is

$$C_r(z) = \lim_{\delta\to 0}\left[\sup_{0<\|u\|<\delta} \frac{\|f(z+u)-f(z)\|/\|f(z)\|}{\|u\|/\|z\|}\right] , \qquad (2.2)$$

In words, for input data near z the relative change in the output is bounded by $C_r(z)$ times the relative change in the data. In Problem 2.26, $C_r(z)$ is computed for the problem of calculating the roots of a polynomial.

In the special case when f is a linear one-to-one map of Z onto X then, for $\|z\| \neq 0$, $C_r(z)$ does not depend upon z and $C_r(z) = C_r$ where (Problem 2.29),

$$C_r = \left[\sup_{\|u\|=1} \frac{\|f(u)\|}{\|u\|}\right]\left[\sup_{\|x\|=1} \frac{\|f^{-1}(x)\|}{\|x\|}\right] \tag{2.3}$$

*PROBLEMS**

2.1 Given two topologies τ_1 and τ_2 for a space X with neighbourhood bases $\beta^{(1)} = \{\beta_x^{(1)}\}$ and $\beta^{(2)} = \{\beta_x^{(2)}\}$, show that: τ_1 is stronger than τ_2 $\Longleftrightarrow$ for each $x \in X$ and each $N_x^{(2)} \in \beta_x^{(2)}$, there exists $N_x^{(1)} \in \beta_x^{(1)}$ satisfying $N_x^{(1)} \subset N_x^{(2)}$. {13}

2.2 Let X be a topological space with topology τ, and let Y be a subset of X. Show that Y is a topological space with respect to the *induced topology* or *relative topology* τ_Y: $U \in \tau_Y$ iff $U = G \cap Y$ where $G \in \tau$. {13}

2.3 Let $K(R^n)$ denote the collection of closed bounded sets in R^n. For $S \in K(R^n)$ let

$$U(S;\delta) = \{x \in R^n : \|x-y\|_2 < \delta \text{ for some } y \in S\},$$

where $\|\cdot\|_2$ denotes Euclidean distance. For $S_1, S_2 \in K(R^n)$ let

$$\delta(S_1,S_2) = \inf\{\delta : U(S_1;\delta) \supset S_2\},$$

$$\rho(S_1,S_2) = \delta(S_1,S_2) + \delta(S_2,S_1).$$

Show that $K(R^n)$ is a metric space with distance function ρ. (Eggleston [1963, p. 60]). {15, 206}

2.4 Suppose that one is given a set X with n elements and a collection of m subsets, τ. By forming $G_1 \cup G_2$ and $G_1 \cap G_2$ for all possible $G_1, G_2 \in \tau$, and checking whether these sets belong to τ, one can determine whether τ endows X with a topology. But this requires $O(m^2)$ checks. Is there a more efficient way? {13}

*If a problem is referred to in the text, the page number of the reference is given in brackets.

2.5 Let X be the set of continuous real functions on $[0,1]$. Consider the families $\beta_x = \{N_x(n) : n = 1,2,\ldots\}$,
$$N_x(n) = \{y : |y(t)-x(t)| \le 1/n, \text{ for } 0 \le t \le 1\} .$$
Show that this is a base of neighbourhoods for a topology. {13}

2.6 Let X be the set of continuous real functions on $[0,1]$. Let $t_1,\ldots,t_n$ be n fixed points in $[0,1]$. If $H_1,\ldots,H_n$ are sets in R^1 and $f \in X$ let
$$U_f(H_1,\ldots,H_n) = \{g \in X : (f(t_k)-g(t_k)) \in H_k \text{ for } 1 \le k \le n\} .$$
Let $\beta_f = \{U_f(G_1,\ldots,G_n) :\ G_1,\ldots,G_n$ are open sets in R^1 which contain $0\}$. Show that $\beta = \{\beta_f : f \in X\}$ is a base of neighbourhoods for a topology on the space X. {13}

2.7 If topologies are defined as in Problems 2.5 and 2.6, which is stronger? {13}

2.8 If f is a real (single-valued) function on $[0,1]$ we will say that f has property D if for all $a,b \in [0,1]$ and every β between $f(a)$ and $f(b)$ there exists α between a and b such that $f(\alpha) = \beta$. Show that every continuous function has property D, but that there also exist discontinuous functions with property D. (Property D is the *Darboux property* (Halperin [1950])).

2.9 Show that the composition of continuous mappings is continuous. {25}

2.10 Show that $(f \circ g)^{-1} = g^{-1} \circ f^{-1}$.

2.11 Let x_0 be a point in a topological vector space X. Use the concept of a projection map on product spaces to show that the map $f: x \in X \to (x,x_0) \in X \times X$ is continuous. Hence show that the map $g: x \in X \to x + x_0 \in X$ is continuous.

2.12 An *interval number* is an ordered pair of real numbers $[a,b]$ with $a \le b$. Let X denote the set of interval numbers. Define vector addition and scalar multiplication by
$$[a,b] + [c,d] = [a+c,b+d] ,$$
$$\lambda[a,b] = \begin{cases} [\lambda a,\lambda b], & \text{if } \lambda > 0 , \\ [\lambda b,\lambda a], & \text{if } \lambda \le 0 . \end{cases}$$

Let

$$\rho([a,b],[c,d]) = \max(|a-c|,|b-d|) .$$

Show that X is a metric space with distance function ρ. Is X a vector space? (Moore [1966, 1979]). {15, 27}

2.13 On computers it is only possible to represent numbers within a certain range. On the CDC 6600 computer, apart from 'normal' real numbers there are special representations for four 'extended arithmetic' numbers, namely $+\infty$, $-\infty$, +'indefinite', and -'indefinite'. The arithmetic rules used are shown in the tables below, in which X and Y denote normal real numbers, P and Q denote strictly positive normal real numbers, and IND = indefinite.

ADDITION: U+V

U	V: Y	V: +∞	V: -∞	V: ±IND
X	X+Y	+∞	-∞	IND
+∞	+∞	+∞	IND	IND
-∞	-∞	IND	-∞	IND
±IND	IND	IND	IND	IND

SUBTRACTION: U-V

U	V: Y	V: +∞	V: -∞	V: ±IND
X	X-Y	-∞	+∞	IND
+∞	+∞	IND	+∞	IND
-∞	-∞	-∞	IND	IND
±IND	IND	IND	IND	IND

MULTIPLICATION: UV

U	V: +Q	V: -Q	V: +0	V: -0	V: +∞	V: -∞	V: ±IND
+P	+PQ	-PQ	0	0	+∞	-∞	IND
-P	-PQ	+PQ	0	0	-∞	+∞	IND
+0	0	0	0	0	IND	IND	IND
-0	0	0	0	0	IND	IND	IND
+∞	+∞	-∞	IND	IND	+∞	-∞	IND
-∞	-∞	+∞	IND	IND	-∞	+∞	IND
±IND	IND	IND	IND	IND	IND	IND	IND

DIVISION: U/V

U	V: +Q	-Q	+0	-0	+∞	-∞	±IND
+P	+P/Q	-P/Q	+∞	-∞	0	0	IND
-P	-P/Q	+P/Q	-∞	+∞	0	0	IND
+0	0	0	IND	IND	0	0	IND
-0	0	0	IND	IND	0	0	IND
+∞	+∞	-∞	+∞	-∞	IND	IND	IND
-∞	-∞	+∞	-∞	+∞	IND	IND	IND
±IND	IND	IND	IND	IND	IND	IND	IND

Is the space $X = R^1 \cup \{+\infty, -\infty, +\mathrm{IND}, -\mathrm{IND}\}$ a vector space? {27}

2.14 In many computers all floating point numbers x are represented in the form $\pm m\beta^e$ where the integer β is the *radix*, the integer e is the *exponent*, and the *mantissa* m is of the form

$$m = \cdot\beta_1 \ldots \beta_r ,$$

with $1 \le \beta_1 < \beta$ and $0 \le \beta_k < \beta$ for $2 \le k \le r$. Determine, for a particular computer, which of the eleven conditions for a vector space given in Definition 2.4 are valid. (Sterbenz [1974], Kulisch and Miranker [1981]) {27}

2.15 Let X be the real line and define a topology as in Example 2.5(3) with the bases of neighbourhoods

$\beta_x = \{N_x(n) : n = 1,2,\ldots\}$,

$$N_x(n) = \{y : x \le y < x + 1/n\} .$$

Show that this is a vector space and a topological space but not a topological vector space.

2.16 Using Theorems 2.2 and 2.4 prove Theorem 2.5 which characterizes local bases. {31}

2.17 Use Theorem 2.5 to show that the space ℓ^p, $0 < p < 1$, defined in Example 2.18 is a topological vector space. {32}

2.18 Let X be the set of all real sequences $x = (x_1, x_2 \ldots)$. Let

$$p_k(x) = |x_k| , \quad k = 1,2,\ldots .$$

(a) Show that p_k is a seminorm.

(b) Show that the topology on X induced by the family $\{p_k\}$ is the same as the topology on X induced by the metric ρ of Example 2.8. {15}

2.19 Let $\|\cdot\|_1$, $\|\cdot\|_2$, and $\|\cdot\|_\infty$ be defined as in Example 2.21. Show that

(a) $\|\cdot\|_1$ is neither strictly convex nor uniformly convex.

(b) $\|\cdot\|_\infty$ is neither strictly convex nor uniformly convex.

(c) $\|\cdot\|_2$ is both strictly convex and uniformly convex. {39}

2.20 Prove that the conditions for an inner product (Definition 2.9) imply the parallelogram identity:

$$\|x+y\|^2 + \|x-y\|^2 = 2[\|x\|^2 + \|y\|^2] \ . \qquad \{41\}$$

2.21 Let X be the set of real continuous functions on [0,1] . Show that X can be normed using either

$$\|x\|_\infty = \max_{[0,1]} |x(t)| \ ,$$

or

$$\|x\|_1 = \int_0^1 |x(t)|dt \ ,$$

but that these norms are not equivalent.

2.22 Let X be the topological vector space of continuous functions on [0,1] with the topology of pointwise convergence (Example 2.6). Show that X does not have a countable local base so that X is not metrisable. {41}

2.23 Let X be the topological vector space of continuous functions on the interval $(-\infty, +\infty)$ with the topology induced by the seminorms

$$p_n(x) = \max_{-n\le t\le n} |x(t)|, \quad n = 1,2,\ldots \ .$$

Show that X is metrisable but not normable. {41}

2.24 Let X be the normed space C[0,1] of continuous functions on [0,1] with the norm

$$\|x\|_\infty = \max_{[0,1]} |x(t)| \ .$$

Show that X is not an inner-product space. {41}

2.25 If E is a set in a normed linear space we say that E is *bounded in norm* if

$$\sup_{x \in E} \|x\| < \infty .$$

Prove the following

Lemma

In a normed linear space: a set E *is bounded topologically* $\Leftrightarrow$ *E is bounded in norm.* □ {41}

2.26 Let $p(t;z)$ be the real polynomial of degree n,

$p(t;z) = a_0 + a_1 t + \ldots + a_{k-1} t^{k-1} + z t^k + a_{k+1} t^{k+1} + \ldots + a_n t^n$,

where z is a parameter and the remaining coefficients a_i are fixed. The n roots of $p(t;z)$ depend continuously upon z. Let x_0 be one root of p when $z = z_0$, and assume that x_0 is a simple root. Then there is a neighbourhood $N \subset R^1$ of z_0 and a continuous mapping $f: N \to \mathbb{C}$ such that $f(z_0) = x_0$ and $p(f(z);z) = 0$. Show that the condition number for this problem is

$$C_r(z_0) = |z_0||x_0|^{k-1}|\frac{\partial p}{\partial t}(x_0;z_0)|^{-1} . \qquad (*)$$

Show that the 'Wilkinson polynomial' $p(t) = \prod_{i=1}^{20} (t-i)$ has some well-conditioned roots and some ill-conditioned roots. {42}

2.27 Prove 'directly' that the topology on R^n defined using the Euclidean norm $\|\cdot\|_2$ is the same as the product topology for $\prod_{i=1}^{n} R^1$. {195}

2.28 Prove that the norm $\|x\| = (x,x)^{\frac{1}{2}}$ in an inner-product space is strictly convex. {40}

2.29 Verify eqn (2.3), which gives an expression for the condition number for linear mappings. {43}

3
Limits and convergence

In numerical analysis one often computes a sequence of elements $x_1, x_2, \ldots, x_n, \ldots$ in a topological space X ; such a sequence is usually denoted by $\{x_n\}$. The question arises: Does x_n converge to an element $x \in X$ and, if so, in what sense?

3.1 *CONVERGENT SEQUENCES*

DEFINITION 3.1. A sequence $\{x_n\}$ in a topological space X *converges* to $x \in X$ if for every neighbourhood V of x there exists an integer $m(V)$ (depending upon V) such that $x_n \in V$ for all $n \geq m(V)$. The sequence $\{x_n\}$ is *convergent* in X if it converges to some x in X . If $\{x_n\}$ converges to x , then x is a *limit* of the sequence $\{x_n\}$. The fact that $\{x_n\}$ converges to x is expressed in many ways: x_n converges to x (as n tends to infinity); $x_n \to x$ as $n \to \infty$; $x_n \to x$; $\lim x_n = x$; $\lim\limits_{n\to\infty} x_n = x$. □

The limit of a sequence need not be unique: in the indiscrete topology (Example 2.5) every sequence converges to every point. However, if X is a Hausdorff space then every convergent sequence has a unique limit: for, if $x_n \to x'$ and $x_n \to x''$ where x' and x'' are distinct then, for large n , x_n must belong to disjoint neighbourhoods of x' and x'' , which is impossible.

THEOREM 3.1. *In a metric space* X *with metric* ρ , *the following are equivalent:*

(1) $x_n \to x$ *as* $n \to \infty$.

(2) $\rho(x_n, x) \to 0$ *as* $n \to \infty$.

Proof. (1) ⇒ (2): Choose $\varepsilon > 0$. Then $B = B[x;\varepsilon] = \{y \in X : \rho(x,y) \leq \varepsilon\}$ is a neighbourhood of x . Thus, there exists an integer $m(\varepsilon)$ such that $x_n \in B$ for

$n \geq m(\varepsilon)$; that is, $\rho(x_n,x) \leq \varepsilon$ for $n \geq m(\varepsilon)$. Since ε was arbitrary, $\rho(x_n,x) \to 0$.

(2) ⇒ (1): Assume that $\rho(x_n,x) \to 0$, and choose any neighbourhood V of x . Since X is a metric space the balls $B[x;1/k] = \{y \in X : \rho(x,y) \leq 1/k\}$, $k = 1,2,\ldots$, form a base of neighbourhoods of x , so that $B[x;1/i] \subset V$ for some i . Choosing an integer $m(V)$ so that $\rho(x_n,x) \leq 1/i$ for $n \geq m(V)$, we see that $x_n \in B[x;1/i] \subset V$ when $n \geq m(V)$ and conclude that $x_n \to x$. □

Let X be a topological space and let $E \subset X$. Then $x \in X$ is a *closure point* of E if, for every neighbourhood V of x, $V \cap E \neq \emptyset$; that is, every neighbourhood of x contains at least one point of E . The set of closure points of E is called the *closure* of E and is denoted by $\overline{E}$. Some of the properties of the closure operation are given in Problem 3.2. If there exists a sequence $\{x_n\}$ in E which converges to a point $x \in X$ then x is a *limit point* of E .

THEOREM 3.2. *Let* X *be a metric space and let* $E \subset X$. *Then*

(1) $x \in \overline{E} \Leftrightarrow x$ *is a limit point of* E .

(2) E *is closed* $\Leftrightarrow E = \overline{E}$. *(This holds in any topological space).*

(3) E *is closed* $\Leftrightarrow E$ *contains all its limit points.*

Proof. *(1) ⇒:* Consider the family of spherical neighbourhoods $B[x;1/n]$. For every n , there must exist $x_n \in E$ such that $x_n \in B[x;1/n]$. That is, $\rho(x,x_n) \leq 1/n$ so that $x_n \to x$ and hence x is a limit point of E .

(1) ⇐: Since x is a limit point of E there exists a sequence $\{x_n\}$ in E which converges to x , so that if V is any neighbourhood of x there exists at least one x_n (and, in fact, infinitely many x_n) belonging to V . Thus, $V \cap E \neq \emptyset$, and hence $x \in \overline{E}$.

(2) ⇒: If E is closed then $X \setminus E$ is open. Let $x \in X \setminus E$. From Theorem 2.1, there exists a neighbourhood V of x with $V \subset X \setminus E$. Thus $V \cap E = \emptyset$ and $x \notin \overline{E}$. Since x was an arbitrary point in $X \setminus E$, we conclude that $\overline{E} \subset E$. On the

other hand, if $x \in E$ and V is a neighbourhood of x then $V \cap E$ is not empty because it contains x. Thus $E \subset \overline{E}$. Combining the above results we see that $E = \overline{E}$.

(2) $\Leftarrow$: Let $x \in X \setminus E$. Since $x \notin E = \overline{E}$, there exists a neighbourhood V of x such that $V \cap E = \emptyset$. That is, $V \subset X \setminus E$. Since x was an arbitrary point in $X \setminus E$, we conclude from Theorem 2.1 that $X \setminus E$ is open and hence that E is closed.

(3): In the course of the above arguments we showed that E is always a subset of $\overline{E}$. Therefore, using *(1)*, we see that $(E = \overline{E}) \Leftrightarrow$ (E contains all its limit points). Applying *(2)*, we obtain *(3)*. □

Theorem 3.2 is not true for every topological space as is shown by:

EXAMPLE 3.1. Let X be the space of continuous real functions on $[0,1]$ with the topology of pointwise convergence (Example 2.6). Let E be the set of functions f in X such that: (1) $0 \le f(t) \le 1$ for $t \in [0,1]$; and (2) the set of points at which $f(t) \ge 1/2$ has length (more precisely, measure) at least 1/2.

In X the zero element $\underline{0}$ is the function which is identically zero. We assert that $\underline{0} \in \overline{E}$. To see this, let V be any neighbourhood of $\underline{0}$. Then, there exists a finite subset T of $[0,1]$ and $\varepsilon > 0$ such that

$$B_{\underline{0}}[T;\varepsilon] \equiv \{f \in X : |f(t)| \le \varepsilon \quad \text{for} \quad t \in T\} \subset V .$$

We can clearly construct $f \in E$ such that $f(t) = 0$ for $t \in T$, so that $f \in B_{\underline{0}}[T;\varepsilon]$; thus, $V \cap E \neq \emptyset$, and $\underline{0} \in \overline{E}$.

If a sequence $\{f_n\}$ in E converges to $\underline{0}$ then $f_n(t) \to 0$ as $n \to \infty$ for all $t \in [0,1]$. It follows from the Lebesgue dominated convergence theorem (see page 104) that

$$\lim_{n\to\infty} \int_0^1 f_n(t)dt = \int_0^1 \lim f_n(t)dt = \int_0^1 0 \, dt = 0 .$$

But this is impossible because $f_n \in E$ and hence

$$\int_0^1 f_n dt \geq 1/4 .$$

Thus, no sequence in E can converge to $\underline{0}$.

We have thus shown that $\underline{0} \in \overline{E}$ but that $\underline{0}$ is not a limit point of E . □

It is possible to generalize the concept of a convergent sequence so that Theorem 3.2 is true for every topological space. This can be done as follows.

DEFINITION 3.2. A set A is *partially ordered* with respect to an *order relation* $\geq$, which is defined for some, but not necessarily all, pairs of elements from A , if:

(1) $a_1 \in A \Rightarrow a_1 \geq a_1$

(2) $a_1, a_2, a_3 \in A$, $a_1 \geq a_2$, and $a_2 \geq a_3 \Rightarrow a_1 \geq a_3$.

If furthermore,

(3) $a_1 \geq a_2$ and $a_2 \geq a_1 \Rightarrow a_1 = a_2$,

then A is *well-ordered*.

We write $a_2 \leq a_1$ if $a_1 \geq a_2$. □

EXAMPLE 3.2. Let A be the set of pairs of integers $a = (n,m)$. A is partially ordered with respect to the order relation $\geq$ defined by: $(n_1,m_1) \geq (n_2,m_2)$ if n_1 is not less than n_2 and m_1 is not less than m_2 . □

A partially ordered set A is a *directed set* or *net* if given $a_1, a_2 \in A$ there exists an *upper bound* $a_3 \in A$ such that $a_3 \geq a_1$ and $a_3 \geq a_2$. Let A be a directed set and X be a topological space, and let $x_\alpha \in X$ for each $\alpha \in A$. Then $\{x_\alpha\}$ is a *generalized sequence*, and $\{x_\alpha\}$ *converges to* $x \in X$ if, for every neighbourhood V of x , there exists $\beta \in A$ such that $x_\alpha \in V$ if $\alpha \geq \beta$; we write $x_\alpha \to x$ or $\lim x_\alpha = x$ or $x_\alpha \to x(\alpha \in A)$. A sequence $\{x_n\}$ is a special case of a generalized sequence, the set A being the set of positive integers. We obtain the following generalization of Theorem 3.2:

THEOREM 3.3. *Let* X *be a topological space, and let* $E \subset X$. *Then* $x \in \overline{E}$ *iff there exists a generalized sequence* $\{x_\alpha : \alpha \in A\}$ *such that* $x_\alpha \in E$ *and* $x_\alpha \to x$.

Outline of Proof. Use the directed set A consisting of all neighbourhoods of the point x; if $U \in A$ and $V \in A$ then $U \geq V$ if $U \subset V$. □

The concept of a generalized sequence may seem rather esoteric, but it does in fact occur quite often in numerical analysis.

EXAMPLE 3.3. Consider the problem of computing the integral of a continuous real function $f(t)$ from 0 to 1. One approach is to use the midpoint rule. Let Δ be any *partition* of $[0,1]$,

$$\Delta = \{0 = t_0 < t_1 < \ldots < t_n = 1\} .$$

We say that $\Delta_2 \geq \Delta_1$ if all the points t_i of Δ_1 belong to Δ_2. The set A of partitions is a partially ordered set with respect to $\geq$. A is also directed since if $\Delta_1, \Delta_2 \in A$ and Δ_3 is the partition containing the points of both Δ_1 and Δ_2 then Δ_3 is an upper bound for Δ_1 and Δ_2. For each Δ we define $x_\Delta \in R^1$ by

$$x_\Delta = \sum_{i=1}^{n} (t_i - t_{i-1})\, f((t_i + t_{i-1})/2) .$$

It follows from the theory of Riemann integration that

$$x_\Delta \to \int_0^1 f(t)dt \quad (\Delta \in A) . \quad □$$

3.2 *CAUCHY SEQUENCES AND COMPLETE SPACES*

We now introduce an important criterion for determining whether a given sequence is convergent. A sequence $\{x_n\}$ in a metric space X is a *Cauchy sequence* if $\rho(x_m, x_n) \to 0$ as $m,n \to \infty$; that is, given $\varepsilon > 0$ there exists an integer $N(\varepsilon)$ (depending on ε) such that if $m,n \geq N(\varepsilon)$, then

$\rho(x_m,x_n) < \varepsilon$. Every convergent sequence is a Cauchy sequence, because if $x_n \to x$ then

$$\rho(x_n,x_m) \le \rho(x_n,x) + \rho(x_m,x) \to 0 \ .$$

A metric space is *complete* if every Cauchy sequence is convergent. Hence, in a complete metric space a sequence $\{x_n\}$ is convergent iff it is a Cauchy sequence. A complete normed space is a *Banach space*. A complete inner-product space is a *Hilbert space*. Finally, a *Fréchet space* is a complete linear metric space with an invariant metric. The concepts of a Cauchy sequence and completeness can also be extended to generalized sequences in linear topological spaces (Kelly and Namioka [1976, p. 56]): a generalized sequence $\{x_\alpha : \alpha \in A\}$ in a linear topological space is a Cauchy sequence if for every neighbourhood V of $\underline{0}$ there exists $\gamma \in A$ such that if $\alpha \ge \gamma$ and $\beta \ge \gamma$ then $x_\alpha - x_\beta \in V$.

The property of completeness is of great importance in numerical analysis. Suppose that we have an algorithm for generating a sequence $\{x_n\}$ in a metric space X . If we suspect that $x_n \to x$ and know something about x then we can check whether $\rho(x_n,x) \to 0$. In many situations however we do not know x explicitly, and we may even not know whether x exists. In such cases we can check whether $\{x_n\}$ is a Cauchy sequence. If so, and if X is complete, then the sequence must converge to some $x \in X$.

Not every metric space is complete as is shown by:

EXAMPLE 3.4. The rational numbers form a metric space with metric $\rho(x,y) = |x-y|$, but this space is not complete. □

EXAMPLE 3.5. Let X be the normed space of continuous real functions on $[0,1]$ with the norm

$$||x|| = \left[\int_0^1 [x(t)]^2 dt\right]^{\frac{1}{2}} ,$$

and let $x_n(t)$ be the function which is equal to one for $0 \le t \le 1/2 - 1/2n$, equal to zero for $1/2 + 1/2n \le t \le 1$, and is linear elsewhere. Direct computation shows that

$||x_n - x_m|| \to 0$ so that $\{x_n\}$ is a Cauchy sequence.

We claim that if the sequence $\{x_n\}$ converges to $x \in X$ then $x(t)$ is equal to one for $0 < t < 1/2$. To see this, suppose that $t \in (0,1/2)$ and $x(t) \neq 1$. Since x is continuous, there is an interval $I = [t-\varepsilon, t+\varepsilon] \subset (0,1/2)$ and a number $\eta > 0$ such that $|x(s)-1| \geq \eta$ for $s \in I$. However, for large n, $x_n(s) = 1$ for $s \in I$ and hence

$$||x - x_n|| \geq \left[\int_I [x(s)-1]^2 ds\right]^{\frac{1}{2}} \geq [2\varepsilon\eta^2]^{\frac{1}{2}} ,$$

which contradicts the assumption that $x_n \to x$.

We have thus shown that if $x_n \to x$ then $x(t) = 1$ for $t \in (0,1/2)$. Similarly, $x(t) = 0$ for $t \in (1/2,1)$. Thus x is discontinuous at $t = 1/2$, $x \notin X$, and X is not complete. □

3.3 *COMPLETION OF SPACES*

The real line R^1 is a complete space which is constructed by defining the real numbers to be the limits of Cauchy sequences of rational numbers. For example, we may think of a real number as an infinite decimal $\cdot a_1 a_2 a_3 \ldots a_n \ldots$ which is the limit of the finite decimals $r_n = \cdot a_1 a_2 \ldots a_n$ each of which is a rational number. There are two aspects of this construction which require clarification. Firstly, the rational numbers X can be identified with a subset, $\tilde{X}$ say, of the real numbers. Secondly, there is more than one construction of the real numbers. The above construction is due to Cantor. One can, however, construct the real numbers as *Dedekind sections* in which a real number x is represented as the set of all rational numbers which are less than x.

Two topological vector spaces X and Y are *isomorphic* if there exists a bicontinuous linear mapping f of X onto Y.

If X and Y are normed spaces then (Problem 3.5) they are isomorphic iff there is a linear mapping f of X onto Y and constants $\alpha_1 > 0$, $\alpha_2 > 0$ with

$$\alpha_1\|x\|_X \le \|f(x)\|_Y \le \alpha_2\|x\|_X \ . \tag{3.1}$$

If eqn (3.1) holds with $\alpha_1 = \alpha_2 = 1$ then X and Y are *isometrically isomorphic*. The rational numbers are isometrically isomorphic to a subset of the real numbers, and the 'Cantor' real numbers are isometrically isomorphic to the 'Dedekind' real numbers.

Before giving a generalization of the Cantor construction of the real numbers, we need two definitions. If E and F are subsets of a topological space X and $F \subset \overline{E}$ then E is *dense* in F. A topological space X is *separable* if it has a denumerable dense subset.

THEOREM 3.4. *Let* X *be a topological vector space. Then*
(1) There exists a complete topological vector space $\hat{X}$ *with a dense linear subspace* $\tilde{X}$ *such that* $\tilde{X}$ *is isomorphic to* X *.* $\hat{X}$ *is a* *<u>completion</u>* *of* X *.*
(2) If X *is also isomorphic to a dense subspace* $\tilde{Y}$ *of a complete topological vector space* $\hat{Y}$*, then* $\hat{X}$ *and* $\hat{Y}$ *are isomorphic.*
(3) Let X *be a locally convex topological vector space of type: (a) Hausdorff; (b) metric with invariant metric; (c) normed; (d) inner-product. Then* $\hat{X}$ *is a space of type: (a) Hausdorff, (b) Fréchet; (c) Banach; (d) Hilbert. In cases (c) and (d),* $\hat{X}$ *may be chosen so that* $\tilde{X}$ *is isometrically isomorphic to* X .

Proof. The proof in the general case is quite sophisticated and we refer the reader to Kelley and Namioka [1976, p. 63]. Here, we only prove (1) in the case when X is a real normed space; the proof is rather lengthy but we give it in some detail because it illustrates the concept of a topological vector space. Also, the present theorem is used to construct the L^p spaces (Section 4.11) and the Sobolev spaces, which are of great importance in numerical functional analysis, so that it is important for the reader to understand clearly the process of construction.

We assume that X is a normed space and construct $\hat{X}$ using Cauchy sequences in X. Two Cauchy sequences $\{x_n\}$ and $\{x_n'\}$ are *equivalent* if $\|x_n - x_n'\| \to 0$; we write $\{x_n\} \sim \{x_n'\}$. If $\{x_n\} \sim \{x_n'\}$ and $\{x_n'\} \sim \{x_n''\}$ then

$$\|x_n - x_n''\| \le \|x_n - x_n'\| + \|x_n' - x_n''\| \to 0 \ ,$$

so that $\{x_n\} \sim \{x_n''\}$. The set of all Cauchy sequences can be divided into equivalence classes, where two Cauchy sequences belong to the same equivalence class $\hat{x}$ iff they are equivalent. $\hat{X}$ is defined to be the set of all equivalence classes $\hat{x}$. If $\{x_n\} \in \hat{x}$ we will say that $\{x_n\}$ is a *representative* of $\hat{x}$.

Before continuing with the proof, we observe that the construction of $\hat{X}$ from X parallels the construction of the real numbers from the rational numbers: the two sequences

$$3{\cdot}0,\ 3{\cdot}1,\ 3{\cdot}14,\ 3{\cdot}141,\ 3{\cdot}1415,\ 3{\cdot}14159,\dots$$
$$3,\ 22/7,\ 311/99,\ 355/113,\ 3195/1017$$

belong to the equivalence class of all rational sequences converging to the real number π .

To prove (1) we must do the following: (a) define vector addition and scalar multiplication in $\hat{X}$; (b) show that $\hat{X}$ is a vector space; (c) define a norm $\|\cdot\|_\wedge$ for $\hat{X}$; (d) show that X is isomorphic to a subspace $\tilde{X}$ of $\hat{X}$; (e) show that $\hat{X}$ is complete; and (f) show that $\tilde{X}$ is dense in $\hat{X}$.

Step a: Define vector addition and scalar multiplication in $\hat{X}$. Let $\hat{x}$ and $\hat{y}$ belong to $\hat{X}$. Choose $\{x_n\} \in \hat{x}$ and $\{y_n\} \in \hat{y}$. Then,

$$\|(x_n+y_n)-(x_m+y_m)\| \le \|x_n-x_m\| + \|y_n-y_m\| \to 0 \ ,$$

as $n, m \to \infty$. Thus $\{x_n+y_n\}$ is a Cauchy sequence and defines an equivalence class, $\hat{u}$ say, in $\hat{X}$. The equivalence class $\hat{u}$ is independent of the choice of representatives of $\hat{x}$ and $\hat{y}$, because if $\{x_n'\}$ and $\{y_n'\}$ are two other representatives then $\|x_n'-x_n\| \to 0$ and $\|y_n'-y_n\| \to 0$ so that

$$\|(x_n'+y_n')-(x_n+y_n)\| \le \|x_n'-x_n\| + \|y_n'-y_n\| \to 0 \ ;$$

thus $\{x_n'+y_n'\} \sim \{x_n+y_n\}$ and hence $\{x_n'+y_n'\} \in \hat{u}$. We define the vector sum $\hat{x}+\hat{y}$ to be $\hat{u}$.

Let $\hat{x} \in \hat{X}$ and $\lambda \in R^1$. Choose $\{x_n\} \in \hat{x}$. Then

$$||\lambda x_n - \lambda x_m|| \le |\lambda| \; ||x_n - x_m|| ,$$

so that $\{\lambda x_n\}$ is a Cauchy sequence which defines an equivalence class, $\hat{u}$ say. As in the case of vector addition it can be shown that $\hat{u}$ is independent of the choice of $\{x_n\} \in \hat{x}$, and we define $\lambda\hat{x} = \hat{u}$.

Step b: $\hat{X}$ *is a vector space.* We show that $\hat{X}$ is a vector space by verifying the conditions of Definition 2.4. Consider condition (1) of Definition 2.4. Let $\hat{x}, \hat{y}, \hat{z} \in \hat{X}$. Choose representatives $\{x_n\}$, $\{y_n\}$, and $\{z_n\}$. Then $(\hat{x}+\hat{y}) + \hat{z}$ is the equivalence class containing $\{(x_n+y_n)+z_n\}$ while $\hat{x} + (\hat{y}+\hat{z})$ is the equivalence class containing $\{x_n+(y_n+z_n)\}$. Since X is a vector space, $(x_n+y_n) + z_n = x_n + (y_n+z_n)$ so that $(\hat{x}+\hat{y}) + \hat{z} = \hat{x} + (\hat{y}+\hat{z})$ and condition (1) has been proved for $\hat{X}$. The remaining conditions of Definition 2.4 may be proved in a similar way, and we conclude that $\hat{X}$ is indeed a vector space. The zero element $\hat{0}$ in $\hat{X}$ is the equivalence class of sequences converging to $\underline{0} \in X$, and one representative of $\hat{0}$ is the sequence $\underline{0}, \underline{0}, \underline{0}, \ldots$.

Step c: Define a norm in $\hat{X}$. Let $\hat{x} \in \hat{X}$ and choose a representative $\{x_n\}$. Then,

$$||x_n|| \le ||x_n - x_m|| + ||x_m|| ,$$

so that

$$||x_n|| - ||x_m|| \le ||x_n - x_m|| .$$

This inequality also holds if m and n are interchanged, so that

$$|\,||x_n|| - ||x_m||\,| \le ||x_n - x_m|| \to 0 ,$$

as $m,n \to \infty$, and hence $\{||x_n||\}$ is a Cauchy sequence in R^1. We may thus define

$$||\hat{x}||_\wedge = \lim_{n\to\infty} ||x_n|| ,$$

after checking, as usual, that the limit does not depend upon the choice of representatives.

To show that $||\cdot||_\wedge$ is a norm we must verify conditions (1) to (3) of Definition 2.7. Since $||x_n|| \ge 0$ we have

$\|\hat{x}\|_\wedge \geq 0$, while if $\|\hat{x}\|_\wedge = 0$ then $\{x_n\} \sim \underline{0}, \underline{0}, \underline{0}, \ldots$ and $\hat{x} = \hat{0}$; thus condition (1) holds. Condition (2) is obviously true. To prove condition (3) we note that if $\hat{x}, \hat{y}, \hat{z} \in \hat{X}$ have representatives $\{x_n\}$, $\{y_n\}$, and $\{z_n\}$, then

$$\begin{aligned} \|\hat{x}-\hat{y}\|_\wedge &= \lim \|x_n - y_n\| , \\ &\leq \lim[\|x_n - z_n\| + \|z_n - y_n\|] , \\ &= \lim \|x_n - z_n\| + \lim \|z_n - y_n\| , \\ &= \|\hat{x}-\hat{z}\|_\wedge + \|\hat{z}-\hat{y}\|_\wedge . \end{aligned}$$

We recall that a vector space with a norm is a topological vector space (with the norm topology), and conclude that $\hat{X}$ is a topological vector space.

Step d: Construct $\tilde{X}$. Define $f : X \to \hat{X}$ by $f : x \to \hat{x}$ where $\hat{x}$ is the equivalence class containing the sequence $x, x, \ldots, x, \ldots$ consisting of copies of x . Let $\tilde{X}$ be the range of f . Clearly $f(x+y) = f(x) + f(y)$ and $f(\lambda x) = \lambda f(x)$ so that f is a linear mapping of X onto $\tilde{X}$. If $\tilde{x}$ and $\tilde{y}$ belong to $\tilde{X}$ then so do $\tilde{x} + \tilde{y}$ and $\lambda\tilde{x}$, so that $\tilde{X}$ is a subspace of $\hat{X}$. Finally, using Problem 3.5 and noting that

$$\|f(x) - f(y)\|_\wedge = \lim_{n\to\infty} \|x-y\| = \|x-y\| ,$$

we conclude that X and $\tilde{X}$ are isometrically isomorphic.

Step e: $\hat{X}$ *is complete.* Let $\{\hat{x}^{(k)}\} = \hat{x}^{(1)}, \hat{x}^{(2)}, \ldots,$ be a Cauchy sequence in $\hat{X}$. Let

$$\{x_n^{(k)}\} = x_1^{(k)}, x_2^{(k)}, \ldots, x_n^{(k)}, \ldots ,$$

where $x_n^{(k)} \in X$, be a representative of $\hat{x}^{(k)}$. Since $\{x_n^{(k)}\}$ is a Cauchy sequence in X , there is an integer $N_k \geq k$ such that $\|x_n^{(k)} - x_m^{(k)}\| \leq 2^{-k}$ when $n, m \geq N_k$. Set $y_k = x_{N_k}^{(k)} \in X$. Then

$$\|y_k - y_\ell\| \leq \|x_{N_k}^{(k)} - x_n^{(k)}\| + \|x_n^{(k)} - x_n^{(\ell)}\| + \|x_n^{(\ell)} - x_{N_\ell}^{(\ell)}\| ,$$

for any n . Letting $n \to \infty$ we conclude that

$$\|y_k - y_\ell\| \leq 2^{-k} + \|\hat{x}^{(k)} - \hat{x}^{(\ell)}\|_\wedge + 2^{-\ell} ,$$

so that $\{y_k\}$ is a Cauchy sequence in X and defines an equivalence class $\hat{y} \in \hat{X}$. Furthermore,

$$\begin{aligned} \|\hat{x}^{(k)} - \hat{y}\|_\wedge &= \lim_{n\to\infty} \|x_n^{(k)} - y_n\| , \\ &\le \sup_{n \ge N_k} (\| x_n^{(k)} - y_k \| + \|y_k - y_n\|) , \\ &= \sup_{n \ge N_k} (\| x_n^{(k)} - x_{N_k}^{(k)} \| + \|y_k - y_n\|) , \\ &\le 2^{-k} + \sup_{n \ge k} \|y_k - y_n\| . \end{aligned}$$

Remembering that $\{y_k\}$ is a Cauchy sequence, we conclude that $\{\hat{x}^{(k)}\} \to \hat{y}$ in $\hat{X}$. Since $\{\hat{x}^{(k)}\}$ was an arbitrary Cauchy sequence, $\hat{X}$ is complete.

Step f: $\tilde{X}$ *is dense in* $\hat{X}$. Let $\hat{x} \in \hat{X}$, and choose a representative $\{x_n\}$. If $f : X \to \hat{X}$ as in Step d above, we have that

$$\|\hat{x} - f(x_k)\|_\wedge = \lim_{n\to\infty} \|x_n - x_k\| \le \sup_{n \ge k} \|x_n - x_k\| .$$

Since $\{x_n\}$ is a Cauchy sequence it follows that $f(x_k) \to \hat{x}$ in $\hat{X}$ as $k \to \infty$. Remembering that $\tilde{X}$ is the range of f, we conclude that $\hat{x}$ is in the closure of $\tilde{X}$ and thus $\tilde{X}$ is dense in $\hat{X}$. □

The process of completion described in Theorem 3.4 permits us to assume that the spaces with which we are working are complete. This assumption brings many advantages. For example, the assumption that the underlying space is complete provides us with very simple existence proofs for certain boundary value problems. Unfortunately, the process of completion also brings certain disadvantages:

(1) The elements in $\hat{X} \setminus \tilde{X}$ are often 'abnormal'. One need think only of the everyday meaning of the word 'irrational'. It will be recalled that the Greeks were confounded by irrational numbers, and that a satisfactory treatment of irrational numbers was first given in the nineteenth century. In the theory of elliptic differential equations, and elsewhere, one often first establishes the existence

of a 'weak' solution $\hat{x} \in \hat{X}$ and then tries to prove that $\hat{x}$ has nice properties so that in fact $\hat{x}$ is a 'classical' solution and belongs to $\tilde{X}$. Furthermore, the elements of $\hat{X} \setminus \tilde{X}$ are often less 'accessible' numerically than the elements of $\tilde{X}$, because they have been constructed using a limit process which is beyond the capabilities of a computer; we can represent rational numbers exactly in a computer but we can only approximate irrational numbers.

(2) The construction used in Theorem 3.4 does not provide a 'concrete' representation of the elements in $\hat{X}$. Such a representation can sometimes be obtained by other methods. For example, in Definition 4.1 we introduce the space $L^2[0,1]$ as a completion of the continuous functions on $[0,1]$, but $L^2[0,1]$ can also be defined as the space of measurable functions which are square-integrable (see the Appendix to Chapter 4), and the latter definition allows us to think of an element of $L^2[0,1]$ as a function and not just a Cauchy sequence. The 'concrete' characterization of spaces obtained by completing spaces of functions has been considered in some generality by Aronszajn and Smith [1956] and Fuglede [1957].

At the beginning of Chapter 2 we pointed out that it is necessary to achieve a balance between the assumptions made about a space and the number of possible applications. This balance is apparently best achieved by complete normed spaces, that is, Banach spaces, which have a rich theory and the simplicity of a normed metric, but also have many applications. In the remainder of this book we will nearly always use Banach spaces, but the reader may be sure that as the subject develops and as new applications arise, other types of spaces will come to play an important rôle.

3.4 *THE RÔLE OF SEQUENCES IN NUMERICAL ANALYSIS*

Only in a very few instances can one compute the solution x to a problem. Usually, one computes a sequence $\{x_n\}$ of approximate solutions. The following possibilities occur:

1. *Convergence proofs*

Here we wish to show that $x_n \to x$. We may do this in several ways:

(a) By estimating $\|x-x_n\|$ and showing that it converges to zero. (e.g. Theorem 9.6)

(b) By showing that $\{x_n\}$ is a Cauchy sequence in a complete space. Example: contraction mappings.

(c) By showing that every subsequence $\{x_{n_i}\}$ contains a sub-subsequence $\{x_{n_{i_j}}\}$ which converges to x. (Problem 3.10 and Section 6.3.3)

2. *Error estimates*

Here we wish to obtain estimates of the form

$$\|x-x_n\| \le F(n;x) ,$$

where F is a function of n and x which may involve constants that depend non-trivially upon x, such as bounds for the derivatives of x. Example: finite element approximations.

3. *Rates of convergence*

Typically, we show that, for some constants K, p, and n_0,

$$\|x_{n+1}-x\| \le K\|x_n-x\|^p , \quad \text{for } n \ge n_0 .$$

If $p = 1$ and $K < 1$ the convergence is *linear*, if $p = 2$ the convergence is *quadratic*. If $p = 1$ and

$$\limsup_{n\to\infty} \frac{\|x_{n+1}-x\|}{\|x_n-x\|} = 0 ,$$

the convergence is *superlinear*. Example: Newton's method. None of these conditions need hold (Prob. 3.9).

4. *Weak convergence*

In some cases $\{x_n\}$ does not converge in the given topology. It is often possible to introduce a weaker topology in which $\{x_n\}$ does converge.

5. *Acceleration of convergence*

If the sequence converges slowly then we may try to transform the sequence so as to improve the convergence. Example: Euler's transformation (Section 5.4.1).

PROBLEMS

3.1 Let X be a space with two topologies τ_1 and τ_2, where τ_1 is stronger than τ_2. Show that if $x_n \to x$ in the topology τ_1, then $x_n \to x$ in the topology τ_2.

3.2 Prove that the closure operation has the following properties in a topological space X:

(1) $\overline{\emptyset} = \emptyset$;

(2) $E \subset \overline{E}$;

(3) If $E_1 \subset E_2$ then $\overline{E}_1 \subset \overline{E}_2$;

(4) $\overline{E}$ is a closed set, that is, $X \setminus \overline{E}$ is open;

(5) E closed $\Rightarrow \overline{E} = E$.

(6) $\overline{E_1 \cup E_2} = \overline{E}_1 \cup \overline{E}_2$;

(7) $\overline{E}$ is the intersection of all the closed sets that contain E. {50}

3.3 Let E be a subset of R^2 with polygonal boundary and let f be a real continuous function defined for $t \in E$. For each triangulation Δ of E define

$$x_\Delta = \sum_{T \in \Delta} |T| \; f(C_T) \; .$$

Here, $|T|$ is the area of the triangle T, and $C_T = (P_1+P_2+P_3)/3$ where P_1, P_2, and P_3 are the vertices of T. Show that, in an appropriate sense,

$$x_\Delta \to \int_E f(t)dt \; .$$

3.4 If A is a directed set then $A_0 \subset A$ is *cofinal* in A if, for every $a \in A$, there exists $a_0 \in A_0$ with $a_0 \geq a$. (This generalizes the concept of a subsequence). Show that if A_0 is cofinal in A and $\{x_\alpha : \alpha \in A\}$ is a generalized sequence in a topological space such that $x_\alpha \to x \; (\alpha \in A)$, then

$$x_\alpha \to x \; (\alpha \in A_0) \; .$$

3.5 Let X and Y be normed linear spaces. Show that X and Y are isomorphic $\Longleftrightarrow$ there exists a linear mapping f of X onto Y and constants $\alpha_1, \alpha_2 > 0$

satisfying

$$\alpha_1 \|x\|_X \leq \|f(x)\|_Y \leq \alpha_2 \|x\| \quad . \qquad (*)$$

[Hint: Recall that if $f : X \to Y$ is continuous then there exists a ball $B_Y[\underline{0};\varepsilon] \subset X$ which is mapped by f into the unit ball $B_Y[\underline{0};1]$ in Y.] {55, 59}

3.6 Prove that if ρ is an invariant metric and $\|\cdot\|$ is defined by $\|x\| = \rho(x,\underline{0})$ then:

(a) $\|x\| \geq 0$, and $\|x\| = 0$ iff $x = \underline{0}$,
(b) $\|-x\| = \|x\|$,
(c) $\|x+y\| \leq \|x\| + \|y\|$;

that is, $\|\cdot\|$ is a *quasinorm*. Conversely, show that if $\|\cdot\|$ is a quasinorm and ρ is defined by $\rho(x,y) = \|x-y\|$, then ρ is an invariant metric.

3.7 Let C_0 be the space of continuous real functions on $(-\infty,+\infty)$ with *compact support* that is, for each $f \in C_0$ there is a finite interval $[-a,+a]$ (depending upon f) outside which f is zero. Let C_v be the space of continuous real functions on $(-\infty,+\infty)$ which 'vanish at infinity'; that is, for each $f \in C_v$ and each $\varepsilon > 0$ there exists a finite interval $[-a,+a]$ (depending upon f) outside which $|f| \leq \varepsilon$. Let C_0 and C_v be each equipped with the maximum norm, $\|f\| = \sup_{-\infty<t<+\infty} |f(t)|$.

(a) Show that C_0 and C_v are normed spaces.
(b) Show that C_v is a Banach space but that C_0 is not complete.
(c) Show that C_0 is dense in C_v .

3.8 Let $BF[0,1]$ be the space of real bounded functions on $[0,1]$ with norm

$$\|x\| = \sup_{[0,1]} |x(t)| .$$

Show that $BF[0,1]$ is a Banach space, but that $BF[0,1]$ is not separable.

[Hint: The functions which are either 0 or 1 at each point of $[0,1]$ belong to X, and the set of all such functions is not denumerable.]

3.9 Show that the real sequence $\{x_n\}$, $x_n = (\sin(n))/n^2$, is neither linearly convergent, nor quadratically convergent, nor superlinearly convergent.
[Hint: According to a theorem of Hurwitz (Niven and Zuckerman [1964, p. 149]) given any irrational number ζ there exist infinitely many rational numbers n/k such that

$$|\zeta - \frac{n}{k}| \leq \frac{1}{\sqrt{5}k^2}\,.]$$

{62}

3.10 Let the sequence $\{x_n\}$ in a topological space X be such that every subsequence has a sub-subsequence which converges to $x \in X$. Show that $\{x_n\}$ converges to x. {62}.

3.11 Let X be a real (complex) topological vector space with subsets V and W. Prove:
(a) $\overline{\alpha V} = \alpha\overline{V}$, for any scalar α.
(b) $\overline{V+w} = \overline{V}+w$, for all $w \in W$.
(c) $\overline{V} + \overline{W} \subset \overline{V+W}$.
(d) Let $X = R^2$, $V = \{(s,t): st = 1 \text{ and } s,t \geq 0\}$, and $W = \{(s,t): s = 0\}$. Then V and W are closed, $\overline{V} + \overline{W}$ is the open right half plane, and $\overline{V+W}$ is the closed right half plane.
{222}

3.12 Let $X = \prod_{i=1}^{\ell} X_i$ where each X_i is a Banach space. Show that X is a Banach space with respect to each of the norms,

$$\|x\|_1 \equiv \|x;X\|_1 = \|(x_1,\ldots,x_\ell)\|_1 = \sum_{i=1}^{\ell} \|x_i;X_i\|\,,$$

$$\|x\|_2 \equiv \|x;X\|_2 = \|(x_1,\ldots,x_\ell)\|_2 = [\sum_{i=1}^{\ell} (\|x_i;X_i\|)^2]^{\frac{1}{2}}\,,$$

$$\|x\|_\infty \equiv \|x;X\|_\infty = \|(x_1,\ldots,x_\ell)\|_\infty = \max_{1\leq i\leq \ell} \|x_i;X_i\|\,.$$

{159, 287}

4
Basic spaces and problems

In this chapter we introduce most of the spaces which are used in numerical analysis.

4.1 *PRELIMINARY REMARKS*

The spaces that we discuss are considered in many texts. The concise presentation in Chapter 4 of Dunford and Schwartz [1966], is a particularly useful source of information.

For each space we mention some of the problems in numerical analysis which can be formulated in that space. Most of these problems have been intensively studied and have an extensive literature; we merely give a few key references.

Unless explicitly stated to the contrary:

(1) Each space considered is a real separable Banach space.

(2) Addition and multiplication are defined in the obvious way:

(a) If $z = (z_i)$ and $w = (w_i)$ are n-tuples or sequences, and λ is a scalar, then

$$\begin{aligned} z + w &= (z_i + w_i) \ , \\ \lambda z &= (\lambda z_i) \ . \end{aligned} \tag{4.1}$$

(b) If $f = f(t)$ and $g = g(t)$ are functions of t defined on some set T and λ is a scalar then

$$\begin{aligned} (f+g)(t) &= f(t) + g(t) \ , \\ (\lambda f)(t) &= \lambda f(t) \ . \end{aligned} \tag{4.2}$$

Some knowledge of Lebesgue measure and integration is assumed; but the essential facts are summarized in an Appendix at the end of this chapter.

4.2 *THE INEQUALITIES OF HÖLDER AND MINKOWSKI*

For each space X we must prove that if $x, y \in X$ then $x+y \in X$ and $||x+y|| \leq ||x|| + ||y||$. To prove this the inequalities of

Hölder and Minkowski are frequently used and we begin by discussing these very useful inequalities. Let $z = (z_i)$ and $w = (w_i)$ be n-tuples or sequences. Let $f(t)$ and $g(t)$ be functions that are Lebesgue measurable on some set T. We say that (z_i) and (w_i) are *proportional* if there exist numbers α, β, with $|\alpha| + |\beta| > 0$ such that $\alpha z_i + \beta w_i = 0$ for all i. We say that f and g are *effectively proportional* if there exist numbers α, β, with $|\alpha| + |\beta| > 0$ such that $\alpha f(t) + \beta g(t) = 0$ a.e. (almost everywhere).

Hölder's inequalities: Let p and q be strictly positive numbers satisfying

$$\frac{1}{p} + \frac{1}{q} = 1 .$$

If

$$\sum_i |z_i|^p < \infty , \quad \text{and} \quad \sum_i |w_i|^q < \infty ,$$

then

$$\sum_i |z_i w_i| \le [\sum_i |z_i|^p]^{\frac{1}{p}} [\sum_i |w_i|^q]^{\frac{1}{q}} .$$

There is inequality unless $((z_i)^p)$ and $((w_i)^q)$ are proportional.

If

$$\int_T |f(t)|^p dt < \infty , \quad \text{and} \quad \int_T |g(t)|^q dt < \infty ,$$

then

$$\int_T |f(t)g(t)| dt \le [\int_T |f(t)|^p dt]^{\frac{1}{p}} [\int_T |g(t)|^q dt]^{\frac{1}{q}} .$$

There is inequality unless f^p and g^q are effectively proportional.

There are special names for Hölder's inequality in the case $p = q = 2$: for sequences it is called the *Cauchy inequality* and for integrals the *Schwarz* or *Bunjakowski inequality* (these names can be combined or interchanged).

Minkowski's inequalities: Let $p \geq 1$ be a real number. If

$$\sum_i |z_i|^p < \infty , \quad \text{and} \quad \sum_i |w_i|^p < \infty ,$$

then

$$[\sum_i |z_i + w_i|^p]^{\frac{1}{p}} \leq [\sum_i |z_i|^p]^{\frac{1}{p}} + [\sum_i |w_i|^p]^{\frac{1}{p}} .$$

If $p > 1$ then there is inequality unless (z_i) and (w_i) are proportional.

If

$$\int_T |f(t)|^p dt < \infty , \quad \text{and} \quad \int_T |g(t)|^p dt < \infty ,$$

then

$$[\int_T |f(t) + g(t)|^p dt]^{\frac{1}{p}} \leq [\int_T |f(t)|^p dt]^{\frac{1}{p}} + [\int_T |g(t)|^p dt]^{\frac{1}{p}} .$$

If $p > 1$ there is inequality unless f and g are effectively proportional.

Detailed discussion of the Hölder and Minkowski inequalities will be found in the book of Hardy, Littlewood, and Polya [1967]. It is possible to prove both inequalities as special cases of general theorems due to Riesz and Jensen.

4.3 *THE SPACES* E^n *AND* ℓ_p^n $(p \geq 1)$

E^n denotes the space, introduced in Example 2.22, of real n-tuples

$$x = (x_1,\ldots,x_n) \equiv (x_i)$$

with inner-product $(\cdot,\cdot)$ defined by

$$(x,y) = \sum x_i y_i \equiv \sum_{i=1}^{n} x_i y_i \equiv x^T y .$$

E^n is an inner-product space--the *real n-dimensional Euclidean space.*

As shown in Section 2.8, the inner product $(\cdot,\cdot)$ gives rise to a norm

$$\|x\|_2 = [\sum_{i=1}^{n} x_i^2]^{\frac{1}{2}} ,$$

and an invariant metric

$$\rho(x,y) = [\sum_{i=1}^{n} (x_i - y_i)^2]^{\frac{1}{2}} .$$

E^n is a locally convex topological vector space. For each $x \in E^n$ the family $\beta_x = \{B(x;1/n) : n=1,2,...\}$,

$$B(x;1/n) = \{y \in E^n : \|x-y\|_2 < 1/n\} ,$$

is a base of convex open neighbourhoods of x . The family $\beta = \{\beta_x : x \in E^n\}$ is a base of neighbourhoods for E^n . The family $\beta_{\underline{0}} = \{B(\underline{0};1/n) : n=1,2,...\}$ is a local base for E^n .

Since E^n is an inner-product space the parallelogram identity holds, namely

$$(\|x+y\|_2)^2 + (\|x-y\|_2)^2 = 2((\|x\|_2)^2 + (\|y\|_2)^2) .$$

If $\|x\|_2 = \|y\|_2 = 1$ and $\|x-y\|_2 \geq \mu$, then it follows from the parallelogram identity that

$$(\|x+y\|_2)^2 \leq 4 - \mu^2 ,$$

so that

$$\|x+y\|_2 \leq 2(1-\mu^2/4)^{\frac{1}{2}} ,$$

$$\leq 2(1-\mu^2/8) .$$

Consequently, $\|\cdot\|_2$ is uniformly convex.

If $\|x+y\|_2 = \|x\|_2 + \|y\|_2$ then it follows from Minkowski's inequality for n-tuples with $p = 2$ that x and y are proportional; thus, $\|\cdot\|_2$ is also strictly convex.

Let e^i , $1 \leq i \leq n$, be the unit vectors

$$e^i = \{\delta_{i1}, \delta_{i2}, ..., \delta_{in}\} ,$$

where δ_{ij} is the *Kronecker delta*

$$\delta_{ij} = \begin{cases} 1, & \text{if } i=j \,, \\ 0, & \text{otherwise.} \end{cases}$$

Then the e^i form a basis in E^n since, for any $x \in E^n$,

$$x = (x_i) = \sum_{i=1}^{n} x_i e^i \,.$$

Clearly $\sum_{i=1}^{n} \lambda_i e^i = 0$ iff $\lambda_i = 0$ for all i, so the vectors e^i are linearly independent. Thus E^n has dimension n.

Next, we show that E^n is complete. Let $\{x^{(k)}\}$, $k = 1,2,\ldots$, be a Cauchy sequence in E^n, that is,

$$\|x^{(k)} - x^{(\ell)}\|_2 \to 0 \quad \text{as} \quad k,\ell \to \infty \,.$$

For each i, $1 \le i \le n$, we have $|x_i^{(k)} - x_i^{(\ell)}| \le \|x^{(k)} - x^{(\ell)}\|_2$, so that

$$|x_i^{(k)} - x_i^{(\ell)}| \to 0 \quad \text{as} \quad k,\ell \to \infty \,.$$

Hence, $\{x_i^{(k)}\}$ is a Cauchy sequence in R^1, the real line. Since R^1 is known to be complete, there exists $x_i \in R^1$ such that

$$x_i^{(k)} \to x_i \quad \text{as} \quad k \to \infty \,.$$

Set $x = (x_1, x_2, \ldots, x_n)$. Choose $\varepsilon > 0$. Then there exist integers N_i such that $|x_i^{(k)} - x_i| \le \varepsilon/\sqrt{n}$ for $k \ge N_i$. Let $N = \max_{1 \le i \le n} N_i$. For $k \ge N$,

$$\|x^{(k)} - x\|_2 = [\sum_{i=1}^{n} |x_i^{(k)} - x_i|^2]^{\frac{1}{2}} \le [\sum_{i=1}^{n} (\varepsilon/\sqrt{n})^2]^{\frac{1}{2}} = \varepsilon \,.$$

Thus, $x^{(k)} \to x$ as $k \to \infty$. But $\{x^{(k)}\}$ was an arbitrary Cauchy sequence, so E^n is complete.

Finally, let S be the set of points of E^n with rational coefficients. Since the rational numbers are dense in R^1, each $x \in E^n$ can be approximated arbitrarily closely by points $s \in S$. Hence $\overline{S} = E^n$ so that S is dense in E^n. But, as will be shown below, S is denumerable; that is, the points of S can be put into a one-to-one correspondence with a subset of the integers. Therefore, E^n is separable.

S IS DENUMERABLE:

(1) Each of the sets

$$T_1 = \{0, +1, -1, +2, -2, +3, -3, +4, -4, \ldots\} ,$$

$$T_2 = \{0, +\tfrac{1}{2}, -\tfrac{1}{2}, +\tfrac{2}{2}, -\tfrac{2}{2}, +\tfrac{3}{2}, -\tfrac{3}{2}, +\tfrac{4}{2}, -\tfrac{4}{2}, \ldots\} ,$$

$$\ldots$$

$$T_k = \{0, +\tfrac{1}{k}, -\tfrac{1}{k}, +\tfrac{2}{k}, -\tfrac{2}{k}, +\tfrac{3}{k}, -\tfrac{3}{k}, +\tfrac{4}{k}, -\tfrac{4}{k}, \ldots\} ,$$

$$\ldots$$

is denumerable.

(2) Since there are denumerably many T_k, $T = \cup T_k$ is denumerable. To see this we first count diagonally,

T_1	$0^{(1)}$	$+1^{(2)}$	$-1^{(4)}$	$+2^{(7)}$	$-2^{(11)}$	$+3 \ldots$
T_2	$0^{(3)}$	$+\frac{1}{2}^{(5)}$	$-\frac{1}{2}^{(8)}$	$+\frac{2}{2}^{(12)}$	$-\frac{2}{2}$	$+\frac{3}{2} \ldots$
T_3	$0^{(6)}$	$+\frac{1}{3}^{(9)}$	$-\frac{1}{3}^{(13)}$	$+\frac{2}{3}$	$-\frac{2}{3}$	$+\frac{3}{3} \ldots$
T_4	$0^{(10)}$	$+\frac{1}{4}^{(14)}$	$-\frac{1}{4}$	$+\frac{2}{4}$	$-\frac{2}{4}$	$+\frac{3}{4} \ldots$

to obtain the denumerable set

$$\{0, +1, 0, -1, +\tfrac{1}{2}, 0, +2, -\tfrac{1}{2}, +\tfrac{1}{3}, 0, -2, +\tfrac{2}{2}, -\tfrac{1}{3}, +\tfrac{1}{4}, \ldots\},$$

and then throw away numbers which appear more than once, to obtain

$$T = \{0, +1, -1, +\tfrac{1}{2}, 2, -\tfrac{1}{2}, +\tfrac{1}{3}, -2, -\tfrac{1}{3}, +\tfrac{1}{4}, \ldots\} . \qquad (*)$$

T is of course the set of rational numbers so we have just shown that the rational numbers are denumerable.

(3) Finally, S consists of n-tuples, each component of which belongs to T . If T is ordered,

$$T = \{t_1, t_2, \ldots\} ,$$

we can order S as follows:

$$S = \bigcup_{k=1}^{\infty} S_k ,$$

where S_k is the finite set,

$$S_k = \{(t_{i_1}, t_{i_2}, \ldots, t_{i_n}) : \sum_{j=1}^{n} i_j = k\} .$$

For example, if T is ordered as in eqn (*) and $n = 3$, then

$$\begin{aligned} S_1 &= \{(t_1,t_0,t_0),\ (t_0,t_1,t_0),\ (t_0,t_0,t_1)\} , \\ &= \{(1,0,0),\ (0,1,0),\ (0,0,1)\} \end{aligned}$$

It follows that S is denumerable. □

ℓ_p^n will denote the space of real n-tuples (x_i) with norm $\|\cdot\|_p$, where $p \geq 1$ and

$$\|x\|_p = [\sum_{i=1}^{n} |x_i|^p]^{1/p} , \quad 1 \leq p < \infty ,$$

$$\|x\|_\infty = \max_{1\leq i\leq n} |x_i| , \qquad p = \infty ,$$

so that $\ell_2^n = E^n$. That $\|\cdot\|_p$ is a norm follows from the inequality of Minkowski. The special cases $p = 1$, $p = 2$, and $p = \infty$ have already been encountered in Example 2.21.

As for E^n it can be shown that ℓ_p^n is a separable Banach space of dimension n .

4.4 *THE SPACE* R^n

We have introduced several topologies for the space of real n-tuples: the product topology (Example 2.13), and topologies defined by the norms $\|\cdot\|_p$ (Section 4.3). It is an important property of n-dimensional spaces that all these topologies are the same, and all these norms are equivalent, as is stated by the following theorem which will be proved, for normed spaces, as Theorem 6.8, p. 196.

THEOREM 4.1. *A real (or complex) finite-dimensional vector space has only one topology under which it is a locally convex topological vector space. In particular, all norms in a finite-dimensional vector space are equivalent.* □

R^n will denote the unique real n-dimensional locally convex topological vector space.

Although any two norms in R^n are equivalent, there can still be significant practical differences between norms as we have already observed (page 37). In Figure 4.1 the unit balls in ℓ_p^2 are shown for several values of p. It will be seen that the unit balls for $p = 1$ and $p = \infty$ are polygons, and this is reflected in the fact that $\|\cdot\|_p$ is uniformly convex and strictly convex only when $1 < p < \infty$ (Problem 4.2). In consequence, some problems posed in ℓ_p^n are more difficult in the extreme cases $p = 1$ and $p = \infty$ than in the 'normal' cases $1 < p < \infty$. (See Theorem 6.22, and Examples 7.1 and 7.2.)

We write E^n or ℓ_p^n instead of R^n when the corresponding inner-product or norm plays a significant rôle in the analysis to hand.

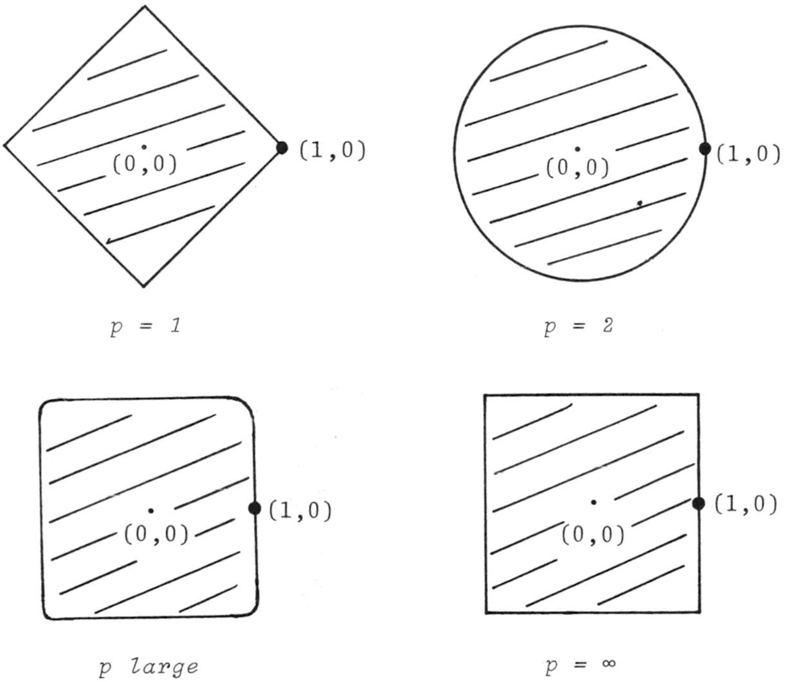

Figure 4.1: Unit balls in ℓ_p^2

PROBLEMS IN R^n

R^n is ubiquitous in numerical analysis because problems in infinite-dimensional spaces must nearly always be approximated by finite-dimensional problems.

Although it would appear very natural to use functional analysis to analyse numerical analysis problems in R^n, this was a relatively late development. The systematic use of norms in matrix analysis is apparently first due to Faddeeva [1952], and even several years later Householder [1958] found it worthwhile to give a detailed introduction to the subject.

Basic problems in R^n include:

(1) The solution of systems of linear algebraic equations,

$$Ax = b \ ,$$

where $x \in R^n$, $b \in R^n$, and A is an $n \times n$ matrix.

See Varga [1962], Wilkinson [1965].

(2) The solution of systems of nonlinear algebraic equations,

$$f(x) = 0 \ ,$$

where $x \in R^n$ and $f : R^n \to R^n$ is a nonlinear mapping.

See Ortega and Rheinboldt [1970], Rheinboldt [1974], Dennis and Schnabel [1979], Wait [1979], and Parlett [1980].

(3) The minimization problem: Compute

$$\min_{x \in C} f(x) \ ,$$

where $f : R^n \to R^1$ and C is a subset of R^n. An example is the *linear programming problem*: Given the real coefficients c_j and a_{ij} find $x = (x_j) \in R^n$ which

$$\text{Minimizes} \quad \sum_{j=1}^{n} c_j x_j \ ,$$

$$\text{Subject to} \quad \sum_{j=1}^{n} a_{ij} x_j \geq 0 \ , \quad \text{for} \quad 1 \leq i \leq m \ .$$

See Section 6.3.1 and Example 7.3.

(4) The approximation problem in ℓ_1^n: Find $x = (x_j)$ in R^n which minimizes

$$f(x) = \sum_{i=1}^{m} |b_i - \sum_{j=1}^{n} a_{ij} x_j| , \qquad (*)$$

where the coefficients b_i and a_{ij} are given. This problem is a discrete version of a continuous problem: Find $x \in R^n$ which minimizes

$$F(x) = \int_0^1 |b(t) - \sum_{j=1}^{n} x_j \phi_j(t)| dt , \qquad (**)$$

where the functions $b(t)$ and $\phi_j(t)$ are given. We obtain problem (*) from (**) by choosing m equidistant points $t_1,\ldots,t_m$ in $[0,1]$, replacing the integral by a sum over the points t_i, and setting $b_i = b(t_i)$, $a_{ij} = \phi_j(t_i)$.

For details about the computational solution of problem (*) see Barrodale and Roberts [1978].

REMARK 4.1. In the preceding discussion of R^n, the value of n did not play a rôle. It should, however, be realized that there are some subtle differences between the spaces R^n of different dimensions:

(1) Among all the spaces R^n, only R^1 is an *ordered topological vector space;* that is, only in R^1 can an order relation $\geq$ be introduced which is defined for all pairs of elements.

If $f : R^1 \to R^1$ is continuous then the equation $f(x) = 0$ can be solved by the *bisection method:*

(a) Find x_p, $x_n \in R^1$ such that $f(x_p) > 0$ and $f(x_n) < 0$.

(b) Compute $x_m = (x_p+x_n)/2$. If $f(x_m) \geq 0$ replace x_p by x_m; otherwise replace x_n by x_m.

(c) Repeat step (b) until x_p and x_n are as close as desired. At each step there exists x lying between x_p and x_n such that $f(x) = 0$.

The bisection method can only be used in R^1 because it depends upon the fact that R^1 is ordered; in particular, for each x_m either $f(x_m) \geq 0$ or $f(x_m) < 0$.

(2) A *continuous vector field* over the *unit sphere* S^n in R^{n+1} ,

$$S^n = \{x \in R^{n+1} : \|x\|_2 = 1\} ,$$

is a continuous mapping $f : S^n \to R^{n+1}$ such that (a) $\|f(x)\|_2 = 1$ and (b) the vector from x to $x+f(x)$ is tangential to S^n at x , that is, $x^T f(x) = 0$. If $n = 1$, $f(x)$ is easily constructed: $f : x = (x_1, x_2)\ (-x_2, x_1)$ (see Figure 4.2).

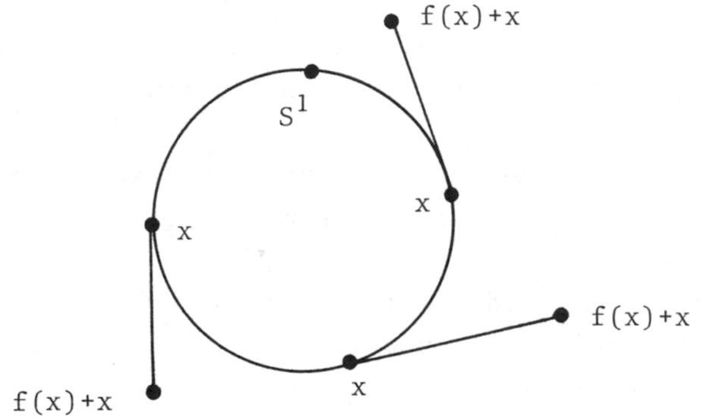

Figure 4.2: A continuous vector field over S^1

However, it can be shown that there exists a continuous vector field over S^n iff n is odd. In particular, one cannot construct a continuous vector field on the surface S^2 of a ball in R^3 . (Hilton and Wylie [1965, p. 219]).

(3) The behaviour of solutions of the wave equation in R^n ,

$$\frac{\partial^2 u}{\partial t^2} = \sum_{i=1}^{n} \frac{\partial^2 u}{\partial x_i^2} ,$$

depends upon n . Disturbances are propagated as sharp signals only if $n \geq 2$ and n is odd *(Huyghen's Principle)*. Furthermore, 'undistorted' signals can be transmitted only if $n = 3$. Thus, human communication is dependent upon the fact that the 'real world' is R^3 . (See Courant and Hilbert [1962, p. 208 and p. 765]).

(4) From a topological point of view, R^3 and R^4 are exceptional spaces. Mandelbaum [1980] concludes his lengthy survey of four-dimensional topology with the words 'It is ironic that the worlds about which we know least are our worlds: the 3-dimensional world of physical geometry and the 4-dimensional world of space-time.'

As these examples show, to describe R^n as a real n-dimensional topological vector space is the truth and nothing but the truth but not the whole truth. □

4.5 *THE SPACE* $\mathbb{C}^n$

$\mathbb{C}^n$ denotes the space of complex n-tuples $z = (z_1, \ldots, z_n)$ over the complex scalars with inner product and norm given by:

$$(z,w) = \sum_{i=1}^{n} z_i \bar{w}_i \ , \quad \|z\| = [\sum_{i=1}^{n} |z_i|^2]^{\frac{1}{2}} \ .$$

Here, $\bar{w}$ denotes the complex conjugate of w, so that if $w = u + iv$ then $\bar{w} = u - iv$. $\mathbb{C}^n$ is a complex separable Hilbert space of dimension n.

PROBLEMS IN $\mathbb{C}^n$

By considering the real and imaginary components separately, it is always possible to reformulate problems in $\mathbb{C}^n$ as problems in R^{2n}, at the expense of losing elegance and simplicity. For an example see Problem 4.4.

Problems in $\mathbb{C}^n$ include:

(1) Computation of complex roots of equations, especially systems of polynomial equations. See Wilkinson [1963], Dejon and Henrici [1969].

(2) Eigenvalue problems for matrices and operators. See Wilkinson [1965], Wilkinson and Reinsch [1971].

(3) Complex numbers, and analytic functions, occur frequently in classical applied mathematics, but seem to occur less frequently in numerical applications.

One reason is that complex function theory is primarily used to handle two-dimensional problems, whereas numerical methods are applicable to a broader class of problems. Garabedian et al (Bauer, Garabedian, and Korn [1977]) have designed supercritical wing sections by, roughly speaking, integrating along a contour in the complex plane so as to avoid singularities on the real axis. For other problems see Henrici [1974] and Garabedian [1964].

(4) Following Henrici [1979] we denote by Π_n the space of bilaterally infinite sequences,

$$x = (x_i)_{i=-\infty}^{+\infty} ,$$

with $x_k \in \mathbb{C}$, which have period n :

$$x_k = x_{k+n} , \quad \text{for all } k .$$

Let $i = \sqrt{-1}$, $w_n = \exp(2\pi i/n)$ and let F_n be the *discrete Fourier transform*

$$F_n : x = (x_k)_{k=-\infty}^{+\infty} \in \Pi_n \to \left(\frac{1}{n}\sum_{k=0}^{n-1} x_k w_n^{-km}\right)_{m=-\infty}^{+\infty} \in \Pi_n \qquad (*)$$

It can be shown (Problem 4.6) that F is a bijective linear map of Π_n onto Π_n . There are n distinct components of each element in Π_n , and the computation of each component of $F_n x$ using eqn (*) requires n complex multiplications, so that the computation of $F_n x$ using eqn (*) requires n^2 complex multiplications. However, there are algorithms, known as *Fast Fourier Transforms*, in which the computation of $F_n x$ is dramatically speeded up. In particular, if $n = 2^m$ then only $\frac{1}{2} n \log_2 n = \frac{1}{2} mn$ complex multiplications are required.

Fast Fourier Transforms have a wide variety of applications including: numerical conformal mapping; fast Poisson solvers; manipulation of power series. (See Henrici [1979]).

4.6 *THE SPACES* ℓ^p, $1 \le p \le \infty$

ℓ^p is the space of real sequences $(x_i)_{i=1}^{\infty}$ with finite norm,

$$\|x\|_p = [\sum_{i=1}^{\infty} |x_i|^p]^{\frac{1}{p}}, \quad 1 \le p < \infty; \quad \|x\|_\infty = \sup_{1 \le i \le \infty} |x_i| .$$

As might be expected, the spaces ℓ^1 and ℓ^∞ are somewhat exceptional: ℓ^p is uniformly convex iff $1 < p < \infty$ (Problem 4.7). ℓ^p is reflexive (this concept is defined in Volume 2) iff $1 < p < \infty$. ℓ^∞ is not separable (Problem 4.8).

PROBLEMS IN ℓ^p

The most important case is $p = 2$ since every real infinite-dimensional separable Hilbert space H is isometrically isomorphic with ℓ^2; that is, there is a one-to-one linear mapping between H and ℓ^2 which preserves norms (see Theorem 8.5). As a result, any problem in an infinite-dimensional separable Hilbert space gives rise to an equivalent problem in ℓ^2. In particular, the spaces $L^2(\Omega)$ (see Section 4.11) and the Sobolev spaces $H^k(\Omega)$ which arise in many problems in partial differential equations, are isometrically isomorphic with ℓ^2.

The use of ℓ^2 instead of L or H^k is especially appropriate in numerical applications where one often seeks to approximate an element $x \in \ell^2$ by "truncated" elements of the form (see Section 8.4)

$$\overline{x} = (x_1, x_2, x_3, \ldots, x_{k-1}, x_k, 0, 0, 0, \ldots) .$$

4.7 *THE SPACE* c

c is the space of real sequences $x = (x_i) = (x_i)_{i=1}^{\infty}$ which are convergent; that is, if $(x_i) \in c$ there is a real number x_∞ such that $x_i \to x_\infty$ as $i \to \infty$. The norm is

$$\|x\| = \sup_{1 \le i < \infty} |x_i| .$$

PROBLEMS IN c

The most important problem in c is the computation of the limit x_∞ of a sequence $x \in c$. There are several methods for transforming slowly convergent sequences. That is, given $(x_i) \in c$ one constructs $(u_i) \in c$ so that $u_i \to u_\infty = x_\infty$ and the convergence of the sequence (u_i) is more rapid than the convergence of the sequence (x_i). See Section 5.4 and Brezinski [1977].

4.8 *THE SPACE* $C[a,b]$

$C[a,b]$ is the space of continuous real-valued functions on $[a,b]$ with the *maximum norm*

$$||x||_\infty = \max_{a \le t \le b} |x(t)| .$$

When no confusion is possible we will write $||\cdot||$ instead of $||\cdot||_\infty$.

A real function $x(t)$ defined on $[a,b]$ is *uniformly continuous* if, given $\varepsilon > 0$, there exists $\delta > 0$ such that $|x(t_1)-x(t_2)| \le \varepsilon$ whenever $|t_1-t_2| \le \delta$. The important point of this definition is that δ depends only upon ε and not upon t_1 and t_2. It is a standard result, which we prove later (Theorem 6.12), that if $x \in C[a,b]$ then x is uniformly continuous on $[a,b]$.

Setting

$$(x+y)(t) = x(t) + y(t) ,$$
$$(\lambda x)(t) = \lambda x(t) ,$$

$C[a,b]$ is clearly a vector space. It is easily checked that $||\cdot||$ is a norm so that $C[a,b]$ is a normed space.

Now let $\{x_n\}$ be a Cauchy sequence in $C[a,b]$ so that

$$\max_{a \le t \le b} |x_n(t)-x_m(t)| \to 0 \quad \text{as } m,n \to \infty .$$

It is a classical result that this implies that there exists a continuous function $x(t)$ such that

$$||x_n - x|| \to 0 \quad \text{as} \quad n \to \infty ,$$

so that $C[a,b]$ is complete. We briefly summarize the arguments leading to the construction of the limit function x. For fixed t, $\{x_n(t)\}$ is a Cauchy sequence in R^1 and so has a limit, $x(t)$ say. That is,

$$|x_n(t)-x(t)| \to 0 \quad \text{as} \quad n \to \infty .$$

We now show that $x(t)$ is continuous and that $x_n \to x$ in $C[a,b]$.

First, choose $\varepsilon > 0$. Then, for some N,

$$\|x_n - x_m\| \leq \varepsilon , \quad \text{for} \quad n,m \geq N ,$$

so that

$$|x_n(t)-x_m(t)| \leq \varepsilon , \quad \text{for} \quad n,m \geq N \quad \text{and} \quad t \in [a,b] .$$

Letting $m \to \infty$,

$$|x_n(t)-x(t)| \leq \varepsilon , \quad \text{for} \quad n \geq N \quad \text{and} \quad t \in [a,b] . \qquad (*)$$

Next, let $t_1, t_2 \in [a,b]$. Then, using eqn (*),

$$|x(t_1)-x(t_2)| \leq |x(t_1)-x_N(t_1)| + |x_N(t_1)-x_N(t_2)| + |x_N(t_2)-x(t_2)| ,$$

$$\leq 2\varepsilon + |x_N(t_1)-x_N(t_2)| .$$

Since $x_N(t)$ is continuous, and hence uniformly continuous, there exists $\delta > 0$ such that $|x_N(t_1)-x_N(t_2)| \leq \varepsilon$ if $|t_1-t_2| \leq \delta$, and so

$$|x(t_1)-x(t_2)| \leq 3\varepsilon , \text{ if } |t_1-t_2| \leq \delta .$$

Combining these results we see that x is continuous and that

$$\|x_n - x\| \to 0 \quad \text{as} \quad n \to \infty ,$$

as asserted. We have thus shown that $C[a,b]$ is a Banach space.

We now show that $C[a,b]$ is separable. To prove this we must construct a denumerable set $\tilde{X} \subset C[a,b]$ which is dense in $C[a,b]$.

Let $\pi[a,b]$ denote the set of all polynomials $p(t)$ defined on $[a,b]$. Then

THEOREM 4.2. (Weierstrass). *$\pi[a,b]$ is dense in $C[a,b]$. That is, given $x \in C[a,b]$ and $\varepsilon > 0$ there exists $p \in \pi[a,b]$ such that*

$$\|x-p\|_\infty = \max_{a\le t\le b} |x(t)-p(t)| \le \varepsilon .$$

Proof. Many proofs are known. We give a proof due to Bernstein.

We begin by observing that it suffices to consider the special case $[a,b] = [0,1]$. For, suppose that the theorem is true in this special case. Given $x \in C[a,b]$ and $\varepsilon > 0$ let $\tilde{x}$ be defined by

$$\tilde{x}(t) = x(a+(b-a)t), \quad t \in [0,1] .$$

Then $\tilde{x} \in C[0,1]$ and so, by assumption, there exists $\tilde{p} \in \pi[0,1]$ such that

$$\|\tilde{x}-\tilde{p}\| \equiv \|\tilde{x}-\tilde{p}\|_{C[0,1]} \le \varepsilon .$$

If p is defined by

$$(u) = \tilde{p}((u-a)/(b-a)), \quad u \in [a,b] ,$$

then $p \in \pi[a,b]$. But

$$\|x-p\|_{C[a,b]} = \|\tilde{x}-\tilde{p}\|_{C[0,1]} \le \varepsilon ,$$

and we conclude that the theorem is true in the general case.

Next we derive an inequality. The binomial expansion

$$\sum_{k=0}^{n} \binom{n}{k} \alpha^k \beta^{n-k} = (\alpha+\beta)^n , \tag{*}$$

holds for any α and β. Differentiating eqn (*) with respect to α and then multiplying by α we find that

$$\sum_{k=0}^{n} k\binom{n}{k} \alpha^k \beta^{n-k} = n\alpha(\alpha+\beta)^{n-1} . \tag{**}$$

Repeating this operation once more we obtain

$$\sum_{k=0}^{n} k^2 \binom{n}{k} \alpha^k \beta^{n-k} = [n(n-1)\alpha^2+n(\alpha+\beta)\alpha](\alpha+\beta)^{n-2} . \tag{***}$$

Setting $\alpha = t$, $\beta = 1 - t$, and forming the linear combination $t^2(*) - 2t/n(**) + 1/n^2(***)$ we obtain the identity

$$\sum_{k=0}^{n} \binom{n}{k} \left(t - \frac{k}{n}\right)^2 t^k (1-t)^{n-k} = t(1-t)/n \,. \qquad \left(\substack{**\\**}\right)$$

Introduce the notation,

$$\Sigma' = \Sigma' \binom{n}{k} t^k (1-t)^{n-k} = \sum_{\substack{k=0\\ |k/n-t| \le n^{-1/4}}}^{n} \binom{n}{k} t^k (1-t)^{n-k} \,,$$

$$\Sigma'' = \Sigma'' \binom{n}{k} t^k (1-t)^{n-k} = \sum_{\substack{k=0\\ |k/n-t| > n^{-1/4}}}^{n} \binom{n}{k} t^k (1-t)^{n-k} \,.$$

Using eqn $\left(\substack{**\\**}\right)$ we see that

$$\begin{aligned} \Sigma'' &\le \sum_{\substack{k=0\\ |k/n-t| \ge n^{-1/4}}}^{n} \binom{n}{k} [(t-k/n)^2 n^{1/2}] t^k (1-t)^{n-k} \,, \\ &\le n^{1/2} \sum_{k=0}^{n} \binom{n}{k} (t-k/n)^2 t^k (1-t)^{n-k} \,. \\ &= t(1-t)/\sqrt{n} \,, \\ &\le 1/(4\sqrt{n}) \,, \end{aligned} \qquad \left(\substack{**\\***}\right)$$

which is the desired inequality.

We now proceed to the proof of the theorem. Let $x \in C[0,1]$, and choose $\varepsilon > 0$. Let $B(n;x)$ denote the *Bernstein polynomial*

$$B(n;x)(t) = \sum_{k=0}^{n} x(k/n) \binom{n}{k} t^k (1-t)^{n-k} \,.$$

Using the inequality $\left(\substack{**\\***}\right)$, we see that

$$|x(t) - B(n;x)(t)| = \left| \sum_{k=0}^{n} (x(t) - x(k/n)) \binom{n}{k} t^k (1-t)^{n-k} \right| \,,$$

$$\le 2 \max_{0\le t\le 1} |x(t)| \sum'' + \max_{\substack{0\le t\le 1 \\ |t-k/n|\le n^{-1/4}}} |x(t)-x(k/n)| \sum' ,$$

$$\le \frac{1}{2\sqrt{n}} \|x\|_\infty + \max_{\substack{0\le t\le 1 \\ |t-k/n|\le n^{-1/4}}} |x(t)-x(k/n)| .$$

Since x is uniformly continuous, there exists $\delta > 0$ such that $|x(t_1)-x(t_2)| \le \varepsilon/2$ if $|t_1-t_2| \le \delta$. Choosing N so that $\|x\|_\infty/2\sqrt{N} < \varepsilon/2$ and $N^{-1/4} \le \delta$, we conclude that $\|x-B(N;x)\| \le \varepsilon$ and the proof of the theorem is complete. □

Now let $\tilde{\pi}[a,b]$ denote the set of all polynomials on $[a,b]$ with rational coefficients. If $p \in \pi[a,b]$ then p is of the form,

$$p(t) = \sum_{i=0}^{n} a_i t^i ,$$

where the a_i are real numbers. For any $\varepsilon > 0$ choose rational numbers $\tilde{a}_i$ satisfying

$$|\tilde{a}_i - a_i| \le \varepsilon/((n+1)[\max(1,|a|,|b|)]^n) ,$$

and set

$$\tilde{p}(t) = \sum_{i=0}^{n} \tilde{a}_i t^i .$$

Then $\tilde{p} \in \tilde{\pi}[a,b]$ and

$$\|\tilde{p}-p\|_\infty \le \sum_{i=0}^{n} |\tilde{a}_i - a_i| \; |t|^i \le \varepsilon .$$

Using Theorem 4.2 we conclude that $\tilde{\pi}[a,b]$ is dense in $C[a,b]$.

We now show that $\tilde{\pi}[a,b]$ is denumerable. To see this, we note that

$$\tilde{\pi}[a,b] = \bigcup_{n=1}^{\infty} \tilde{\pi}_n[a,b] ,$$

where $\tilde{\pi}_n[a,b]$ is the set of polynomials on $[a,b]$ of degree n with rational coefficients. But there is a one-to-one correspondence between $\tilde{\pi}_n[a,b]$ and the set of rational (n+1)-tuples, namely,

$$\sum_{i=0}^{n} \tilde{a}_i t^i \leftrightarrow (\tilde{a}_0, \tilde{a}_1, \ldots, \tilde{a}_n) .$$

In showing that E^n is separable we showed that the set of rational n-tuples is denumerable. Hence $\tilde{\pi}_n[a,b]$ is denumerable, and $\tilde{\pi}[a,b]$ is the countable union of denumerable sets. Using the 'diagonal counting' method used in Section 4.3, it follows that $\tilde{\pi}[a,b]$ is denumerable.

In summary, $\tilde{\pi}[a,b]$ is a denumerable set which is dense in $C[a,b]$; that is $C[a,b]$ is separable.

REMARK 4.2. On a computer the only functions that can be evaluated are polynomials with rational coefficients and simple combinations thereof. The fact that the polynomials with rational coefficients are dense in $C[a,b]$ means that we can actually compute approximations to functions in $C[a,b]$. For this reason, the Weierstrass Theorem (Theorem 4.2) may be thought of as the 'Fundamental Theorem of Numerical Analysis'.

$C[a,b]$ has many similarities with the spaces ℓ_∞^n , and the unit ball in $C[a,b]$ may be thought of as an infinite-dimensional version of the unit ball in ℓ_∞^2 shown in Figure 4.1, that is, an infinite-dimensional cube. It is therefore no surprise that $C[a,b]$ is neither uniformly convex nor strictly convex (Problem 4.11). Furthermore, $C[a,b]$ is not reflexive. □

PROBLEMS IN $C[a,b]$

After R^n, $C[a,b]$ is probably the most widely used space in numerical functional analysis. Not only are many problems formulated in $C[a,b]$, but many other problems involving functions of several variables are solved by solving a succession of problems in $C[a,b]$. Basic problems include:

(1) *Approximation*

Given a function $x \in C[a,b]$ and a subspace $\tilde{X}$ in $C[a,b]$ we wish to determine $\tilde{x} \in \tilde{X}$ which approximates x as well as possible. For example, we may wish to determine the polynomial $\tilde{x}(t) = a_0 + a_1 t^1 + a_2 t^2 + a_3 t^3$ which approximates $x(t) = \sin t$ as well as possible on the interval $[-1,+1]$; here, $\tilde{X}$ is the four-dimensional subspace of X spanned by the functions $1, t, t^2, t^3$.

See Section 6.3.2, Example 7.2, and Cheney [1966].

(2) *Quadrature*

Compute numerically

$$\int_a^b x(t)\,dt \ .$$

See Sections 5.3 and 5.4.2, Davis and Rabinowitz [1975], and Krylov [1962].

(3) *Integral equations*

Given $g \in C[a,b]$ and a continuous kernel $k(s,t)$ defined for $a \le s,\ t \le b$, find $u \in C[a,b]$ such that

$$u(s) + \int_a^b k(s,t)u(t)\,dt = g(s), \quad \text{for} \quad a \le s \le b \ .$$

There are many possible generalizations: the kernel can be discontinuous, have unpleasant singularities, or be a nonlinear function of x

See Chapter 9, as well as Baker [1977], Zabreiko et al [1975], Jaswon and Symm [1977], Delves and Walsh [1974], and Atkinson [1976].

4.9. *THE SPACE* $C(\overline{\Omega})$, Ω *BOUNDED*

Let Ω be an open bounded set in R^n . $C(\overline{\Omega})$ denotes the space of real functions which are continuous on $\overline{\Omega}$, the closure of Ω , with the maximum norm

$$||x|| \equiv ||x;C(\overline{\Omega})|| \equiv ||x||_{\infty,\Omega} = \max_{t \in \overline{\Omega}} |x(t)| \ .$$

The arguments of p. 81 can be extended to show that $C(\overline{\Omega})$ is complete. By an extension of Theorem 4.2, the Stone-Weierstrass Theorem, it can be shown that the polynomials with rational coefficients are dense in $C(\bar{\Omega})$ so that $C(\bar{\Omega})$ is separable. (Rudin [1964, p. 150])

PROBLEMS IN $C(\bar{\Omega})$

The basic problems in $C(\bar{\Omega})$ are the same as those for the special case $C[a,b]$ discussed above. However, the problems are much more difficult. The reasons for this include:

(1) The number of computations required to achieve a given accuracy grows rapidly with the number of spatial dimensions. As an example, consider the problem of integrating a smooth function x over a cube in n dimensions, by the quadrature formula

$$\int_{-1}^{1} dt_1 \cdots \int_{-1}^{1} dt_i \cdots \int_{-1}^{1} x(t_1,\ldots,t_n)\,dt_n$$

$$\doteq \left(\frac{1}{N}\right)^n \sum_{k_1=-N}^{N} \sum_{k_2=-N}^{N} \cdots \sum_{k_n=-N}^{N} x(k_1/N, k_2/N, \ldots, k_n/N) ,$$

where N is a large integer. The error is $O(1/N)$ and the number of computations is $O(N^n)$. On many current computers arithmetic operations take about 1 microsecond = 10^{-6} seconds, and even on the fastest computers at present available arithmetic operations require several nanoseconds (one nanosecond = 10^{-9} seconds). Thus, in one dimension we could take $N = 10^6$ in the above approximation and obtain good accuracy with a few seconds of computer time, but if $n \geq 2$ then we would have to take N to be much smaller and accept lower accuracy.

(2) The theory of the approximation of functions in n spatial dimensions is much more difficult if $n \geq 2$ than if $n = 1$. For example, as shown in Example 7.1, there is more than one polynomial of best approximation of the form $a_0 + a_1 s + a_2 t + a_3 s^2 + a_4 t^2$ to the function $x(s,t) = st$ in $C([0,1] \times [0,1])$. (This is not

unrelated to the fact that R^n is an ordered topological vector space iff $n = 1$.)

(3) In one spatial dimension, problems are usually formulated on an interval $\Omega = (a,b)$. In two or more spatial dimensions, irregular domains Ω arise quite naturally and give rise to many difficulties. For example, it is easy to construct high accuracy quadrature formulae for an integral

$$\int_\Omega x(t)dt$$

if $\Omega = (a,b) \subset R^1$, but it is extremely difficult to do so if Ω is a general region in R^n with $n \geq 2$. (See Theorem 7.13 and Stroud [1971]).

4.10 *THE SPACES* $C^m(\Omega)$, $C^m(\overline{\Omega})$, *and* $C^m(\overline{\Omega};R^\ell)$

Let Ω be an open set in R^n. Points in Ω are denoted by $t = (t_1,\ldots,t_n)$. The closure of Ω is denoted by $\overline{\Omega}$, and the boundary of Ω is denoted by $\partial\Omega = \overline{\Omega}\backslash\Omega$.

Let $x(t) = x(t_1,\ldots,t_n)$ be a real function defined on Ω.

A *multi-index* α is an n-tuple of non-negative integers $\alpha = (\alpha_1,\ldots,\alpha_n)$; we set $|\alpha| = \sum_{i=1}^{n} |\alpha_i|$, and denote by $D^\alpha x$ the partial derivative

$$D^\alpha x = \frac{\partial^{|\alpha|}x}{\partial t_1^{\alpha_1}\ldots\partial t_n^{\alpha_n}},$$

of order $|\alpha|$. For example, if $\alpha = (1,0,5)$ then $|\alpha| = 6$ and

$$D^\alpha x = \frac{\partial^6 x}{\partial t_1\ \partial t_3^5}.$$

If x is defined on Ω then x is said to have *compact support* in Ω if there is a closed bounded subset A of Ω

such that $x(t) = 0$ for $t \in \Omega \backslash A$. If x has compact support in Ω then x 'vanishes near $\partial\Omega$'.

Let m be a non-negative integer. Then:

$C^m(\Omega)$ is the vector space of real functions x defined on Ω which, together with all their partial derivatives $D^\alpha x$ of order $|\alpha| \le m$, are continuous on Ω.

$C^\infty(\Omega)$ is the vector space consisting of functions x such that $x \in C^m(\Omega)$ for all m, $0 \le m < \infty$.

$C_0^m(\Omega)$ is the subspace of $C^m(\Omega)$ consisting of functions $x \in C^m(\Omega)$ with compact support in Ω, for $0 \le m \le \infty$.

We write $C(\Omega)$ and $C_0(\Omega)$ instead of $C^0(\Omega)$ and $C_0^0(\Omega)$.

$C^\infty(\Omega)$ and $C^m(\Omega)$ have some undesirable features:

(1) If $x \in C^\infty(\Omega)$ then x may be unbounded. E.g. $\Omega = (0,1)$ and $x(t) = 1/t$.

(2) It may not be possible to extend $x \in C^\infty(\Omega)$ continuously to $\overline{\Omega}$. E.g. $\Omega = (0,1)$ and $x(t) = \sin(1/t)$.

(3) If $x \in C^\infty(\Omega)$ then x may not be uniformly continuous. E.g. $\Omega = (-\infty,+\infty)$ and $x(t) = \sin(t^2)$.

The space $C_0^\infty(\Omega)$ can be made into a metric space by using the seminorms

$$p_k(x) = \sum_{|\alpha|=k} \sup_{t\in\Omega} |D^\alpha x| , \quad k = 0,1,2,\ldots,$$

but this space is not normable (Problem 4.12).

These undesirable features are eliminated in the following spaces:

$C^m(\overline{\Omega})$ is the Banach space of real functions x defined on $\overline{\Omega}$ such that, for $0 \le |\alpha| \le m$, $D^\alpha x$ exists as a bounded and uniformly continuous function on Ω. $C^m(\overline{\Omega})$ is equipped with the norm

$$\|x;\ C^m(\overline{\Omega})\| = \sum_{|\alpha|\le m} \sup_{t\in\Omega} |D^\alpha x(t)| .$$

$C(\overline{\Omega}) = C^0(\overline{\Omega})$.

$C_0^m(\bar{\Omega})$ is the subspace of $C^m(\bar{\Omega})$ consisting of the functions $x \in C^m(\bar{\Omega})$ with compact support in Ω.

$C^m(\bar{\Omega};R^\ell)$ is the Banach space consisting of mappings x of $\bar{\Omega}$ into R^ℓ,

$$x(t) = \begin{bmatrix} x_1(t) \\ \cdots \\ \cdots \\ \cdots \\ x_\ell(t) \end{bmatrix} \in R^\ell, \quad \text{for} \quad t \in \bar{\Omega},$$

where each $x_i : \bar{\Omega} \to R^1$ belongs to $C^m(\bar{\Omega})$. If $x \in C^m(\bar{\Omega};R^\ell)$ then

$$\|x\| = \|x;\ C^m(\bar{\Omega};R^\ell)\| = \sum_{i=1}^{\ell} \|x_i;\ C^m(\bar{\Omega})\| .$$

Notice that $C(\overline{R^n}) \neq C(R^n)$.

It is left as an exercise for the reader (Problem 4.14) to show that $C^m(\bar{\Omega})$ and $C^m(\bar{\Omega};R^\ell)$ are Banach spaces. From the Stone-Weierstrass theorem (see page 86) the polynomials with rational coefficients are dense in $C(\bar{\Omega})$ when Ω is a bounded set. It can be shown that the polynomials with rational coefficients are dense in $C^m[a,b]$. (Problem 4.15). See also Problem 4.16.

REMARK 4.3. It is of interest to compare possible alternative definitions of $C^m(\bar{\Omega})$. The definition given above will be called *Definition 1.*

Definition 2: We could define $C^m(\bar{\Omega})$ to be those functions $u \in C^m(\Omega)$ such that, for $1 \leq |\alpha| \leq m$, $D^\alpha u$ can be extended continuously to $\bar{\Omega}$. That is, for $1 \leq |\alpha| \leq m$, $D^\alpha u$ is the restriction to Ω of a function g_α which is defined and continuous on $\bar{\Omega}$. (This definition is used, for example, by Smirnov [1964, p. 304].).

Definition 3: We could define $C^m(\bar{\Omega})$ to consist of functions u which are the restriction to $\bar{\Omega}$ of functions $\tilde{u}$ which are m-times continuously differentiable on some open set $\tilde{\Omega}$ containing $\bar{\Omega}$. (This definition is frequently used in the theory of differential manifolds.)

As shown in Problem 4.13, if $u \in C^m(\bar{\Omega})$ (Def 1) then $u \in C^m(\bar{\Omega})$ (Def 2). A continuous function g_α defined on a closed bounded set in R^n is bounded and uniformly continuous (Theorem 6.12), so that Definitions 1 and 2 are equivalent if Ω is bounded (which Smirnov [1964] assumes). In particular, the definition of $C(\overline{\Omega})$ given in this section coincides with the definition previously given in Section 4.9 when Ω is bounded. If Ω is not bounded then Definitions 1 and 2 differ. (Problem 4.17).

If Ω is not bounded, then Definitions 1 and 3 differ as shown by Problem 4.17. If Ω is bounded then $u \in C^m(\overline{\Omega})$ (Def 3) $\Rightarrow u \in C^m(\overline{\Omega})$ (Def 1), but the converse is not always true. If Ω is smooth and $u \in C^m(\overline{\Omega})$ (Def 1) then (Problems 4.18 and 4.19) we can extend u smoothly across $\partial\Omega$, so that $u \in C^m(\overline{\Omega})$ (Def 3). If $\partial\Omega$ is not smooth, then such an extension may not be possible (Problem 4.20). □

EXAMPLE 4.1. $C^m[a,b]$ is the space of real-valued functions x defined on $[a,b]$ which, together with their first m derivatives, are uniformly continuous and bounded on (a,b). If $x \in C^m[a,b]$, then x is m-times continuously differentiable on $[a,b]$, with the convention that the derivatives at a and b are the right and left derivatives, respectively (see Problem 4.18). The norm is

$$\|x\| = \|x;\ C^m[a,b]\| = \sum_{j=0}^{m} \sup_{t\in(a,b)} |x^{(j)}(t)| ,$$

where $x^{(0)}(t) = x(t)$ and $x^{(j)}(t) = \dfrac{d^j x(t)}{dt^j}$. Of course, $C^{(0)}[a,b] = C[a,b]$.

$C^m([a,b];\ R^\ell)$ is the space of m-times continuously differentiable functions mapping the interval $[a,b]$ into R^ℓ. That is, if $x \in C^m([a,b];\ R^\ell)$ then

$$x(t) = \begin{pmatrix} x_1(t) \\ \cdots \\ \cdots \\ \cdots \\ x_\ell(t) \end{pmatrix} \in R^\ell .$$

where $x_i(t) \in C^m[a,b]$ for $1 \le i \le \ell$. We set

$$\|x\| = \sum_{i=1}^{\ell} \sum_{j=0}^{m} \sup_{a<t<b} |x_i^{(j)}(t)| .$$

For example, $C^2([a,b]; R^3)$ consists of triplets $x = (x_1,x_2,x_3)$, where each x_i is a twice continuously differentiable mapping of $[a,b]$ into R^1, with norm

$$\|x\| = \sum_{i=1}^{3} [\sup_{t\in(a,b)} |x_i(t)| + \sup_{t\in(a,b)} |\dot{x}_i(t)| + \sup_{t\in(a,b)} |\ddot{x}_i(t)|]. \quad \square$$

PROBLEMS IN $C^m(\overline{\Omega})$ AND $C^m(\overline{\Omega}; R^\ell)$

The spaces $C^m(\overline{\Omega})$ arise in the theory of ordinary differential equations, partial differential equations, integrodifferential equations, etc. Here, we describe two applications to ordinary differential equations.

(1) *Initial value problems.* The initial value problem for a system of ℓ ordinary differential equations is: Given a continuous mapping $f: [a,b] \times R^\ell \to R^\ell$, find $x \in C^1([a,b]; R^\ell)$ satisfying:

(a) $x(a) = \eta \in R^\ell$ (the given initial values).

(b) $\dot{x}(t) = f(t,x(t))$, $a < t < b$; that is,

$\frac{dx_i}{dt}(t) = f_i(t,x_1(t),\ldots,x_\ell(t))$, for $1 \le i \le \ell$, and $a < t < b$.

Set

$$X = C^1([a,b]; R^\ell) \qquad (*)$$
$$Y = C([a,b]; R^\ell) ,$$
$$Z = R^\ell$$

and define $A(t): Z \to Z$ and $\gamma: X \to R^1$ by

$$A(t)z = f(t,z) , \quad \text{for } a \le t \le b ,$$
$$\gamma x = x(a) .$$

Then the problem (*) can be written in the form: Find $x \in X$ such that

$$\frac{dx}{dt} = A(t)x(t) , \quad a < t < b \; ; \; \gamma x = \eta . \qquad (**)$$

References to the numerical solution of such problems include Henrici [1962], and Lambert [1973]. In Section 6.3.3 we prove the existence of the solution of this problem.

(2) *Boundary value problems.* Consider the two-point boundary value problem for the real function $x(t)$:

$$a_0\, x(t) + a_1\, \dot{x}(t) + a_2\, \ddot{x}(t) = f(t) , \quad 0 \le t \le 1 ,$$
$$x(0) = \alpha , \; x(1) = \beta , \qquad (***)$$

where a_0, a_1, a_2, α, and β are given constants.

Set

$$v(t) = \alpha + (\beta-\alpha)t \ , \ y(t) = x(t) - v(t)$$

then

$$\begin{aligned} a_0\, y(t) + a_1\, \dot{y}(t) + a_2\, \ddot{y}(t) \\ &= f(t) - [a_0\, v(t) + a_1\, \dot{v}(t) + a_2\, \ddot{v}(t)] \ , \\ &= g(t) \ , \text{ say,} \end{aligned}$$

and

$$y(0) = 0 \ , \quad y(1) = 0 \ .$$

Set

$$Y = \{u \in C^2[0,1] : u(0) = u(1) = 0\} \ ,$$
$$Z = C[0,1] \ ,$$

and define $T: Y \to Z$ by $(Ty)(t) = a_0\, y(t) + a_1 \dot{y}(t) + a_2\, \ddot{y}(t)$. Then the problem $(*_*)$ takes the form: Find $y \in Y$ such that

$$Ty = g \ . \qquad ({}^{**}_{**})$$

References to the numerical solution of boundary value problems for ordinary differential equations include Keller [1968]. □

REMARK 4.4. The reader may feel that we are just 'playing with symbols' and that there is no point to using eqns $(**)$ and $({}^{**}_{**})$ instead of eqns $(*)$ and $(*_*)$, respectively. The new forms contain less information than the old ones, but this is an advantage since it allows us to concentrate on those properties of the operators and spaces which are important.

For example, in the case of eqn $({}^{**}_{**})$, the most important properties of the operator T are that, as is readily verified, T is linear and continuous. As one application of these facts, assume that the coefficients a_i in eqn $(*_*)$ are such that T is one-to-one and surjective. Then it follows from the open mapping theorem (see Theorems 8.6 and 8.7) that the solution x of eqn $(*_*)$ depends continuously upon f. (see Problem 8.27) □

4.11 *THE SPACES* $L^p(\Omega)$

DEFINITION 4.1. Let $1 \leq p < \infty$. Let Ω be an open set in R^n. Let X be the normed linear space consisting of the real-valued continuous functions in $C(\Omega)$ with finite norm,

$$\|x\|_p = [\int_\Omega |x(t)|^p \, dt]^{\frac{1}{p}} , \quad 1 \le p < \infty .$$

(That $\|\cdot\|_p$ is a norm follows readily with the help of the inequality of Minkowski). According to Theorem 3.4 there exists a Banach space $\hat{X}$ with a dense subspace $\tilde{X}$ such that X is isometrically isomorphic to $\tilde{X}$. We denote $\hat{X}$ by $L^p(\Omega)$. □

DEFINITION 4.2. Let $1 \le p < \infty$. Let Ω be an open set in R^n. $L^p(\Omega)$ is the space of real measurable functions x defined on Ω for which $|x|^p$ is summable, with norm

$$\|x\|_p = \|x; L^p(\Omega)\| = [\int_\Omega |x(t)|^p \, dt]^{\frac{1}{p}} .$$

$L^\infty(\Omega)$ is the space of real measurable functions x which are *essentially bounded* on Ω: that is, $x \in L^\infty(\Omega)$ if

$$\begin{aligned} \|x\|_\infty &= \operatorname*{ess\,sup}_{t \in \Omega} |x(t)| , \\ &= \inf \{K: |x(t)| \le K \text{ a.e. on } \Omega\} , \\ &< \infty . \end{aligned}$$

Two elements of $L^p(\Omega)$ are identified if they are equal a.e. (almost everywhere), so that the elements of $L^p(\Omega)$ consist of equivalence classes. (A corollary is that if $x \in L^p(\Omega)$ and $t \in \Omega$ then the value of $x(t)$ is not known since x can be changed arbitrarily on sets of measure zero.) □

REMARK 4.5. The two definitions of $L^p(\Omega)$ given above are equivalent, and each has advantages and disadvantages. The first definition is useful when we wish to approximate elements of $L^p(\Omega)$ by continuous functions. The second definition has the advantage that it provides a concrete representation for the elements of $L^p(\Omega)$, but has the disadvantage that it requires knowledge of the theory of Lebesgue measure and integration. (The necessary material is summarized in the Appendix). Definition 4.1 cannot be used for the case $p = \infty$ since $C(\Omega)$ is itself complete with respect to the supremum norm (Problem 4.17). □

The basic properties of $L^p(\Omega)$ are listed below (see, for example, Adams [1975, Chapter 2]):

Properties of the spaces

(1) $L^p(\Omega)$ is a Banach space for $1 \leq p \leq \infty$.

(2) $L^p(\Omega)$ is separable iff $1 \leq p < \infty$.

(3) $L^p(\Omega)$ is a Hilbert space iff $p = 2$.

(4) $L^p(\Omega)$ is reflexive iff $1 < p < \infty$.

(5) $\|\cdot\|_p$ is uniformly convex iff $1 < p < \infty$.

(6) If $1 \leq p \leq q \leq \infty$, $\mu(\Omega) = \int_\Omega 1\,dx < \infty$, and $u \in L^q(\Omega)$, then $u \in L^p(\Omega)$ and $\|u\|_p \leq (\mu(\Omega))^{(\frac{1}{p}-\frac{1}{q})}\|u\|_q$. If $u \in L^\infty(\Omega)$ then $\lim_{p\to\infty}\|u\|_p = \|u\|_\infty$. If $u \in L^p(\Omega)$ for $1 \leq p < \infty$, and if there is a constant K such that $\|u\|_p \leq K$ for $1 \leq p < \infty$, then $u \in L^\infty(\Omega)$ and $\|u\|_\infty \leq K$.

(7) *Hölder's inequality.* If $1 \leq p \leq \infty$, $\frac{1}{p} + \frac{1}{q} = 1$, $u \in L^p(\Omega)$, and $v \in L^q(\Omega)$ then

$$\int_\Omega |u(x)v(x)|\,dx \leq \|u\|_p\|v\|_q .$$

Properties of the elements

(8) *Translation is continuous.* That is, if $1 \leq p < \infty$ and $x \in L^p(\Omega)$ then

$$\int_\Omega |x(t) - \tilde{x}(t+h)|^p dt \to 0 \text{ , as } \|h\| \to 0 .$$

Here, $h \in R^n$ and

$$\tilde{x}(t+h) = \begin{cases} x(t+h), & \text{if } t+h \in \Omega , \\ 0, & \text{otherwise.} \end{cases}$$

(9) If $1 \leq p \leq \infty$ then a Cauchy sequence in $L^p(\Omega)$ has a subsequence converging pointwise a.e. in Ω. That is, if the sequence $\{x_k\}$ is a Cauchy sequence in $L^p(\Omega)$ then there is a subsequence $\{x_{k_i}\}$ such that $\{x_{k_i}(t)\}$ converges in R^1 for almost all $t \in \Omega$.

Approximation by continuous functions

(10) $C_0^\infty(\Omega)$ is dense in $L^p(\Omega)$ iff $1 \leq p < \infty$. One proof uses mollifiers, which are described below.

For $t \in R^n$ let

$$J(t) = \begin{cases} k \exp[-1/(1-\|t\|^2)], & \text{if } \|t\| < 1 , \\ 0, & \text{otherwise,} \end{cases}$$

where k is a constant chosen so that $\int_{R^n} J(t)dt = 1$.

If $\varepsilon > 0$ then $J_\varepsilon(t) = \varepsilon^{-n}J(t/\varepsilon)$ is non-negative, belongs to $C_0^\infty(R^n)$, and satisfies

$$J_\varepsilon(t) = 0 , \quad \text{if } \|t\| \geq \varepsilon ; \quad \int_{R^n} J_\varepsilon(t)dt = 1 .$$

Any function with those properties is called a *mollifier.*

Let $x \in L^p(\Omega)$, $1 \leq p < \infty$, and define $\tilde{x} \in L^p(R^n)$

$$\tilde{x}(t) = \begin{cases} x(t), & \text{if } t \in \Omega , \\ 0, & \text{otherwise.} \end{cases}$$

If J_ε is a mollifier, the function $J_\varepsilon * x$ defined by

$$(J_\varepsilon * x)(t) = \int_{R^n} J_\varepsilon(t-s)\tilde{x}(s)ds , \quad \text{for } t \in R^n ,$$

is called a *mollified function corresponding to* x .
It can be shown that, for $1 \leq p < \infty$:

(a) $J_\varepsilon * x \in L^p(\Omega)$;

(b) $\|J_\varepsilon * x\|_p \leq \|x\|_p$;

(c) $J_\varepsilon * x \in C^\infty(R^n)$;

(d) $\|J_\varepsilon * x - x\|_p \to 0$ as $\varepsilon \to 0$.

(11) Although $C_0(\Omega)$ is not dense in $L^\infty(\Omega)$, we have *Lusin's Theorem:* If $x \in L^\infty(\Omega)$ and $\mu(\Omega) < \infty$, then for each $\varepsilon > 0$ there exists $\psi \in C_0(\Omega)$ such that:

(a) $\|\psi\|_\infty \leq \|x\|_\infty$,

(b) $\mu(\{t \in \Omega: x(t) \neq \psi(t)\}) < \varepsilon$,

where $\mu(e)$ denotes the Lebesgue measure of e . Note, however, that, although ψ and x coincide except on a set of measure ε , $\|\psi - x\|_\infty$ need not be small.

PROBLEMS IN $L^p(\Omega)$

For $1 \le p < \infty$, $L^p(\Omega)$ is obtained by taking the completion, with respect to the norm $\|\cdot\|_p$, of the continuous functions (Definition 4.1). Such a process of completion occurs frequently in the theory of partial differential equations, where $\|x\|_p$ can often be interpreted as the energy associated with x. There are, therefore, many applications of the spaces $L^p(\Omega)$ to partial differential equations (see Chapter 5). The spaces $L^p(\Omega)$ also form the 'building blocks' of the Sobolev spaces, and these in turn play an important rôle in the theory of finite elements.

4.12 *SUMMARY OF APPLICATIONS*

To conclude, the following table summarizes the areas of numerical analysis in which the spaces discussed in this chapter play a rôle. When considering a new problem, the choice of the spaces in which the problem should be formulated is often crucial. Table 4.1 provides a guide to the usual choices of spaces, but sometimes other choices are desirable

Matrix theory: R^n and $\mathbb{C}^n$

Summation of series: c

Approximation theory: All spaces but especially $C(\bar{\Omega})$ and $L^p(\Omega)$

Integral equations: $C(\bar{\Omega})$ and $L^p(\Omega)$

Initial value problems for ordinary differential equations: $C^m(\bar{\Omega})$

Initial value problems for hyperbolic partial differential equations: $L^p(\Omega)$

Initial value problems for parabolic partial differential equations: $C(\bar{\Omega})$ and $L^p(\Omega)$

Boundary value problems for ordinary differential equations: $C^m(\bar{\Omega})$ and $L^p(\Omega)$

Boundary value problems for elliptic partial differential equations: $C^m(\bar{\Omega})$, $L^p(\Omega)$, and $H^k(\Omega)$

Table 4.1: Typical applications of the common Banach spaces

APPENDIX: MEASURE THEORY AND LEBESGUE INTEGRATION

We briefly summarize the basic properties of Lebesgue measure and integration. For details, see Halmos [1950], Smirnov [1964], Rudin [1966] or Dunford and Schwartz [1966, Chapter 3]. Much of this material will be needed in Volume 2.

Lebesgue Measure

Lebesgue measure is an extension of the concept of the 'area' or 'volume' of a set.

A (*semi-open*) *interval* I in R^n is a 'rectangle' of the form

$$I = \{t = (t_i) \in R^n : -\infty < a_i \le t_i < b_i \le +\infty \ , 1 \le i \le n\}$$

The Lebesgue measure of such an interval is defined to be

$$\mu(I) = \prod_{i=1}^{n} (b_i - a_i) \le +\infty \ .$$

That is, $\mu(I)$ is the 'area' or 'volume' of I in the ordinary sense.

Let a set E in R^n be covered by a finite or denumerable number of semi-open intervals $\{I_k\}$. That is, $E \subset \cup I_k$. The *exterior measure* $\mu^*(E)$ of E is the infimum of the values of the sums $\sum \mu(I_k)$. That is

$$\mu^*(E) = \inf_{E \subset \cup I_k} \sum \mu(I_k) \ .$$

$\mu^*(E)$ may be equal to $+\infty$.

A set E in R^n is (Lebesgue) *measurable* if for every $\varepsilon > 0$ there exists an open set G containing E such that

$$\mu^*(G \backslash E) < \varepsilon \ .$$

The *measure* of a measurable set E is denoted by $\mu(E)$, and is equal to $\mu^*(E)$.

The following properties hold:

(M1) Every open set is measurable.

(M2) Every closed set is measurable.

(M3) Every interval is measurable.

(M4) If E is measurable then $R^n \setminus E$ is measurable.

(M5) If $E_1, E_2, \ldots$ is a finite or denumerable sequence of measurable sets then

(a) $E = \cup E_i$ is measurable and $\mu(E) \le \sum \mu(E_i)$.

(b) If the E_i are disjoint then $\mu(\cup E_i) = \sum \mu(E_i)$.

(c) $\cap E_i$ is measurable.

(M6) If E_1 and E_2 are measurable then $E_1 \setminus E_2$ is measurable. If $E_2 \subset E_1$ and $\mu(E_2) < \infty$ then $\mu(E_1 \setminus E_2) = \mu(E_1) - \mu(E_2)$.

(M7) If $E_1, E_2, \ldots$ is a non-increasing sequence of measurable sets such that $\mu(E_1) < \infty$ then

$$\mu(\cap E_i) = \lim \mu(E_i) .$$

(M8) If $E_1, E_2, \ldots$ is a non-decreasing sequence of measurable sets then

$$\mu(\cup E_i) = \lim \mu(E_i) .$$

(M9) E is measurable with measure zero iff $\mu^*(E) = 0$.

(M10) E is measurable iff for every $\varepsilon > 0$ there exists an open set G and a closed set F such that $F \subset E \subset G$ and

$$\mu(G \setminus F) < \varepsilon .$$

(M11) E is measurable iff for every set $A \subset R^n$,

$$\mu^*(A) = \mu^*(A \cap E) + \mu^*(A \setminus E) .$$

(M12) If E_1 is measurable and E_2 is a translation or rotation of E_1 then E_2 is measurable and $\mu(E_2) = \mu(E_1)$.

Although the class of measurable sets is very large, there are many sets in R^n which are not measurable. Indeed, every measurable set of positive measure contains a non-measurable subset. (Natanson [1954, p. 79]). It is natural to inquire whether it is possible to define the measure of a set in such a way that every set is measurable. It can be proved that even in R^1 there is no measure μ

satisfying properties M5b and M12 such that every set is measurable and $\mu([0,1]) = 1$. (See also Sullivan [1981])

A property which holds at all the points of a set E with the exception of the points in a subset E_1 with measure $\mu(E_1) = 0$ is said to hold *a.e.* (*almost everywhere*) on E.

Measurable Functions

Let f be an extended real function f defined on a measurable set $E \subset R^n$; that is, f maps E into the extended real line $\overline{R}^1 = [-\infty, +\infty]$. Then f is *measurable* if the sets

$$f^{-1}[a,+\infty] = \{t \in E : f(t) \ge a\} ,$$
$$f^{-1}[-\infty,a) = \{t \in E : f(t) < a\} ,$$
$$f^{-1}(a,+\infty] = \{t \in E : f(t) > a\} ,$$
$$f^{-1}[-\infty,a] = \{t \in E : f(t) \le a\} ,$$

are measurable for all $a \in R^1$.

The following properties hold:

(F1) f is measurable on E iff *one* of the four families of sets $f^{-1}[a,+\infty]$, $f^{-1}[-\infty,a)$, $f^{-1}(a,+\infty]$, or $f^{-1}[-\infty,a]$ is measurable for all $a \in R^1$.

(F2) If $|f|$ and g are measurable then

(a) f is measurable.

(b) The set $\{t \in E : f(t) > g(t)\}$ is measurable.

(c) If c is a finite constant, $c + f$ and cf are measurable.

(d) $f - g$, $f + g$, fg, and f/g are measurable, provided that they are defined (that is, $\infty - \infty$ and $0/0$ are excluded).

(e) $f^+(t) = \begin{cases} f(t), & \text{if } f(t) \ge 0 , \\ 0, & \text{otherwise} , \end{cases}$

is measurable.

(F3) If $\{f_k\}$ is a sequence of measurable functions on E then

$$f_1 = \inf_k f_k , \quad f_2 = \lim_k \inf f_k ,$$

$$f_3 = \sup_k f_k \quad f_4 = \lim_k \sup f_k ,$$

are measurable.

(F4) If $\{f_k\}$ is a sequence of measurable functions on E which is convergent almost everywhere, then the limit function f is measurable. (The limit function may not be defined on a set of measure zero.).

Lebesgue Integration

Let $E \subset R^n$ be a measurable set. A measurable function s on E is a *simple function* if s is of the form

$$s = \sum_{i=1}^{m} \alpha_i \psi_{A_i}$$

where the α_i are finite constants, the A_i are disjoint measurable subsets of E, and ψ_{A_i} is the *characteristic function* of A_i,

$$\psi_{A_i}(t) = \begin{cases} 1, & \text{if } t \in A_i , \\ 0, & \text{otherwise.} \end{cases}$$

The Lebesgue integral of a non-negative simple function s over E is defined to be

$$\int_E s \, dt = \sum_{i=1}^{m} \alpha_i \, \mu(A_i) ,$$

it being understood in the sum on the right that $0 \cdot \infty = 0$.

The Lebesgue integral of a non-negative measurable function f over E is defined to be

$$\int_E f \, dt = \sup_{0 \le s \le f} \int_E s \, dt ,$$

where the supremum is taken over all non-negative simple functions s which are less than or equal to f.

We cannot in general define the integral of an arbitrary measurable function because of the possibility of cancellation between $+\infty$ and $-\infty$. A function f is *summable* on E if f is measurable on E and

$$\int_E |f|\,dt < \infty .$$

If f is summable, we define

$$\int_E f\,dt = \int_E f^+\,dt - \int_E f^-\,dt .$$

where

$$f^+(t) = \max\{0, f(t)\} ,$$

and

$$f^-(t) = -\min\{0, f(t)\} .$$

The basic properties of the Lebesgue integral are:

(I1) If $f_1,\ldots,f_m$ are summable on E and $\alpha_1,\ldots,\alpha_m$ are finite real numbers then $f = \sum \alpha_i f_i$ is summable and

$$\int_E f\,dt = \sum_{i=1}^{m} \alpha_i \int_E f_i\,dt .$$

(I2) If f_1 and f_2 are summable on E and $f_1 \geq f_2$ then

$$\int_E f_1\,dt \geq \int_E f_2\,dt .$$

(I3) If f is summable on E then

$$\left|\int_E f\,dt\right| \leq \int_E |f|\,dt .$$

(I4) If f is summable on E then f is summable on any measurable subset of E.

(I5) If f is summable on E and

$$E = \bigcup_{k=1}^{\infty} E_k$$

where the E_k are disjoint measurable subsets of E then

$$\int_E f\,dt = \sum_{k=1}^{\infty} \int_{E_k} f\,dt .$$

(I6) If $E = \bigcup_{k=1}^{\infty} E_k$, f is summable on E_k , and

$$\sum_{k=1}^{\infty} \int_{E_k} |f| dt < \infty ,$$

then f is summable on E and

$$\int_E f\, dt = \sum_{k=1}^{\infty} \int_{E_k} f\, dt .$$

(I7) If f is summable on E then for every $\varepsilon > 0$ there exists $\delta > 0$ such that

$$\left|\int_e f\, dt\right| \le \varepsilon$$

for every measurable set $e \subset E$ satisfying $\mu(e) \le \delta$.

(I8) If $\mu(E) = 0$ then, for any function f ,

$$\int_E f\, dt = 0 .$$

(I9) If f is summable on E and $g = f$ a.e. then g is summable on E and

$$\int_E f\, dt = \int_E g\, dt .$$

(I10) If f is measurable on E and $|f| \le F$ where F is summable on E then f is summable and

$$\left|\int_E f\, dt\right| \le \int_E F\, dt .$$

(I11) If f is measurable and non-negative on E , and satisfies

$$\int_E f\, dt = 0 ,$$

then $f = 0$ a.e. in E .

(I12) If, for every measurable $e \subset E$, f is summable on e and

$$\int_e f \, dt = 0 ,$$

then $f = 0$ a.e. in E.

(I13) *Fatou's lemma*

If $\{f_k\}$ is a sequence of non-negative measurable functions on E then

$$\int_E [\liminf_{k\to\infty} f_k(t)] dt \le \liminf_{k\to\infty} \int_E f_k(t) dt .$$

(I14) *Lebesgue's monotone convergence theorem*

If $\{f_k\}$ is a sequence of real measurable functions converging a.e. on E to a function f, such that

$$0 \le f_1(t) \le f_2(t) \le \ldots \le f_k(t) \le \ldots , \text{ for } t \in E$$

then

$$\lim_{k\to\infty} \int_E f_k \, dt = \int_E f \, dt .$$

(I15) *Dominated convergence theorem*

Let $\{f_k\}$ be a sequence of summable functions which converges a.e. on E to a function f. Assume that there exists a summable function g such that $|f_k| \le g$ a.e. on E. Then f is summable and

$$\int_E f \, dt = \lim \int_E f_k \, dt .$$

(I16) *Fubini's theorem*

Let f be measurable on R^{n+m} and suppose that at least one of the integrals

$$\int_{R^{n+m}} |f(s,t)| \, ds \, dt , \quad \int_{R^m} (\int_{R^n} |f(s,t)| ds) dt , \quad \int_{R^n} (\int_{R^m} |f(s,t)| dt) ds , \tag{*}$$

exists and is finite. Then

(a) $f(\cdot,t) \in L^1(R^n)$, for almost all $t \in R^m$.

(b) $f(s,\cdot) \in L^1(R^m)$, for almost all $s \in R^n$.

(c) $\int_{R^n} |f(s,\cdot)|ds \in L^1(R^m)$.

(d) $\int_{R^m} |f(\cdot,t)|dt \in L^1(R^n)$.

(c) The three integrals in (*) exist and are equal.

PROBLEMS

4.1 Give two alternative spellings of Bunjakowski that are used in the literature.

4.2 Show that the norm $\|\cdot\|_p$ in ℓ_p^n is strictly convex and uniformly convex iff $1 < p < \infty$.
[Hint. See Problem 2.19. If $a,b \in R^1$, and $2 \le p < \infty$, then

$$\left|\frac{a+b}{2}\right|^p + \left|\frac{a-b}{2}\right|^p \le \frac{1}{2}|a|^p + \frac{1}{2}|b|^p \; . \qquad (*)$$
]

{73}

4.3 Let $1 \le p \le q \le \infty$. Show that for all real n-tuples x ,

$$\frac{1}{\alpha_1}\|x\|_p \le \|x\|_q \le \alpha_2\|x\|_p \; , \qquad (*)$$

where $\|\cdot\|_p$ is the ℓ_p^n norm, $\alpha_1 = n^{\frac{1}{p}-\frac{1}{q}}$, and $\alpha_2 = 1$. Show that α_1 and α_2 are the smallest constants for which eqn (*) holds for all x .

4.4 Consider the solution of the problem

$$Ax = b \qquad (*)$$

where $x,b \in \mathbb{C}^1$ and A is an $n \times n$ complex matrix.

If A, x, and b are broken into their real and complex components, $A = A_r + iA_c$, $x = x_r + ix_c$, $b = b_r + ib_c$, eqn (*) becomes

$$\begin{pmatrix} A_r & -A_c \\ A_c & A_r \end{pmatrix} \begin{pmatrix} x_r \\ x_c \end{pmatrix} = \begin{pmatrix} b_r \\ b_c \end{pmatrix} . \qquad (**)$$

Assume that a complex multiplication is equivalent to four real multiplications and two real additions, and that a complex addition is equivalent to two real additions.

Show that, for large n, if Gaussian elimination is used to solve eqns (*) and (**), the latter computation requires twice as much work as the former. {77}.

4.5 If $f(z): \mathbb{C}^1 \to \mathbb{C}^1$ is an analytic function which is regular within the unit circle C in the complex plane $\mathbb{C}^1$, the derivatives of $f(z)$ at $z = 0$ are given by

$$f^{(k)}(0) = \frac{k!}{2\pi i} \int_C \frac{f(\zeta)}{(\zeta - z)^{k+1}} d\zeta , \quad k = 0,1,2,\ldots, \qquad (*)$$

where $i = \sqrt{-1}$.

Let $f(z) = (z+1)^4$. Compute $f^{(1)}(0)$ approximately by first making the substitution $\zeta = e^{i\theta}$ in the integral (*), and then approximating the resulting integral using the trapezoidal rule with the points $\theta = 0$, $\pi/2$, π, $3\pi/2$, and 2π.

4.6 Show that the discrete Fourier transform F_n defined on page 78 is a bijective linear map of Π_n onto Π_n. {78}.

4.7 Show that ℓ^p is uniformly convex iff $1 < p < \infty$. [Hint: See Problem 4.2] {78}.

4.8 Show that ℓ^∞ is not separable. [Hint: Let $\{x^k\}$, $k = 1,2,\ldots,$ be a denumerable set of points which is dense in ℓ^∞. Let $x^k = (x_i^k)$. Construct $x = (x_i)$ in ℓ^∞ such that $|x_i - x_i^i| \geq 1$.] {79}.

4.9 Show that the term $t^k(1-t)^{n-k}$ in the expression for the Bernstein polynomial $B(n;x)$ attains its maximum on the interval $[0,1]$ when $t = k/n$.

4.10 We denote by $\$^1_{n+1}[a,b]$ the set of functions $p \subset C[a,b]$ which are *piecewise linear on* $[a,b]$ with n+1 *breakpoints*

$$t_i^n = a + i(b-a)/n, \quad 0 \leq i \leq n. \tag{*}$$

That is,

$$p(t) = p(t_i^n)\frac{t-t_i^n}{h} + p(t_{i+1}^n)\frac{t_{i+1}^n - t}{h},$$

$$\text{for } t_i^n \leq t \leq t_{i+1}^n \text{ and } 0 \leq i \leq n-1, \quad h = (b-a)/n.$$

If $f \in \$^1_{n+1}$, f is a *linear spline*.

Let $\tilde{\$}^1[a,b] = \{p \in C[a,b]:$ for some n, $p \in \$^1_{n+1}$ and $p(t_i^n)$ is rational, $0 \leq i \leq n\}$. Show that $\tilde{\$}^1[a,b]$ is denumerable and dense in $C[a,b]$.

4.11 Show that $C[a,b]$ is neither uniformly convex nor strictly convex. {85}.

4.12 Use Theorems 2.6 and 2.7 to show that the seminorms

$$p_k(x) = \sum_{|\alpha|=k} \max_{t\in\Omega} |D^\alpha x(t)|, \quad k = 0,1,2,\ldots,$$

induce a metric on $C_0^\infty(\Omega)$, but that the resulting linear metric space is not normable. {89}

4.13 Show that if $u \in C^m(\overline{\Omega})$ and $|\alpha| \leq m$, then $D^\alpha u$ can be extended continuously to $\overline{\Omega}$. That is, $D^\alpha u = g_\alpha|_\Omega$ where g_α is defined and continuous on $\overline{\Omega}$. Show also that g_α is unique and that g_α is uniformly continuous on $\overline{\Omega}$. {90}.

4.14 Prove that $C^m(\bar{\Omega})$ and $C^m(\bar{\Omega};R^\ell)$ are Banach spaces. [Hint: Use Problem 3.12]. {90}.

4.15 Let $x \in C^k[a,b]$. Choose the polynomial p_k in the space $\pi[a,b]$ of polynomials on $[a,b]$ such that

$$\sup_{a<t<b} |p_k(t)-x^{(k)}(t)| \le \varepsilon . \qquad (*)$$

Show that by integrating p_k k times one obtains a polynomial $p \in \pi[a,b]$ satisfying

$$\|x-p;C^k[a,b]\| \le \varepsilon \sum_{\ell=0}^{k} [(b-a)/2]^\ell . \qquad (**)$$

Hence, show that $C^k[a,b]$ is separable. {90}.

4.16 Let $x \in C^1(\bar{\Omega})$, where $\Omega = \prod_{i=1}^{n} (a,b) \subset R^n$. From the Stone-Weierstrass theorem it is known that, given $\varepsilon > 0$ there exist polynomials p_α such that

$$\sup_{t\in\Omega} |D^\alpha x(t)-p_\alpha(t)| \le \varepsilon , \text{ for all } \alpha \text{ such that } |\alpha| = 1 . \qquad (*)$$

Does this imply that the polynomials are dense in $C^1(\bar{\Omega})$? {90}

4.17 Let X be the space of real valued functions x which are bounded and continuous on $\Omega = \bar{\Omega} = R^1$. Show that:

(1) X is a Banach space with respect to the norm $\|x\| = \sup_{t\in\Omega} |x(t)|$.

(2) X strictly contains $C(\bar{\Omega})$.

(3) X is not separable.

4.18 Let $u \in C^1[a,b]$. Show that u has a right derivative at a,

$$D_+u(a) \equiv \lim_{\substack{h\to 0\\ h>0}} \frac{u(a+h)-u(a)}{h} ,$$

and that

$$D_+u(a) = \lim_{t\to a} \dot{u}(t) . \qquad \{91\}$$

4.19 Assume that $u \in C^1(\overline{H}_+)$ where H_+ is the right half plane $\{(x_1,x_2) \in R^2 : x_1 > 0\}$. Let v be defined on R^2 by

$$v(x_1,x_2) = \begin{cases} u(x_1,x_2), & \text{if } x_1 \geq 0, \\ 3u(-x_1,x_2) - 2u(-2x_1,x_2), & \text{if } x_1 < 0. \end{cases}$$

Show that $v \in C^1(\overline{R^2})$. {91, 234}.

4.20 Let $\overline{\Omega} \subset R^2$ be defined by

$$\overline{\Omega} = \{(x_1,x_2) : x_1^2 + x_2^2 \leq 1 \text{ AND (EITHER } x_1 \leq 0 \text{ OR } |x_2| \geq x_1^4)\}.$$

Let $f: \overline{\Omega} \to R^1$ be given by:

$$f(x_1,x_2) = \begin{cases} 0, & \text{if } x_1 \leq 0, \\ x_1^2 \operatorname{sign}(x_2), & \text{otherwise.} \end{cases}$$

Show that $f \in C^1(\overline{\Omega})$ but that f cannot be extended to a continuously differentiable function $\tilde{f}$ defined on an open set $\tilde{\Omega} \supset \overline{\Omega}$. {91}.

4.21 Let $\mathcal{D}[a,b]$ denote the space of functions x defined on $[a,b]$ such that:

(1) $x(t+0) = \lim_{h\to 0} x(t+h)$, $h > 0$, exists for $a \leq t < b$.

(2) $x(t-0)$ exists, for $a < t \leq b$.

(3) $x(t) = \frac{1}{2}[x(t+0) + x(t-0)]$, for $a < t < b$.

Show that

(a) $\mathcal{D}$ is a Banach space with respect to the maximum norm.

(b) The piecewise constant functions are dense in $\mathcal{D}$.

{Problem 9.54}

5
The principle of uniform boundedness

In this chapter we consider a basic principle, the principle of uniform boundedness, which is relatively easy to prove and which has many applications in numerical analysis in situations where one wishes to determine conditions under which a sequence of linear operators T_n, obtained by some approximation procedure, converges to a linear operator T.

5.1 *OPERATOR NORMS*

Let T be a linear mapping from a normed linear space X to a normed linear space Y. Then

$$\|T\| = \sup_{\|x\|_X \le 1} \|Tx\|_Y , \tag{5.1}$$

is called the *norm* of T. T is *bounded* if $\|T\| < \infty$. There are two other definitions of $\|T\|$ which are often useful and which are equivalent to eqn (5.1) (Problem 5.1):

$$\|T\| = \sup_{\|x\|_X = 1} \|Tx\|_Y , \tag{5.2}$$

$$\|T\| = \inf\{\lambda : \|Tx\|_Y \le \lambda \|x\|_X \text{ for all } x \in X\} . \tag{5.3}$$

In the preceding definitions of $\|T\|$ we have carefully distinguished between $\|\cdot\|_X$ and $\|\cdot\|_Y$; when no confusion is possible the subscripts will be dropped and we will write $\|\cdot\|$ for both norms.

To prove that $\|T\| = M$ one must show:

(1) If $\|x\| \le 1$ then $\|Tx\| \le M$.

(2) Given $\varepsilon > 0$ there exists x_ε with $\|x_\varepsilon\| \le 1$ such that $\|Tx_\varepsilon\| \ge M-\varepsilon$. (5.4)

Let us now compute the norms of a few operators:

EXAMPLE 5.1. $X = Y = \ell_\infty^n$, and A is a real $n\times n$ matrix with components a_{ij}. Then

$$||A|| = ||A||_\infty = \max_i \sum_{j=1}^{n} |a_{ij}| , \tag{5.5}$$

the *maximum (absolute) row-sum norm.*

Proof. We use eqn (5.1). If $x = (x_1,\ldots,x_n) \in \ell_\infty^n$ satisfies $||x||_\infty \le 1$, then each component x_i satisfies $|x_i| \le 1$ and so

$$\begin{aligned} ||Ax||_\infty &= \max_i |(Ax)_i| , \\ &= \max_i \Big| \sum_{j=1}^{n} a_{ij}x_j \Big| , \\ &\le \max_i \sum_{j=1} |a_{ij}| , \\ &= M , \quad \text{say}, \end{aligned}$$

from which it follows that

$$||A|| \le M . \tag{*}$$

On the other hand, for some k,

$$\sum_{j=1}^{n} |a_{kj}| = M .$$

Let $x = (x_j) \in \ell_\infty^n$ be defined by

$$x_j = \begin{cases} 1, & \text{if } a_{kj} \ge 0, \\ -1, & \text{otherwise.} \end{cases}$$

Then $||x||_\infty = 1$ and

$$\begin{aligned} ||Ax||_\infty &= \max_i \Big| \sum_{j=1} a_{ij}x_j \Big| , \\ &\ge \Big| \sum_{j=1}^{n} a_{kj}x_j \Big| , \\ &= M , \end{aligned}$$

since, by construction, $a_{kj}x_j = |a_{kj}|$. Hence

$$||A|| \ge M . \tag{**}$$

We conclude from eqns (*) and (**) that $||A|| = M$. □

EXAMPLE 5.2. $X = Y = \ell_2^n = E^n$ and A is the $n \times n$ real matrix with components a_{ij}. Then

$$||A|| = ||A||_2 = [\rho(A^TA)]^{\frac{1}{2}} , \tag{5.6}$$

where A^T is the transpose of A and $\rho(A^TA)$ is the *spectral radius* of A^TA, that is, the largest eigenvalue of A^TA in absolute value. $||A||_2$ is the *Euclidean norm.*

Proof. We recall that if $x \in E^n$ then

$$||x||_2^2 = x^Tx ,$$

so that

$$||Ax||_2^2 = (Ax)^TAx ,$$
$$= x^TBx ,$$

where $B = A^TA$.

Now, B is a real symmetric matrix and so B has n distinct real eigenvectors $u_1,\ldots,u_n$ which are orthogonal, which correspond to n real eigenvalues $\lambda_1,\ldots,\lambda_n$, and which form a basis in E^n. Every $x \in E^n$ has a unique expansion in terms of the u_i :

$$x = \sum_{i=1}^n \alpha_i u_i , \quad \text{say.} \tag{*}$$

Using eqn (*) and the fact that $Bu_i = \lambda_i u_i$ we find that

$$x^Tx = \sum_{i=1}^n \alpha_i^2 ,$$

$$x^TA^TAx = \sum_{i=1}^n \lambda_i \alpha_i^2 ,$$

so that

$$||A|| = \sup_{x^Tx \le 1} [x^TA^TAx]^{\frac{1}{2}} ,$$

$$= \sup_{\sum_{i=1}^n \alpha_i^2 \le 1} [\sum_{i=1}^n \lambda_i \alpha_i^2]^{\frac{1}{2}} ,$$

$$= \max_i |\lambda_i|^{\frac{1}{2}} ,$$

$$= [\rho(A^TA)]^{\frac{1}{2}} . \qquad \square$$

For an alternative proof see Problem 5.2.

EXAMPLE 5.3. $X = Y = \ell_1^n$ and A is the $n\times n$ real matrix with components a_{ij}. Then

$$\|A\| = \|A\|_1 = \max_j \sum_{i=1}^{n} |a_{ij}| , \tag{5.7}$$

the *maximum (absolute) column-sum norm.*

The proof is given as an exercise (Problem 5.3). □

EXAMPLE 5.4. $X = C[0,1]$, $Y = \ell_\infty^n$ and

$$Tx = \sum_{i=1}^{n} x(t_i)a_i , \tag{5.8}$$

where the $a_i \in Y$ and $t_i \in [0,1]$ are fixed. Then

$$\|T\| = \sum_{i=1}^{n} |a_i| . \tag{5.9}$$

The proof is given as an exercise (Problem 5.5). □

EXAMPLE 5.5. $X = Y = C[0,1]$ and

$$(Tx)(s) = \int_0^1 k(s,t)x(t)dt , \tag{5.10}$$

where $k(s,t)$ is a continuous function of s and t. Then

$$\|T\| = \max_{0\le s\le 1} \int_0^1 |k(s,t)|dt . \tag{5.11}$$

Proof. Let

$$M = \max_{0\le s\le 1} \int_0^1 |k(s,t)|dt .$$

It is readily seen that

$$\|T\| \le M . \tag{*}$$

Since k is continuous, there exists $s_0 \in [0,1]$ such that

$$\int_0^1 |k(s_0,t)|dt = M .$$

Let

$$f(t) = \begin{cases} +1, & \text{if } k(s_0,t) \geq 0, \\ -1, & \text{if } k(s_0,t) < 0. \end{cases}$$

Then, using Lebesgue integrals,

$$\int_0^1 k(s_0,t)f(t)dt = \int_0^1 |k(s_0,t)|dt = M . \qquad (**)$$

Of course, in general $f \notin X$. However, it follows from Lusin's theorem (see Section 4.11) that given $\varepsilon > 0$ we can find $x_\varepsilon \in X$ such that $||x_\varepsilon|| \leq 1$ and $x_\varepsilon(t) = f(t)$ except on a set of points of measure less than ε . Thus,

$$\left|\int_0^1 k(s_0,t)[x_\varepsilon(t)-f(t)]dt\right| \leq 2\varepsilon \max_{0\leq t\leq 1} |k(s_0,t)| . \qquad (\overset{*}{_*}{}^*)$$

From eqns (**) and ($\overset{*}{_*}{}^*$) we conclude that

$$||Tx_\varepsilon|| \geq M - 2\varepsilon \max_{0\leq t\leq 1} |k(s_0,t)| ,$$

where $\varepsilon > 0$ is arbitrary. Together with eqn (*) this shows that $||T|| = M$.

A more elementary proof which does not use Lusin's theorem is as follows. Since $k(s_0,t)$ is continuous for $0 \leq t \leq 1$, it is uniformly continuous. Choose $\varepsilon > 0$. Then there exists $\delta > 0$ such that

$$|k(s_0,t_1) - k(s_0,t_2)| \leq \varepsilon , \quad \text{if} \quad |t_1-t_2| \leq \delta . \qquad (\overset{**}{_{**}})$$

Let V_ε denote the set of points t where $|k(s_0,t)| \leq \varepsilon$. For each $t \in V_\varepsilon$ let

$$A_t = (t-\delta, t+\delta) \cap [0,1] .$$

Since k is continuous, V_ε is closed. The collection of open sets $\{A_t : t \in V_\varepsilon\}$ covers V_ε and thus, by the classical Heine-Borel theorem (Theorem 6.7), there exist a finite number of disjoint sets A_{t_i} which cover V_ε . Hence

$$V_\varepsilon \subset T_\varepsilon \equiv [\bigcup_{i=1}^n A_{t_i}] \cap [0,1] .$$

From eqn (**) and the definition of V_ε we see that $|k(s_0,t)| \le 2\varepsilon$ for $t \in T_\varepsilon$. The set $U_\varepsilon = [0,1] \setminus T_\varepsilon$ consists of a finite number of closed intervals on which $|k(s_0,t)| \ge \varepsilon$. We set

$$x_\varepsilon(t) = k(s_0,t)/|k(s_0,t)| , \quad t \in U_\varepsilon ,$$

and define x_ε piecewise linearly on T_ε, so that x_ε is continuous on $[0,1]$ and satisfies $||x_\varepsilon|| \le 1$. Then,

$$\begin{aligned}
||Tx_\varepsilon|| &\ge \left|\int_0^1 k(s_0,t)x_\varepsilon(t)dt\right| , \\
&= \left|\int_{U_\varepsilon} k(s_0,t)x_\varepsilon(t)dt + \int_{T_\varepsilon} k(s_0,t)x_\varepsilon(t)dt\right| , \\
&\ge \int_{U_\varepsilon} |k(s_0,t)|dt - \int_{T_\varepsilon} |k(s_0,t)|dt , \\
&= \int_0^1 |k(s_0,t)|dt - 2\int_{T_\varepsilon} |k(s_0,t)|dt , \\
&\ge M - 4\varepsilon .
\end{aligned}$$

As before, it follows that $||T|| = M$. □

REMARK 5.1. In the above examples it was possible to give an explicit expression for $||T||$. In many cases we must be satisfied with less, namely bounds for $||T||$. And even when an explicit expression for $||T||$ is known it may be impossible to compute it exactly. For instance, in Example 5.2 above, one cannot in general compute $\rho(A^TA)$ exactly and so one cannot in general compute $||A||_2$ exactly. □

REMARK 5.2. As the reader may have noticed, in Examples 5.1 through 5.4 we were able to find an $x \in X$ such that $||Tx|| = ||T|| \cdot ||x||$. In Example 5.5 on the other hand this is not always possible (see Problem 5.6). □

EXAMPLE 5.6. Let

$$T : C^1[a,b] \to R^1 ,$$

be given by

$$Tx = \sum_{k=1}^{n} a_k x(t_k) , \tag{5.12}$$

where $a \le t_1 \le \ldots \le t_n \le b$, and where $a_k \in R^1$. Then

$$K/(2+b-a) \le ||T|| \le K , \tag{5.13}$$

where

$$K = [\sum_{k=1}^{n} (t_{k+1}-t_k) |\sum_{j=1}^{k} a_j|] + |\sum_{j=1}^{n} a_j| , \tag{5.14}$$

and

$$t_{n+1} = b .$$

Proof. We recall (Example 4.1) that the norm in $C^1[a,b]$ is defined by

$$||x|| = \max|x(t)| + \max|\dot{x}(t)| .$$

We begin by observing that for any real numbers $a_1,\ldots,a_n$ and $b_1,\ldots,b_n$, we have the identity

$$\sum_{k=1}^{n} a_k b_k = \sum_{k=1}^{n} b_k (\sum_{j=1}^{k} a_j - \sum_{j=1}^{k-1} a_j) ,$$

$$= \sum_{k=1}^{n} b_k \sum_{j=1}^{k} a_j - \sum_{k=0}^{n-1} b_{k+1} \sum_{j=1}^{k} a_j , \tag{*}$$

$$= -[\sum_{k=1}^{n} (b_{k+1}-b_k) \sum_{j=1}^{k} a_j] + b_{n+1} \sum_{j=1}^{n} a_j ,$$

where b_{n+1} can be defined arbitrarily and where we have used the convention that $\sum_{j=1}^{0} a_j = 0$.

From eqns (5.12) and (*) with $b_k = x(t_k)$ we obtain

$$Tx = -[\sum_{k=1}^{n} (x(t_{k+1})-x(t_k)) \sum_{j=1}^{k} a_j] + x(b) \sum_{j=1}^{n} a_j . \tag{**}$$

Now

$$|x(b)| \leq \|x\| \ ,$$

and

$$|x(t_{k+1})-x(t_k)| \leq \max|\dot{x}(t)| \cdot (t_{k+1}-t_k) \leq \|x\|(t_{k+1}-t_k) \ ,$$

so that it follows immediately from eqn (**) that

$$\|Tx\| \leq K\|x\| \ . \qquad (\substack{**\\*})$$

Define the continuous piecewise linear function y so that

(1) $y(b) = \operatorname{sign} \sum_{j=1}^{n} a_j$.

(2) $y(t_{k+1}) - y(t_k) = -(t_{k+1}-t_k) \operatorname{sign} \sum_{j=1}^{k} a_j$.

(3) y is linear in each interval (t_k, t_{k+1}) .

Consequently, from eqns (**) and (5.14),

$$Ty = K \ . \qquad (\substack{**\\**})$$

In eqn $(\substack{**\\**})$ we have abused our notation slightly since y may have corners at the points t_k and, therefore, may not belong to $C^1[a,b]$. By replacing any such corners in y by segments of small circular arcs we obtain, for any $\varepsilon > 0$, a function $x_\varepsilon \in C^1[a,b]$ such that $|\dot{x}_\varepsilon| \leq 1$, $x_\varepsilon(b) = \pm 1$, and

$$Tx_\varepsilon \geq K - \varepsilon \ . \qquad (\substack{***\\**})$$

But

$$\begin{aligned} \|x_\varepsilon\| &= \max|x_\varepsilon| + \max|\dot{x}_\varepsilon| \ , \\ &\leq [1 + (b-a)] + 1 \ , \\ &= 2 + b - a \ . \qquad (\substack{***\\***}) \end{aligned}$$

Combining eqns $(\substack{***\\**})$ and $(\substack{***\\***})$, we see that

$$\begin{aligned} \|T\| &\geq (K-\varepsilon)/\|x_\varepsilon\| \ , \\ &\geq (K-\varepsilon)/(2+b-a) \ . \end{aligned}$$

Noting eqn $(\substack{**\\*})$ we see that eqn (5.13) has been proved. □

There is a simple but important relationship between continuity and norms for linear operators:

THEOREM 5.1. *Let* T *be a linear mapping from a normed linear space* X *to a normed linear space* Y . *Then the following are equivalent:*

(1) T *is continuous; that is, if* $x_n \to x$ *then* $Tx_n \to Tx$, *for all* $x \in X$.

(2) T *is continuous at some point; that is, there exists* $\bar{x} \in X$ *such that if* $x_n \to \bar{x}$ *then* $Tx_n \to T\bar{x}$.

(3) $\|T\| < \infty$.

(4) *For some real* $M < \infty$, $\|Tx\| \le M\|x\|$ *for all* x .

Proof. We prove only that (2) ⇒ (3), leaving the remainder of the proof to the reader (Problem 5.8).

Assume (3) does not hold. Then there exists a sequence $\{u_n\}$ in X such that

$$\|Tu_n\| \ge n \quad \text{and} \quad \|u_n\| = 1 .$$

Let $\bar{x}$ be any point in X and set $x_n = \bar{x} + (1/n)u_n$. Then, $x_n \to \bar{x}$ because $\|\bar{x}-x_n\| = \|u_n\|/n \to 0$. However, $Tx_n \not\to T\bar{x}$ because

$$\|T\bar{x}-Tx_n\| = \|Tu_n\|/n \ge 1 .$$

Thus NOT (3) ⇒ NOT (2), and hence (2) ⇒ (3) . □

5.2 *THE PRINCIPLE OF UNIFORM BOUNDEDNESS*

THEOREM 5.2. *(Baire category theorem).*

If a complete metric space X *is the denumerable union of closed subsets then at least one of the closed subsets contains a non-empty open ball.*

Proof. Let

$$X = \bigcup_{k=1}^{\infty} X_k ,$$

where each X_k is closed. We assume that no X_k contains a non-empty open ball and show that this leads to a contradiction.

From the assumptions, the complement $X_k^c = X \setminus X_k$ of X_k is open and non-empty for all k. In particular, X_1^c contains an open ball $B_1 = B(p_1;\varepsilon_1)$ with centre p_1 and radius $\varepsilon_1 \leq 1/3$ such that $B_1 \cap X_1 = \emptyset$. By assumption, the open ball $B(p_1;\varepsilon_1/3)$ is not contained in X_2. Thus the open set $X_2^c \cap B(p_1;\varepsilon_1/3)$ is non-empty and hence contains a ball $B_2 = B(p_2;\varepsilon_2)$ with $\varepsilon_2 \leq \varepsilon_1/3 \leq (1/3)^2$. Repeating the argument we obtain a sequence of balls

$$B_n = B(p_n;\varepsilon_n)$$

satisfying

$$0 < \varepsilon_n \leq 1/3^n;\ \varepsilon_{n+1} \leq \varepsilon_n/3;\ B_{n+1} \subset B(p_n;\varepsilon_n/3);\ B_n \cap X_n = \emptyset .$$

Since, for $n < m$,

$$\begin{aligned}\rho(p_n,p_m) &\leq \rho(p_n,p_{n+1}) + \ldots + \rho(p_{m-1},p_m) ,\\ &\leq \frac{\varepsilon_n}{3} + \ldots + \frac{\varepsilon_{m-1}}{3} \leq \frac{\varepsilon_n}{2} ,\end{aligned}$$

the centres p_n form a Cauchy sequence. By assumption, X is complete and so the sequence $\{p_n\}$ must converge to a point $p \in X$.

Because

$$\begin{aligned}\rho(p_n,p) &\leq \rho(p_n,p_m) + \rho(p_m,p) ,\\ &\leq \varepsilon_n/2 + \rho(p_m,p) ,\\ &\to \varepsilon_n/2 , \quad \text{as } m \to \infty ,\end{aligned}$$

it follows that $p \in B_n$ for all n. Thus $p \notin X_n$ for all n, and this contradicts the assumption that $X = \cup X_n$. □

THEOREM 5.3. *(Principle of uniform boundedness)*

For each a *in an index set* A *let* T_a *be a continuous linear map from a Fréchet space* X *into a Fréchet space* Y. *If for each* $x \in X$ *the set* $\{T_a x;\ a \in A\}$ *is topologically bounded then* $\lim_{x \to \underline{0}} T_a x = 0$ *uniformly for* $a \in A$. *That is, given* $\varepsilon > 0$ *there exists* $\delta > 0$ *such that*

$$\rho_Y(T_a x, \underline{0}) < \varepsilon \quad \textit{for all} \quad a \in A ,$$

provided that

$$\rho_X(x, \underline{0}) < \delta .$$

Proof. For given $\varepsilon > 0$ and $k = 1,2,\ldots,$ let

$$X_k = \{x \in X\colon \rho_Y(\tfrac{1}{k} T_a x, \underline{0}) \le \varepsilon/2 \quad \text{for all} \quad a \in A\} . \qquad (*)$$

Since T_a is a continuous mapping and since the mapping $y \to y/k$ is continuous the set

$$X_k(a) = \{x \in X\colon \rho_Y(\tfrac{1}{k} T_a x,\ \underline{0}) \le \varepsilon/2\} ,$$

is closed. Since $X_k = \bigcap_{a \in A} X_k(a)$, X_k is also closed.

Let $x_o \in X$. By assumption,

$$E_o = \{T_a x_o\colon a \in A\}$$

is topologically bounded, so there exists $\mu_o = 1/k_o$ such that

$$\mu_o E_o = \{\tfrac{1}{k_o} T_a x_o\colon a \in A\} \subset B_Y[\underline{0};\ \varepsilon/2] .$$

That is, $x_o \in X_{k_o}$. Since x_o was arbitrary,

$$X = \cup X_k .$$

By the Baire category theorem (Theorem 5.2), at least one X_k, X_{k_o} say, contains an open ball, $B(x_o;\delta_o)$ say. Multiplication is continuous, so there exists $\delta > 0$ such that if

$$\rho_X(x, \underline{0}) < \delta \quad \text{then} \quad \rho_X(k_o x, \underline{0}) < \delta_o . \qquad (**)$$

Now let x satisfy $\rho_X(x, \underline{0}) < \delta$. Then, from eqn (**),

$$x_o + k_o x \in B(x_o;\delta_o) \subset X_{k_o} .$$

Hence, from eqn (*), for any $a \in A$,

$$\rho_Y(\frac{1}{k_o}T_a x_o,\underline{0}) \le \varepsilon/2 , \quad \text{and} \quad \rho_Y(\frac{1}{k_o}T_a(x_o+k_o x),\underline{0}) \le \varepsilon/2 .$$

Thus, by the triangle inequality, for any $a \in A$,

$$\begin{aligned} \rho_Y(T_a x,\underline{0}) &= \rho_Y(\frac{1}{k_o}T_a k_o x,\underline{0}) , \\ &\le \rho_Y(\frac{1}{k_o}T_a x_o,\underline{0}) + \rho_Y(\frac{1}{k_o}T_a(x_o + k_o x),\underline{0}) , \\ &\le \varepsilon/2 + \varepsilon/2 = \varepsilon . \end{aligned}$$

□

The most important applications in numerical functional analysis of the principle of uniform boundedness concern the question of whether or not a sequence of linear operators has a limit. The basic result is the following:

THEOREM 5.4. *(Banach-Steinhaus)*

Let X *and* Y *be Banach spaces and* $\{T_n\}$ *a sequence of bounded linear operators from* X *into* Y *. Then:*
(1) $\Longleftrightarrow$ *(2) AND (3), where*

(1) The limit $\lim_{n\to\infty} T_n x$ *exists for every* $x \in X$.

(2) The limit $\lim_{n\to\infty} T_n x$ *exists for each* x *in a dense subset* E *of* X .

(3) The T_n *are uniformly bounded; that is,* $||T_n|| \le M$ *for all* n *and some constant* $M < \infty$.

If (1) holds, let $T\colon X \to Y$ *be defined by*

$$Tx = \lim_{n\to\infty} T_n x , \quad x \in X .$$

Then

(4) T *is a bounded linear operator and* $||T|| \le \liminf_{n\to\infty} ||T_n|| \le M < \infty$.

Proof (1) ⇒ (2): Trivial. (1) ⇒ (3): For each $x \in X$ the sequence $\{T_n x\}$ is convergent and hence bounded. Thus, by the principle of uniform boundedness, there exists $\delta > 0$ such that if $||x|| = \rho(x,\underline{0}) < \delta$ then $||T_n x|| = \rho(T_n x,\underline{0}) < 1$. for all n. That is, $||T_n|| \le M = \delta^{-1}$.

(2) AND (3) ⇒ (1). Choose any $x \in X$ and any $\varepsilon > 0$. Pick $u \in E$ satisfying $||u-x|| \le \varepsilon$. Since $u \in E$ the sequence $\{T_n u\}$ converges and there exists N such that $||T_n u - T_m u|| \le \varepsilon$ when $m,n > N$. Thus, if $m,n > N$ we have

$$||T_n x - T_m x|| \le ||T_n u - T_m u|| + ||T_n(x-u)|| + ||T_m(x-u)|| ,$$
$$\le \varepsilon + M\varepsilon + M\varepsilon ,$$

from which we conclude that $\{T_n x\}$ is a Cauchy sequence. By assumption, Y is a Banach space and hence the limit of $T_n x$ exists as $n \to \infty$.

(1) ⇒ (4): For any scalars λ,μ

$$\begin{aligned} T(\lambda x+\mu y) &= \lim T_n(\lambda x+\mu y) , \\ &= \lambda \lim T_n x + \mu \lim T_n y , \\ &= \lambda Tx + \mu Ty , \end{aligned}$$

so that T is linear. Furthermore, as previously shown, (1) ⇒ (3). Hence, for any $x \in X$,

$$||Tx|| = \lim_{n\to\infty} ||T_n x|| \le \liminf_{n\to\infty} ||T_n||\cdot||x|| \le M||x|| ,$$

so that T is a bounded continuous operator (see Theorem 5.1, part (4)). □

REMARK 5.3. The principle of uniform boundedness is so useful that it is of interest to determine the class of spaces for which it holds. For remarks on this see Robertson and Robertson [1964, p. 65], and Dunford and Schwartz [1966, p. 80].

REMARK 5.4. A set F in a Banach space X is *fundamental* if the linear subspace spanned by F is dense in X. For example, the set of functions t^k, $k = 0,1,2,\ldots$, is fundamental in $C[a,b]$ because the functions t^k span the subspace consisting of the polynomials, which is dense in $C[a,b]$. Condition (2) of the Banach-Steinhaus theorem may be replaced by the condition

(2') The limit Tx exists for each x in a fundamental set F.

(The proof is left to the reader - see Problem 5.10). □

REMARK 5.5. If $L: X \to Y$ is a given bounded linear operator, and $T_n: X \to Y$ is a sequence of bounded linear operators, then

$$\lim_{n\to\infty} T_n x = Lx ,$$

for all $x \in X$ iff conditions (2) and (3) of Theorem 5.4 hold and, furthermore,

$$Lx = Tx$$

for all x in a fundamental subset F of X, where X and Y are Banach spaces (Problem 5.11). □

REMARK 5.6. Can the conditions of the Banach-Steinhaus theorem be relaxed?

There are many examples of sequences of operators T_n such that $T_n x$ converges for all x in a dense set, but not for all $x \in X$.

For example, let $X = C[0,1]$ and

$$T_n x = [x(\tfrac{1}{n}) - x(0)]n .$$

Then $T_n x \to \frac{dx(0)}{dt}$ for every x which is differentiable at $t = 0$, and, in particular, for every polynomial. But the sequence $\{T_n x\}$ is not convergent for all $x \in X$.

Given a sequence of bounded linear operators $\{T_n\}$ mapping a Banach space X into a normed linear space Y, the set $\{x \in X: \limsup \|T_n x\| < \infty\}$ either coincides with X or is a set of the first category in X. (For a definition of sets of the first category see Problem 5.9.) See Yosida [1968, p. 73], and Dunford and Schwartz [1966, p. 81]. □

REMARK 5.7. To apply the Banach-Steinhaus theorem we must check condition (2) of the theorem namely that $T_n x \to Tx$ for all x in a dense subset E, and it might seem that little has been gained. In applications, however, the fact that $T_n x \to Tx$ for $x \in E$ is often trivially true because of the way in which the operators T_n have been constructed. In quadrature formulae for example we often have $T_n x = Tx$ for n sufficiently large if x is any polynomial. □

We now give some applications of the Banach-Steinhaus theorem.

5.3 *NUMERICAL QUADRATURE*

Let $X = C[a,b]$ and $Y = R^1$. Let $T: X \to Y$,

$$Tx = \int_a^b x(t)dt .$$

Then T is a bounded linear operator with $||T|| = b-a$. Let T_n be a sequence of quadrature formulae

$$T_n x = \sum_{k=0}^{n} w_k^{(n)} x(t_k^{(n)}), \quad n = 1,2,\ldots , \qquad (5.15)$$

where $a \le t_0^{(n)} < \ldots < t_n^{(n)} \le b$, and where the *weights* $w_k^{(n)}$ and *nodes* $t_k^{(n)}$ are real numbers which do not depend upon x.

THEOREM 5.5. *(Steklov)*

The quadrature formulae (5.15) converge for any $x \in C[a,b]$ *iff:*

(1) The formulae converge for every polynomial.

(2) $\sum_{k=0}^{n} |w_k^{(n)}| \le M$ *for some constant* M *and all* n.

Proof. The theorem follows from the Banach-Steinhaus theorem since, from Example 5.4,

$$||T_n|| = \sum_{k=0}^{n} |w_k^{(n)}| . \qquad \square$$

For Gaussian quadrature formulae the weights $w_k^{(n)}$ are always positive. Also, such formulae are exact for constants so that

$$\sum_{k=0}^{n} |w_k^{(n)}| = \sum_{k=0}^{n} w_k^{(n)} = b-a .$$

Hence, if T_n denotes an (n+1)-point Gaussian quadrature formula, then $T_n x \to Tx$ as $n \to \infty$ for all $x \in C[a,b]$.

EXAMPLE 5.7. The following table gives the approximations to the integral

$$Tx = \int_{-1}^{+1} \frac{dt}{1+25t^2} = \frac{2}{5} \text{ arc tan } 5 \doteq .54936 \qquad (*)$$

obtained using the (n+1)-point closed Newton-Cotes formulac, the (n+1)-point trapezoidal rule, and the (n+1)-point Gauss-Legendre formulae. It is clear that the Newton-Cotes approximations are diverging.

n	(n+1)-point closed Newton-Cotes	(n+1)-point Trapezoidal	(n+1)-point Gauss-Legendre
2	1.358	1.03846	.95833
4	0.474	0.65716	.70694
6	0.774	0.57767	.61612
8	0.300	0.55689	.57870
10	0.934	0.55122	.56245
12	-0.062	0.54968	.55524
14	1.579	0.54929	.55201
16	-1.248	0.54922	.55055
18	3.775	0.54922	.54990
20	-5.369	0.54924	.54960

Table 5.1: Numerical integration of (*)

For $n \le 7$ and $n = 9$ the weights $w_k^{(n)}$ of the (n+1)-point Newton-Cotes quadrature formulae are all positive so that $\|T_n\| = 2$. For all other values of n negative weights occur. For large n very large positive and negative weights occur, and $\|T_n\| \to \infty$ as $n \to \infty$, which explains the divergence of the Newton-Cotes approximations in Table 5.1.

It is known (Krylov [1962, p. 86]) that, for large m ,

$$\left|w_m^{(2m)}\right| \sim \frac{(2m)!}{m!m!\ m^2 \ell n^2(2m)} ,$$

$$\sim \frac{2^{2m}}{\sqrt{\pi m}\ m^2 \ell n^2(2m)} \qquad (*)$$

For small values of n, $\|T_n\|$ can be computed directly and some values are given in Table 5.2. (Problem 5.12)

n	$\|T_n\|$
2	2.00
4	2.00
6	2.00
8	2.00
10	6.12
12	15.06
14	40.68
16	116.91
18	350.92
20	1088.35

Table 5.2: $\|T_n\|$ *for (n+1)-point closed Newton-Cotes quadrature formulae*

□

5.4 *TRANSFORMATION OF SEQUENCES*

Let $X = Y = c$, the space of convergent real sequences; if $x \in c$, $x = (x_1, x_2, \ldots, x_n, \ldots)$, $x_\infty = \lim_{n\to\infty} x_n$ exists, and $\|x\| = \sup |x_n|$.

We define $L: c \to R^1$ by

$$L: x \in c \to x_\infty = \lim_{n\to\infty} x_n \in R^1 .$$

As is readily checked, L is a continuous linear map.

Let A be the infinite matrix

$$A = \begin{bmatrix} a_{11}, & a_{12}, & \dots, & a_{1k}, & \dots \\ a_{21}, & a_{22}, & \dots, & a_{2k}, & \dots \\ \cdot & & & & \\ \cdot & & & & \\ \cdot & & & & \\ a_{n1}, & a_{n2}, & \dots, & a_{nk}, & \dots \\ \text{-----} & \text{-----} & \text{-----} & \text{-----} & \text{-----} \end{bmatrix} . \qquad (5.16)$$

Let $T_n^{(m)}: c \to R^1$ be given by

$$T_n^{(m)} x = \sum_{k=1}^{m} a_{nk} x_k . \qquad (5.17)$$

If it exists, we consider $T_n: c \to R^1$,

$$T_n x \equiv (Ax)_n = \lim_{m\to\infty} T_n^{(m)} x = \sum_{k=1}^{\infty} a_{nk} x_k . \qquad (5.18)$$

and, if it exists, we consider $T: c \to R^1$,

$$Tx \equiv Ax = \lim T_n x = \lim_{n\to\infty} \left(\sum_{k=1}^{\infty} a_{nk} x_k \right) . \qquad (5.19)$$

DEFINITION 5.1. If

(α) $T_n x$ exists for all $x \in c$; that is, the series (5.18) converges for all $x \in c$,

(β) Tx exists for all $x \in c$; that is, the limit (5.19) exists for all $x \in c$,

(γ) $Tx = Lx \equiv \lim_{n\to\infty} x_n$, for all $x \in c$; that is, Tx is equal to the limit Lx of the original sequence x,

we say that A defines a *regular method of summability*. □

THEOREM 5.6. *(Toeplitz)*

The infinite matrix A *defines a regular method of summability iff*

(1) $\lim_{n\to\infty} a_{nk} = 0$, for $k = 1,2,\ldots$,

(2) $\lim_{n\to\infty} \sum_{k=1}^{\infty} a_{nk} = 1$,

(3) $\sum_{k=1}^{\infty} |a_{nk}| \le M$, for $n = 1,2,\ldots$, *and some fixed constant* M .

Proof. We begin by observing that the linear subspace spanned by the elements

$u_0 = (1,1,1,\ldots)$, (all ones),
$u_1 = (1,0,0,\ldots)$,
$u_r = (0,0,\ldots,0,1,0,\ldots)$, (all zeros except for a one in the r-th place)

is dense in c . That is, the elements $u_0, u_1, \ldots$, are fundamental in c . The proof is left to the reader (Problem 5.15)

We must prove that conditions (α), (β), and (γ) of Definition 5.1 are equivalent to conditions (1), (2), and (3) above. We begin by reformulating (α) .

Applying the Banach-Steinhaus theorem and Remark 5.4 we see that, *for fixed* n , $T_n x = \lim_{m\to\infty} T_n^{(m)} x$ exists for all $x \in c$ iff

(a) $\lim_{m\to\infty} T_n^{(m)} u_r$ exists for $r = 0,1,2,\ldots$.

and

(b) $\|T_n^{(m)}\| \le M_n$, for all m and some constant $M_n < \infty$.

But,

$$T_n^{(m)} u_0 = \sum_{k=1}^{m} a_{nk} ,$$

$$T_n^{(m)} u_r = \begin{cases} 0, & \text{if } m < r , \\ a_{nr}, & \text{if } m \ge r , \end{cases}$$

and (see Example 5.4)

$$\|T_n^{(m)}\| = \sum_{k=1}^{m} |a_{nk}| .$$

Thus, $\lim_{m\to\infty} T_n^{(m)}x$ exists for all x iff

(c) $\sum_{k=1}^{\infty} a_{nk}$ exists,

and

(d) $\sum_{k=1}^{\infty} |a_{nk}| < M_n < \infty$.

That is, $(\alpha) \Leftrightarrow (c,d)$. Obviously, $(d) \Rightarrow (c)$, and so $(\alpha) \Leftrightarrow (d)$.

We now assume (α) and apply the Banach-Steinhaus theorem and Remarks 5.4 and 5.5 to the sequence of operators T_n . It follows that $Tx = \lim_{n\to\infty} T_n x$ exists and is equal to $Lx = \lim_{m\to\infty} x_m$ for all x iff

(e) $\lim_{n\to\infty} T_n u_r = u_r = Lu_r$ for $r = 0,1,\ldots$,

(f) $\|T_n\| \le M$, for some constant M .

But

$$T_n u_0 = \sum_{k=1}^{\infty} a_{nk} \; ; \quad Lu_0 = 1 \; ;$$

$$T_n u_k = a_{nk} , \quad Lu_k = 0 , \quad \text{for } k > 0 \; ;$$

and

$$\|T_n\| = \sum_{k=1}^{\infty} |a_{nk}| .$$

Thus, $\lim_{n\to\infty} T_n x$ exists and is equal to $Lx = \lim_{m\to\infty} x_m$ for all x iff

(g) $\lim_{n\to\infty} \sum_{k=1}^{\infty} a_{nk} = 1$,

(h) $\lim_{n\to\infty} a_{nk} = 0$, for $k = 1,2,\ldots$,

(i) $\sum_{k=1}^{\infty} |a_{nk}| \le M$.

That is, if (α) is true then $(\beta,\gamma) \Leftrightarrow (g,h,i)$. By inspection we see that $(g,h,i) \Rightarrow (d) \Leftrightarrow (\alpha)$. Thus $(\alpha,\beta,\gamma) \Leftrightarrow (g,h,i)$ and the theorem is proved. □

REMARK 5.6. It is possible to obtain some insight into the conditions of Toeplitz's theorem using simple arguments. Recall that

$$(Ax)_n = \sum_{k=1}^{\infty} a_{nk} x_k \; .$$

(1) If condition (1) of Theorem 5.6 does not hold, then for some k and some ε there exist arbitrarily large n such that $|a_{nk}| \geq \varepsilon > 0$. For these values of n , $(Ax)_n$ will contain a non-trivial multiple of x_k , and so $(Ax)_n$ will not converge to x_∞ for general x .

(2) Consider the sequence $\bar{x} = (\bar{x}_k)$ with $\bar{x}_k = \bar{x}_\infty$ for all k . Then $(A\bar{x})_n = \bar{x}_\infty \sum_{k=1}^{\infty} a_{nk}$. Thus condition (2) of Theorem 5.6 is necessary to ensure that

$$\lim_{n\to\infty} (A\bar{x})_n = \lim_{n\to\infty} \bar{x}_n \; .$$

□

We now apply the theorem of Toeplitz to two transformations: Euler's transformation and Romberg's transformation.

5.4.1 *EULER'S TRANSFORMATION*

Euler's transformation was originally developed to sum alternating series of the form

$$\sum_{s=0}^{\infty} (-1)^s u_s = u_0 - u_1 + u_2 + \ldots \; , \tag{5.20}$$

with given terms u_s .

If the *shift operator* E and *forward difference operator* Δ are defined by

$$Eu_s = u_{s+1} \; , \quad \Delta u_s = u_{s+1} - u_s \; ,$$

we have that

$$E = 1 + \Delta \; .$$

Thus, formally,

$$\sum_{s=0}^{\infty}(-1)^s u_s = \sum_{s=0}^{\infty}(-E)^s u_0 = \frac{1}{1+E}u_0 ,$$

$$= \frac{1}{2}\frac{1}{1+\Delta/2}u_0 = \frac{1}{2}\sum_{s=0}^{\infty}(-\tfrac{1}{2}\Delta)^s u_0 ,$$

$$= \sum_{s=0}^{\infty} v_s , \qquad (5.21)$$

where

$$v_s = \tfrac{1}{2}(-\Delta/2)^s u_0 = \frac{1}{2^{s+1}}(1-E)^s u_0 ,$$

$$= \frac{1}{2^{s+1}}(u_0 - su_1 + \ldots + (-1)^j \binom{s}{j} u_j + \ldots + (-1)^s u_s) . \qquad (5.22)$$

Euler's transformation consists of computing the terms v_s defined by eqn (5.22) and summing the series (5.21) instead of the original series (5.20).

One expects the Euler transformation to be effective when the terms u_s decrease smoothly to zero since then the differences $\Delta^s u_0$ will decrease rapidly.

When applying the Euler transformation in practice, one often sums the first few terms of the series directly and then applies the Euler transformation to the remainder.

EXAMPLE 5.8. Consider the series

$$S = 1 - \tfrac{1}{3} + \tfrac{1}{5} \ldots = \sum_{k=0}^{\infty} \frac{(-1)^k}{2k+1} .$$

We have that

$$S = \sum_{k=0}^{5} \frac{(-1)^k}{2k+1} + \sum_{k=6}^{\infty} \frac{(-1)^k}{2k+1} ,$$

$$\doteq .74401 + \sum_{s=0}^{\infty} (-1)^s u_s ,$$

with $u_s = \dfrac{1}{2s+13}$,

Applying Euler's transformation,

s	u_s	Δ	Δ^2	Δ^3	Δ^4
0	.07692				
1	.06666	-1026			
2	.05882	- 784	242		
3	.05263	- 619	165	-77	
4	.04761	- 502	117	-48	29

and

$$v_0 = \frac{1}{2} u_0 = \frac{.07692}{2} = .03846$$

$$v_1 = \frac{1}{2} (-\Delta/2) u_0 = \frac{.01026}{4} = .00256$$

$$v_2 = \frac{1}{2} (-\Delta/2)^2 u_0 = \frac{.00242}{8} = .00030$$

$$v_3 = \frac{1}{2} (-\Delta/2)^3 u_0 = \frac{.00077}{16} = .00005$$

$$v_4 = \frac{1}{2} (-\Delta/2)^4 u_0 = \frac{.00029}{32} = .00001$$

so that

$$S \doteq .74401 + \sum_{s=0}^{4} v_s = .78539 .$$

The true value is

$$S = \pi/4 = .785398 \ldots$$

The sum has thus been determined to five decimal places with relatively little work. The whole computation was performed by the author using a hand calculator in about fifteen minutes. To obtain comparable accuracy, about one million terms of the original series must be summed. □

REMARK 5.7. There is a computationally useful modification of Euler's method due to van Wijngaarden (Modern Computing Methods [1961, p. 125]). We rewrite Euler's transformation (5.21), (5.22) in the form

$$\sum_{s=0}^{\infty}(-1)^s u_s = \sum_{s=0}^{n-1}(-1)^s u_s + \sum_{s=0}^{\infty}(-1)^{n+s}u_{n+s} ,$$

$$= \sum_{s=0}^{n-1}(-1)^s u_s + \frac{1}{2}(-1)^n \sum_{s=0}^{\infty}\frac{(1-E)^s}{2^s}u_n ,$$

$$= \sum_{s=0}^{n-1}\tilde{u}_s + \frac{1}{2}\sum_{s=0}^{\infty}M^s\tilde{u}_n , \qquad (*)$$

where

$$\tilde{u}_s = (-1)^s u_s ,$$

and where M is the operator

$$M\tilde{u}_s = \tfrac{1}{2}(\tilde{u}_s+\tilde{u}_{s+1}) = \tfrac{1}{2}(1+E)\tilde{u}_s = \tfrac{1}{2}(-1)^s(1-E)u_s .$$

If the series in eqn (*) is truncated after (p+1) terms we obtain the approximation $S_{n,p}$ given by

$$S_{n,p} = \sum_{s=0}^{n-1}\tilde{u}_s + \frac{1}{2}\sum_{s=0}^{p}M^s\tilde{u}_n . \qquad (**)$$

The approximations $S_{n,p}$ satisfy the recurrence relations

$$S_{n,p+1} = S_{n,p} + \tfrac{1}{2}M^{p+1}\tilde{u}_n ,$$

$$S_{n+1,p} = S_{n,p} + M^{p+1}\tilde{u}_n . \qquad (\overset{**}{*})$$

The first of these formulae is obvious. The second formula follows from the observation that $E = 2M-1$ so that

$$S_{n+1,p} = \sum_{s=0}^{n}\tilde{u}_s + \frac{1}{2}\sum_{s=0}^{p}M^s\tilde{u}_{n+1} ,$$

$$= \sum_{s=0}^{n}\tilde{u}_s + \frac{1}{2}\sum_{s=0}^{p}M^s(2M-1)\tilde{u}_n ,$$

$$= \sum_{s=0}^{n}\tilde{u}_s + \frac{1}{2}\sum_{s=0}^{p}M^s\tilde{u}_n + [M^{p+1}\tilde{u}_n - M^0\tilde{u}_n] ,$$

$$= S_{n,p} + M^{p+1}\tilde{u}_n .$$

Starting with $n = p = 0$ and $S_{n,p} = S_{0,0} = 0 + \frac{1}{2}M^0\tilde{u}_0 = u_0/2$, the van Wijngaarden algorithm recursively constructs a sequence of approximations $S_{n,p}$ using the relations $(\overset{*}{{}_{*}}{}^{*})$ and the criterion:

$$\text{if } |M^{p+1}\tilde{u}_n| < |M^p\tilde{u}_{n+1}| \text{ then } p \leftarrow p+1 \text{ else } n \leftarrow n+1 \,. \tag{$\substack{**\\**}$}$$

The justification for $(\substack{**\\**})$ is that, from eqn (**), $\frac{1}{2}M^{p+1}\tilde{u}_n$ and $\frac{1}{2}M^p\tilde{u}_{n+1}$ are, respectively, the last terms in the approximations $S_{n,p+1}$ and $S_{n+1,p}$, so that at each stage the approximation with the smallest last term is chosen.

The algorithm is readily implemented. At each stage only the $p+1$ quantities

$$\tilde{u}_{n+p},\ M\tilde{u}_{n+p-1},\ldots,M^p\tilde{u}_n \,,$$

need be stored. During the updating of $S_{n,p}$ one computes the new $p+2$ quantities

$$\tilde{u}_{n+p+1},\ M\tilde{u}_{n+p},\ldots,M^{p+1}\tilde{u}_n \,,$$

using the formula

$$M^k\tilde{u}_{n+p+1-k} = \tfrac{1}{2}[M^{k-1}\tilde{u}_{n+p+1-k} + M^{k-1}\tilde{u}_{n+p+2-k}] \,, \quad k = 1,\ldots,p+1;$$

when $M^k\tilde{u}_{n+p+1-k}$ has been computed, the quantity $M^{k-1}\tilde{u}_{n+p+1-k}$ is no longer needed and may be over-written by $M^{k-1}\tilde{u}_{n+p+2-k}$. Finally, if p is not increased, the term $M^{p+1}\tilde{u}_n$ is not retained.

For the problem of Example 5.8, the first few approximations obtained with van Wijngaarden's modification of Euler's transformation are:

n	p	$S_{n,p}$
0	0	.50000
1	0	.83333
1	1	.80000
1	2	.79047
2	2	.78412
2	3	.78499
2	4	.78525
3	4	.78543
3	5	.78541

By chance, $M^2\tilde{u}_2 = M^3\tilde{u}_1$, with the consequence that some computer programmes will not perform the test (**) correctly for this problem when $n = 1$ and $p = 2$. □

We conclude this section by showing that Euler's transformation is a regular method of summability.

Let (x_n) and (y_n) be the sequences defined by

$$x_0 = 0 ,$$

$$x_n = \sum_{s=0}^{n-1} (-1)^s u_s , \quad \text{if} \quad n \geq 1 , \tag{*}$$

$$y_n = \sum_{s=0}^{n-1} v_s , \quad \text{for} \quad n \geq 1 .$$

where the terms u_s are terms in the infinite series (5.20) and v_s is defined by eqn (5.22). Thus, x_n and y_n are, respectively, the n-th partial sums of the original series and the transformed series.

From eqn (*) we conclude that

$$\begin{aligned} u_n &= (-1)^n (x_{n+1} - x_n) , \\ &= (-1)^n (E-1) x_n , \end{aligned} \tag{**}$$

Using eqns (5.22) and (**) we obtain

$$\begin{aligned} y_n &= \sum_{s=0}^{n-1} v_s , \\ &= \sum_{s=0}^{n-1} \frac{1}{2^{s+1}} (u_0 - su_1 + \ldots + (-1)^j \tbinom{s}{j} u_j + \ldots + (-1)^s u_s) , \\ &= \sum_{s=0}^{n-1} \frac{1}{2^{s+1}} (E-1)(x_0 + sx_1 + \ldots + \tbinom{s}{j} x_j + \ldots + x_s) , \\ &= \sum_{s=0}^{n-1} \frac{1}{2^{s+1}} (E-1)(E+1)^s x_0 , \\ &= \sum_{s=0}^{n-1} \left[\frac{(E+1)}{2} - 1\right] \left[\frac{E+1}{2}\right]^s x_0 . \end{aligned}$$

This latter sum telescopes, and we obtain

$$y_n = \left[\left(\frac{E+1}{2}\right)^n - 1\right] x_0 ,$$

$$= \left(\frac{E+1}{2}\right)^n x_0 ,$$

$$= \frac{1}{2^n}\left(x_0 + nx_1 + \ldots + \binom{n}{j} x_j + \ldots + x_n\right) . \qquad (*\!*\!*)$$

Comparing eqns (5.16), (5.17), and (5.18) with eqn (***), we see that Euler's transformation transforms the sequence (x_n) into the sequence $((Ax)_n)$ where the coefficients a_{nk} are given by

$$a_{nk} = \begin{cases} \binom{n}{k} 2^{-n} , & \text{if } k \le n , \\ 0 , & \text{otherwise} . \end{cases}$$

The conditions of Toeplitz's theorem are readily seen to be satisfied, so that Euler's transformation is indeed a regular method of summability.

REMARK 5.8. For a further discussion of Euler's transformation see Knopp [1954, p. 262] and Brezinski [1977].

5.4.2 *ROMBERG'S TRANSFORMATION*

Romberg's transformation is applicable to a convergent sequence denoted by $T_0^{(0)}, T_0^{(1)}, T_0^{(2)}, \ldots$ and generates a triangular array

$$\begin{array}{llll} T_0^{(0)} & & & \\ T_0^{(1)} & T_1^{(0)} & & \\ T_0^{(2)} & T_1^{(1)} & T_2^{(0)} & \\ T_0^{(3)} & T_1^{(2)} & T_2^{(1)} & T_3^{(0)} \\ \cdot & \cdot & \cdot & \cdot \end{array} \qquad (5.23)$$

according to the formula

$$T_{m+1}^{(k)} = \frac{\beta^{m+1} T_m^{(k+1)} - T_m^{(k)}}{\beta^{m+1} - 1} , \qquad (5.24)$$

where the constant $\beta > 1$ is chosen using information about the sequence $T_0^{(k)}$.

Romberg's transformation is based on the hypothesis that there exists an expression of the form

$$T_0^{(v)} = \sum_{s=0}^{\infty} \alpha_s \left(\frac{1}{\beta^v}\right)^s , \tag{5.25}$$

for some constant $\beta > 1$.

The transformation is constructed so that $T_m^{(k)}$ is equal to the limit as $v \to \infty$ of the given sequence $T_0^{(v)}$ provided that

$$T_0^{(v)} = \sum_{s=0}^{m} \alpha_s \left(\frac{1}{\beta^v}\right)^s , \quad \text{for} \quad k \leq v \leq k + m . \tag{5.26}$$

One could, therefore, compute $T_m^{(k)}$ by first solving the system of $m + 1$ linear equations (5.26) for $\alpha_0, \ldots, \alpha_m$ and then setting $T_m^{(k)} = \alpha_0$. The recurrence relation (5.24) achieves the same goal more efficiently (see Problem 5.19).

EXAMPLE 5.9. Consider the trapezoidal rule with steplength $h = 1/n$ for approximating the integral

$$I = \int_0^1 f(t)dt ,$$

namely,

$$I(h) = \frac{h}{2} \sum_{i=1}^{n} [f(ih) + f([i-1]h)] .$$

If f is smooth, the Euler-Maclaurin expansion for $I(h)$ (see e.g. Isaacson and Keller [1966, p.287]) leads to an asymptotic expansion of the form

$$I(h) \sim I + \sum_{s=1}^{\infty} \alpha_s h^{2s} .$$

If $T_0^{(k)} = I(2^{-k})$ then

$$T_0^{(k)} \sim I + \sum_{s=1}^{\infty} \alpha_s (4^{-k})^s ,$$

so that eqn (5.25) holds with $\beta = 4$.

Let $T_0^{(k)}$ be obtained by applying the trapezoidal rule with step length $h = 1/2^k$ to the integral $I = \int_0^1 \frac{1}{1+t}\,dt$. Romberg's rule with $\beta = 4$ and $0 \le k \le 4$ gives:

.750 000 000				
.708 333 333	.694 444 444			
.697 023 809	.693 253 968	.693 174 603		
.694 121 850	.693 154 530	.693 147 901	.693 147 477	
.693 391 202	.693 147 652	.693 147 194	.693 147 183	.693 147 181 .

The true value is

$$I = \int_0^1 \frac{1}{1+t}\,dt = \ln 2 \doteq .693\ 147\ 180\ 559\ . \qquad \square$$

When considering the triangular array $T_m^{(k)}$ there are two natural questions which arise:

(1) For fixed m, what is $\lim_{k\to\infty} T_m^{(k)}$?

(2) What is $\lim_{m\to\infty} T_m^{(0)}$?

In case (1) we are considering a fixed column, column m say. It is readily shown using elementary analysis (Problem 5.20) that

$$\lim_{k\to\infty} T_m^{(k)} = \lim_{k\to\infty} T_0^{(k)}\ .$$

In case (2) it is necessary to use the theorem of Toeplitz. Romberg's transformation defines a transformation from the sequence $(T_0^{(k)})$ to the sequence $(T_m^{(0)})$ with transformation matrix C of the form

$$C = \begin{pmatrix} c_{00} & 0 & 0 & \cdot & \cdot \\ c_{11} & c_{10} & 0 & \cdot & \cdot \\ c_{22} & c_{21} & c_{20} & \cdot & \cdot \\ \cdot & \cdot & \cdot & \cdot & \cdot \\ \cdot & \cdot & \cdot & \cdot & \cdot \end{pmatrix}\ .$$

For example, when $\beta = 4$,

$$C = \begin{pmatrix} 1 & 0 & 0 & 0 & \cdots \\ -\frac{1}{3} & +\frac{4}{3} & 0 & 0 & \cdots \\ \frac{1}{45} & -\frac{20}{45} & +\frac{64}{45} & 0 & \cdots \\ -\frac{1}{2835} & +\frac{84}{2835} & -\frac{1344}{2835} & \frac{4096}{2835} & \cdots \\ \cdots & \cdots & \cdots & \cdots & \cdots \end{pmatrix}$$

Romberg's transformation is analysed for the case $\beta = 4$ by Bauer, Rutishauser, and Stiefel [1963]. In particular, Bauer, Rutishauser, and Stiefel show that:

$$\text{(1)} \quad \sum_{k=0}^{m} c_{m,k} = 1 .$$

$$\text{(2)} \quad \sum_{k=0}^{m} |c_{m,k}| < 2 , \quad \text{for all } m . \tag{5.27}$$

$$\text{(3)} \quad \lim_{m\to\infty} c_{m,m-s} = 0 .$$

(The method of proof is outlined in Problem 5.21.) It follows from the theorem of Toeplitz, that the transformation from $(T_0^{(k)})$ to $(T_m^{(0)})$ is a regular method of summability.

Romberg's transformation can also be applied to processes other than quadrature, for example numerical differentiation (Problem 5.22).

5.5 *INTERPOLATION BY POLYNOMIALS*

In this section we show how the Banach-Steinhaus theorem can sometimes be used to prove that a certain method *cannot exist*.

Let $X = C[-1,1]$. Let π_n be the subspace of X consisting of all polynomials of degree less than or equal to n . A continuous linear operator $T_n: X \to X$ is called a *polynomial operation of order n* if

(1) $T_n(x) \in \pi_n$ for $x \in X$,

(2) $T_n(x) = x$ if $x \in \pi_n$.

(In language which will be introduced in Chapter 8, T_n is a projection of X onto π_n) . A particular example of a polynomial operation of order n is the operation which associates the interpolation polynomial of order n (at some preassigned set of nodes) with each $x \in X$.

Let $\tilde{X} = \tilde{C}[0,2\pi]$, be the space of functions $x(t)$ which are continuous for $t \in [0,2\pi]$ and satisfy the restraint $x(0) = x(2\pi)$; $\tilde{C}$ is equipped with the maximum norm. If $x \in \tilde{X}$, then x can be extended as a periodic function on the real axis. Let $\tilde{\pi}_n$ denote the subspace of $\tilde{X}$ consisting of *trigonometric polynomials* of degree n ; that is, if $x \in \tilde{\pi}_n$

$$x(t) = a_0/2 + \sum_{k=1}^{n} (a_k \cos kt + b_k \sin kt) \ , \quad t \in [0,2\pi] \ ,$$

for some constants $a_0,\dots,a_n,b_1,\dots,b_n$. If T_n is a continuous linear mapping of $\tilde{X}$ into $\tilde{X}$ such that

(1) $T_n x \in \tilde{\pi}_n$ for $x \in \tilde{X}$,

(2) $T_n x = x$ for $x \in \tilde{\pi}_n$,

then T_n is a *trigonometric operation of order* n.

There is a close relationship between $C[-1,1]$ and $\tilde{C}[0,2\pi]$. For example, if $f \in C[-1,1]$ let $\tilde{f} : t \to f(\cos(t))$; then $\tilde{f} \in \tilde{C}[0,2\pi]$, and if $f \in \pi_n$ then $\tilde{f} \in \tilde{\pi}_n$. Because of the relationship between $C[-1,1]$ and $\tilde{C}[0,2\pi]$ it is possible to prove results about one space and then deduce them for the second space. Since it is technically easier to work with $\tilde{C}[0,2\pi]$ it is usual to work with this space and we shall also do so.

If $x \in \tilde{C}[0,2\pi]$ the Fourier series of x is

$$x \sim \frac{a_0}{2} + \sum_{k=1}^{\infty} (a_k \cos kt + b_k \sin kt) \ ,$$

where,

$$a_k = \frac{1}{\pi} \int_0^{2\pi} x(t) \cos kt \, dt ,$$

$$b_k = \frac{1}{\pi} \int_0^{2\pi} x(t) \sin kt \, dt .$$

Let $S_n : \tilde{C}[0,2\pi] \to \tilde{C}[0,2\pi]$ where $S_n x$ is the sum of the first n terms in the Fourier series for x. It is easily proved that

$$(S_n x)(s) = \frac{1}{2\pi} \int_0^{2\pi} x(t) \frac{\sin[(2n+1)(t-s)/2]}{\sin[(t-s)/2]} dt .$$

Now, if $K(s,t)$ is continuous and has period 2π with respect to s then

$$(Tx)(s) = \int_0^{2\pi} x(t) K(s,t) \, dt$$

maps $\tilde{C}[0,2\pi]$ into $\tilde{C}[0,2\pi]$ and

$$||T|| = \max_{0 \le s \le 2\pi} \int_0^{2\pi} |K(s,t)| dt .$$

A series of elementary manipulations (Problem 5.23) yields the result that

$$||S_n|| = \frac{1}{2\pi} \max_{0 \le s \le 2\pi} \int_0^{2\pi} \left| \frac{\sin((2n+1)(t-s)/2)}{\sin((t-s)/2)} \right| dt ,$$

$$\ge \frac{4}{\pi^2} \ln\left(\frac{n+1}{2}\right) . \tag{5.28}$$

It follows from the Banach-Steinhaus theorem, that there must be a function $\tilde{x} \in \tilde{C}[0,2\pi]$ such that the sequence $\{S_n \tilde{x}\}$ does not converge.

It is possible to extend this result to general trigonometric polynomial operators. The key result needed is the following identity. If $x \in \tilde{C}[0,2\pi]$ and $\tau \in R^1$ we denote by x^τ the function obtained by translation by τ :

$$x^\tau(s) = x(s+\tau) , \quad \text{for } 0 \le s \le 2\pi . \tag{*}$$

Since x has period 2π so does x^{τ}. If T_n is any trigonometric polynomial operation of order n then the following identity holds for every $x \in \tilde{C}[0,2\pi]$:

$$(S_n x)(s) = \frac{1}{2\pi}\int_0^{2\pi} (T_n x^{\tau})(s-\tau)d\tau . \tag{5.29}$$

This identity is proved in three steps:

(1) Eqn (5.29) is trivially true if $x \in \tilde{\pi}_n$.

(2) Using the orthogonality relationships among the trigonometric polynomials it follows that eqn (5.29) holds for any trigonometric polynomial x .

(3) It can be shown that, for fixed s , the operators on both sides of eqn (5.29) are continuous linear mappings from $\tilde{C}[0,2\pi]$ into R^1 .

The details of the proof of eqn (5.29) are left to the reader (Problem 5.24).

It follows from eqn (5.29) that if T_n is any trigonometric polynomial operation of order n then

$$\max_{0\le s\le 2\pi} |(S_n x)(s)| \le \frac{1}{2\pi}\int_0^{2\pi} \max_{0\le s\le 2\pi} |(T_n x^{\tau})(s-\tau)|d\tau ,$$

$$\le \frac{1}{2\pi}\int_0^{2\pi} ||T_n x^{\tau}|| \, d\tau ,$$

$$\le \frac{1}{2\pi}\int_0^{2\pi} ||T_n||\cdot||x|| \, d\tau ,$$

so that, using eqn (5.28),

$$||T_n|| \ge ||S_n|| \ge \frac{4}{\pi^2}\,\ell n\left(\frac{n+1}{2}\right) . \tag{5.30}$$

Thus we obtain the following theorem (for the case of trigonometric polynomials):

THEOREM 5.7. *(Lozinskii-Kharshiladze)*

Let T_n *be a sequence of (trigonometric) polynomial operations,* T_n *being an operation of order* n . *Then*

$\|T_n\| \to \infty$ as $n \to \infty$. *In particular, there must exist* $x \in C[-1,+1]$ $(x \in \tilde{C}[0,2\pi])$ *such that the sequence* $\{T_n x\}$ *is not convergent.* □

The Lozinskii-Kharshiladze theorem shows that if we wish to construct operators T_n such that $T_n x \to x$ as $n \to \infty$, for all x , then we cannot require that T_n reproduces polynomials.

REMARK 5.9. Of course, there are many mappings $T_n : C[a,b] \to \pi_n[a,b]$ such that $T_n x \to x$ for all x . Two examples are:

(1) $B_n : x \in C[0,1] \to B(n;x) \in \pi_n$ where $B(n;x)$ is the Bernstein polynomial (see Section 4.8):

$$B(n;x)(t) = \sum_{j=0}^{n} \binom{n}{j} t^j (1-t)^{n-j} x(j/n) .$$

(2) $H_n : x \in C[-1,+1] \to H_n x \in \pi_{2n-1}$ where

$$(H_n x)(t) = \frac{2^{2n-2}[p_n(t)]^2}{n^2} \sum_{k=1}^{n} \frac{x(t_k^{(n)})(1-tt_k^{(n)})}{(t-t_k^{(n)})^2} .$$

Here $p_n(t)$ is the Chebyshev polynomial.

$$p_n(t) = \frac{1}{2^{n-1}} \cos (n \arccos t) ,$$

and the $t_k^{(n)}$ are the zeros of p_n . The function $H_n x$ satisfies the conditions,

$$(H_n x)(t_k^{(n)}) = x(t_k^{(n)}), \quad (d/dt\ H_n x)(t_k^{(n)}) = 0 , \quad 1 \le k \le n ,$$

so that $H_n x$ is the Hermite interpolant to x at the Chebyshev points with the modification that the interpolant has zero slope at the points of interpolation.

For arbitrary $x \in C[0,1]$ it was shown in Section 4.8 that $B_n x \to x$. Fejer showed that $H_n x \to x$ as $n \to \infty$ (Natanson [1955, p. 397]). So neither B_n nor H_n is a polynomial operator of order n ; indeed, neither B_n nor H_n reproduces polynomials of degree greater than one (Problem 5.25). □

REMARK 5.10. Although Theorem 5.7 is of theoretical importance, its practical importance is relatively small. The most obvious class of polynomial operations is obtained by letting T_n map x onto the polynomial $T_n x$ of degree n which interpolates x at prescribed points $t_0^{(n)},\ldots,t_n^{(n)}$. If the points $t_k^{(n)}$ are equidistant on $[-1,+1]$ then $\|T_n\|$ grows rapidly, as is shown by the well-known Runge example where interpolating the 'bell-shaped' function

$$x(t) = \frac{1}{1+25t^2}$$

leads to polynomials $T_n x$ which oscillate wildly. However, if the points $t_k^{(n)}$ are chosen to be the zeros of the Chebyshev polynomial of the first kind, that is,

$$t_k^{(n)} = \cos\left(\frac{2k+1}{2n+2}\pi\right), \quad k = 0,1,\ldots,n,$$

then

$$\|T_n\| \le \frac{2}{\pi}\,\ell n(n+1) + 4 .$$

so that $\|T_n\|$ grows quite slowly as a function of n. See de Boor [1978, p. 26]. □

REMARK 5.11. For further information about polynomial operators see Korovkin [1960] and Natanson [1955]. There is also a very pretty theory concerning sequences of *positive operators*, that is, operators $T_n: C[a,b] \to C[a,b]$ such that $(T_n x)(t) \ge 0$ for all $t \in [a,b]$, whenever $x(t) \ge 0$ for all $t \in [a,b]$; see Korovkin [1960] and de Vore [1972]. □

5.6 *CUBIC SPLINE INTERPOLATION*

Let $x \in C[0,1]$, and let $t_i^{(n)} = i/n \equiv ih$, for $0 \le i \le n$. The *cubic spline* which interpolates x at the *knots* $t_i^{(n)}$ will be taken to be the function $s_n = T_n x$ which satisfies the following conditions:

(1) s_n is a cubic polynomial between any two knots.
(2) $s_n(t_i^{(n)}) = x(t_i^{(n)})$, $0 \le i \le n$.
(3) s_n is twice continuously differentiable for $0 < t < 1$. (5.31)
(4) $\dot{s}_n(0) = \dot{s}_n(1) = 0$.

We will show that $s_n = T_n x \to x$ as $n \to \infty$ for all $x \in C[0,1]$.

REMARK 5.12. The definition of $T_n x$ given above is quite restrictive. In the general case the knots $t_i^{(n)}$ need not be distributed uniformly on [0,1] . Furthermore, condition (4) is arbitrary; a more accurate condition might be the 'not-a-knot' condition (de Boor [1978, p. 55]). We use this restrictive definition of cubic splines so as to be able to give a simple presentation of the basic ideas. □

(1) <u>Derivation of explicit formulae for</u> s_n: To simplify the notation we write t_i for $t_i^{(n)}$ and s for s_n . We set $I_i = [t_i, t_{i+1}]$. We denote by s_i, s_i', and s_i'' the values of s and its derivatives at the knots t_i . x_i denotes the value of x at t_i .

We begin by noting that $\ddot{s}(t)$ is linear on I_i :

$$\ddot{s}(t) = [s_i''(t_{i+1}-t) + s_{i+1}''(t-t_i)]/h \text{ , on } I_i . \qquad (*)$$

Integrating twice,

$$s(t) = \frac{1}{6h}[s_i''(t_{i+1}-t)^3 + s_{i+1}''(t-t_i)^3] + a_i(t_{i+1}-t) + b_i(t-t_i) \text{ ,}$$

$$\text{on } I_i . \qquad (**)$$

where a_i and b_i are integration constants. Since $s_i = x_i$, and $s_{i+1} = x_{i+1}$ we have that

$$x_i = \frac{h^2}{6} s_i'' + ha_i \text{ ,}$$

$$x_{i+1} = \frac{h^2}{6} s_{i+1}'' + hb_i \text{ .}$$

Solving for a_i and b_i and substituting into eqn (**) we obtain

$$s(t) = \frac{1}{6h}[s''_i(t_{i+1}-t)^3 + s''_{i+1}(t-t_i)^3] +$$

$$+ \frac{1}{h}[x_i(t_{i+1}-t) + x_{i+1}(t-t_i)] - \qquad (5.32)$$

$$-\frac{h}{6}[s''_i(t_{i+1}-t) + s''_{i+1}(t-t_i)] \text{ , on } I_i .$$

Differentiating,

$$\dot{s}(t) = \frac{1}{2h}[-s''_i(t_{i+1}-t)^2 + s''_{i+1}(t-t_i)^2] +$$

$$+ \frac{1}{h}[x_{i+1}-x_i] - \frac{h}{6}[s''_{i+1}-s''_i] \text{ , on } I_i . \quad (***)$$

When $t = 0$ or $t = 1$ it follows from condition (4) of (5.31) that $\dot{s}(t) = 0$. Thus,

$$\dot{s}(0) = \frac{1}{2h}[-s''_0 \; h^2+0] + \frac{1}{h}[x_1-x_0] - \frac{h}{6}[s''_1-s''_0] = 0 \; ,$$

or,

$$2s''_0 + s''_1 = 6(x_1-x_0)/h^2 \; .$$

Similarly,

$$\dot{s}(1) = \frac{1}{2h}[-0+s''_n \; h^2] + \frac{1}{h}(x_n-x_{n-1}) - \frac{h}{6}(s''_n-s''_{n-1}) = 0 \; ,$$

or,

$$2s''_n + s''_{n-1} = -6(x_n-x_{n-1})/h^2 \; .$$

Finally, for $0 < i < n$ we know that s is continuous at t_i , so that $\dot{s}(t_i+0) = \dot{s}(t_i-0)$. Thus,

$$\frac{1}{2h}[-s''_i \; h^2+0] + \frac{1}{h}[x_{i+1}-x_i] - \frac{h}{6}[s''_{i+1}-s''_i]$$

$$= \frac{1}{2h}[0+s''_i \; h^2] + \frac{1}{h}[x_i-x_{i-1}] - \frac{h}{6}[s''_i-s''_{i-1}] \; ,$$

or,

$$4s''_i + s''_{i-1} + s''_{i+1} = 6[x_{i+1}-2x_i+x_{i-1}]/h^2 \; .$$

Combining the above equations we see that

$$Au = \frac{6}{h^2} b \; , \qquad (5.33)$$

where

$$u = \begin{bmatrix} s''_0 \\ \cdot \\ \cdot \\ \cdot \\ s''_n \end{bmatrix}, \qquad b = \begin{bmatrix} x_1 - x_0 \\ x_2 - 2x_1 + x_0 \\ \cdots \\ x_{i+1} - 2x_i + x_{i-1} \\ \cdots \\ x_n - 2x_{n-1} + x_{n-2} \\ x_{n-1} - x_n \end{bmatrix}, \tag{5.34}$$

and

$$A = (a_{ij}) = \begin{bmatrix} 2 & 1 & & & & \\ 1 & 4 & 1 & & \bigcirc & \\ & 1 & 4 & 1 & & \\ & & \cdot & \cdot & \cdot & \\ & & & \cdot & \cdot & \cdot \\ & & & & \cdot & \cdot & \cdot \\ & \bigcirc & & & 1 & 4 & 1 \\ & & & & & 1 & 2 \end{bmatrix}. \tag{5.35}$$

The $(n+1) \times (n+1)$ matrix A is strictly diagonally dominant and hence non-singular. Therefore, there exists a unique spline $s = T_n x$ satisfying conditions (1) to (4) of eqn (5.31), and s is defined by eqns (5.32) through (5.35).

(2) *Estimation of* $\|T_n\|$: Let v be any vector in R^{n+1} and set $w = A^{-1}v \in R^{n+1}$. Then

$$\|w\|_\infty = \max_i |w_i| = |w_k|, \quad \text{say.}$$

From eqn (5.35),

$$\begin{aligned} \|v\|_\infty &= \|Aw\|_\infty, \\ &\geq |(Aw)_k| \\ &\geq |a_{kk}w_k| - \sum_{\substack{j=1 \\ j\neq k}}^{n+1} |a_{kj}w_j|, \\ &\geq [|a_{kk}| - \sum_{j\neq k} |a_{kj}|]|w_k|, \\ &\geq |w_k|, \\ &= \|w\|_\infty, \\ &= \|A^{-1}v\|_\infty. \end{aligned}$$

Since v was arbitrary,

$$||A^{-1}||_\infty \le 1 . \tag{*}$$

Combining eqns (5.33) and (*) we have that

$$||u||_\infty \le ||A^{-1}||_\infty \; ||6b/h^2||_\infty \le 6||b||_\infty/h^2 . \tag{5.36}$$

But, from eqn (5.34), $||b||_\infty \le 4||x||_\infty$, so that, using eqn (5.36),

$$\max_i |s_i''| = ||u||_\infty \le 24||x||_\infty/h^2 . \tag{5.37}$$

Substituting into eqn (5.32) we find that

$$\begin{aligned}
||T_n x||_\infty &= ||s||_\infty , \\
&= \max_{a\le t\le b} |s(t)| , \\
&\le \frac{1}{6h} \max_i |s_i''| \cdot \max_{0\le t\le h} [(h-t)^3+t^3] + \\
&\quad + \frac{1}{h} \max_i |x_i| \cdot \max_{0\le t\le h} [(h-t)+t] + \\
&\quad + \frac{h}{6} \max_i |s_i''| \cdot \max_{0\le t\le h} [(h-t)+t] , \\
&\le \frac{2h^2}{6} \max_i |s_i''| + \max_i |x_i| , \\
&\le 8||x||_\infty + ||x||_\infty , \\
&= 9||x||_\infty ,
\end{aligned}$$

so that

$$||T_n|| \le 9 . \tag{5.38}$$

(3) <u>*Proof that* $T_n x \to x$ *, as* $n \to \infty$ *, for every continuous function* x</u>. Assume first that the derivative $\dot{x}$ exists and is continuous. Applying the mean value theorem to eqn (5.34) we obtain

$$||b||_\infty \le 2h||\dot{x}||_\infty ,$$

so that, from eqn (5.36),

$$\max_i |s_i''| \le 12||\dot{x}||_\infty/h . \tag{*}$$

Now consider the expression (5.32) for $s(t)$. It will be noted that the second term in this expression is a piecewise linear function, p say, which interpolates x at the knots t_i. On the other hand, the first and third terms in the expression can be bounded as was done in the derivation of eqn (5.38). We obtain

$$\|T_n x - x\|_\infty \le \frac{h^2}{3} \max |s_i''| + \|x-p\|_\infty . \qquad (**)$$

Since x and p coincide at the knots, applying the mean value theorem on the smaller of the intervals $[t_i, t]$, $[t, t_{i+1}]$ shows that for $t \in I_i$,

$$|(x-p)(t)| = \min(|t-t_i|, |t-t_{i+1}|)|(\dot{x}-\dot{p})(\zeta_i)| ,$$

for some $\zeta_i \in I_i$. Thus,

$$\|x-p\|_\infty \le \frac{h}{2}[\|\dot{x}\|_\infty + |x_{i+1}-x_i|/h] \le h\|\dot{x}\|_\infty . \qquad (\overset{*}{*}{*})$$

Combining eqns (*), (**), and ($\overset{*}{*}{*}$) it follows that

$$\|T_n x - x\| \le 5h\|\dot{x}\|_\infty . \qquad (5.39)$$

In particular, $T_n x \to x$ for every polynomial x.

Noting the uniform bound (5.38) on $\|T_n\|$, it follows from the Banach-Steinhaus theorem that $T_n x \to x$ for every $x \in C[a,b]$.

5.7 *NUMERICAL SOLUTION OF INITIAL VALUE PROBLEMS FOR PARTIAL DIFFERENTIAL EQUATIONS*

This section is organized as follows: (1) Introduction to the theory; (2) The abstract initial-value problem and the Lax theory; (3) Analysis of stability using Fourier transforms; (4) A detailed example for the heat equation.

(1) INTRODUCTION

The theory discussed in this section is applicable to many hyperbolic and parabolic initial-value problems.

To motivate the theory, and to provide concrete examples, consider the problem of determining the temperature $u(x,t)$ in an infinitely long bar, $-\infty < x < +\infty$, for time t in the the interval $0 \le t \le T < \infty$, given the initial temperature distribution $u_0(x)$. That is, we wish to find a function $u(x,t)$ such that:

(1) u satisfies the *(one-dimensional) heat equation*

$$\frac{\partial u}{\partial t} = \frac{\partial^2 u}{\partial x^2}, \tag{5.40}$$

for $-\infty < x < +\infty$, $0 < t \le T$.

(2) u satisfies the *initial condition*

$$u(x,0) = u_0(x), \quad -\infty < x < \infty .$$

It is known (Courant and Hilbert [1962, p. 198]) that if u_0 is continuous and bounded then problem (5.40) has a unique solution

$$u(x,t) = \frac{1}{2\sqrt{\pi t}} \int_{-\infty}^{+\infty} \exp[-(x-z)^2/4t]\, u_0(z)dz , \quad t > 0 \tag{5.41}$$

It can be verified by direct computation that if $u(x,t)$ is given by eqn (5.41) then u satisfies the conditions (5.40). (Problem 5.28).

We now put the problem into a more abstract framework by writing eqn (5.41) in the form

$$u(t) = E(t)u_0 , \tag{5.42}$$

where $E(t)$ is the *solution operator*, and $u(t) = u(\cdot,t)$ and $u_0 = u_0(\cdot)$ represent, respectively, the solution at time t and the initial values.

For $t \in [0,T]$, $E(t)$ is an operator acting upon the space X of initial values u_0. How should X be chosen? Since the differential equation involves second partial derivatives with respect to x, one possible choice for X would be $C^2(\overline{R^1})$. However, this is too restrictive for the following reasons:

(1) It is quite reasonable physically to specify an initial temperature distribution which has discontinuities. For example, u_0 could be the Heaviside step function $H(x)$ (see Figure 2.2).

(2) We wish to develop a theory which will also be applicable to hyperbolic partial differential equations. The space $C^2(\overline{R^1})$ is quite inappropriate for initial value problems for hyperbolic equations, where smooth initial data can give rise to solutions with discontinuities.

We therefore choose X to be the Banach space $X = L^2(-\infty,+\infty)$. That is, $u_0 \in X$ if $u_0(x)$ is Lebesgue-measurable and

$$\int_{-\infty}^{+\infty} [u_0(x)]^2 dx < \infty . \tag{5.43}$$

It can be verified (Problem 5.29) that if $E(t)$ is defined by eqns (5.41) and 5.42) and $X = L^2(-\infty,+\infty)$ then $E(t)$ has the following properties:

(1) $E(t): X \to X$ for $0 \le t \le T$. (5.44)

(2) $E(t)$ is linear; that is,

$$E(t)[\lambda u_0 + \mu v_0] = \lambda E(t)u_0 + \mu E(t)v_0 . \tag{5.45}$$

(3) The solution at time $t + s$ is obtained if we first apply $E(t)$ and then apply $E(s)$. That is,

$$E(s+t) = E(s)\, E(t) . \tag{5.46}$$

This is the *semigroup property*.

(4) $E(t)$ is a continuous function of t. That is, given $\varepsilon > 0$ there exists $\delta > 0$ such that if $|\Delta t| < \delta$ then

$$||E(\Delta t) - I|| \le \varepsilon , \tag{5.47}$$

where I is the identity operator.

(5) Finally, the problem is *properly-posed*. That is, if u_0 is 'small' and $0 \le t \le T$ then $E(t)u_0$ is 'small'. More precisely, for some constant $K_E > 0$,

$$||E(t)|| \le K_E \quad \text{for} \quad 0 \le t \le T . \tag{5.48}$$

Let $X_0 = C_0^\infty(-\infty,+\infty)$ so that X_0 is the subset of X consisting of the 'smooth' functions in X. X_0 is dense in X. (Section 4.11, Property (10)). If $u_0 \in X_0$ then $E(t)u_0$

satisfies the heat equation in the usual sense (such solutions are often called *classical solutions*). We set

$$X_c = \{E(t)u_0 : u_0 \in X_0, \quad 0 \le t \le T\} , \qquad (5.49)$$

so that X_c is the subspace of X consisting of those points $u \in X$ which can be reached by solving the initial value problem (5.40) with 'smooth' initial values $u_0 \in X_0$.

REMARK 5.13. The heat equation is a parabolic equation and smoothes initial data, so that $E(t)u_0 \in C^\infty(-\infty,+\infty)$ even if u_0 is discontinuous. $u(x,t) = (E(t)u_0)(x)$ is also a *weak solution* of eqn (5.40) since (Problem 5.30):

$$\int_0^T \int_{-\infty}^{+\infty} [u(x,t)\psi_t(x,t) + u(x,t)\psi_{xx}(x,t)]dxdt = 0 ,$$

$$\text{for all } \psi \in C_0^\infty[(-\infty,+\infty) \times (0,T)] .$$

For hyperbolic equations, $E(t)u_0$ is often only a weak solution. □

(2) ABSTRACT THEORY

We now forget about the origins of the problem and consider the question of approximating a family of operators $\{E(t): 0 \le t \le T\}$ which map a Banach-space X into itself and satisfy eqns (5.45) through (5.48). X_0 will denote a dense subspace of X , while X_c will be defined by eqn (5.49).

To compute $u(t) = E(t)u_0$ numerically the interval $[0,T]$ is divided into subintervals of length Δt , and the approximate solution is computed at the points $t = n\Delta t$, $0 \le n\Delta t \le T$. For some $\delta_C > 0$ a family of *approximation operators,*

$$\{C(\Delta t): 0 < \Delta t \le \delta_C\}$$

is constructed so that each $C(\Delta t)$ is a bounded linear mapping of X into X which 'approximates' $E(\Delta t)$. Then $u(n\Delta t)$ is approximated by $U(n\Delta t)$:

$$u(n\Delta t) = [E(\Delta t)]^n u_0 \approx [C(\Delta t)]^n u_0 = U(n\Delta t) . \qquad (5.50)$$

We observe that $E(t)$ may first be defined on X_0 and then extended to X (Remark 7.3, p. 236).

To illustrate how approximation operators $C(\Delta t)$ can be constructed, we return briefly to the initial-value problem (5.40) for the heat equation. Every finite difference approximation gives rise to a corresponding operator $C(\Delta t)$. Suppose, for example, that the heat equation is approximated by the explicit finite difference approximation

$$\frac{U(x,t+\Delta t)-U(x,t)}{\Delta t}=\frac{U(x+\Delta x,t)-2U(x,t)+U(x-\Delta x,t)}{(\Delta x)^2}. \tag{5.51}$$

Then the corresponding $C(\Delta t)$ is given by:

$$[C(\Delta t)v](x) = v(x) + \frac{\Delta t}{(\Delta x)^2}[v(x+\Delta x)+v(x-\Delta x)-2v(x)] ,$$

for any $v \in X$, and any Δt. In this example, $\delta_C = T$.

For given initial data u_0, we can compute approximate solutions using eqn (5.50). If the computations are repeated with a sequence of time steps $\Delta t = \Delta_j t$, where $\Delta_j t \to 0$ as $j \to \infty$, it is desired that the approximate solutions converge to the exact solution. We introduce a number of definitions, and then prove the theorem of Lax which gives necessary and sufficient conditions for convergence for properly posed problems. (An example of a problem which is not properly posed is given in Problem 5.31).

DEFINITION 5.2. The family of operators $C(\Delta t)$, $0 < \Delta t \le \delta_C$, approximating $E(t)$ is

(1) A *consistent approximation* if given $v_0 \in X_0$ and $\varepsilon > 0$ there exists $\delta = \delta(v_0,\varepsilon) > 0$ such that

$$\| [C(\Delta t) - E(\Delta t)]v(t) \| \le \varepsilon \Delta t , \tag{5.52}$$

for $0 < \Delta t \le \delta$ and $0 \le t \le T$, where $v(t) = E(t)v_0$.

(2) *Stable* if there exists $K_C > 0$ such that

$$\| [C(\Delta t)]^n \| \le K_C \quad \text{for} \quad 0 < n\Delta t \le T , \tag{5.53}$$

and $\Delta t \in (0,\delta_C]$.

(3) *Convergent* if given any $t \in [0,T]$, any $u_0 \in X$, and any sequences $\Delta_j t \downarrow 0$, $n_j \to \infty$, such that $n_j \Delta_j t \to t$ as $j \to \infty$, then

$$||[C(\Delta_j t)]^{n_j} u_0 - E(t)u_0|| \to 0, \quad \text{as } j \to \infty. \quad \square \qquad (5.54)$$

THEOREM 5.8. *(Lax) For a consistent approximation to a properly posed initial-value problem satisfying conditions (5.44) through (5.48), stability is necessary and sufficient for convergence.*

Proof. Necessity: Assume that the approximation is convergent. Let F be the family of operators:

$$F = \{[C(\Delta t)]^n : \ 0 \le n\Delta t \le T, \ 0 < \Delta t \le \delta_C\}.$$

Choose $u_0 \in X$. We assert that $||Lu_0|| \le K(u_0)$ for some constant $K(u_0)$ and all $L \in F$. If not, then there exist sequences $\{n_j\}$ and $\{\Delta_j t\}$ such that $||[C(\Delta_j t)^{n_j} u_0|| \to \infty$ as $j \to \infty$; since $0 \le n_j \Delta_j \le T$ we may assume by the Bolzano-Weierstrass theorem (Theorem 6.5), extracting a subsequence if necessary, that $n_j \Delta_j t \to t$ for some $t \in [0,T]$. But this contradicts the convergence condition (5.54). Hence, as asserted, $||Lu_0|| \le K(u_0)$ for $L \in F$. By the principle of uniform boundedness (Theorem 5.3), $||L|| \le K_C < \infty$ for some constant K_C and all $L \in F$. That is, the approximation is stable.

Sufficiency: Assume that the approximation is stable. Choose $\varepsilon > 0$ and $v_0 \in X_0$, and set $v(t) = E(t)v_0$. Then, using the semigroup property of $E(t)$ (condition (5.46)),

$$||\{[C(\Delta_j t)]^{n_j} - E(n_j \Delta_j t)\} v_0||$$

$$= ||\sum_{k=0}^{n_j - 1} [C(\Delta_j t)]^k [C(\Delta_j t) - E(\Delta_j t)] E([n_j - 1 - k]\Delta_j t) v_0||,$$

$$\le \sum_{k=0}^{n_j - 1} ||C(\Delta_j t)^k|| \ ||[C(\Delta_j t) - E(\Delta_j t)] v([n_j - 1 - k]\Delta_j t)||.$$

Thus, using the assumptions of stability (condition (5.53)), consistency (condition (5.52)), and properly posedness (condition (5.48)), we obtain

$$\|\{[C(\Delta_j t)]^{n_j} - E(n_j\Delta_j t)\}v_0\| \leq n_j K_C \,\varepsilon\, \Delta_j t \leq K_C T\,\varepsilon\,, \qquad (*)$$

for $0 < \Delta_j t \leq \delta = \delta(v_0,\varepsilon)$, and $0 \leq n_j\Delta_j t \leq T$.

Now let $u_0 \in X$. Since X_0 is dense in X, there is a $v_0 \in X_0$ such that $\|u_0 - v_0\| \leq \varepsilon$. Let δ be such that eqns (5.47) and (*) hold. Let $\Delta_j t \downarrow 0$ and $n_j\Delta_j t \to t$, and choose N such that $|\Delta_j t| \leq \delta$ and $|t - n_j\Delta_j t| \leq \delta$ for $j \geq N$. Then, for $j > N$,

$$\|\{[C(\Delta_j t)]^{n_j} - E(t)\}u_0\|$$

$$\leq \|\{[C(\Delta_j t)]^{n_j} - E(n_j\Delta_j t)\}v_0\| + \|[C(\Delta_j t)]^{n_j}(u_0 - v_0)\| +$$

$$+ \|E(n_j\Delta_j t)(v_0 - u_0)\| + \|\{E(n_j\Delta_j t) - E(t)\}u_0\|\,.$$

The four terms on the right may be bounded using, respectively, eqns (*), (5.53), (5.48), and (5.47). We obtain

$$\|\{[C(\Delta_j t)]^{n_j} - E(t)\}u_0\| \leq K_C T\varepsilon + K_C\varepsilon + K_E\varepsilon + K_E\|u_0\|\varepsilon\,.$$

Consequently, the method is convergent. □

REMARK 5.14. An approximation $C(\Delta t)$ is consistent if it is 'convergent for smooth solutions', and stable if 'successive applications of $C(\Delta t)$ do not magnify the error'.

Results linking convergence and stability have been proved in other contexts. For example, for ordinary differential equations it can be proved (see Henrici [1962]) that a linear multistep method is convergent iff it is stable and consistent. This result is valid for *non-linear* ordinary differential equations, but of course ordinary differential equations are much easier to handle than partial differential equations. □

REMARK 5.15. The concept of semigroups of operators is a powerful tool for handling linear and nonlinear problems. As shown by Crandall and others, the solution operator

$E(t)$ can often be constructed using difference approximations. See Evans [1977].

REMARK 5.16. In applying Theorem 5.8 to a particular finite-difference approximation, we must check for consistency and stability.

Checking condition (5.52) for consistency is a straightforward, though tedious, procedure which can even be automated on a computer. One expands both $C(\Delta t)v(t)$ and $E(\Delta t)v(t)$ in Taylor series, and then uses the differential equation to simplify the resulting expressions. An example is given in Example 5.10 at the end of this section.

Checking condition (5.53) for stability is much more difficult. An approach which simplifies the work is described below. □

(3) STABILITY ANALYSIS USING FOURIER TRANSFORMS

The determination of whether or not a family of finite difference approximations is stable, is considerably simplified by the use of Fourier transforms as we now show.

Preliminary Comments

We begin by making some comments to justify the framework within which we will be working:

(1) The model which we have used so far is the heat equation, where the solution $u(t)$ is a real-valued function in $L^2(-\infty,+\infty)$. It is of importance to be able to handle problems where the solution $u(t)$ is a vector-valued function with p components, since:

   (a) Problems involving higher derivatives of t, such as,

$$\frac{\partial^2 w}{\partial t^2} = \frac{\partial^2 w}{\partial x^2}\,, \tag{5.55}$$

   can be reformulated as systems of first order equations. For example, eqn (5.55) is equivalent to

$$\frac{\partial u_1}{\partial t} = \frac{\partial u_2}{\partial x} , \quad \frac{\partial u_2}{\partial t} = \frac{\partial u_1}{\partial x} ,$$

where $u_1 = \frac{\partial w}{\partial t}$, $u_2 = \frac{\partial w}{\partial x}$.

(b) Many physical problems involve systems of coupled equations for the temperature, velocity components, etc.

(c) Some finite difference approximations are multi-step methods which involve several time levels. For example, the DuFort-Frankel approximation to the heat equation is

$$\frac{U(x,t+\Delta t)-U(x,t-\Delta t)}{2\Delta t} = \frac{U(x+\Delta x,t)+U(x-\Delta x,t)-U(x,t+\Delta t)-U(x,t-\Delta t)}{(\Delta x)^2} . \tag{5.57}$$

Such methods can be brought into the present framework by regarding them as one-step methods for vector-valued functions. For example, the DuFort-Frankel method expresses $(U(t+\Delta t), U(t))$ in terms of $(U(t), U(t-\Delta t))$. (Problem 5.32).

(2) It is clearly important to consider problems in more than one space variable.

(3) For technical reasons it is convenient to consider complex-valued solutions rather than real solutions. As will be shown below, an approximation $C(\Delta t)$ is stable for real-valued functions iff it is stable for complex-valued functions.

Fourier Transforms

With the above comments in mind, let p and d be given integers. We will use the spaces $\mathbb{C}^p$ (Section 4.5) and R^d with the norms,

$$\begin{aligned} \|x\|^2 &= \|(x_j)\|^2 = \|x;R^d\|^2 = \sum_{j=1}^{d} x_j^2 , \quad x \in R^d , \\ \|z\|^2 &= \|(z_j)\|^2 = \|z;\mathbb{C}^p\|^2 = \sum_{j=1}^{p} |z_j|^2 , \quad z \in \mathbb{C}^p . \end{aligned} \tag{5.58}$$

We define B to be the Banach space of complex-valued p-vectors in the d real variables $x_1,\ldots,x_d$. If $w \in B$ then $w\colon R^d \to \mathbb{C}^p$,

$$w(x) = \begin{pmatrix} w_1(x) \\ \cdot \\ \cdot \\ \cdot \\ w_p(x) \end{pmatrix} = \begin{pmatrix} w_1(x_1,\ldots,x_d) \\ \cdots \\ \cdots \\ \cdots \\ w_p(x_1,\ldots,x_d) \end{pmatrix} \in \mathbb{C}^p , \text{ for } x \in R^d .$$

The elements of B are of course required to have finite norm, the norm being defined by

$$\|w\|^2 = \|w;B\|^2 = \int_{R^d} \|w(x);\mathbb{C}^p\|^2 dx . \tag{5.59}$$

For example, if $p = 2$ and $d = 3$,

$$\|w;B\|^2 = \int_{-\infty}^{+\infty} dx_1 \int_{-\infty}^{+\infty} dx_2 \int_{-\infty}^{+\infty} dx_3 [|w_1(x_1,x_2,x_3)|^2 + |w_2(x_1,x_2,x_3)|^2] .$$

(When using norms, we will usually suppress the names of the spaces and the dependence upon p and d.)

Let $\hat{B}$ denote the Banach space of complex-valued p-vectors in the d real variables $k_1,\ldots,k_d$. That is, if $\hat{w} \in \hat{B}$ then $\hat{w}\colon R^d \to \mathbb{C}^p$,

$$\hat{w}(k) = \begin{pmatrix} \hat{w}_1(k) \\ \cdot \\ \cdot \\ \cdot \\ \hat{w}_1(k) \end{pmatrix} \in \mathbb{C}^p , \quad \text{for } k \in R^d .$$

We set

$$\|\hat{w}\|^2 = \|\hat{w};\hat{B}\|^2 = \int_{R^d} \|\hat{w}(k);\mathbb{C}^p\|^2 \, dk . \tag{5.60}$$

Of course, B and $\hat{B}$ are copies of the same space, but it is convenient to distinguish between them.

REMARK 5.17. It follows from eqns (5.58) and (5.59) that

$$\|w;B\|^2 = \sum_{j=1}^{p} \| |w_j|;L^2(R^d)\|^2 ,$$

$$= \sum_{j=1}^{p} [\|\text{Real } (w_j);L^2(R^d)\|^2 + \|\text{Imag } (w_j);L^2(R^d)\|^2 .$$

Consequently, B and $\hat{B}$ are Banach spaces because they are isomorphic to the space

$$\prod_{j=1}^{2p} L^2(R^d) .$$

See Problem 3.12. □

We now define a map between B and $\hat{B}$. Given $w \in B$ we define its *Fourier transform* $\hat{w} = Fw \in \hat{B}$ by

$$\hat{w}(k) = (Fw)(k) = \frac{1}{(2\pi)^{d/2}} \int_{R^d} w(x)\, e^{-i\langle k,x\rangle} dx , \qquad (5.61)$$

where $i = \sqrt{-1}$ and

$$\langle k,x\rangle = \sum_{j=1}^{d} k_j x_j .$$

Then it can be shown (Yosida [1968, p. 154]) that F is a continuous linear map of B *onto* $\hat{B}$ which preserves norms. That is

$$\|\hat{w}\| = \|Fw\| = \|w\| . \qquad (5.62)$$

The inverse of F is given explicitly by

$$w(x) = (F^{-1}\hat{w})(x) = \frac{1}{(2\pi)^{d/2}} \int_{R^d} \hat{w}(k)\, e^{+i\langle k,x\rangle} dk . \qquad (5.63)$$

Let $T^{\beta}: B \to B$ be the translation operator

$$(T^{\beta}w)(x) = w(x_1 + \beta_1\Delta x_1,\ldots,x_d + \beta_d\Delta x_d) , \qquad (5.64)$$

where the constant $\beta \in R^d$ has integer coefficients and where $\Delta x_1,\ldots,\Delta x_n$ are real constants. Then

$$(FT^{\beta}w)(k) = \frac{1}{(2\pi)^{d/2}} \int_{R^d} w(x+y)\, e^{-i\langle k,x\rangle}\, dx ,$$

where $y = (\beta_1 \Delta x_1, \ldots, \beta_d \Delta x_d)$. Hence, setting $x + y = z$,

$$(FT^{\beta}w)(k) = \frac{1}{(2\pi)^{d/2}} \int_{R^d} w(z)\, e^{-i\langle k,z-y\rangle}\, dz ,$$

$$= e^{+i\langle k,y\rangle}\, \hat{w}(k) ,$$

$$= \hat{w}(k) \exp \{i \sum_{j=1}^{d} k_j \beta_j \Delta x_j\} . \tag{5.65}$$

Stability Analysis

We now return to finite difference approximations. Let $C(\Delta t)$ be an approximation to an initial value problem in d spatial dimensions, with a solution $u(t)$,

$$u(t): x \in R^d \to u(t,x_1,\ldots,x_d) \in R^p . \tag{5.66}$$

We take the space X to be

$$X = \prod_{j=1}^{p} L^2(R^d) , \tag{5.67}$$

with norm

$$\|x;X\|^2 = \|(x_j);X\|^2 = \sum_{j=1}^{p} \|x_j;L^2(R^d)\|^2 . \tag{5.68}$$

From eqn (5.53), $C(\Delta t)$ is stable iff

$$\|[C(\Delta t)]^n v;X\| \le K_C \|v;X\| , \tag{5.69}$$

for $0 \le n\Delta t \le T$, and all $v \in X$. It follows from eqn (5.67) and Remark 5.17 (page 159) that $w \in B$ iff

$$w = v_1 + iv_2$$

where v_1 , $v_2 \in X$. Furthermore,

$$\|v_1+iv_2;B\|^2 = \|v_1;X\|^2 + \|v_2;X\|^2 .$$

We extend $C(\Delta t)$ to a map of B into B by setting

$$C(\Delta t)w = C(\Delta t)\mathrm{Real}(w) + iC(\Delta t)\,\mathrm{Imag}(w), \text{ for } w \in B . \tag{5.70}$$

Again using eqn (5.67) and Remark 5.17 it easily follows (Problem 5.33) that $C(\Delta t)$ satisfies eqn (5.69) iff

$$\|[C(\Delta t)]^n w;B\| \le K_C \|w;B\| , \text{ for } 0 \le n\Delta t \le T , \text{ and all } w \in B . \tag{5.71}$$

In the remainder of this discussion, we regard $C(\Delta t)$ as a map from B to B as in eqn (5.70) and take eqn (5.71) as the definition of stability. We define $\hat{C}(\Delta t): \hat{B} \to \hat{B}$ by

$$\hat{C}(\Delta t) = \mathcal{F}C(\Delta t)\mathcal{F}^{-1} . \tag{5.72}$$

A typical finite difference operator $C(\Delta t)$ is defined by an equation of the form

$$\sum_{\beta \in N_1} A_1^{(\beta)}[(T^\beta C(\Delta t)w)(x)] = \sum_{\beta \in N_0} A_0^{(\beta)}[(T^\beta w)(x)] , \quad x \in R^d \tag{5.73}$$

where N_0 and N_1 are finite sets and where $A_1^{(\beta)}$ and $A_0^{(\beta)}$ are $p \times p$ matrices. For example, when solving the heat equation using the DuFort-Frankel method given by eqn (5.57), we have (Problem 5.32):

$$d = 1 , \quad p = 2 , \quad N_1 = \{-1,0,1\} , \quad N_0 = \{0\} ,$$

and

$$A_1^{(0)} = \begin{pmatrix} 1+\mu & 0 \\ 0 & 1 \end{pmatrix} ; \quad A_1^{(1)} = A_1^{(-1)} = \begin{pmatrix} 0 & -\mu \\ 0 & 0 \end{pmatrix} ; \quad A_0^{(0)} = \begin{pmatrix} 0 & 1-\mu \\ 1 & 0 \end{pmatrix} \tag{5.74}$$

where $\mu = 2\Delta t/(\Delta x)^2$.

REMARK 5.18. The matrices $A_0^{(\beta)}$ and $A_1^{(\beta)}$ in eqn (5.73) may depend upon both Δt and Δx , but in any finite difference computation Δx will be specified as a function of Δt , so that $A_0^{(\beta)}$ and $A_1^{(\beta)}$ will depend only on Δt . □

Applying $\mathcal{F}$ to eqn (5.73) and using eqns (5.65) and (5.72) we find that $\hat{C}(\Delta t)$ satisfies:

$$H_1[(\hat{C}(\Delta t)\hat{w})(k)] = H_0[\hat{w}(k)] , \quad k \in R^d , \tag{5.75}$$

or

$$(\hat{C}(\Delta t)\hat{w})(k) = G(k,\Delta t)[\hat{w}(k)] , \quad k \in R^d , \tag{5.76}$$

where $G(k,\Delta t)$, H_0, and H_1 are complex $p \times p$ matrices:

$$G(k,\Delta t) = H_1^{-1} H_0 ,$$

$$H_0 = \sum_{\beta\in N_0} A_0^{(\beta)} \exp\{i \sum_{j=1}^{d} k_j \beta_j \Delta x_j\} , \tag{5.77}$$

$$H_1 = \sum_{\beta\in N_1} A_1^{(\beta)} \exp\{i \sum_{j=1}^{d} k_j \beta_j \Delta x_j\} .$$

$G(k,\Delta t)$ is the *amplification matrix* corresponding to $C(\Delta t)$. As an example for the explicit method (5.51) we obtain (Problem 5.34)

$$\lambda = \Delta t/(\Delta x)^2 .$$

$$H_1 = 1 ,$$

$$\begin{aligned} H_0 &= \lambda \exp(-ik\Delta x) + (1 - 2\lambda) + \lambda \exp(+ik\Delta x) , \\ &= 2\lambda \cos(k\Delta x) + (1 - 2\lambda) , \\ &= 1 - 4\lambda \sin^2(k\Delta x/2) , \end{aligned} \tag{5.78}$$

$$G(\Delta t,k) = 1 - 4\lambda \sin^2(k (\tfrac{\Delta t}{\lambda})^{\frac{1}{2}}/2) .$$

We see that $\hat{C}(\Delta t)$ is much easier to handle than $C(\Delta t)$ since it merely involves pointwise matrix multiplication by complex $p \times p$ matrices. The following lemma is, therefore, of help in computing the norm of $\hat{C}(\Delta t)$.

LEMMA 5.9. *Let* $\hat{B}$ *be as defined on page 158 and let* $\hat{L}: \hat{B} \to \hat{B}$,

$$(\hat{L}\hat{w})(k) = A(k)\hat{w}(k) , \quad k \in R^d ,$$

where $A(k): \mathbb{C}^p \to \mathbb{C}^p$ *is a* $p \times p$ *complex matrix with coefficients which depend continuously upon* $k \in R^d$. *Then*

$$||\hat{L}|| = \sup_{k\in R^d} ||A(k)|| ,$$

where

$$||A(k)|| = \sup_{||z;\mathbb{C}^p|| \le 1} ||A(k)z;\mathbb{C}^p|| . \tag{5.79}$$

Proof.

$$\|\hat{L}\| = \sup_{\|\hat{w};\hat{B}\|\le 1} \|\hat{L}\hat{w};\hat{B}\| ,$$

$$= \sup_{\|\hat{w}\|\le 1} \left[\int_{R^d} \|A(k)\hat{w}(k);\mathbb{C}^p\|^2 dk\right]^{\frac{1}{2}} ,$$

$$\le \sup_{\|\hat{w}\|\le 1} \left[\int_{R^d} \|A(k)\|^2 \|\hat{w}(k);\mathbb{C}^p\|^2 dk\right]^{\frac{1}{2}} ,$$

$$\le \sup_{\|\hat{w}\|\le 1} \left\{\left[\sup_{k\in R^d} \|A(k)\|\right]\left[\int_{R^d} \|\hat{w}(k);\mathbb{C}^p\|^2 dk\right]^{\frac{1}{2}}\right\} ,$$

$$= \sup_{k\in R^d} \|A(k)\| ,$$

where, in the last step, we used the definition (5.60) of $\|\hat{w}\|$.

On the other hand, since the coefficients of $A(k)$ are continuous functions of k , given $\varepsilon > 0$ there exist k_0 and δ such that

$$\|A(k_0)\| \ge \sup_k \|A(k)\| - \varepsilon ,$$

and such that $\|A(k)-A(k_0)\| \le \varepsilon$ if $\|k-k_0\| \le \delta$. Let $z \in \mathbb{C}^p$ be such that $\|z\| = 1$ and $\|A(k_0)z\| \ge (\|A(k_0)\|-\varepsilon)\|z\|$ and set

$$\hat{w}(k) = \begin{cases} z , & \text{if } \|k-k_0\| \le \delta , \\ 0 , & \text{otherwise} . \end{cases}$$

Then $\hat{w} \in \hat{B}$ and, if $3\varepsilon < \sup \|A(k)\|$,

$$\|\hat{L}\hat{w}\| = \left[\int_{R^d} \|A(k)\hat{w}(k);\mathbb{C}^p\|^2 dk\right]^{\frac{1}{2}},$$

$$= \left[\int_{\|k-k_0\|\le\delta} \|[A(k_0)+(A(k)-A(k_0))]z\|^2 dk\right]^{\frac{1}{2}},$$

$$\ge \left[\int_{\|k-k_0\|\le\delta} [(\|A(k_0)\|-2\varepsilon)\|z\|]^2 dk\right]^{\frac{1}{2}},$$

$$\ge \left[\sup_{k\in R^d} \|A(k)\|-3\varepsilon\right]\|\hat{w}\|,$$

and the lemma follows. □

We can now prove

THEOREM 5.10. *The difference approximation* $C(\Delta t)$ *defined by eqn (5.73) is stable iff there exist* $\tau > 0$ and $K_C > 0$ *such that*

$$\|[G(k,\Delta t)]^n\| \le K_C,$$

for all $k \in R^d$, *and all* Δt *such that* $0 < n\Delta t \le T$ *and* $0 < \Delta t < \tau$. *Here,* $G(k,\Delta t)$ *is the* $p \times p$ *complex matrix defined by eqns (5.77), while the norm of* G *is defined by eqn (5.79).*

Proof. By definition, $\hat{C}(\Delta t) = FC(\Delta t)F^{-1}$. Since F is an isometry, (see eqn (5.62)), we conclude that

$$\|C(\Delta t)^n\| = \|\hat{C}(\Delta t)^n\|.$$

Using Lemma 5.9 and eqn (5.76) we obtain

$$\|C(\Delta t)^n\| = \sup_{k\in R^d} \|(G(k,\Delta t))^n\|.$$

It is only necessary to consider small Δt, since if $\Delta t \geq \tau$ for some constant τ then $n \leq T/\tau$. It is assumed, as part of the definition of $C(\Delta t)$, that $C(\Delta t)$ is bounded for each Δt, and hence the finite collection of numbers

$$\|[C(\Delta t)]^n\| \leq \|C(\Delta t)\|^n \;, \quad \Delta t \geq \tau \;, \quad n \leq T/\tau \;,$$

is bounded.

The theorem now follows from the definition of stability (see eqn (5.71)). □

Although Theorem 5.10 reduces the question of stability to the analytic problem of computing the supremum for $k \in R^d$ of the amplification matrix $G(k,\Delta t)$, this is often not an easy problem. If $p = 1$ or $G(k,\Delta t)$ has a very simple structure, then it is possible to use Theorem 5.10 to obtain necessary and sufficient conditions for stability. A theorem due to Kreiss gives several different equivalent algebraic conditions for stability. (Richtmyer and Morton [1967, p. 74]).

There is also a very useful *necessary* condition for stability which follows from Theorem 5.10. Let $R(k,\Delta t)$ denote the spectral radius of $G(k,\Delta t)$. Then (Problem 5.40)

$$[R(k,\Delta t)]^n \leq \|[G(k,\Delta t)]^n\| \;.$$

Thus, if the method is stable with constant K_C,

$$\begin{aligned} R(k,\Delta t) &\leq \exp[\tfrac{1}{n}\,\ell n(\|G(k,\Delta t)\|^n)] \;, \\ &\leq \exp[\tfrac{1}{n}\,\ell n\; K_C] \;, \\ &\leq \exp[\tfrac{1}{n}\,(\ell n(\max(1,K_C)))] \;, \end{aligned}$$

for all $k \in R^d$, and $0 \leq n\Delta t \leq T$. The smallest value in the expression on the right occurs when $n\Delta t = T$. Thus, a necessary condition for stability is that

$$\begin{aligned} R(k,\Delta t) &\leq \exp[\alpha\Delta t] \;, \\ &= 1 + O(\Delta t) \;, \end{aligned} \tag{5.80}$$

for all $k \in R^d$, where

$$\alpha = \frac{1}{T} \ln(\max(1, K_C)) .$$

This is the celebrated *necessary condition of von Neumann*. If $p=1$ it is also a sufficient condition for stability (Problem 5.35).

(4) AN EXAMPLE

EXAMPLE 5.10. We consider yet again the scalar initial-value problem (5.40):

$$\frac{\partial u}{\partial t} = \frac{\partial^2 u}{\partial x^2} , \quad -\infty < x < +\infty , \quad 0 < t \le T ,$$

$u(x,0) = u_0(x)$.

For some constant $\theta \in [0,1]$ let $C(\Delta t)$ be the operator corresponding to the finite difference approximation

$$U(x,t+\Delta t) - U(x,t) = \lambda\theta[U(x+\Delta x,t+\Delta t) - 2U(x,t+\Delta t) + U(x-\Delta x,t+\Delta t)] +$$
$$+ \lambda(1-\theta)[U(x+\Delta x,t) - 2U(x,t) + U(x-\Delta x,t)] . \tag{5.81}$$

Here,

$$\lambda = \Delta t/(\Delta x)^2 . \tag{5.82}$$

We assume that Δt and Δx vary in such a way that the ratio λ remains constant.

(1) Existence of $C(\Delta t)$

If $\theta = 0$ then eqn (5.81) defines $U(t+\Delta t)$ explicitly in terms of $U(t)$, and a simple application of the triangle inequality shows that

$$||U(t+\Delta t)|| \le (1+4\lambda)||U(t)|| ,$$

so that $C(\Delta t)$ is defined and is a bounded linear operator.

If $\theta > 0$ then eqn (5.81) defines $U(t+\Delta t)$ implicitly, and it is not at all clear that $U(t+\Delta t)$ exists and belongs to X . If $U(t+\Delta t)$ did exist then, applying the Fourier transform F, we would obtain, as a special case of eqns (5.75) and (5.76),

$$H_1 = 1 - \lambda\theta[e^{+ik\Delta x} - 2 + e^{-ik\Delta x}] ,$$
$$= 1 - \lambda\theta[2 \cos k\Delta x - 2] ,$$
$$= 1 + 4\lambda\theta[\sin (k\Delta x/2)]^2 ,$$

and

$$H_0 = 1 + \lambda(1-\theta)[e^{ik\Delta x} - 2 + e^{-ik\Delta x}] ,$$
$$= 1 - 4\lambda(1-\theta)[\sin (k\Delta x/2)]^2 ,$$

so that

$$\hat{U}(k,t+\Delta t) = (\hat{C}(\Delta t)\hat{U}(t))(k) = G(k,\Delta t)\hat{U}(k,t) , \qquad (5.83)$$

with

$$G(k,\Delta t) = H_1^{-1}H_0 = \frac{1-(1-\theta)s}{1+\theta s} ,$$

where

$$s = 4\lambda[\sin (k\Delta x/2)]^2 , \qquad (5.84)$$
$$= 4\lambda\left[\sin \left(\frac{k}{2}\left(\frac{\Delta t}{\lambda}\right)^{\frac{1}{2}}\right)\right]^2 .$$

From (5.84) we see that the scalar $G(k,\Delta t)$ is bounded, so that, by Lemma 5.9, eqn (5.83) defines a bounded linear operator $\hat{C}(\Delta t)$ such that

$$\hat{C}(\Delta t)\hat{U}(t) = \hat{U}(t+\Delta t) .$$

Setting

$$C(\Delta t) = \mathcal{F}^{-1}\hat{C}(\Delta t)\mathcal{F} , \qquad (5.85)$$

it can be verified, using eqn (5.65), that if

$$U(t) \in X \quad \text{and} \quad U(t+\Delta t) = C(\Delta t)U(t) = \mathcal{F}^{-1}\hat{C}(\Delta t)\mathcal{F}U(t)$$

then $U(t+\Delta t) \in X$ and eqn (5.81) is satisfied.

(2) Stability

From eqn (5.84),

$$\|[G(k,\Delta t)]^n\| = \left|\frac{1-(1-\theta)s}{1+\theta s}\right|^n .$$

As k ranges over R^1 , s takes on all values in $I = [0,4\lambda]$.

If $\theta \in [\frac{1}{2},1]$ then

$$\sup_k \| [G(k,\Delta t)]^n \| = \left[\max_{s\in I}\left|\frac{1-(1-\theta)s}{1+\theta s}\right|\right]^n \le 1 ,$$

so that, from Theorem 5.10, the method is stable for all values of λ .

If $\theta \in [0,\frac{1}{2})$, then

$$\sup_k \| [G(k,\Delta t)]^n \| = \left[\max_{s\in I}\left|\frac{1-(1-\theta)s}{1+\theta s}\right|\right]^n ,$$

$$= \left[\max\left(1, \left|\frac{1+4\lambda\theta-4\lambda}{1+4\lambda\theta}\right|\right)\right]^n ,$$

which is bounded for all n iff

$$\lambda \le \frac{1}{2(1-2\theta)} . \tag{5.86}$$

This is, therefore, from Theorem 5.10 and Problem 5.35, a necessary and sufficient condition for stability.

(3) Consistency

To establish the consistency of $C(\Delta t)$ we must show that, given $\varepsilon > 0$ and $v_0 \in X_0$ there exists $\delta > 0$ such that

$$\| [C(\Delta t)-E(\Delta t)]v(t) \| \le \varepsilon \, \Delta t , \tag{5.87}$$

for all $\Delta t \le \delta$ and $t \in [0,T]$, where $v(t) = E(t)v_0$.

In principle, it is very simple to check eqn (5.87): one substitutes the exact solution $v(x,t)$ into the difference approximation (5.81) and expands every term in a Taylor series about one point. Expanding about the point (x,t) we obtain (Problem 5.36):

$$\begin{aligned} v(x,t+\Delta t) - v(x,t) = \tau &+ \\ &+ \lambda\theta[v(x+\Delta x,t+\Delta t)-2v(x,t+\Delta t)+v(x-\Delta x,t+\Delta t)] + \\ &+ \lambda(1-\theta)[v(x+\Delta x,t)-2v(x,t)+v(x-\Delta x,t)] . \end{aligned} \tag{5.88}$$

Here, the local truncation error τ is given by

$$\tau = \tau(v_0, x, t) ,$$

$$= \Delta t\, v_t + \frac{(\Delta t)^2}{2!} v_{tt} + \frac{(\Delta t)^3}{3!} v_{ttt} + \frac{(\Delta t)^4}{4!} \bar{v}_{tttt} -$$

$$-2\lambda \left[\frac{(\Delta x)^2}{2!} v_{xx} + \frac{(\Delta x)^4}{4!} v_{xxxx} + \frac{(\Delta x)^6}{6!} v_{xxxxxx} + \frac{(\Delta x)^8}{8!} \bar{v}_{xxxxxxxx} \right] -$$

$$-2\lambda\theta \left[\frac{(\Delta x)^2}{2!} [\Delta t\, v_{xxt} + \frac{(\Delta t)^2}{2!} v_{xxtt} + \frac{(\Delta t)^3}{3!} \bar{v}_{xxttt}] + \right. \qquad (5.89)$$

$$\left. + \frac{(\Delta x)^4}{4!} [\Delta t\, v_{xxxxt} + \frac{(\Delta t)^2}{2!} \bar{v}_{xxxxtt}] + \frac{(\Delta x)^6}{6!} [\Delta t \bar{v}_{xxxxxxt}] \right] ,$$

where the terms without bars, such as v_{xx}, are evaluated at the point (x,t) and the terms with bars, such as $\bar{v}_{tttt}$, are evaluated at different nearby points.

Since $\lambda = \Delta t/(\Delta x)^2$, and $v_t = v_{xx}$, it follows from eqn (5.89) that $\tau = O((\Delta t)^2)$. Specific choices of λ and θ lead to even greater accuracy (Problem 5.36).

Now set

$$W = [C(\Delta t) - E(\Delta t)]v(t) .$$

If $U(t) = v(t)$ then $U(t+\Delta t) = C(\Delta t)U(t)$ satisfies eqn (5.81). Subtracting eqns (5.81) and (5.88) from one another we obtain an equation for W:

$$W(x) = \lambda\theta[W(x+\Delta x) - 2W(x) + W(x-\Delta x)] - \tau . \qquad (5.90)$$

At this point we encounter some (minor) technical difficulties:

(a) Given a bound for $||\tau||$, we must use equation (5.90) to obtain a bound for $||W||$. However, eqn (5.90) is an equation for W which must be solved, unless $\theta = 0$ in which case the difference approximation (5.81) is explicit.

(b) We can use eqn (5.89) to bound τ in terms of the derivatives of $v(t)$. However, the bound (5.87) must hold for all $t \in [0,T]$, so that we need bounds for the derivatives of $v(t)$ in terms of v_0.

(c) $X = L^2(-\infty,+\infty)$, so that to bound $\|\tau\|$ using eqn (5.89) it is not sufficient to know that $|\tau(x,t)| \le K\,\varepsilon\,\Delta t$. Instead we need to have $|\tau(x,t)| \le K\,\varepsilon\,\Delta t\, g(x)$, for some $g \in X$.

These difficulties are only minor, and are readily overcome. However, it is easier and more elegant to use the Fourier transform.

In eqn (5.70) we extended $C(\Delta t)$ to a map of B into B. We now do the same for $E(t)$:

$$E(t)w = E(t)\,\mathrm{Real}(w) + iE(t)\mathrm{Imag}(w)\,, \quad \text{for} \quad w \in B\,.$$

If X_0 is a dense subset of X, then

$$B_0 = \{v_1+iv_2\colon\ v_1,\ v_2 \in X_0\}\,,$$

is a dense subset of B. We set $\hat{B}_0 = \mathcal{F}B_0 = \{\hat{w} = \mathcal{F}w\colon\ w \in B_0\}$.

Since the Fourier transform $\mathcal{F}$ is an isometry, the consistency condition (5.87) is equivalent to the condition: Given $\varepsilon > 0$ and $\hat{w}_0 \in \hat{B}_0 = \mathcal{F}B_0$, there exists $\delta > 0$ such that

$$\|[\hat{C}(\Delta t)-\hat{E}(\Delta t)]\hat{w}(t)\| \le \varepsilon\,\Delta t\,, \tag{5.91}$$

for all $\Delta t \le \delta$ and $t \in [0,T]$, where $\hat{w}(t) = \hat{E}(t)\hat{w}_0$, and

$$\hat{E}(t) = \mathcal{F}E(t)\mathcal{F}^{-1}\colon\ \hat{B} \to \hat{B}\,. \tag{5.92}$$

The space $\hat{B}_0$ is dense in $\hat{B} = \mathcal{F}B$ iff $B_0 = \mathcal{F}^{-1}\hat{B}_0$ is dense in B. We may, therefore, choose $\hat{B}_0$ to be any convenient dense subspace of $\hat{B}$. For the present problem, $p = d = 1$ and, since $C_0^\infty(-\infty,+\infty)$ is dense in $L^2(-\infty,+\infty)$ (Section 4.11, property (10)) we set

$$\hat{B}_0 = \{v_1+iv_2\colon\ v_1,v_2 \in C_0^\infty(-\infty,+\infty)\}\,.$$

Applying the Fourier transform to the heat equation we find that

$$\frac{d}{dt}\hat{w}(k,t) = -k^2\hat{w}(k,t)\,, \tag{5.93}$$

so that $\hat{E}(t)$ is given by

$$\hat{w}(k,t) = (\hat{E}(t)\hat{w}_0)(k) = e^{-k^2t}\hat{w}_0(k)\,. \tag{5.94}$$

We now prove condition (5.91). To do so, we choose $\varepsilon > 0$ and $\hat{w}_0 \in \hat{B}_0$. $\hat{w}_0$ has compact support, so that $\hat{w}_0(k) = 0$ if $|k| \geq A$, where $A = A(w_0)$ is a constant.

The growth factor $G(k,\Delta t)$ is given by eqn (5.84). Expanding as a Taylor series in powers of Δt we obtain that there are constants K_G and δ (depending on A) such that

$$|G(k,\Delta t)-e^{-k^2\Delta t}| \leq K_G(\Delta t)^2, \quad \text{for } |k| \leq A \text{ and } \Delta t \leq \delta . \tag{5.95}$$

From eqns (5.76) and (5.94),

$$([\hat{C}(\Delta t)-\hat{E}(\Delta t)]\hat{w}(t))(k) = (G(k,\Delta t)-e^{-k^2\Delta t})e^{-k^2 t}\hat{w}_0(k) .$$

Remembering that $w_0(k)$ vanishes when $|k| \geq A$, we conclude, using eqn (5.95), that

$$\|[\hat{C}(\Delta t)-\hat{E}(\Delta t)]\hat{w}(t)\| \leq K_G(\Delta t)^2\|w_0\| ,$$

for $t \in [0,T]$ and $t \leq \delta$. Picking $\Delta t \leq \varepsilon/(K_G\|w_0\|)$, we satisfy eqn (5.91), and the proof of consistency is complete.

(4) Numerical results

In practice, one solves problems over a finite interval. As an illustration, consider the initial-boundary value problem

$$\frac{\partial u}{\partial t} = \frac{\partial^2 u}{\partial x^2}, \quad -1 \leq x \leq 1, \quad t > 1,$$

$$u(x,1) = \exp(-x^2/4), \quad -1 \leq x \leq 1, \tag{*}$$

$$u(+1,t) = u(-1,t) = \frac{1}{\sqrt{t}}\exp(-1/4t), \tag{**}$$

with exact solution

$$u(x,t) = \frac{1}{\sqrt{t}}\exp(-x^2/4t) . \tag{***}$$

The interval $[-1,+1]$ is divided into m subintervals, so that

$$\Delta x = 2/m ,$$

$$\Delta t = \lambda(\Delta x)^2 .$$

If $U(-1+j\Delta x,1+n\Delta t)$ is denoted by U_j^n then we obtain from eqn (5.81) the following tri-diagonal system of equations:

$$U_j^{n+1} - U_j^n = \lambda\theta[U_{j+1}^{n+1}-2U_j^{n+1}+U_{j-1}^{n+1}] + \lambda(1-\theta)[U_{j+1}^n-2U_j^n+U_{j-1}^n], \quad (\text{**}_{**})$$

$$\text{for } 1 \le j \le m-1 \text{ and } n \ge 0 .$$

The values of U_0^n and U_m^n are given by the boundary conditions (**) while the values of U_j^0 are given by the initial condition (*).

The problem was solved, with $m = 10$, for various values of λ and θ. (The tri-diagonal system ($^{**}_{**}$) is easily solved, using Gaussian elimination, for the vector U^{n+1} once the vector U^n is known).

The necessity of the stability condition (5.86) is readily demonstrated. If $\theta = 0$ the stability condition becomes $\lambda \le \frac{1}{2}$. In Table 5.3 we compare the numerical results at time $t = 2.296$ obtained using $\lambda = \cdot 4$ (stable) and $\lambda = \cdot 6$ (unstable). (The odd choice of t is necessary because the time must be an integer multiple of different stepsizes).

j	0	1	2	3	4	5
$\lambda = \cdot 6$	.5919	.9156	.0631	.4341	-.2676	.6310
$\lambda = \cdot 4$	.5919	.6154	.6344	.6483	.6568	.6596
exact	.5919	.6155	.6346	.6486	.6571	.6600

Table 5.3: Numerical and exact results at $t = 2.296$ *with* $\theta = 0$, $\lambda = \cdot 4$ *and* $\lambda = \cdot 6$. $(U_{10-j}^n = U_j^n)$.

If $\lambda = 5$, the stability condition (5.86) becomes

$$\theta \ge \cdot 45 .$$

In Table 5.4 we compare numerical results for $\theta = \cdot 4$ (unstable) and $\theta = \cdot 6$ (stable).

j	0	1	2	3	4	5
$\theta = \cdot 4$	.2325	.4558	-.1514	.7040	-.2563	.7290
$\theta = \cdot 6$	.2325	.2336	.2345	.2352	.2356	.2357
exact	.2325	.2336	.2345	.2352	.2356	.2357

Table 5.4: Numerical and exact results at $t = 18$ *with* $\lambda = 5$, $\theta = \cdot 4$, *and* $\theta = .6$. $(U^n_{10-j} = U^n_j)$.

PROBLEMS

5.1 Prove the equivalence of the three expressions (5.1), (5.2), and (5.3) for the norm of a linear operator. {110}

5.2 $A = (a_{ij})$ is a real $n \times n$ matrix. As shown in Example 5.2,

$$||A||_2 = [\rho(A^TA)]^{\frac{1}{2}} .$$

Give an alternative proof by using the fact that every real symmetric matrix B can be factorized as $B = U^TDU$ where U is an orthogonal matrix and D is a diagonal matrix. {112}

5.3 Prove that $||A||_1 = \max_j \sum_{i=1}^n |a_{ij}|$, where A is the $n \times n$ matrix (a_{ij}) . (Example 5.3, p. 113).

5.4 Compute the $||\cdot||_1$, $||\cdot||_2$, and $||\cdot||_\infty$ norms of the following matrices

$$A_1 = \begin{bmatrix} 1 & 2 \\ 3 & -4 \end{bmatrix}, \quad A_2 = \begin{bmatrix} 2 & -1 & & & & \bigcirc \\ -1 & 2 & -1 & & & \\ & \ddots & \ddots & \ddots & & \\ & & \ddots & \ddots & \ddots & \\ \bigcirc & & & -1 & 2 & -1 \\ & & & & -1 & 2 \end{bmatrix}$$

where A_2 is an $n \times n$ tridiagonal matrix.
[Hint: If $v = (v_1, \ldots, v_n)$ is an eigenvector of A_2 corresponding to the eigenvalue λ then the v_i satisfy the three-term recurrence relation

$$-v_{i-1} + 2v_i - v_{i+1} = \lambda v_i \ , \ 2 \le i \le n-1 \ ,$$

the general solution of which is

$$v_i = a\nu^i + b\mu^i$$

where a and b are arbitrary constants, and μ and ν are the two solutions of the quadratic equation

$$-1 + 2t - t^2 = \lambda t \ .]$$

5.5 Verify equation (5.9) of Example 5.4.

5.6 Let $T: C[0,1] \to R^1$ be given by

$$Tx = \int_0^1 (t-\tfrac{1}{2})x(t)dt \ .$$

Show that $||T|| = 1/4$. Also show that there is no $x \in C[0,1]$ satisfying $||Tx|| = ||T|| \ ||x||$. {115}

5.7 There are a large number of elegant results on the norms of certain operators in ℓ^p and L^p. Two examples are:
(1) For $1 < p < \infty$, and $x = (x_i) \in \ell^p$ let $Tx = y = (y_i)$ be defined by

$$y_n = \frac{1}{n} \sum_{i=1}^{n} x_i \ .$$

Then $y \in \ell^p$, so that $T: \ell^p \to \ell^p$. Furthermore, $||T|| = p/(p-1)$. □

(2) For $1 < p < \infty$, and $x \in L^p(0,\infty)$ let $Tx = y$ be defined by

$$y(t) = \int_0^\infty \frac{x(s)}{s+t}\, ds \,, \quad \text{for} \quad 0 < t < \infty \,.$$

Then $y \in L^p(0,\infty)$, so that $T: L^p(0,\infty) \to L^p(0,\infty)$. Furthermore,

$$||T|| = \pi/\sin(\pi/p) \,. \quad \Box$$

For (1) above and $p = 2$ prove that $||T|| \leq 2$ by showing the following:

(a) $y_i^2 - 2x_i y_i = (1-2i)y_i^2 + 2(i-1)y_i y_{i-1}$. (*)

where $y_i = (\sum_{j=1}^{i} x_j)/i$.

(b) $y_i^2 - 2x_i y_i \leq (i-1)y_{i-1}^2 - iy_i^2$. (**)

(c) $\sum_{i=1}^{n} y_i^2 \leq 2 \sum_{i=1}^{n} x_i y_i$. (***)

(d) Apply the Cauchy-Schwarz inequality.

See Dunford and Schwartz [1966, p. 532] and Hardy, Littlewood, and Polya [1967, p. 240].

5.8 Complete the proof of Theorem 5.1 (which gives different equivalent conditions for the continuity of a linear operator). {118}

5.9 Let X be a metric space, and E a subset of X . E is *nowhere dense* in X if every ball in X contains a subball which has no points in common with E . E is of the *first category in X* if E is the union of a countable number of nowhere dense sets. E is of the *second category in X* if it is not of the first category in X .

Example

Let X be the Euclidean plane $-\infty < x,y < \infty$. Let $E_z = \{(x,z): -\infty < x < \infty\}$, so that E_z is the line parallel to the x-axis with y-coordinate z . Then E_z is nowhere dense. If $\{z_i\}$ is any denumerable sequence then

$$E = \cup E_{z_i}$$

is of the first category. (Note that if $\{z_i\}$ is an ordered sequence of all the rational numbers then E is dense in X.) □

Use Theorem 5.2 to prove the following version of the Baire category theorem:

Theorem

A complete metric space X is a set of the second category in itself.

[Hint: Assume that $X = \cup X_i$, where each X_i is nowhere dense in X , and derive a contradiction.]

5.10 Prove that condition (2) of the Banach-Steinhaus theorem may be replaced by a condition requiring convergence on a fundamental set. (Remark 5.4, p. 122).

5.11 Determine necessary and sufficient conditions such that

$$\lim_{n\to\infty} T x = Lx ,$$

where $\{T_n\}$ is a sequence of linear operators between Banach spaces. (Remark 5.5, p. 123).

5.12 Compute the norms of the Newton-Cotes quadrature rules on $C[-1,1]$. (Table 5.2, p. 126)

5.13 Given an $(m+1)$-point quadrature formula

$$\int_0^1 x(t)dt \doteq \sum_{k=0}^{m} w_k x(t_k) ,$$

one can construct a composite rule by dividing $[0,1]$ into n subintervals and applying the original quadrature rule in each interval:

$$\int_0^1 x(t)dx = \sum_{r=1}^{n} \int_{(r-1)/n}^{r/n} x(t)dt ,$$

$$\doteq \sum_{r=1}^{n} \sum_{k=0}^{m} \frac{1}{n} w_k x\left(\frac{r-1+t_k}{n}\right) . \qquad (*)$$

Show that:

as $n \to \infty$ for any $x \in C[0,1]$ $\iff$ $\sum_{k=0}^{m} w_k = 1$.

5.14 Use the results of Example 5.6 to prove the following relative of Steklov's theorem (Theorem 5.5):

THEOREM

The quadrature formulae

$$T_n x = \sum_{k=1}^{n} w_k^{(n)} x(t_k^{(n)}) \doteq \int_a^b x(t)dt = Tx , \qquad (*)$$

converge for all continuously differentiable functions x $\iff$

(1) The formulae converge for each polynomial.

AND (2) There exists a constant M *such that*

$$\sum_{k=1}^{n} (t_{k+1}^{(n)} - t_k^{(n)}) \left| \sum_{j=1}^{k} w_j^{(n)} \right| \le M ,$$

where $a \le t_1^{(n)} < t_2^{(n)} < \ldots < t_n^{(n)} \le t_{n+1}^{(n)} = b$.

5.15 Prove that

$$u_0 = (1,1,1,\ldots) ,$$
$$u_1 = (1,0,0,\ldots) ,$$
$$u_2 = (0,1,0,\ldots) ,$$
$$\ldots$$

form a fundamental set in c, as asserted in the proof of Theorem 5.6. {128}

5.16 Compute

$$\sum_{n=0}^{\infty} \frac{(-1)^n}{2n+1} = \pi/4$$

to eight decimal places using van Wijngaarden's modification of Euler's transformation. (Remark 5.7, p. 134).

5.17 Prove that van Wijngaarden's modification of Euler's transformation maps every convergent sequence into a convergent sequence.

5.18 Show that Euler's transformation transforms

$$(1) \quad \sum_{n=0}^{\infty} (-1)^n 2^{-n} \quad \text{into} \quad \frac{1}{2} \sum_{n=0}^{\infty} 4^{-n} ,$$

$$(2) \quad \sum_{n=0}^{\infty} (-1)^n 3^{-n} \quad \text{into} \quad \frac{1}{2} \sum_{n=0}^{\infty} 3^{-n} ,$$

and

$$(3) \quad \sum_{n=0}^{\infty} (-1)^n 4^{-n} \quad \text{into} \quad \frac{1}{2} \sum_{n=0}^{\infty} \left(\frac{3}{8}\right)^n ,$$

so that the transformed series may converge faster, at the same speed, or slower than the original series.

5.19 Let $T_0^{(0)}, T_0^{(1)}, \ldots,$ be given. Let the constants $\beta > 1$, $\alpha_s^{(k,m)}$, and $\alpha_s^{(k+1,m)}$ be such that

$$T_0^{(v)} = \sum_{s=0}^{m} \alpha_s^{(k,m)} \left(\frac{1}{\beta^v}\right)^s , \quad \text{for } k \le v \le k+m , \qquad (*)$$

$$T_0^{(v)} = \sum_{s=0}^{m} \alpha_s^{(k+1,m)} \left(\frac{1}{\beta^v}\right)^s , \quad \text{for } k+1 \le v \le k+m+1 . \qquad (**)$$

By forming a weighted combination of eqns (*) and (**) show that

$$\left[\frac{1}{\beta^k} - \frac{1}{\beta^{k+m+1}}\right] T_0^{(v)} = \left[\sum_{s=0}^{m} \alpha_s^{(k,m)} \left(\frac{1}{\beta^v}\right)^s\right]\left[\frac{1}{\beta^v} - \frac{1}{\beta^{k+m+1}}\right] + \qquad (*{*\atop *})$$

$$+ \left[\sum_{s=0}^{m} \alpha_s^{(k+1,m)} \left(\frac{1}{\beta^v}\right)\right]\left[\frac{1}{\beta^k} - \frac{1}{\beta^v}\right] ,$$

for $k \le v \le k+m+1$.

Since eqns (*), (**), and (***) are of the form (5.26), deduce eqn (5.24), namely

$$T_{m+1}^{(k)} = \alpha_0^{(k,m+1)} = \frac{\beta^{m+1}\alpha_0^{(k+1,m)} - \alpha_0^{(k,m)}}{\beta^{m+1}-1} ,$$

$$= \frac{\beta^{m+1}T_m^{(k+1)} - T_m^{(k)}}{\beta^{m+1}-1} . \quad \{137\}$$

5.20 Let $T_m^{(k)}$ be elements of a triangular Romberg array. Show that, for fixed m, there are constants c_s such that

$$T_m^{(k)} = \sum_{s=0}^{m} c_s T_0^{(k+s)} , \quad k = 0,1,2,\ldots \quad . \qquad (*)$$

Hence show that

$$\lim_{k\to\infty} T_m^{(k)} = \lim_{k\to\infty} T_0^{(k)} . \quad \{138\}$$

5.21 Prove the properties (5.27) for the coefficients of Romberg's method in the case $\beta = 4$, using the approach outlined below.

Let $t_m(z)$ be the generating polynomial for the coefficients c_{mk}, namely,

$$t_m(z) = \sum_{k=0}^{m} c_{mk} z^k .$$

Prove the following:

(a) $t_m(z) = (\beta^m - z)\, t_{m-1}(z)/(\beta^m - 1)$,

$$t_m(z) = \prod_{k=1}^{m} (1 - z\beta^{-k})/(1-\beta^{-k}) .$$

(b) $\sum_{k=0}^{m} c_{mk} = t_m(1) = 1$.

(c) If $\beta > 1$, $t_m(-z)$ has strictly positive coefficients,

$(-1)^k c_{mk} > 0$, and $\sum_{k=0}^{m} |c_{mk}| = t_m(-1)$.

(d) If $\varepsilon \le 4^{-1}$ and $\beta = 4$ then

$$\ell n\left(\frac{1+\varepsilon}{1-\varepsilon}\right) \le 2\varepsilon\left(1+\frac{1}{3}\sum_{i=1}^{\infty}\varepsilon^{2i}\right) \le \frac{92}{45}\cdot\varepsilon .$$

$$\ell n\ t_m(-1) = \sum_{k=1}^{m}\ell n\left(\frac{1+4^{-k}}{1-4^{-k}}\right) < \frac{92}{45}\cdot\frac{1}{3} .$$

$$t_m(-1) < \exp\left(\frac{92}{45}\cdot\frac{1}{3}\right) < 2 .$$

(e) If $\beta = 4$, $|c_{m,m-s}|$ is less than the coefficient of z^{m-s} in

$$\prod_{k=1}^{m}\frac{(1+\frac{z}{4})}{(1-4^{-k})} .$$

Hence $c_{m,m-s} \to 0$ as $m \to \infty$. {139}

5.22 Let $x \in C^{\infty}[-1,+1]$. Consider the approximations to $x(0)$:

(1) $$T_0^{(k)} = \frac{x(2^{-k}) - x(-2^{-k})}{2.2^{-k}} .$$

(2) $$T_0^{(k)} = \frac{x(2^{-k}) - x(0)}{2^{-k}} .$$

Show how Romberg's method can be applied. Compute the Romberg $T_m^{(k)}$ arrays for (1) and (2) if $x(t) = \sin t$ and $0 \le m + k \le 4$. {139}

5.23 Prove that

$$\max_{0\le s\le 2\pi}\int_0^{2\pi}\left|\frac{\sin[(2n+1)(t-s)/2]}{\sin[(t-s)/2]}\right|dt \ge \frac{8}{\pi}\,\ell n(\frac{n+1}{2}) \qquad (*)$$

[Hint: Choose $s = \pi$. Then the integrand is symmetric about $t = \pi$. Setting $u = (2n+1)(t-\pi)/2$, it follows that eqn (*) will hold provided that

$$I_n = \int_0^{(2n+1)\pi/2}\frac{|\sin u|}{\sin(u/(2n+1))}\,du \ge \frac{2(2n+1)}{\pi}\,\ell n(\frac{n+1}{2}) .$$

But,

$$I_n > \sum_{k=1}^{n-1} \int_{k\pi}^{(k+1)\pi} \frac{|\sin u|}{\sin(u/(2n+1))} du ,$$

$$> \sum_{k=1}^{n-1} \int_{k\pi}^{(k+1)\pi} \frac{(2n+1)}{u} |\sin u| du .] \quad \{141\}$$

5.24 Prove the identity (5.29) for trigonometric polynomial operators. {142}

5.25 Prove that if $x(t) = t^2$ and B_n and H_n are the 'Bernstein' and 'Modified Hermite' operators defined in Remark 5.9, then neither $B_n x$ nor $H_n x$ is equal to x . {143}

5.26 Prove that if the knots $t_i^{(n)}$ of a cubic spline s_n are not equidistant but satisfy

$$\frac{1}{\alpha} \min_{0\le i\le n-1} |t_i^{(n)} - t_{i-1}^{(n)}| \le \max_{0\le i\le n-1} |t_i^{(n)} - t_{i-1}^{(n)}| ,$$

$$\le \alpha \min_{0\le i\le n-1} |t_i^{(n)} - t_{i-1}^{(n)}| ,$$

for some constant α , then we still have that $s_n = T_n x \to x$ as $n \to \infty$ for all $x \in C[a,b]$. (See Section 5.6).

5.27 Modify the definition (5.31) of a cubic spline $s_n = T_n x$ in such a way that

$$||T_n x - x||_\infty = O(h^4)$$

if $x \in C^4[a,b]$.

5.28 Verify that if $u_0 \in C(\overline{R^1})$ then $u(x,t)$ given by eqn (5.41) satisfies the initial value problem (5.40) for the heat equation. {150}

5.29 Prove that if $E(t)$ is defined by eqns (5.41) and (5.42) and $X = L^2(-\infty,+\infty)$ then $E(t)$ satisfies conditions (5.44) through (5.48) on page 151.

[Hints: Let

$$k(x,t) = \frac{1}{2\sqrt{\pi t}} \exp(-x^2/4t) .$$

Then

$$\int_{-\infty}^{+\infty} k(x,t)dx = 1 .$$

Thus, if $u_0 \in X$ then, from the Hölder inequalities,

$$|u(x,t)| = \left| \int_{-\infty}^{+\infty} k(x-z,t)u_0(z)dz \right| ,$$

$$= \left| \int_{-\infty}^{+\infty} [k(x-z,t)]^{\frac{1}{2}}[k(x-z,t)]^{\frac{1}{2}}u_0(z)dz \right| ,$$

$$\leq \left[\int_{-\infty}^{+\infty} k(x-z,t)dz \right]^{\frac{1}{2}} \left[\int_{-\infty}^{+\infty} k(x-z,t)(u_0(z))^2dz \right]^{\frac{1}{2}} ,$$

$$= \left[\int_{-\infty}^{+\infty} k(x-z,t)(u_0(z))^2dz \right]^{\frac{1}{2}} . \qquad (*)$$

Using the theorem of Fubini, we see that,

$$\|u(t)\|_X^2 \leq \int_{-\infty}^{+\infty} dx \int_{-\infty}^{+\infty} k(x-z,t)(u_0(z))^2dz ,$$

$$= \int_{-\infty}^{+\infty} (u_0(z))^2dz \int_{-\infty}^{+\infty} k(x-z,t)dx ,$$

$$= \|u_0\|_X^2 . \text{]}$$

5.30 Verify that any smooth solution u of the heat equation satisfies the weak form of the heat equation. (Remark 5.13, p. 152).

5.31 Consider the *backwards heat equation*,

$$\frac{\partial u}{\partial t} = -\frac{\partial^2 u}{\partial x^2}, \quad -\infty < x < \infty, \quad 0 < t < T, \qquad (*)$$

with initial condition $u(x,0) = u_0(x)$.
(This is not a properly-posed problem.)

(1) Show that if $u_0 = \sin kx$ then $u(x,t) = O(e^{k^2 t})$. Compare this with the behaviour of solutions of the usual heat equation.

(2) Show that if the numerical method of Example 5.10 is adapted to the backwards heat equation, the resulting numerical method is unstable for all values of $\theta \in [0,1]$. {42}

5.32 Show that the DuFort-Frankel method (5.57) can be written in the form

$$\begin{pmatrix} 1+\mu & 0 \\ 0 & 1 \end{pmatrix} V(x,t+\Delta t) + \begin{pmatrix} 0 & -\mu \\ 0 & 0 \end{pmatrix} [V(x+\Delta x,t+\Delta t) + V(x-\Delta x,t+\Delta t)]$$

$$= \begin{pmatrix} 0 & 1-\mu \\ 1 & 0 \end{pmatrix} V(x,t), \qquad (*)$$

where $V(x,t) = \begin{pmatrix} U(x,t) \\ U(x,t-\Delta t) \end{pmatrix}$ and $\mu = \dfrac{2\Delta t}{(\Delta x)^2}$.

Show that the amplification matrix for eqn (*) is

$$G(k,\Delta t) = \begin{pmatrix} \frac{2\mu}{1+\mu}\cos k\Delta x & \frac{1-\mu}{1+\mu} \\ 1 & 0 \end{pmatrix}.$$

{157,161}

5.33 Prove that $C(\Delta t)$ satisfies the stability condition (5.69) $\Longleftrightarrow$ the extended operator $C(\Delta t)$ defined by eqn (5.70) satisfies the stability condition (5.71). {161}

5.34 Verify the expression (5.78) for the amplification matrix for the explicit method for the heat equation. {162}

5.35 Prove that the von Neumann condition (5.80) is necessary and sufficient for stability when $p = 1$, i.e. the solution u has only one component. {166,168}

5.36 Verify the expansion of eqn (5.89) for the truncation error of the finite difference approximation (5.81) for the heat equation.

Show that:

(1) $\tau = O((\Delta t)^3)$ if $\theta = \frac{1}{2} - \frac{1}{12\lambda}$,

(2) $\tau = O((\Delta t)^4)$ if $\theta = \frac{1}{2} - \frac{1}{12\lambda}$ and $\lambda = 1/\sqrt{20}$.

{168,169}

5.37 Consider the initial value problem

$$u_t + u_x = 0\ , \quad -\infty < x < \infty$$
$$u(x,0) = u\ (x) \in X = L^2(-\infty,+\infty)\ ,$$

and the finite difference approximations

(1) $$\frac{U(x,t+\Delta t) - U(x,t)}{\Delta t} + \frac{U(x+\Delta x,t) - U(x-\Delta x,t)}{2\Delta x} = 0\ .$$

(2) $$\frac{U(x,t+\Delta t) - U(x,t)}{\Delta t} + \frac{U(x+\Delta x,t) - U(x,t)}{\Delta x} = 0\ .$$

(3) $$\frac{U(x,t+\Delta t) - U(x,t)}{\Delta t} + \frac{U(x,t) - U(x-\Delta x,t)}{\Delta x} = 0\ .$$

(4) $$\frac{U(x,t+\Delta t) - U(x,t-\Delta t)}{2\Delta t} + \frac{U(x+\Delta x,t) - U(x-\Delta x,t)}{2\Delta x} = 0\ .$$

Assuming that Δt and Δx vary in such a way that $\lambda = \Delta t/\Delta x$ is constant, show that all four methods are consistent but that only methods (3) and (4) can be stable.

5.38 Suppose that T is a bounded linear operator mapping a Banach space X into itself and that we attempt to solve the equation $x = Tx$ by iteration: $x_{n+1} = Tx_n$. If the sequence $\{x_n\}$ does not converge we can try to obtain convergence by averaging. That is, we choose $\alpha \in [0,1)$ and consider the iteration:

$$y_1 = x_1 \, ,$$
$$\tilde{y}_{n+1} = Ty_n \, , \qquad (*)$$
$$y_{n+1} = \alpha y_n + (1-\alpha)\tilde{y}_{n+1} \, , \quad \text{for } n \geq 1 \, .$$

Show that $\{y_n\}$ is convergent whenever $\{x_n\}$ is.

5.39 Let $X = C^1[0,1]$. There is a formal expansion for the differentiation operator D in terms of the difference operator Δ_h , $\Delta_h x(t) = x(t+h) - x(t)$, namely,

$$hD = \Delta_h - \frac{1}{2}\Delta_h^2 + \frac{1}{3}\Delta_h^3 \cdots \qquad (*)$$

Truncating the series after n terms and setting $h = 1/n$ we obtain a sequence of operators $T_n : X \to R^1$, beginning with

$$T_1 x = \Delta_1 x(0) = x(1) - x(0) \, ,$$
$$T_2 x = 2[\Delta_{\frac{1}{2}} - \tfrac{1}{2}\Delta_{\frac{1}{2}}^2]x(0) = -x(1) + 4x(\tfrac{1}{2}) - 3x(0) \, .$$

Show that the sequence $\{T_n x\}$ does not converge to $Dx(0) = \dot{x}(0)$ for all $x \in X$.

5.40 Let A be a complex $n \times n$ matrix, with spectral radius $\rho(A)$, considered as an operator on $\mathbb{C}^n$.

(1) Prove that, for any norm $\|\cdot\|$ on $\mathbb{C}^n$, $\rho(A) \leq \|A\|$.

(2) Prove that, for any $\varepsilon > 0$, there exists a norm $\|\cdot\|_\wedge$ on $\mathbb{C}^n$ such that $\|A\|_\wedge \leq \rho(A) + \varepsilon$.

{165}

6
Compactness

Compactness is a property of certain sets and mappings. It is of importance in numerical analysis for several reasons:

(1) Compact operators have useful properties which make them easy to analyse and approximate.

(2) Compactness is a basic tool in establishing the convergence of sequences of approximations because it makes it possible to extract convergent subsequences.

(3) Compactness is in several ways equivalent to 'nearly finite dimensional'.

(4) A continuous function attains its supremum and infimum on a compact set.

6.1 *DEFINITIONS AND BASIC PROPERTIES*

Let S be a subset of a topological space X. An *open cover* or *open covering* of S is any family κ of open sets in X whose union contains S, so that

$$S \subset \bigcup_{V \in \kappa} V .$$

A cover is *finite* if it contains only a finite number of sets. Given an open cover κ of S, it is sometimes possible to choose a subset κ_1 of κ which is also an open cover of S; such a subset will sometimes be called a *subcover* to emphasize that it is a subset of the original cover.

A set S in a topological space X is *compact* if every open cover of S contains a finite subcover (of S). S is *precompact* or *conditionally compact* if its closure $\overline{S}$ is compact. X is a *compact space* if every open cover of X contains a finite subcover. S is *locally compact* if every point of X has a precompact neighbourhood.

REMARK 6.1. Let X be a topological space with topology τ. Let $S \subset X$. On S the induced topology τ_S is defined by (see Problem 2.2)

$$\tau_S = \{S \cap G: G \in \tau\} .$$

S is compact (as a subset of X) iff S is a compact space with respect to the induced topology τ_S (Problem 6.1). □

THEOREM 6.1.

(1) A closed subset of a compact set is compact.

(2) A continuous image of a compact space is compact.

(3) A compact subset of a Hausdorff space is closed.

(4) A compact set in a topological vector space is bounded topologically.

Proof. (1): Let S be a compact set and let T be a closed subset of S. Given an open cover κ_T of T, we construct κ by adjoining the complement T^c to κ_T. κ is an open cover of S. Since S is compact, κ contains a finite subcover κ'. Then, $\kappa_T' = \{G \in \kappa': G \neq T^c\}$ is a finite cover of T which is a subcover of κ_T.

(2) Let $f: X \to Y$ be a continuous mapping of a compact topological space X into a topological space Y. We wish to prove that $f(X)$ is compact. Let κ_Y be an open cover of $f(X)$, and set

$$\kappa_X = \{f^{-1}(G): G \in \kappa_Y\} .$$

f is continuous, and so $f^{-1}(G)$ is open whenever G is open (Theorem 2.3). Furthermore, for each $x \in X$, $f(x) \in G$ for some $G \in \kappa_Y$. Consequently, κ_X is an open cover of the compact space X. Choosing a finite subcover κ_X' of X, it follows that

$$\kappa_Y' = \{G \in \kappa_Y: f^{-1}(G) \in \kappa_X'\}$$

is a finite subcover of $f(X)$.

It may be remarked that the continuous image of a closed set is not always closed. (Problem 6.30).

(3) Let S be a compact subset of a Hausdorff space X. We show that S is closed by proving that $X \setminus S$ is open. Choose any $x \in X \setminus S$. For each $y \in S$ there exist disjoint open neighbourhoods $N_x(x,y)$ of x and $N_y(x,y)$ of y. Since $\{N_y(x,y): y \in S\}$ is an open cover of S it contains a finite subcover $\{N_{y_i}(x,y_i): 1 \le i \le n\}$. Then $N_x = \bigcap_{1 \le i \le n} N_x(x,y_i)$ is an open neighbourhood of x. Furthermore, $N_x \cap N_{y_i}(x,y_i) = \emptyset$ for $1 \le i \le n$ so that $N_x \subset X \setminus S$. Thus, $X \setminus S$ is the union of the open sets N_x, and is open.

(4) Let S be a compact subset of a topological vector space X. Let V be a neighbourhood of $\underline{0} \in X$. Since multiplication is continuous, for each $x \in S$ there exists a neighbourhood N_x of x and a $\delta_x > 0$ such that $[0,\delta_x]N_x \subset V$. But $\{N_x : x \in S\}$ is an open cover of S, so that there is a finite subcover $\{N_{x_i} : 1 \le i \le n\}$. Then $S \subset \delta^{-1}V$ where $\delta = \min \delta_{x_i} > 0$. Since V was an arbitrary neighbourhood of $\underline{0}$, we conclude that S is topologically bounded. □

A set S in a topological space X is *sequentially compact* if for every sequence $\{x_n\}$ in S there exists a subsequence $\{x_{n_i}\}$ which converges to a point $x \in X$. Note that the limit point x need not belong to S.

REMARK 6.2. Let X be a metric space containing a point x and a sequence $\{x_n\}$. Then there exists a subsequence $\{x_{n_i}\}$ which converges to x iff every neighbourhood of x contains infinitely many of the points x_n.

The proof, which is left to the reader (Problem 6.2) uses the fact that every point in a metric space has a countable local base. It is, therefore, of interest that a Hausdorff topological vector space is metrisable iff it has a countable local base. (Theorem 2.7) □

THEOREM 6.2. *Let* S *be a subset of a metric space* X *with metric* ρ . *Then*

(1) S *is compact* $\Leftrightarrow$ S *is closed and sequentially compact.*

(2) S *is sequentially compact* $\Rightarrow$ S *is separable, that is, there is a denumerable set* S_o *which is dense in* S .

Proof. *(1)*$\Rightarrow$: Since S is a compact subset of a Hausdorff space it follows from Theorem 6.1 (part 3) that S is closed.

To prove that S is sequentially compact we assume the opposite and obtain a contradiction. If S is not sequentially compact then some sequence $\{x_n\}$ in S contains no convergent subsequence. Hence, by Remark 6.2, each point $y \in S$ has an open neighbourhood N_y containing at most a finite number of the x_n . The family $\kappa = \{N_y : y \in S\}$ is an open cover of S , so there is a finite subcover, $\kappa' = \{N_{y_i} : 1 \le i \le n\}$ say. But each N_{y_i} contains only a finite number of the x_n , so the infinite sequence $\{x_n\}$ has only a finite number of terms. This is the desired contradiction.

(2): We begin by observing that S is bounded. [Otherwise there exists a sequence $\{s_n\}$ in S satisfying $\rho(\underline{0},s_n) \to \infty$. Since S is sequentially compact, we can find $s_{n_i} \to x \in X$. By Theorem 3.1, $\rho(s_{n_i},x) < 1$ for large i . Thus, for large i ,

$$\rho(\underline{0},s_{n_i}) \le \rho(\underline{0},x) + \rho(x,s_{n_i}) < \rho(\underline{0},x) + 1 ,$$

which contradicts the assumption that $\rho(\underline{0},s_{n_i}) \to \infty$.]

Choose $p_o \in S$, and let

$$d_o = \sup_{q \in S} \rho(p_o,q) .$$

Then $d_o < \infty$ because S is bounded. We construct a sequence $p_1,p_2,\ldots$ in S as follows. Let

$$d_n = \sup_{q \in S} \min_{0 \le i \le n} \rho(q, p_i) \ . \qquad (*)$$

Choose $p_{n+1} \in S$ such that

$$\min_{0 \le i \le n} \rho(p_{n+1}, p_i) \ge d_n/2 \ .$$

Clearly, $d_0 \ge d_1 \ge d_2 \ \dots$. Since S is sequentially compact, there exists a subsequence $\{p_{n_j}\}$, $p_{n_j} \to p \in X$. But,

$$\begin{aligned} \frac{1}{2} d_{n_j} &\le \min_{0 \le i \le n_j} \rho(p_{n_j+1}, p_i) \ , \\ &\le \rho(p_{n_j+1}, p_{n_j}) \\ &\to 0 \ , \text{ as } \ j \to \infty \ . \end{aligned}$$

Consequently, $d_n \to 0$ as $n \to \infty$. Recalling the definition (*) of d_n , we see that the denumerable set $\{p_n\}$ is dense in S . That is, S is separable.

(1)⇐: We continue the arguments which proved (2), with the additional assumption that S is closed.

To prove that S is compact we proceed by contradiction and assume that S is not compact. Then there is an open cover of S, $\kappa = \{G_\alpha : \alpha \in A\}$, with no finite subcover. Let the sequence $\{p_n\}$ be as in the proof of (2). With each p_n we associate the open balls $B(p_n; 1/m)$, for $m = 1, 2, \dots$. The set of all such balls is denumerable. Construct the denumerable family $\sigma = \{G_{\alpha_k} : k = 1, 2, \dots\} \subset \kappa$ as follows: If $B(p_n; 1/m)$ is contained in one or more $G_\alpha \in \kappa$, adjoin *one* such G_α to σ . Then σ is denumerable. We assert that σ is also an open cover of S . For, let $p \in S$. Then $p \in G_\beta$ for some $\beta \in A$. Thus, $B(p; \varepsilon) \subset G_\beta$ for some $\varepsilon > 0$. Since $\{p_n\}$ is dense in S , there exist m and n such that $p \in B(p_n; 1/m) \subset B(p; \varepsilon) \subset G_\beta$. It follows that $p \in B(p_n; 1/m) \subset G_{\alpha_k}$ for some k , so that σ is indeed an open subcover of S .

For $n = 1,2,\ldots$ let

$$S_n = \bigcup_{k=1}^{n} G_{\alpha_k} .$$

Since no finite subcover of κ covers S, we know that $S \not\subset S_n$. For $n = 1,\ldots,$ choose $x_n \in S$ such that $x_n \notin S_n$. Since S is sequentially compact and closed, there exists a subsequence $\{x_{n_i}\}$ such that $x_{n_i} \to x \in S$. But $x \in G_{\alpha_r}$ for some r because σ covers S, and thus $x_{n_i} \in G_{\alpha_r}$ for i sufficiently large. Consequently, $x_{n_i} \in S_r \subset S_{n_i}$ for i sufficiently large. This contradicts the definition of x_{n_i}, from which we conclude that S is indeed compact. □

Let S be a subset of metric space X. $M \subset X$ is an *ε-net for S* if for every $x \in S$ there is a $z \in M$ such that $\rho(x,z) < \varepsilon$. S is *totally bounded* if for every $\varepsilon > 0$ there exists a finite ε-net for S. The next theorem links two aspects of compactness which are of importance in numerical analysis: convergence of subsequences and approximation by finite sets.

THEOREM 6.3. *If* S *is a set in a metric space* X *the following are equivalent:*

(1) S *is sequentially compact.*
(2) S *is precompact; that is,* $\overline{S}$ *is compact.*
(3) S *is totally bounded and* $\overline{S}$ *is complete (every Cauchy sequence in* $\overline{S}$ *converges to a point in* $\overline{S}$*).*

Furthermore, a compact metric space is complete and separable.

Proof. (1) ⇒ (2): Let $\{y_n\}$ be a sequence in $\overline{S}$. Choose $x_n \in S$ so that $\rho(x_n,y_n) < 1/2^n$. Since S is sequentially compact, some subsequence $\{x_{n_i}\}$ converges, $x_{n_i} \to x$, and it is clear that $y_{n_i} \to x$ also. Thus,

$\overline{S}$ is sequentially compact. By Theorem 6.2 (part 1), it follows that $\overline{S}$ is compact.

(2) ⇒ (3): If $\varepsilon > 0$ the open balls $B(x;\varepsilon)$ with $x \in S$ cover $\overline{S}$. Since $\overline{S}$ is compact, a finite number of these balls cover $\overline{S}$ and hence S; that is, S is totally bounded. Let $\{y_n\}$ be a Cauchy sequence in $\overline{S}$. By Theorem 6.2 (part 1) $\overline{S}$ is sequentially compact, so that a convergent subsequence $\{y_{n_i}\}$ exists, $y_{n_i} \to y$. Clearly $y \in \overline{S}$ and $y_n \to y$, and we conclude that $\overline{S}$ is complete.

(3) ⇒ (1): Let $\{x_n\}$ be a sequence in S. Since S is totally bounded, there exists for each k a finite cover of S with balls of radius $1/k$. Thus, some subsequence of $\{x_n\}$, $\{x_n^{(1)}\}$ say, is contained in a ball of radius 1. Repeating the process we obtain a series of sequences $\{x_n^{(k)}\}$, $k = 1,2,\ldots$. Each sequence $\{x_n^{(k)}\}$ is a subsequence of its predecessor $\{x_n^{(k-1)}\}$ and is contained in a ball of radius $1/k$. Set $s_n = x_n^{(n)}$. Then $\{s_n\}$ is a Cauchy sequence in $\overline{S}$ and, since $\overline{S}$ is complete, $s_n \to s \in \overline{S}$. Since $\{s_n\}$ is a subsequence of $\{x_n\}$, which was arbitrary, S is sequentially compact.

Finally, let X be a compact metric space. By Theorem 6.2 (part 2), X is separable. As a space, X is closed, and it follows from (3) above that $\overline{X} = X$ is complete. □

REMARK 6.3. As shown in Theorem 6.3, in a metric space compactness can be defined using three different approaches: open covers, sequences, and ε-nets. In the subsequent discussion of compactness we are free to use any of these approaches. As an analyst, the author prefers to use sequences even though this sometimes leads to a proliferation of subscripts and superscripts. However, we do try to indicate the alternative approaches. □

In Section 2.4 the concept of the Cartesian product of spaces was introduced. We conclude this section by proving, in part, an extremely useful property of Cartesian products.

THEOREM 6.4. *(Tychonoff)*

The Cartesian product of compact spaces is compact with respect to the product topology.

Proof. Let X_α be a compact space for $\alpha \in A$, and set

$$X = \prod_{\alpha \in A} X_\alpha .$$

When A is not a finite set, the proof that X is compact requires Zorn's lemma, and we refer the reader to Dunford and Schwartz [1966, p. 32] or Yosida [1968, p. 6].

When A is a finite set, $A = \{1,2,\ldots,m\}$ say, then each $x \in X$ can be regarded as an m-tuple $x = (x_i)$, where $x_i \in X_i$ for $i = 1,\ldots,m$. Let $\{x^{(n)}\}$ be a sequence in X. Then $\{x_1^{(n)}\}$ is a sequence in X_1 and so, since X_1 is compact, there exists a subsequence, $\{x_1^{(n)'}\}$ say, such that $x_1^{(n)'} \to y_1 \in X_1$. Repeating this process m times, we obtain a subsequence of $\{x^{(n)}\}$, $\{\tilde{x}^{(n)}\}$ say, such that $\tilde{x}_i^{(n)} \to y_i \in X_i$ as $n \to \infty$ for $1 \le i \le m$.

Now let G be any neighbourhood (in the product topology) of $y = (y_i) \in X$. From the definition of the product topology, there exist neighbourhoods $G_i \subset X_i$ of y_i, for $1 \le i \le m$, satisfying

$$\prod_{i=1}^{m} G_i \subset G .$$

Furthermore, there are integers N_i such that $\tilde{x}_i^{(n)} \in G_i$ whenever $n \ge N_i$, so that $x^{(n)} \in G$ whenever $n \ge \max N_i$.

We conclude that the subsequence $\{\tilde{x}^{(n)}\}$ converges to $y \in X$. Since $\{\tilde{x}^{(n)}\}$ is a subsequence of the arbitrary sequence $\{x^{(n)}\}$, it follows that X is sequentially compact. But every topological space, and X in particular, is closed. Thus, using Theorem 6.2 (part 1), we see that X is compact. □

REMARK 6.4. Let Y_α be a compact subset of X_α for $\alpha \in A$. From Remark 6.1, Y_α is a compact space (with the induced topology). Let

$$Y = \prod_{\alpha \in A} Y_\alpha , \quad X = \prod_{\alpha \in A} X_\alpha .$$

It is readily shown that Y is a compact subset of X. (Problem 6.3). □

6.2 *CRITERIA FOR COMPACTNESS*

We now derive criteria for the compactness of subsets of specific spaces X. We recall (Theorem 6.1) that for a set S to be compact it is necessary that S be closed and bounded.

6.2.1 *COMPACTNESS IN FINITE-DIMENSIONAL SPACES*

We begin by reminding the reader of two basic theorems for the real line:

THEOREM 6.5. *(Bolzano-Weierstrass)*

Let $S \subset R^1$. *Then:* S *is bounded* $\Leftrightarrow S$ *is sequentially compact.*

Proof $\Rightarrow$*:* Let $\{x_n\}$ be a sequence in S. We construct the sequence of intervals $[a_i,b_i]$, $i = 1,2,\dots$ such that

(a) $[a_i,b_i]$ is either the left half or the right half of $[a_{i-1},b_{i-1}]$;

(b) Each $[a_i,b_i]$ contains infinitely many of the points x_n.

By construction

$$|b_i - b_j| \le |b_0 - a_0|/2^i , \qquad \text{for } j \ge i ,$$

so that $\{b_i\}$ is a Cauchy sequence in R^1 and, since R^1 is complete, $b_i \to b \in R^1$. If $G \subset R^1$ is a neighbourhood of b, then $[a_i,b_i] \subset G$ for i sufficiently large, and so G also contains infinitely many of the points x_n.

It follows from Remark 6.2 that there exists a subsequence of $\{x_n\}$ which converges to b.

$\Leftarrow$: Left to the reader (Problem 6.4). □

The following theorem is an immediate consequence of Theorems 6.2 (part 1) and Theorem 6.5. (An alternative proof is outlined in Problem 6.5):

THEOREM 6.6. *A set* $S \subset R^1$ *is compact iff it is closed and bounded.* □

Consider the real n-dimensional space E^n. As shown in Problem 2.27 the topology on E^n may be taken to be the product topology. Consequently, a subset of E^n is compact iff it is compact in the product topology. Now let S be a bounded and closed subset of E^n. Then S is contained in some bounded n-dimensional interval I,

$$I = \prod_{i=1}^{n} Y_i \equiv \prod_{i=1}^{n} [a_i, b_i] \subset \prod_{i=1}^{n} R^1 = R^n .$$

It follows from Remark 6.4 and Theorem 6.6, that I is a compact subset of E^n. Since S is a closed subset of I, we conclude from Theorem 6.1 (part 1) that S is compact. In summary, we have established the following generalization of Theorem 6.6:

THEOREM 6.7. *(Heine-Borel)*

A subset of E^n *is compact iff it is closed and bounded.* □

REMARK 6.5. We chose to prove the Heine-Borel theorem using Tychonoff's theorem, as a way of illustrating Tychonoff's theorem. Two other possible proofs are suggested in Problems 6.6 and 6.7. □

REMARK 6.6. The Heine-Borel theorem is non-constructive: given an open cover of a bounded closed set in E^n, we are told that there exists a finite subcover but we are not told how to construct it. This is typical of compactness arguments.

There are many interesting results and problems connected with the construction of finite open covers for compact sets in E^n, and with the related problem of partitioning a set in E^n. In numerical computations, one often wishes to partition a set S in E^2 or E^3 into a large number of smaller sets S_i of a certain type. For example, in finite element computations, one might require the S_i to be small triangles without sharp corners (since these reduce the accuracy of the finite element approximation). See also Rogers [1964]. □

We now consider finite dimensional subspaces. We recall that a subspace X_0 of a vector space X is finite dimensional if there exists a finite number of points $x_i \in X_0$, $1 \le i \le n$, such that every $x \in X_0$ can be expressed as a linear sum of the x_i,

$$x = \sum_{i=1}^{n} \lambda_i x_i . \tag{*}$$

The x_i are linearly independent if $\sum_{i=1}^{n} \lambda_i x_i = 0$ iff $\lambda_i = 0$ for all i. In this case X_0 is said to have dimension n and each $x \in X_0$ has a unique representation of the form (*).

As promised earlier, we prove Theorem 4.1 (in part):

THEOREM 6.8. *Let* X_0 *be a real n-dimensional normed space. Then* X_0 *is isomorphic to* E^n*; that is, there exists a bicontinuous linear mapping of* X_0 *onto* E^n.

Consequently, all norms on X_0 *are equivalent. In particular, all norms on* R^n *are equivalent.*

Proof. Since X_0 is n-dimensional, there exist n linearly independent elements x_i such that every $x \in X_0$ has a *unique* representation

$$x = \sum_{i=1}^{n} \lambda_i(x)x_i , \tag{6.1}$$

where $\lambda_i(x) \in R^1$. Let

$$\lambda: x \in X_0 \to \lambda(x) = \begin{pmatrix} \lambda_1(x) \\ \vdots \\ \lambda_n(x) \end{pmatrix} \in E^n . \tag{6.2}$$

It is readily checked (Problem 6.8) that λ is a linear one-to-one mapping of X_0 onto E^n with continuous inverse

$$\lambda^{-1}: \mu = (\mu_i) \in E^n \to \sum_{i=1}^{n} \mu_i x_i \in X_0 . \tag{6.3}$$

It remains only to show that λ is continuous. Since λ is linear, it suffices to show, by Theorem 5.1, that there exists a constant α such that

$$||\lambda(y);E^n|| \le \alpha \quad \text{if} \quad ||y;X_0|| \le 1 . \tag{*}$$

The proof that eqn (*) holds for some α proceeds by contradiction. If there is no α such that eqn (*) holds, then there must exist a sequence $\{y_k\}$ in X_0 such that $||y_k;X_0|| \le 1$ but $||\lambda(y_k);E^n|| \to \infty$ as $k \to \infty$. Let

$$z_k = y_k/||\lambda(y_k);E^n|| \in X_0 .$$

Then

$$||z_k;X_0|| \to 0 \quad \text{and} \quad ||\lambda(z_k);E^n|| = 1 , \quad \text{as} \quad k \to \infty .$$

Thus $\lambda(z_k)$ belongs to the closed unit ball B in E^n . Since B is a closed and bounded subset of E^n , B is compact, by the Heine-Borel theorem, and there exists a subsequence, $\{\tilde{z}_k\}$ say, such that

$$\lambda(\tilde{z}_k) \to \mu \in E^n ,$$

where $||\mu;E^n|| = 1$. Set

$$z = \lambda^{-1}(\mu) = \sum_{i=1}^{n} \mu_i x_i \in X_0 .$$

Then,

$$\|z;X_0\| \le \|\tilde{z}_k;X_0\| + \|z-\tilde{z}_k;X_0\| ,$$

$$\le \|\tilde{z}_k;X_0\| + \sum_{i=1}^{n} |\mu_i-\lambda_i(\tilde{z}_k)| \ \|x_i;X_0\| ,$$

$$\le \|\tilde{z}_k;X_0\| + (n \max_i \|x_i;X_0\|) \|\mu-\lambda(\tilde{z}_k);E^n\| .$$

The terms on the right can be made arbitrarily small by choosing k sufficiently large. Consequently, $\|z;X_0\| = 0$. Thus,

$$\sum_{i=1}^{n} \mu_i x_i = 0 \quad \text{and} \quad \|\mu;E^n\| = 1 .$$

This contradicts the assumption that the x_i are linearly independent. □

REMARK 6.7. A simple, but important, corollary of Theorem 6.8 is that every real finite-dimensional normed space is complete, and thus a Banach space. Furthermore, every finite-dimensional subspace of a normed space is closed. (Problem 6.11). □

As another application of Theorem 6.8, let S be a bounded and closed subset of a finite-dimensional space X_0 . Let $\lambda: X_0 \to R^n$ be linear and bicontinuous, and set $T = \lambda(S)$. Since S is closed and T is the inverse image of S with respect to the continuous mapping $\lambda^{-1}: R^n \to X_0$, we conclude that T is closed. Furthermore, T is bounded. It follows from the Heine-Borel theorem that T is compact, and hence, from Theorem 6.1 (part 2), $S = \lambda^{-1}(T)$ is also compact. We have thus established

THEOREM 6.9. *Let* X_0 *be a finite-dimensional subspace of a normed space* X . *Then* $S \subset X_0$ *is compact iff* S *is closed and bounded.* □

6.2.2 *COMPACTNESS IN INFINITE-DIMENSIONAL SPACES*

Theorem 6.9 provides a very satisfactory characterization of compact sets in finite-dimensional spaces. The purpose of the next two theorems is to show that this characterization of compact sets does not extend to infinite-dimensional spaces.

THEOREM 6.10. *Let* X *be a normed space. Let* Y *be a finite dimensional subspace of* X *with* $Y \neq X$. *Then for every* $\varepsilon > 0$ *there exists* $x_0 \in X$ *such that* $\|x_0\| = 1$ *and*

$$\|x_0 - y\| \geq 1 - \varepsilon$$

for all $y \in Y$.

Proof: Y is closed, by Remark 6.7.

Let $\varepsilon \in (0,1)$. Since Y is closed and $Y \neq X$ there is a point $x_1 \in X \setminus Y$ such that

$$\inf_{z \in Y} \|x_1 - z\| = d > 0 \ , \qquad (*)$$

and a point $y_1 \in Y$ such that

$$\|x_1 - y_1\| \leq d/(1-\varepsilon) \ . \qquad (**)$$

Set

$$x_0 = (x_1 - y_1)/\|x_1 - y_1\| \ .$$

Then, $\|x_0\| = 1$. Furthermore, if $y \in Y$, then, using eqns (*) and (**),

$$\begin{aligned} \|x_0 - y\| &= \|(x_1 - y_1) - (\|x_1 - y_1\| y)\| \,/\, \|x_1 - y_1\| \ , \\ &= \|x_1 - z\| \,/\, \|x_1 - y_1\| \ , \\ &\geq d \,/\, \|x_1 - y_1\| \ , \\ &\geq 1 - \varepsilon \ , \end{aligned}$$

where $z = y_1 + \|x_1 - y_1\| y \in Y$. □

THEOREM 6.11. *A necessary and sufficient condition for every bounded set in a normed space* X *to be sequentially compact is that* X *be finite-dimensional.* □

Proof: Assume that X is infinite dimensional. Choose $x_1 \in X$ with $\|x_1\| = 1$ and let X_1 be the subspace spanned by x_1. By Theorem 6.10, there exists $x_2 \in X$ such that $\|x_2\| = 1$ and $\|x_1 - x_2\| \geq 1/2$. In general let X_n be the subspace spanned by $x_1, \ldots, x_n$. Then there exists $x_{n+1} \in X$ such that $\|x_{n+1}\| = 1$ and $\|x_{n+1} - x_m\| \geq 1/2$ for $1 \leq m \leq n$. The sequence $\{x_i\}$ is bounded, but no subsequence is convergent because $\|x_n - x_m\| \geq 1/2$ if $n \neq m$.

On the other hand, if X is finite-dimensional, it follows readily from Theorems 6.3 and 6.9 that every bounded subset of X is sequentially compact. □

We see from Theorem 6.11 that in an infinite-dimensional space a compact set S must be closed and bounded, but must also satisfy some other requirements; loosely speaking, these additional requirements can be thought of as 'regularity' conditions upon the elements of S.

We now consider some specific infinite dimensional spaces:

(1) *COMPACT SETS IN* $C(\bar{\Omega})$; $\bar{\Omega}$ *A COMPACT SET IN* R^n.

We first obtain some preliminary results on uniform continuity. It will be recalled that a function $x \in C(\bar{\Omega})$ is said to be uniformly continuous if, given $\varepsilon > 0$ there exists $\delta > 0$ such that

$$|x(t_1) - x(t_2)| \leq \varepsilon \qquad (*)$$

for all $t_1, t_2 \in \bar{\Omega}$ satisfying

$$\|t_1 - t_2\| \leq \delta . \qquad (**)$$

THEOREM 6.12. *If* x *is a real continuous function on* $\bar{\Omega}$, *a compact subset of* R^n, *then* x *is uniformly continuous on* $\bar{\Omega}$.

Proof. Choose $\varepsilon > 0$. Since, by assumption, x is continuous on $\bar{\Omega}$, for each $v \in \bar{\Omega}$ there exists $\delta(v) > 0$ such that

$$|x(v)-x(t)| \leq \varepsilon/2 \;, \quad \text{if} \quad t \in \bar{\Omega} \quad \text{and} \quad \|v-t\| \leq \delta(v) \;. \qquad (*)$$

The family of neighbourhoods $\kappa = \{N_v : v \in \bar{\Omega}\}$,

$$N_v = B(v;\ \delta(v)/2) = \{t \in R^n : \|t-v\| < \delta(v)/2\} \;,$$

is an open cover of $\bar{\Omega}$. Since $\bar{\Omega}$ is compact, there exists a finite subcover κ' of $\bar{\Omega}$. Set

$$\delta = \frac{1}{2} \min \{\delta(v) : N_v \in \kappa'\} \;.$$

Now let $t_1, t_2 \in \bar{\Omega}$ satisfy $\|t_1 - t_2\| \leq \varepsilon$. Since κ' is a cover of $\bar{\Omega}$, $t_1 \in N_v$ for some $N_v \in \kappa'$. As a consequence, we have that

$$\|t_1 - v\| \leq \delta(v)/2 \;,$$

and

$$\|t_2 - v\| \leq \|t_2 - t_1\| + \|t_1 - v\| \leq \delta + \delta(v)/2 \leq \delta(v) \;.$$

Using eqn (*) it follows that

$$|x(t_1)-x(t_2)| \leq |x(t_1)-x(v)| + |x(v)-x(t_2)| \;,$$

$$\leq \varepsilon/2 + \varepsilon/2 \;. \quad \square$$

REMARK 6.8. The compactness of the 'underlying space' $\bar{\Omega} \subset R^n$ is essential in the proof of Theorem 6.12. If $\bar{G} \subset R^n$ is not compact (that is, not bounded) then it is trivial to give examples of functions which are continuous on $\bar{G}$ but not uniformly continuous; for example, if $x(t) = \sin(t^2)$, then x is continuous on $(-\infty, +\infty)$ but x is not uniformly continuous.

Problems on unbounded domains in R^n are usually substantially more difficult than similar problems on bounded domains. For instance, the Fredholm integral equation of the second kind,

$$x(s) + \int_a^b k(s,t)x(t)dt = f(s) \ , \quad a \le s \le b \ ,$$

is usually readily amenable to numerical methods when a and b are finite (see Atkinson [1976]), but is far more intractable when either a or b is infinite. As Baker [1977, p. 578] remarks, '...the treatment of [these] equations is not completely understood at the present.'

As another example, let Γ be a smooth closed curve in R^2 with interior Ω and exterior G. The exterior Dirichlet problem

$$u_{xx} + u_{yy} = 0 \ , \quad \text{in } G \ ,$$
$$u = g \ , \quad \text{on } \partial G = \Gamma \ ,$$

is more difficult than the interior Dirichlet problem

$$u_{xx} + u_{yy} = 0 \ , \quad \text{in } \Omega \ ,$$
$$u = g \ , \quad \text{on } \partial\Omega = \Gamma \ .$$

These difficulties can be ascribed to the lack of compactness of the underlying space $\overline{G}$. □

Let S be a subset of $C(\overline{\Omega})$. Then S is *equicontinuous* if, given $\varepsilon > 0$, there exists $\delta > 0$ such that

$$|x(t_1)-x(t_2)| \le \varepsilon \ , \tag{6.4}$$

for all $x \in S$ and all t_1, $t_2 \in \overline{\Omega}$ satisfying

$$\|t_1-t_2\| \le \delta \ . \tag{6.5}$$

THEOREM 6.13. *(Ascoli-Arzela)*

Let $\overline{\Omega} \subset R^n$ *be compact, and let* $S \subset C(\overline{\Omega})$.

Then: (1) S *is sequentially compact* $\Leftrightarrow$ *(2)* S *is bounded AND (3)* S *is equicontinuous.*

Proof: *(1) ⇒ (3).* By Theorem 6.3, (part 3) S is totally bounded. For $\varepsilon > 0$ let $x_1, \ldots, x_r$ be points in S forming an ε-net; that is, each $x \in S$ is within a distance ε of at least one x_k. Each of the finitely many functions x_k is continuous on the compact set $\bar{\Omega}$, and hence uniformly continuous. Thus, there exists $\delta > 0$ such that

$$|x_k(t_1) - x_k(t_2)| \le \varepsilon$$

for $1 \le k \le r$ whenever $\|t_1 - t_2\| \le \delta$. If $x \in S$ then

$$|x(t_1) - x(t_2)| \le 2 \min_{1 \le k \le r} \|x - x_k\| + \max_{1 \le k \le r} |x_k(t_1) - x_k(t_2)| ,$$

$$\le 3\varepsilon ,$$

whenever $\|t_1 - t_2\| \le \delta$. That is, S is equicontinuous.

(1) ⇒ (2). Theorem 6.1 (part 4).

(2) AND (3) ⇒ (1). Take $\varepsilon > 0$ and choose δ so as to satisfy the 'equicontinuity conditions' (6.4) and (6.5). Choose a δ-net $t_1, \ldots, t_m$ for $\bar{\Omega}$. Choose $\eta_0, \ldots, \eta_N$ such that

$$-K = \eta_0 < \eta_1 < \eta_2 < \ldots < \eta_N = K, \text{ and } |\eta_{k+1} - \eta_k| \le \varepsilon ,$$

where

$$K = \sup_{x \in S} \|x\| .$$

The case $n = 1$, $\bar{\Omega} = [0,1]$, $m = 3$, and $N = 4$ is illustrated in Figure 6.1.

Each $x \in S$ passes through at most two of the intervals $I_i = [\eta_i, \eta_{i+1}]$ at each point t_k. We construct a finite subset P of S as follows. Given a combination $I_{i_1} \ldots I_{i_k} \ldots I_{i_m}$ of intervals, $0 \le i_k < N$, there may be one or more $x \in S$ such that $x(t_k) \in I_{i_k}$ for $1 \le k \le m$; if so, we adjoin *one* such x to P. (For example, the function x shown in Figure 6.1 could be adjoined to P for the combination $I_1\ I_2\ I_0$.) .

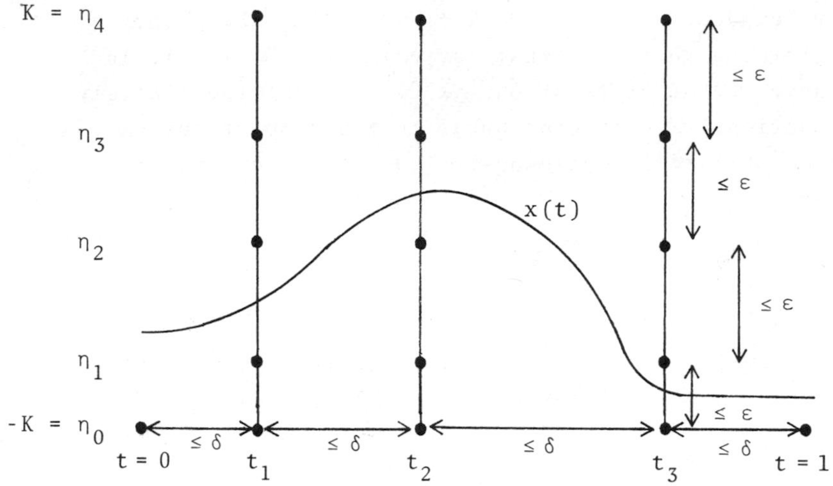

Figure 6.1: The ε-net in [-K,+K] and δ-net in $\overline{\Omega}$ = [0,1] as constructed in the proof of Theorem 6.13

Doing this for each possible combination of intervals, we find that P has at most N^m elements, and if $x \in S$ then there exists $y \in P$ such that

$$\max_{1 \le k \le m} |x(t_k)-y(t_k)| \le \varepsilon \; . \qquad (*)$$

For any $x \in S$ choose $y \in P$ satisfying eqn (*), and let $t \in \overline{\Omega}$. Then there exists t_k such that $\|t-t_k\| \le \delta$.

$$|x(t)-y(t)| \le |x(t)-x(t_k)| + |y(t)-y(t_k)| + |x(t_k)-y(t_k)| \; . \qquad (**)$$

The first and second terms on the right of eqn (**) are less than ε by equicontinuity (eqn(6.4)) while the last term is less than ε by virtue of eqn (*). Since t was arbitrary, we conclude that $\|x-y\| \le 3\varepsilon$. That is, P is a finite 3ε-net for S, so that S is totally bounded.

Finally, $\overline{S}$ is complete because it is a closed subset of the complete space $C(\overline{\Omega})$. Hence, by Theorem 6.3, S is sequentially compact. □

The Ascoli-Arzela theorem can be used to obtain compactness criteria for sets in $C^m(\overline{\Omega})$ (Problems 6.13 and 6.14). The theorem has many applications to the solution of differential and integral equations. (see Section 6.3.3).

(2) *COMPACT SETS IN* $L^p(\Omega)$

The next theorem, the proof of which is given by Adams [1975, p. 31] and outlined by Dunford and Schwartz [1966, p. 301], characterizes compact sets in $L^p(\Omega)$. This theorem will be used when considering integral equations (see Chapter 9).

THEOREM 6.14. *Let* $\Omega \subset R^n$, *and* $1 \leq p < \infty$. *A subset* $S \subset L^p(\Omega)$ *is sequentially compact iff*

(1) S *is bounded.*

(2) For every $\varepsilon > 0$ *there exists* $\delta > 0$ *and a bounded set* $G \subset \Omega$ *such that*

(a) $\overline{G} \subset \Omega$.

(b) For every $x \in S$, *and every* $h \in R^n$ *satisfying* $\|h\| < \delta$,

$$\int_{\Omega \setminus \overline{G}} |x(s)|^p ds < \varepsilon^p, \text{ and } \int_{\Omega} |\tilde{x}(s) - \tilde{x}(s+h)|^p ds < \varepsilon^p, \tag{*}$$

where

$$\tilde{x}(s) = \begin{cases} x(s), & \text{if } t \in \Omega, \\ 0, & \text{otherwise.} \end{cases}$$ □

(3) *COMPACT SETS IN* ℓ^p *AND* c

The following criteria for compactness in ℓ^p and c can be proved by using the equivalence between sequential compactness and total boundedness. (See Theorem 6.3.) The proofs are left to the reader (Problem 6.17).

THEOREM 6.15. *A subset* S *of* c *is sequentially compact* $\iff$

(1) S *is bounded, AND*

(2) For every $\varepsilon > 0$ *there exists* N *such that if* $x = (x_i) \in S$ *then*

$$\max_{i \geq N} |x_i - x_\infty| \leq \varepsilon .$$

□

THEOREM 6.16. *Let* $1 \leq p < \infty$. *A subset* S *of* ℓ^p *is sequentially compact* $\iff$

(1) S *is bounded, AND*

(2) For every $\varepsilon > 0$ *there exists* N *such that if* $x = (x_i) \in S$ *then*

$$\sum_{i=N}^{\infty} |x_i|^p < \varepsilon .$$

□

(4) *COMPACTNESS, MONOTONICITY, AND CONVEXITY*

The assumption of convexity or monotonicity is sometimes sufficient to ensure compactness. We give two examples.

(a) *Compactness in* $K(R^n)$.

In Problem 2.3 we introduced the space $K(R^n)$ consisting of closed bounded sets in R^n , and defined a metric ρ . We have (Eggleston [1963, p. 64]):

THEOREM 6.17. *Blaschke Selection Theorem*

Every infinite sequence of closed non-empty convex subsets of a bounded portion of R^n *contains an infinite subsequence which converges (in* $K(R^n)$*) to a closed non-empty convex subset.* □

(b) *Compactness in* $\prod_{\alpha\in[0,1]} R^1$.

In Examples 2.6 and 2.15 we introduced the space of real functions defined on $[0,1]$, denoted by $\prod_{\alpha\in[0,1]} R^1$.

THEOREM 6.18. *(Helly)*

Let S *be a collection of real functions* x *defined on* $[0,1]$ *such that*

(1) Each $x \in S$ *is monotone. That is,*

$$x(t_1) \ge x(t_2) \quad \textit{if} \quad t_1 \ge t_2 .$$

(2) S *is bounded. That is, there is a constant* K *such that*

$$|x(t)| \le K \quad \textit{for} \quad x \in S \quad \textit{and} \quad t \in [0,1] .$$

Then, if $\{x_n\}$ *is any sequence of functions in* S *there exists a monotone function* x *and a subsequence* $\{\tilde{x}_n\}$ *such that, for each* $t \in [0,1]$, $\tilde{x}_n(t) \to x(t)$. □

A proof of Theorem 6.18 is given by Natanson [1954, p. 224]. When the functions are continuous, there is an interesting alternative approach based on rotating the coordinate axes by 45° (Problem 6.18).

Theorems 6.17 and 6.18 have applications in the solution of free boundary problems (Garabedian [1964, p. 580]).

(5) *WEAK COMPACTNESS*

Just as life is full of sorrows so is mathematics. There are many sets which we would dearly love to be compact but which are not. In life, the young man seeking a blonde for a wife must sometimes weaken his conditions to allow brunettes and red-heads. Similarly, by weakening the topology we enlarge the class of compact sets (Problem 6.19). The concept of weak compactness will be introduced in Volume 2.

6.3 *APPLICATIONS OF COMPACTNESS*

We discuss briefly a variety of applications.

6.3.1. *OPTIMIZATION*

The basic *minimization problem* in optimization is as follows: Given a set V in a topological space X and a mapping $f : V \to R^1$, find

$$m = \inf_{x \in V} f(x) \in \overline{R}^1 = [-\infty, +\infty] . \qquad (6.6)$$

V is called the *feasible region* and f is called the *objective function*. If $m = f(x_b)$ for some $x_b \in V$, we say that the minimization problem (6.6) has a *solution* x_b, and that f *attains its infimum* at x_b. To maximize f on V, we minimize $-f$.

In certain arguments it is necessary to consider V as a topological space, in which case V is given the topology induced on it by X (see Problem 2.2).

Minimization problems occur in every branch of human endeavour, and the solution of minimization problems in finite-dimensional spaces is the goal of the subject 'linear and nonlinear programming'. Problems in numerical analysis which can be reformulated as minimization problems include: the approximation of functions with minimum error (see Section 6.3.2); the solution of ordinary and partial differential equations using variational methods; and the construction of optimum quadrature formulae.

Given a minimization problem (6.6), there exists a sequence $\{x_n\}$ in V such that

$$f(x_n) \downarrow \inf_{x \in V} f(x) = m , \quad \text{as} \quad n \to \infty \qquad (6.7)$$

It is tempting to conclude from eqn (6.7) that if V is closed and bounded, f is continuous and m is finite then there exists x_b satisfying eqn (6.6). This erroneous conclusion was made by several distinguished nineteenth century mathematicians, including Gauss, Lord Kelvin, and Dirichlet, in attempting to prove the existence of the solution of the Dirichlet problem. As the following example shows, additional assumptions about V are needed:

EXAMPLE 6.1.

$$X = C[0,1] ,$$

$$V = \{x \in X: x(0) = 0, x(1) = 1, ||x|| \le 1\} ,$$

$$f(x) = \int_0^1 [x(t)]^2 dt .$$

Then V is closed, convex, and bounded, f is continuous, and

$$\inf_{x \in V} f(x) = 0 ,$$

but there is no $x_b \in V$ satisfying $f(x_b) = 0$. □

If V is compact and X is a metric space then it follows from eqn (6.7) that there exists a subsequence $\{\tilde{x}_n\}$ converging to a point $x_b \in V$. If, furthermore, f is continuous, then

$$f(x_b) = \lim f(\tilde{x}_n) ,$$

so that

$$\inf_{x \in V} f(x) = f(x_b) > -\infty .$$

We have thus proved (for metric spaces) .

THEOREM 6.19. *A continuous real function attains its infimum and its supremum on a compact set.* □

A mapping f from a topological space V to R^1 is *lower semi-continuous (upper semi-continuous)* if the set

$$f^{-1}(t,\infty) = \{x \in V: f(x) > t\}$$

$$(f^{-1}(-\infty,t) = \{x \in V: f(x) < t\})$$

is open for all real t .

In Volume 2 we will encounter minimization problems for functions which are lower semi-continuous, but not continuous, and this case is covered by the following generalization of Theorem 6.19; the proof of which is left to the reader (Problem 6.21). (An application is given on page 346 - see also Problem 9.46).

THEOREM 6.20. *A real lower (upper) semi-continuous function attains its infimum (supremum) on a compact set.* □

The theory of optimization is simplified if the constraint set V and objective function f are convex. We recall that a subset V of a vector space is *convex* if $\lambda x_1 + (1-\lambda)x_2 \in V$ for all $\lambda \in [0,1]$ and $x_1, x_2 \in V$.

Let V be a convex set and let $f: V \to R^1$. f is *convex* if

$$(1-\lambda)f(x_1) + \lambda f(x_2) \geq f((1-\lambda)x_1 + \lambda x_2), \quad \lambda \in [0,1] \text{ and } x_1, x_2 \in V .$$

f *is concave* if $-f$ is convex. f is *strictly convex at* $x_1 \in V$ if

$$(1-\lambda)f(x_1) + \lambda f(x_2) > f((1-\lambda)x_1 + \lambda x_2), \quad x_2 \neq x_1 ,$$

for all $\lambda \in (0,1)$ and all $x_2 \in V$. f is *strictly concave at* $x_1 \in V$ if $-f$ is strictly convex at $x_1 \in V$.

EXAMPLE 6.2. There is a close connection between convex sets and convex functions: if $V \subset X$ is convex then $f : V \to R^1$ is convex iff its *epigraph*

$$\text{epi } f = \{(x,t) \in X \times R^1 : x \in V, t \geq f(x)\} ,$$

is a convex set in $X \times R^1$.

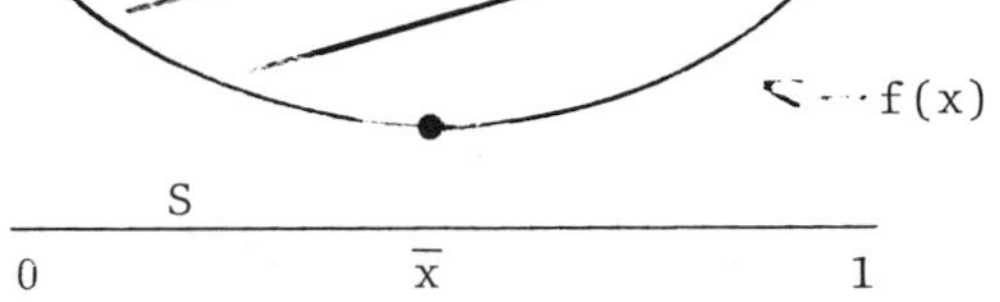

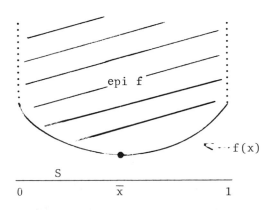

Figure 6.2: A convex function $f : V = [0,1] \to R^1$ *which is strictly convex at* $\bar{x}$. □

The proof of the following theorem is straightforward and is left to the reader (Problem 6.23).

THEOREM 6.21. *Let* V *be a convex set in a topological vector space* X . *Let* S *denote the, possibly empty, set of solutions of the minimization problem (6.6). Then*

(1) If f *is convex then* S *is convex.*

(2) If $x \in S$ *and* f *is strictly convex at* x *, then* x *is the unique solution of the problem (6.6).*

(3) If f *is concave and not constant then* S *is a subset of the boundary points* ∂V *of* V , $\partial V = \overline{V} \cap (\overline{X \backslash V})$.

(4) If f *is convex,* $y_b \in V$ *,* N *is a neighbourhood of* y_b *, and* y_b *solves the local minimization problem,*

$$f(y_b) = \inf_{y \in N \cap V} f(y) ,$$

then $y_b \in S$. □

REMARK 6.9. From part 3 of Theorem 6.21, if V is convex and f is linear, then every solution of the minimization problem lies on the boundary of V, as one might perhaps expect from intuition.

This leads to the following observation. When formulating a problem it is often the case that some aspects of the original problem cannot be incorporated into the minimization problem. If the solution x_b of the minimization problem is on the boundary of the feasible set V, it may very well happen that x_b has undesirable properties. For example, if we construct a numerical method for solving differential equations by choosing V to be a family of numerical methods parameterized by x, letting $f(x)$ be the truncation error, and then determining x so as to minimize f, then the resulting numerical method will have minimum truncation error, but may well be unstable. For example, the Dahlquist theory for linear multistep methods for ordinary differential equations states that k-step methods of order 2k can be constructed, but only methods of order less than k+3 can be stable (Henrici [1962, p. 229]). □

For an introduction the vast literature on optimization see Ekeland and Teman [1976].

6.3.2. *APPROXIMATION*

An important special case of the minimization problem (6.6) considered in the preceding section arises in normed spaces when the objective function $f(y)$ is taken to be the distance from $y \in V$ to a given point $x \in X$. Then the problem takes the form: Given $x \in X$ find $x_b \in V$ such that

$$||x_b - x|| = \inf_{y \in V} ||y - x|| \equiv d(x,V) . \tag{6.8}$$

Any x_b satisfying eqn (6.8) will be called a *(point of) best approximation to* x *(in* V*)*. If X and V are such that the solution x_b of (6.8) exists and is unique, we may introduce the *proximity mapping of* x *onto* V denoted by

$$\text{prox}(V,\cdot) : x \in X \to x_b \in V . \tag{6.9}$$

An elementary, but very important, result is

THEOREM 6.22. *Let* X_0 *be a finite-dimensional linear subspace in a normed space* X. *For each* $x \in X$ *there exists an* $x_b \in X_0$ *(not necessarily unique) which is closest to* x *; that is such that*

$$||x - x_b|| = d(x,X_0) \equiv \inf_{y \in X_0} ||x - y|| .$$

If the norm $||\cdot||$ *is strictly convex then* x_b *is unique.* □

Proof. Let

$$V = \{y \in X_0 : ||y - x|| \le 2d(x,X_0)\} .$$

Then V is a bounded and closed subset of the finite-dimensional subspace X_0. Hence, from Theorem 6.9, V is compact. Applying Theorem 6.19, we conclude that there exists $x_b \in V$ such that

$$||x_b - x|| = \inf_{y \in V} ||x - y|| = \inf_{y \in X_0} ||x - y|| .$$

We denote $||x_0 - x|| = d(x,X_0)$ by d.

To prove the last part of the theorem, we assume that $||\cdot||$ is strictly convex; that is, if

$$||u+v|| = ||u|| + ||v|| , \qquad (*)$$

then either $u = 0$ or $v = \lambda u$ for some λ. If $y_b \in X_0$ and $||x-y_b|| = d$ let $u = (x-x_b)/2$, $v = (x-y_b)/2$; then

$$d = ||x-x_b|| = ||x-y_b|| = 2||u|| = 2||v|| .$$

Set $w = (x_b+y_b)/2$. Then $w \in X_0$ so that

$$||u+v|| = ||x-w|| \geq d = ||u|| + ||v|| \geq ||u+v|| ,$$

which implies that eqn (*) holds.

It now follows readily that $x_b = y_b$ (Problem 6.24). □

EXAMPLE 6.3. Let

$$X = L^p[0,1] , \quad 1 \leq p \leq \infty .$$

$$X_0 = \pi_n = \text{set of real polynomials of degree at most } n .$$

It follows from Theorem 6.22 that for each $x \in X$ there exists an approximation $x_b \in \pi_n$ which minimizes $||x-x_b||$. Remembering that L^p is strictly convex if $1 < p < \infty$, we conclude that the approximation x_b is unique if $1 < p < \infty$. If $p = 1$ then x_b need not be unique as shown by the example: $n = 0$,

$$x(t) = \begin{cases} 0 , & 0 \leq t < \tfrac{1}{2} , \\ 1 , & \tfrac{1}{2} \leq t \leq 1 . \end{cases}$$

The minimum distance d from x to π_0 is $\tfrac{1}{2}$, and this is attained by each of the functions $x_\alpha : t \to \alpha$ for any $\alpha \in [0,1]$. □

The single most important approximation problem in numerical analysis is probably the approximation of a function x in $C[a,b]$ by elements of π_n, the subspace of polynomials of degree at most n . The existence of polynomials of best approximation follows from Theorem 6.22, but $C[a,b]$ is not strictly convex (see Problem 4.11) and hence we cannot infer uniqueness. Indeed, for some finite dimensional subspaces of $C[a,b]$, the best approximation is not unique (Problem 6.25). However, π_n possesses the *Haar property* which ensures that polynomials of best approximation in π_n are unique.

There are many texts on approximation theory where further information can be found: see, for example, Lorentz [1966], Cheney [1966], Natanson [1955].

6.3.3. *SOLUTION OF ORDINARY DIFFERENTIAL EQUATIONS*

The proof of the basic Cauchy-Peano existence theorem for the initial value problem for ordinary differential equations uses compactness in an essential way:

THEOREM 6.23. *(Cauchy-Peano)*

Let $\tau,\xi \in R^1$ *and* $a,b \in R^1_+$ *be given, and set*

$$X = [\xi-b,\xi+b] ,$$

$$\overline{\Omega} = [\tau,\tau+a] \times X .$$

Let

$$f : (t,y) \in \overline{\Omega} \to f(t,y) \in R^1$$

be continuous.

Then there exists $\alpha > 0$ *and* $x \in C^1[\tau,\tau+\alpha]$ *such that*

$$\begin{aligned} \dot{x}(t) &= f(t,x(t)) , \quad \tau \le t \le \tau + \alpha , \\ x(\tau) &= \xi . \end{aligned} \tag{6.10}$$

x *is the limit of a subsequence of the approximations* x_n *obtained by applying Euler's method to the initial value problem (6.10).*

Proof. The proof proceeds in several steps:

Step 1: Let $M = \max_{\overline{\Omega}} |f|$, and set $\alpha = \min[a, b/M]$. Throughout the remainder of the proof it will be assumed that

$$\tau \le t \le \tau + \alpha .$$

Step 2: Construct $x_n \in C[\tau, \tau+\alpha]$ for $n = 2^k$, $k = 0,1,2...$ as follows:

(a) $t_{n,i} = \tau + i\alpha/n$, for $0 \le i \le n$.

(b) $x_n(t_{n,o}) = x_n(\tau) = \xi$.

(c) $x_n(t_{n,i+1}) = x_n(t_{n,i}) + \frac{\alpha}{n} f(t_{n,i}, x_n(t_{n,i}))$, (6.11)

for $0 \le i \le n - 1$.

(d) x_n is piecewise linear on $[\tau, \tau+\alpha]$.

That is, x_n is the numerical solution of the initial value problem (6.10) obtained by first applying *Euler's method* with stepsize $h = \alpha/n$ and then interpolating linearly between the gridpoints.

Step 3: It is necessary to show that x_n can be constructed as described in (2) above. The only difficulty which can arise is that $x(t_{n,i}) \notin [\xi-b, \xi+b]$ since then $f(t_{n,i}, x(t_{n,i}))$ is not defined. However, by induction on i we can see that

$$(t_{n,i}, x_n(t_{n,i})) \in \overline{\Omega} ,$$

$$|f(t_{n,i}, x(t_{n,i}))| \le \max_{\overline{\Omega}} |f| \le M , \qquad (6.12)$$

$$|x_n(t_{n,i+1}) - \xi| \le |x_n(t_{n,i}) - \xi| + \alpha M/n \le (i+1)M\alpha/n \le b ,$$

for $0 \le i \le n - 1$.

Consequently, x_n is defined for $t \in [\tau, \tau+\alpha]$. Furthermore, x_n has absolute slope at most M .

As an illustration, in Figure 6.3 the broken line AD represents x_4 . x_4 lies below AB (which has slope + M) and above AC (which has slope -M).

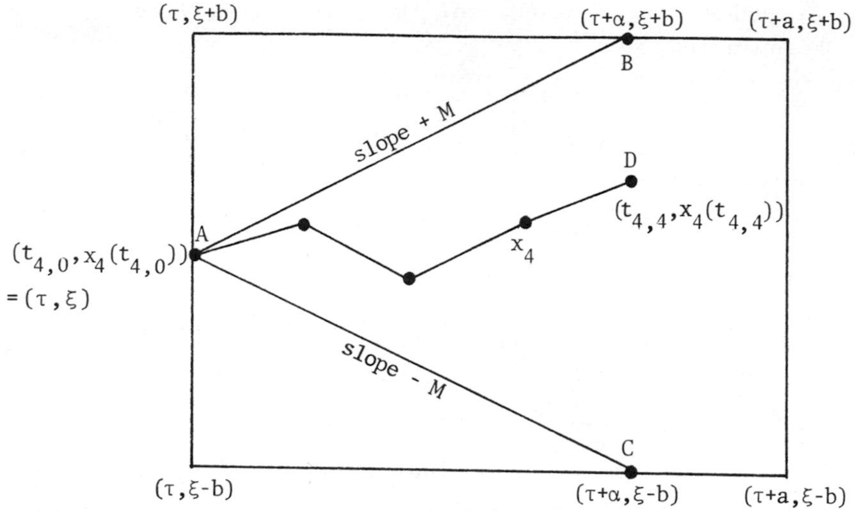

Figure 6.3. Construction of x_4 *in the proof of the Cauchy-Peano theorem (Theorem 6.23).*

Step 4: Let

$$S = \{x_n : n=1,2,\ldots\} \subset C[\tau,\tau+\alpha] \ . \tag{6.13}$$

Since each x_n has slope at most M, S is a bounded and equicontinuous subset of $C[\tau,\tau+\alpha]$. Hence, by the Ascoli-Arzela criterion (Theorem 6.13) S is sequentially compact, so that there exists $x \in C[\tau,\tau+\alpha]$ and a convergent subsequence $\{x_{n_r}\}$ such that

$$\|x_{n_r} - x; C[\tau,\tau+\alpha]\| \to 0 \ , \quad \text{as} \quad r \to \infty \ .$$

Step 5: We assert that

$$x(t) = \xi + \int_\tau^t f(s,x(s))ds \ , \quad \tau \le t \le \tau + \alpha \ . \tag{6.14}$$

To prove eqn (6.14) we first consider the special case when $t = \bar{t}$, where $(\bar{t}-\tau)/\alpha$ can be represented exactly as a binary fraction. Remembering that n_r is a power of 2, it follows, for r sufficiently large, that $\bar{t} = t_{n_r, i_r}$ and we assume henceforth that this is the case. Summing eqn (6.11c) we obtain

$$x_{n_r}(\bar{t}) = \xi + \frac{\alpha}{n_r} \sum_{t_{n_r,i}<\bar{t}} f(t_{n_r,i}, x_{n_r}(t_{n_r,i})) ,$$

$$= \xi + \Big[\frac{\alpha}{n_r} \sum_{t_{n_r,i}<\bar{t}} f(t_{n_r,i}, x(t_{n_r,i}))\Big] +$$

$$+ \Big[\frac{\alpha}{n_r} \sum_{t_{n_r,i}<\bar{t}} f(t_{n_r,i}, x_{n_r}(t_{n_r,i})) -$$

$$- f(t_{n_r,i}, x(t_{n_r,i}))\Big] ,$$

$$= \xi + J_1(n_r) + J_2(n_r) , \text{ say} . \qquad (6.15)$$

We consider the individual terms in eqn (6.15) as $n_r \to \infty$:

(a) On the left, $x_{n_r}(\bar{t}) \to x(\bar{t})$.

(b) On the right, $J_1(n_r)$ is the rectangular rule approximation to

$$J(\bar{t}) = \int_{\tau}^{\bar{t}} f(s,x(s))ds .$$

Since $f(s,x(s))$ is continuous, it follows, as in Section 5.3 (Theorem 5.5), that $J_1(n_r) \to J(\bar{t})$.

(c) Choose $\varepsilon > 0$. By assumption $f(t,y)$ is continuous on $\bar{\Omega} = [\tau,\tau+\alpha] \times [\xi-b,\xi+b]$, so that (by Theorem 6.12) f is uniformly continuous on $\bar{\Omega}$. In particular, there exists $\delta > 0$ such that

$$|f(t,y_1) - f(t,y_2)| \le \varepsilon$$

when $y_1, y_2 \in [\xi-b,\xi+b]$, $|y_1-y_2| \le \delta$, and $t \in [\tau,\tau+\alpha]$. For large r,

$$|x_{n_r}(t)-x(t)| \le \delta , \quad t \in [\tau,\tau+\alpha] ,$$

and so

$$|J_2(n_r)| \le \frac{\alpha}{n_r} \sum_{t_{n_r,i}<\bar{t}} |f(t_{n_r,i},x_{n_r}(t_{n_r,i}))-f(t_{n_r,i},x(t_{n_r,i}))| ,$$

$$\le \frac{\alpha}{n_r} \sum_{i=0}^{n_r-1} \varepsilon ,$$

$$= \alpha\varepsilon .$$

Since ε was arbitrary, we have proved that $J_2(n_r) \to 0$.

Combining (a), (b), and (c) above, we see that eqn (6.14) follows for $t = \bar{t}$ by letting $r \to \infty$ in eqn (6.15).

Both sides of eqn (6.14) are continuous functions of t. The points $\bar{t}$ are dense in $[\tau,\tau+\alpha]$, and so eqn (6.14) holds for all $t \in [\tau,\tau+\alpha]$.

Step 6: The integrand $f(s,x(s))$ on the right of eqn (6.14) is continuous. Therefore, the right hand side (and hence also the left hand side) of eqn (6.14) is continuously differentiable. Differentiating, we conclude that x satisfies the initial value problem (6.10). □

The basic idea in the above proof of the Cauchy-Peano theorem is to construct a sequence of numerical approximations using a very simple numerical method (Euler's method) and then invoke compactness to establish the existence of a convergent subsequence. This approach is readily extended to a variety of problems including (Problem 6.26):

(1) Systems of ordinary differential equations.

(2) Volterra integral equations.

(3) Delay differential equations.

The initial value problem (6.10) can be generalized as follows. Let X be a real Banach space and let

$$f: [\tau,\tau+a] \times X \to X \tag{6.16}$$

be continuous and bounded. Let

$$x: [\tau,\tau+a] \to X . \tag{6.17}$$

We say that x is *differentiable* at the point $t \in (\tau,\tau+a)$ and has *derivative* $\dot{x}(t) \in X$ if

$$\left\| \frac{x(t+h)-x(t)}{h} - \dot{x}(t) \right\| \to 0 \quad \text{as} \quad h \to 0 . \tag{6.18}$$

If x is continuous then we may introduce the integral

$$\int_a^b x(s)ds = \lim_{n\to\infty} h \sum_{i=0}^{n-1} x(\tau+ih) , \tag{6.19}$$

where $h = (b-a)/n$.

We now consider an initial value problem which generalizes problem (6.10); Find $x : [\tau,\tau+a] \to X$ such that

$$\begin{aligned} x(\tau) &= \xi \in X , \\ \dot{x}(t) &= f(t,x(t)) , \quad \tau < t < \tau + a . \end{aligned} \tag{6.20}$$

Given the initial value problem (6.20), we can construct a sequence of approximations x_n , exactly as in the proof of the Cauchy-Peano theorem. However, the proof breaks down at Step 4 because, in general, $S = \{x_n : n=1,2...\}$ is not compact. This is a very significant difference between the initial value problem for ordinary differential equations in R^n and other initial value problems.

For a further discussion of differential equations in Banach spaces see Ladas and Lakshmikantham [1972], Cellina [1972], Martin [1976], and Deimling [1977]. Example 6.4 below illustrates the fact that the problem (6.20) may not have a solution.

EXAMPLE 6.4. Let $X = c_0$, the space of real sequences $x = (x_n)$ with limit 0, so that $x_n \to 0$ as $n \to \infty$. As usual, $\|x\| = \sup_n |x_n|$. Set $\xi = (0) = (0,0,..) \in c_0$, $\tau = 0$.

Define $f : X \to X$ by

$$f((x_n)) = (|x_n|^{\frac{1}{2}} + n^{-1}) . \tag{*}$$

If $x = (x_n)$ and $y = (y_n$ then

$$\|f(x)-f(y)\| = \sup_n \left| |x_n|^{\frac{1}{2}} - |y_n|^{\frac{1}{2}} \right| ,$$

from which it follows easily that (Problem 6.38)

$$\|f(x)-f(y)\| \le \sqrt{\varepsilon} , \quad \text{if} \quad \|x-y\| \le \varepsilon , \tag{6.21}$$

so that f is continuous.

For this choice of X, ξ, τ, and f, the problem (6.20) has no solution. For, if $x(t) = (x_n(t))$ were a solution of problem (6.20) then we would have

$$\begin{aligned} x_n(t) &= |x_n(t)|^{\frac{1}{2}} + \frac{1}{n} , \; t > 0 \\ x_n(0) &= 0 . \end{aligned} \tag{**}$$

for $n = 1,2,...$. It follows from eqn (**) that $x_n(t)$ is increasing and non-negative. Furthermore,

$$\dot{x}_n(t) > [x_n(t)]^{\frac{1}{2}} ,$$

so that

$$\frac{d}{dt} [x_n(t)]^{\frac{1}{2}} > \frac{1}{2} ,$$

or, since $x_n(0) = 0$,

$$x_n(t) > \frac{1}{4} t^2 .$$

Thus, $x_n(t) \not\to 0$ as $n \to \infty$, and so $x(t) \notin c_0$. □

REMARK 6.10. The Cauchy-Peano theorem does not assert that the entire sequence of approximations x_n is convergent, and indeed this is not necessarily the case as we now show.

The scalar problem

$$\begin{aligned} x(0) &= 0 , \\ \dot{x}(t) &= \tfrac{3}{2}[x(t)]^{1/3} , \quad t > 0 , \end{aligned} \tag{*}$$

has three solutions: $x_0(t) = 0$; $x_+(t) = t^{3/2}$; $x_-(t) = -t^{3/2}$, but, if the problem (*) is solved numerically using Euler's method, only the solution x_0 is found.

Consider now the perturbed problem

$$\begin{aligned} x(0) &= 0 , \\ \dot{x}(t) &= \tfrac{3}{2}[x(t)]^{1/3} + g(t) , \end{aligned} \tag{**}$$

where

$$g(t) = \begin{cases} \frac{1}{8} t^{1/2} \sin(\pi/(2t)), & \text{if } t > 0 , \\ 0 , & \text{if } t = 0 . \end{cases}$$

Let x_n denote the approximate solution of eqn (**) obtained using Euler's method with stepsize $h = 1/n$. Then

$$x_n(0) = 0 ,$$

$$x_n(h) = 0 ,$$

$$x_n(2h) = hg(h) = \begin{cases} \frac{1}{8} h^{3/2} , & \text{if } n = 4m+1 , \\ 0 , & \text{if } n = 2m , \\ -\frac{1}{8} h^{3/2} , & \text{if } n = 4m-1 , \end{cases}$$

where m is an integer. Then (Problem 6.27)

$$x_n(ih) \begin{cases} \geq \frac{1}{8}(ih)^{3/2} , & \text{if } n = 4m+1 , \\ = 0 , & \text{if } n = 2m , \\ \leq -\frac{1}{8}(ih)^{3/2} , & \text{if } n = 4m-1 \end{cases} \tag{6.22}$$

□

if $i \geq 3$.

6.4 *COMPACT OPERATORS*

DEFINITION 6.1. An operator T mapping a topological space X into a topological space Y is *compact* if T maps every topologically bounded set U in X into a precompact set $T(U) \subset Y$. If T is continuous and compact it is *completely continuous*. □

In applications, X and Y are usually metric spaces in which case $T : X \to Y$ is compact iff $T(U)$ is sequentially compact for every bounded set U. Furthermore, if T is linear and X is a normed space, then T is compact iff it maps the unit ball in X into a sequentially compact set in Y.

Some examples of compact operators are given in Problems 6.31 to 6.34.

We will encounter many applications of compact operators in Chapter 9. Here, we merely summarize some of the basic 'topological' properties of compact operators. As a preliminary step, we obtain some properties of compact sets in topological vector spaces:

THEOREM 6.24. *Let* X *be a topological vector space. Then*

(1) *If* $V \subset X$ *is (pre) compact then* αV *is (pre) compact for any scalar* α .

(2) *If* $V \subset X$ *is compact and* $W \subset X$ *is closed then* $V + W$ *is closed.*

(3) *If* V *and* W *are (pre) compact subsets of* X *then* $V + W$ *is (pre) compact.*

Proof. We give the proof for the general case. If X is a metric space, then a much simpler proof can be given by using the equivalence between sequential compactness and and precompactness (Problem 6.28).

(1): Since $\overline{\alpha V} = \alpha \overline{V}$ (Problem 3.11) it suffices to show that $\alpha \overline{V}$ is compact. But $\alpha \overline{V}$ is the continuous image

of the compact set $\overline{V}$ and is, therefore, compact (Theorem 6.1 (part 2) and Remark 6.1).

(2): To prove that $V + W$ is closed, we prove that its complement $(V+W)^c$ is open. Let $z \in (V+W)^c$. Let β be a local base of neighbourhoods of $\underline{0}$ in X. Let $v \in V$. Then

$$\{z\} \cap (v+W) = \emptyset .$$

W is closed and so $v + W$ is closed; hence $(v+W)^c$ is open and there exists $M_v \in \beta$ such that

$$(z+M_v) \cap (v+W) = \emptyset . \tag{*}$$

Using Theorem 2.5 (parts (b) and e)), and eqn (*), we can show that there exists $N_v \in \beta$ such that $N_v \subset C_v$ and

$$N_v + N_v \subset C_v + C_v \subset M_v ,$$

where C_v is a circled set. Hence,

$$(z+N_v) \cap ((v+N_v) + W) = \emptyset . \tag{**}$$

Now, $\kappa = \{v+N_v : v \in V\}$ is an open cover of V. Choosing a finite subcover κ' we see from eqn (**) that

$$z + \bigcap_{v+N_v \in \kappa'} N_v \subset (V+W)^c ,$$

so that $(V+W)^c$ is open.

(3): $\overline{V}$ and $\overline{W}$ are compact. Using (2) above and Problem 3.11 we see that

$$\overline{V} + \overline{W} \subset \overline{V + W} \subset \overline{\overline{V} + \overline{W}} = \overline{V} + \overline{W} .$$

It therefore suffices to show that $\overline{V} + \overline{W}$ is compact.

Let κ be an open cover of $\overline{V} + \overline{W}$. Choose $v \in V$. For each $w \in \overline{W}$, we know that

$$v + w \in S(v,w)$$

for some $S(v,w) \in \kappa$. It follows from Theorem 2.5 that there is an open neighbourhood $N(v,w)$ of the origin in

X such that

$$(v+N(v,w)) + (w+N(v,w)) \subset S(v,w) \ .$$

The sets $w + N(v,w)$, $w \in \overline{W}$, form an open cover of $\overline{W}$; choosing a finite subcover, we conclude that

$$(v+N(v)) + \overline{W} \subset S(v) \ ,$$

where $N(v)$ is the intersection of a finite number of the open sets $N(v,w)$, and $S(v)$ is the union of a finite number of the open sets $S(v,w)$. The sets $v + N(v)$, $v \in \overline{V}$, form an open cover of $\overline{V}$; choosing a finite subcover, we conclude that

$$\overline{V} + \overline{W} \subset S$$

where S is the union of a finite number of the $S(v)$, and hence a finite number of the sets $S(v,w) \in \kappa$. □

THEOREM 6.25.

(1) *Let* X *be a topological space and* Y *a topological vector space; if* T_1 *and* T_2 *are compact operators mapping* X *into* Y *, then* $\alpha_1 T_1 + \alpha_2 T_2$ *is compact.*

(2) *If* $T_1 : X \to Y$ *is compact and* $T_2 : Y \to Z$ *is continuous, then* $T_2 T_1 : X \to Z$ *is compact.*

(3) *Let* X *and* Y *be normed spaces. If* $T_1 : X \to Y$ *is a continuous linear operator and* $T_2 : Y \to Z$ *is compact, then* $T_2 T_1 : X \to Z$ *is compact.*

(4) *Let* X *and* Y *be Banach spaces. Let* $\{T_n\}$ *be a sequence of compact linear operators mapping* X *into* Y *. If* $T: X \to Y$ *is continuous and linear, and* $\|T_n - T\| \to 0$ *as* $n \to \infty$ *, then* T *is compact.*

Outline of proof. (1): Follows from Theorem 6.24 (parts 1 and 3).

(2): See Theorem 6.1 (part 2).

(3): If U is bounded then $T_1(U)$ is bounded.

(4): Choose a bounded set U . Choose T_n such that

$$\|T-T_n\| \le \varepsilon \, (\sup_{u\in U} \|u\|)^{-1} ,$$

and choose an ε-net $y_1,\ldots,y_m$ for $T_n(U)$. Then $y_1,\ldots,y_m$ is a 2ε-net for $T(U)$. □

PROBLEMS

6.1 Let S be a subset of a topological space X . Prove: S is compact $\Leftrightarrow$ S is a compact space with respect to its relative topology. (Remark 6.1, p. 187).

6.2 Let $\{x_n\}$ be a sequence in a metric space. Prove: $\{x_n\}$ has a subsequence which converges to a point x $\Leftrightarrow$ every neighbourhood of x contains infinitely many of the points x_n . (Remark 6.2, p. 188)

6.3 Prove that if Y_α is a compact subset of X_α for $\alpha \in A$ then $Y = \prod_{\alpha\in A} Y_\alpha$ is a compact subset of $X = \prod_{\alpha\in A} X_\alpha$. (Remark 6.4, p. 194).

6.4 Prove that, for $S \subset R^1$, if S is sequentially compact then S is bounded (Theorem 6.5, p. 194).

6.5 Prove the one-dimensional Heine-Borel theorem (Theorem 6.6) using the following line of argument. Assume that $S \subset R^1$ is bounded and closed but not compact, so that there is an open cover κ of S with no finite subcover. As in the proof of Theorem 6.5, construct a sequence of intervals $[a_i,b_i]$ such that: (1) $S \cap [a_i,b_i]$ has no finite subcover; (2) $|a_i-b_i| \to 0$; and (3) $b_i \to b$. Show that this leads to a contradiction. {195}

6.6 Taking the proof of Tychonoff's theorem as a model, prove directly that a bounded sequence $\{x^{(k)}\}$ in E^n has a convergent subsequence. (Remark 6.5, p. 195).

6.7 Let κ be an open cover of the unit square $[0,1] \times [0,1]$ in R^2. Use the fact that $[0,1]$ is compact to show that: for each $x \in [0,1]$ there exists a finite subset κ_x of κ such that

$$L_x = \{(x,y) \in E^2 : \ y \in [0,1]\} \subset \bigcup_{G \in \kappa_x} G .$$

Hence show that κ contains a finite subcover of the unit square. (Remark 6.5, p. 195).

6.8 Complete the proof of Theorem 6.8 by showing that if $\lambda : X_0 \to E^n$ is as defined in the proof of the theorem then λ is linear and bijective, and λ^{-1} is continuous. {197}

6.9 Let $x_1,\ldots,x_n$ be linearly independent elements of a real normed space X. Show that: a sequence $\{y^k\}$,

$$y^k = \sum_{i=1}^{n} \lambda_i^{(k)} x_i ,$$

is bounded in $X \iff$ the sequences $\{\lambda_i^{(k)}\}$ are bounded in R^1, $i=1,\ldots,n$.

6.10 Prove that every complex n-dimensional normed space X is isomorphic to $\mathbb{C}^n$.

6.11 Prove that: (a) every finite-dimensional normed space X_0 is a Banach space; (b) every finite-dimensional subspace X_0 of a normed space X is closed. (Remark 6.7, p. 198).

6.12 In the representation (6.1),

$$x = \sum_{i=1}^{n} \lambda_i(x) x_i ,$$

of an element x in terms of basis elements x_i, prove that $\lambda_i : X \to R^1$ is continuous.

6.13 Use the Ascoli-Arzela theorem to show that: a subset S of $C^m[a,b]$ is sequentially compact $\Longleftrightarrow$

(1) S is bounded, AND

(2) $S_m = \{D^m x : x \in S\}$ is equicontinuous.

6.14 Let Ω be an open bounded subset of R^n. Show that: a subset S of $C^m(\overline{\Omega})$ is sequentially compact $\Longleftrightarrow$

(1) S is bounded, AND

(2) the set $S_\alpha = \{D^\alpha x : x \in S\}$ is equicontinuous for $|\alpha| \le m$.

[Hint: Given a sequence $\{x^{(k)}\}$ in S show that there is a subsequence $\{\tilde{x}^{(k)}\}$ such that $D^\alpha \tilde{x}^{(k)} \to F_\alpha \in C(\overline{\Omega})$ for $|\alpha| \le m$. Conclude by showing that $F_\alpha = D^\alpha F$ for $|\alpha| \le m$ so that $F \in C^m(\overline{\Omega})$.]

6.15 Prove that the family of functions $S = \{x_n : t \to t^n;\ n = 1,2,\ldots\}$ is not a sequentially compact subset of $C[0,1]$.

6.16 Consider the family S of functions x_n, $n = 0,1,\ldots$, where

$$x_n : t \in R^1 \to e^{-(n-t)^2} \in R^1 .$$

Show that S is a bounded and equicontinuous subset of $C(-\infty,+\infty)$, but that S is not sequentially compact.

6.17 Prove Theorems 6.15 and 6.16 which give criteria for sequential compactness in c and ℓ^p. {206}

6.18 (1) Let x be a continuous monotonic real function on $[0,1]$, with $x(0) = a$ and $x(1) = b$, so that $y = x(t)$ is a continuous curve, C say, in the yt-plane. Introduce coordinates $Y = y - t$, $T = y + t$. Show that C can be represented as a curve $Y = X(T)$, $a \le T \le b + 1$, where X is Lipschitz continuous,

$$|X(T_1) - X(T_2)| \le |T_1 - T_2| . \qquad (*)$$

(2) Now let $\gamma = \{C_n\}$ be a sequence of continuous monotonic curves in the yt-plane joining the points (a,0) and (b,1) . By transforming to the YT co-ordinates and using the Ascoli-Arzela theorem, show that there exists a convergent subsequence $\{C_{n_i}\}$. {207}

6.19 Let X be a space with topologies τ_1 and τ_2 , τ_1 being stronger than τ_2 . Show that if a subset S of X is compact (τ_1) then it is compact (τ_2) . {207}

6.20 Prove, using Theorem 6.1 that a real continuous function attains its supremum on a compact set (Theorem 6.19, p. 209).

6.21 Prove that a real lower (upper) semi-continuous function attains its infimum (supremum) on a compact set. (Theorem 6.20, p. 210).

6.22 Show that the function $f\colon [0,1] \to R^1$ defined by

$$f(x) = \begin{cases} x , & \text{if } 0 \le x < 1 , \\ \tfrac{1}{2} , & \text{if } x = 1 , \end{cases}$$

is lower semi-continuous, but not upper-semicontinuous and does not attain its supremum on [0,1] .

6.23 Prove Theorem 6.21, which gives properties of the solution set of a minimization problem when the feasible region V is convex and the objective function f is convex or concave. {211}

6.24 Complete the proof of Theorem 6.22 that the approximation x_b is unique if the norm is strictly convex. {213}

6.25 Let $x \in X = C[0,1]$ be given by $x\colon t \to 1$. Let $X_0 \subset C[0,1]$ be the linear subspace spanned by $x_1\colon t \to t$. Show that there exists a point of best approximation to x in X_0 , but that it is not unique. {214}

6.26 As in Theorem 6.23 (Cauchy-Peano) prove that each of the following problems has a solution $x \in C[\tau,\tau+\alpha]$, for some $\alpha > 0$, which can be approximated by Euler's method.

(1) *An initial-value problem for a system of ordinary differential equations* (see Section 4.10).

Given

$$\xi \in R^{\ell} ,$$

$$X = \{u \in R^{\ell} : \|u-\xi\|_{\infty} \le b\} ,$$

$$\overline{\Omega} = [\tau,\tau+a] \times X , \quad f \in C(\overline{\Omega};R^{\ell})$$

find $x \in C^1([\tau,\tau+\alpha]; R^{\ell})$ satisfying

$$x(\tau) = \xi ,$$

$$\dot{x}(t) = f(t,x(t)) , \quad \tau \le t \le \tau + \alpha .$$

(2) *A Volterra integral equation*

Given

$$\xi \in R^1 ,$$

$$X = \{u \in R^1 : |u-\xi| \le b\} ,$$

$$\overline{\Omega} = [\tau,\tau+a] \times [\tau,\tau+a] \times X , \quad f \in C(\overline{\Omega})$$

find $x \in C[\tau,\tau+\alpha]$ satisfying

$$x(t) = \xi + \int_{\tau}^{t} f(t,s,x(s))ds , \quad \tau \le t \le \tau + \alpha .$$

(3) *A scalar delay differential equation*

Given

$$\xi \in R^1 ,$$

$$X = [\xi-b,\xi+b] ,$$

$$\overline{\Omega} = [\tau,\tau+a] \times X \times X , \quad f \in C(\overline{\Omega})$$

and

$$\phi \in C[\tau-1,\tau] ,$$

$$\phi(\tau) = \xi ,$$

$$\phi([\tau-1,\tau]) \subset X ,$$

find $x \in C[\tau,\tau+\alpha]$ satisfying

$$x(\tau) = \xi ,$$
$$x(t) = f(t,x(t),x_1(t)) , \quad \tau \le t \le \tau + \alpha ,$$

where

$$x_1(t) = \begin{cases} x(t-1) , & \text{if } t \ge \tau + 1 , \\ \phi(t-1) , & \text{otherwise.} \end{cases}$$ {218}

6.27 Complete Remark 6.10 by using induction to prove eqn (6.22). {221}

6.28 Prove Theorem 6.24 for the case of a metric space using sequential compactness arguments. {222}

6.29 Let $X = [0,1]$ and $T : X \to R^1$,

$$T : x \in X \to \begin{cases} 0 , & \text{if } 0 \le x \le \frac{1}{2} , \\ 1 , & \text{if } \frac{1}{2} < x \le 1 . \end{cases}$$

Show that T is compact but not continuous.

6.30 (a) Let R^n_+ denote the closed non-negative orthant in R^n . If A is a real $m \times n$ matrix show that $A(R^n_+)$ is closed in R^m . (See also Problem 7.24).

(b) Let $f: R^1_+ \to R^1$, where $f: t \in R^1_+ \to e^{-t}$. Show that f is continuous but that $f(R^1_+)$ is not closed. {21, 187, 250}

6.31 Let $X = Y = C[a,b]$. Let $T : X \to Y$ be defined by

$$(Tx)(s) = \int_a^b K(s,t)F(x(t))dt ,$$

where $K(s,t)$ is continuous for $a \le s,t \le b$ and where $F : R^1 \to R^1$ is continuous. Use the Ascoli-Arzela criterion to show that T is compact.

6.32 Let $X = Y = L^p[a,b]$, $1 \le p < \infty$. Let $T : X \to Y$ be defined by

$$(Tx)(s) = \int_a^b K(s,t)x(t)dt ,$$

where $K(s,t)$ is continuous for $a \le s , t \le b$. Use Theorem 6.14 to show that T is compact.

6.33 Let $X = Y = L^p[a,b]$, $1 < p < \infty$. Let $K(s,t)$ be measurable on $[a,b] \times [a,b]$ and satisfy

$$\sup_{s\in[a,b]} \int_a^b |K(s,t)|^q dt = M < \infty \qquad (*)$$

where $1/p + 1/q = 1$. Then $T : X \to Y$ defined by

$$(Tx)(s) = y(s) = \int_a^b K(s,t)x(t)dt \qquad (**)$$

is compact.

6.34 Consider the mapping

$$(Tx)(s) = \int_0^1 \ell n|s-t| \; x(t)dt .$$

Show that T is a completely continuous map of $C[0,1]$ into $C[0,1]$.

6.35 Let $f\colon \Omega \subset R^n \to R^1$ be a convex function on the open convex set Ω . Prove that f is continuous.

6.36 Let $T: X \to Y$ be a compact linear map between the normed spaces X and Y. Prove that T is continuous.

6.37 Prove eqn (6.21). {220}

6.38 Not only mathematics has jargon. What is a semihemidemisemiquaver?

7
The Hahn-Banach theorem

In Section 2.4 we introduced the concept of the extension of a mapping: given $f : X \to Y$ and $g : Z \to Y$ then g is an extension of f if $Z \supset X$ and $f(x) = g(x)$ for $x \in X$.

In this chapter we consider various results on the extension of mappings, the most important being provided by the Hahn-Banach theorem, which is a basic theorem on the existence of linear functionals. (We recall that a real (complex) linear functional is a linear mapping into R^1 ($\mathbb{C}^1$)). There are two slightly different ways of looking at the Hahn-Banach theorem: (1) as a procedure for extending a linear functional defined on a subspace; (2) as a procedure for separating convex sets. Each point of view has its own advantages, and we consider each point of view in turn.

7.1 *EXTENSIONS OF REAL CONTINUOUS FUNCTIONS*

We state two classical theorems on extensions in normal topological spaces. (A topological space is *normal* if (1) sets consisting of single points are closed and (2) if A and B are closed disjoint sets in X then there exist open disjoint sets U and V such that $A \subset U$ and $B \subset V$.)

THEOREM 7.1. *(Urysohn)*

If A *and* B *are disjoint closed sets in a normal topological space* X, *then there exists a continuous real function* $f : X \to [0,1]$ *such that* $f(A) = \{0\}$ *and* $f(B) = \{1\}$. □

THEOREM 7.2. *(Tietze)*

If f *is a real bounded and continuous function defined on a closed set* A *in a normal topological space* X, *then there exists a continuous real function* $F : X \to R^1$ *such that* $f(x) = F(x)$ *on* A *and*

$$\sup_{x \in X} |F(x)| = \sup_{x \in A} |f(x)| . \qquad (7.1)$$

□

The proofs of the above two theorems (which will be found for example in Dunford and Schwartz [1966, p. 15]) are non-constructive. In special cases it is possible to construct the extensions, a simple example being the case

$$A = \{(x_1, x_2) \in R^2 : x_2 > 0\} ,$$

where any $f \in C(\bar{A})$ can be extended to $F \in C(\overline{R^2})$,

$$F(x_1, x_2) = \begin{cases} f(x_1, x_2) , & \text{if } x_2 \geq 0 , \\ f(x_1, -x_2) , & \text{if } x_2 < 0 . \end{cases}$$

The theorems of Urysohn and Tietze are occasionally useful in numerical functional analysis:

(1) They provide a useful tool for constructing unpleasant functions which demonstrate the fallibility of numerical methods. (Problem 7.4)

(2) The problem of interpolating or extrapolating data occurs frequently in numerical analysis. The theorem of Tietze assures us that the problem has a solution, in much the same way that the theorem of Weierstrass (Theorem 4.2) assures us that one-dimensional polynomial interpolation is possible.

REMARK 7.1. Given $\Omega \subset R^n$, one can inquire about the possibility of extending $f \in C^m(\bar{\Omega})$ to $F \in C^m(R^n)$ in such a way that

$$\| F;\ C^m(\overline{R^n}) \| \leq K \| f;\ C^m(\bar{\Omega}) \| , \qquad (*)$$

where K does not depend on f . This is not possible for general Ω as is shown in Problem 7.5. On the other hand, as shown in Problem 4.19, extensions satisfying eqn (*) are possible for certain domains. See Adams [1975, p. 83], and Whitney [1944]. □

REMARK 7.2. Given a mapping f with certain properties one can try to find an extension F which preserves these properties:

(1) *Linearity.* This is the most important case and is considered in the remainder of this chapter.

(2) *Convexity.* If f is a real convex map from a convex set A in a vector space X, then it is not always possible to extend f to X and preserve convexity. (Problem 7.6). However, it is possible to preserve convexity if f is allowed to take on the extended real values $\pm\infty$. (Problem 7.7)

(3) *Analyticity.* Given an analytic function $f : A \to \mathbb{C}$, where A is a subset of the complex plane $\mathbb{C}$, the theory of analytic continuation (Hille [1962, Chapter 10]) is concerned with analytic extensions of f. The numerical implementation of analytic continuation has applications to a variety of problems including the computation of the transonic flow past aerofoils (Garabedian [1964, Chapter 16]; Bauer, Garabedian, and Korn [1977]).

7.2 *EXTENSION OF DENSELY DEFINED LINEAR OPERATORS*

THEOREM 7.3. *Let* X *be a normed space, and let* Y *be a Banach space. Let* $T_0 : X_0 \to Y$ *be a bounded linear operator defined on a subspace* X_0 *of* X.

For $x \in \overline{X}_0$, *choose a Cauchy sequence* $\{x_i\}$ *in* X_0 *converging to* x. *Then* $\{T_0 x_i\}$ *is a Cauchy sequence. Set*

$$Tx = \lim_{i\to\infty} T_0 x_i . \qquad (*)$$

Then T *is uniquely defined by eqn* (*), *is a bounded linear operator on* $\overline{X}_0$, *agrees with* T_0 *on* X_0, *and satisfies*

$$\|T\| = \sup_{\substack{x\in\overline{X}_0\\ \|x\|\le 1}} \|Tx\| = \|T_0\| = \sup_{\substack{x\in X_0\\ \|x\|\le 1}} \|T_0 x\| . \qquad (**)$$

Proof. To show that T is uniquely defined by eqn (*), let $\{\tilde{x}_i\}$ be any other Cauchy sequence in X_0 converging to $x \in \overline{X}_0$. Then

$$||T_0\tilde{x}_i - T_0x_i|| \leq ||T_0||[||\tilde{x}_i - x|| + ||x - x_i||] \to 0 ,$$

so that

$$\lim_{i\to\infty} T_0\tilde{x}_i = \lim_{i\to\infty} T_0 x_i .$$

To prove that T is linear, let $\{y_i\}$ and $\{x_i\}$ be Cauchy sequences in X_0 converging to y and x, respectively. Then

$$\begin{aligned} T(\lambda x+\mu y) &= \lim T_0(\lambda x_i + \mu y_i) , \\ &= \lambda \lim T_0 x_i + \mu \lim T_0 y_i , \\ &= \lambda Tx + \mu Ty . \end{aligned}$$

Finally, to prove eqn (**), we note that if $x_i \to x$ then

$$||Tx|| = \lim ||T_0x_i|| \leq ||T_0|| \lim ||x_i|| = ||T_0|| \, ||x|| . \quad \square$$

REMARK 7.3. Although Theorem 7.3 is apparently trivial it has important applications of which we mention two.

(1) In Section 5.7 we introduced the idea of a solution operator $E(t)$ (defined on a Banach space X). One way to establish the existence of $E(t)$ is to establish the existence of a solution operator $E_0(t)$ defined on a dense subspace X_0 of X and then invoke Theorem 7.3 to extend $E_0(t)$ to an operator $E(t)$ defined on X. X_0 could for example be a subspace consisting of smooth functions x, so that $E_0(t)x$ would be a classical solution of the initial value problem.

(2) One method of defining the Riemann integral I for functions $x \in X = C[0,1]$ is to define an integral I_0 on a suitable dense subspace X_0 (such as the stepfunctions) and then use Theorem 7.3. (Problem 7.8).

$\square$

REMARK 7.4. There are certain similarities between Theorem 7.3 and results in earlier chapters:

(1) There is a close relationship between Theorem 7.3 and Theorem 3.4 (on the completion of topological vector spaces), and the comments following Theorem 3.4 apply here also.

(2) In the Banach-Steinhaus theorem (Theorem 5.4) an operator $T : X \to Y$ was constructed as the limit of operators T_n defined on the space X. In Theorem 7.3, on the other hand, T_0 and T are defined on different spaces. □

7.3 *THE HAHN-BANACH THEOREM: EXTENSION OF LINEAR FUNCTIONALS*

THEOREM 7.4. *(Hahn-Banach)*

Let the real function p *on the real vector space* X *satisfy*

(1) $p(x+y) \le p(x) + p(y)$,

(2) $p(\alpha x) = \alpha p(x)$, $\alpha \ge 0$.

(These conditions are satisfied, for example, if p *is a seminorm).*

Let f_0 *be a real linear functional on a linear subspace* $X_0 \subset X$ *satisfying*

$$f_0(x) \le p(x) \ , \quad \text{for} \quad x \in X_0 \ . \tag{*}$$

Then there exists a real linear functional f *on* X *which is an extension of* f_0 *and satisfies*

$$f(x) \le p(x), \quad \text{for} \quad x \in X \ .$$

Proof. Suppose that $X_0 \ne X$. Then there exists $x_1 \in X$ such that $x_1 \notin X_0$. Let X_1 denote the linear subspace with elements of the form

$$x = x_0 + \lambda x_1 \ , \quad \text{for} \quad x_0 \in X_0 \quad \text{and} \quad \lambda \quad \text{real.} \tag{**}$$

Every element in X_1 has a unique representation of the form (**). [If $x = \tilde{x}_0 + \tilde{\lambda} x_1$ and $x = x_0 + \lambda x_1$ then

$\tilde{x}_0 - x_0 = (\tilde{\lambda}-\lambda)x_1$. Since $\tilde{x}_0 - x_0 \in X_0$ and $x_1 \notin X_0$, it follows that $\tilde{\lambda} = \lambda$ and $\tilde{x}_0 = x_0$.]

We define a linear functional f_1 on X_1 by means of

$$f_1(x) = f_1(x_0+\lambda x_1) = f_0(x_0) + \lambda c ,$$

where $c = f_1(x_1)$ is a constant which will be chosen so that

$$f_1(x_0+\lambda x_1) \le p(x_0+\lambda x_1) , \quad \text{for all } x_0 \in X_0 \text{ and } \lambda \in R^1 . \tag{$^{**}_{*}$}$$

To find c we begin by observing that, for any $u, v \in X_0$ it follows from eqn (*) and condition (1), that

$$f_0(u) - f_0(v) = f_0(u-v) \le p(u-v) \le p(u+x_1) + p(-v-x_1) ,$$

so that

$$f_0(u) - p(u+x_1) \le f_0(v) + p(-v-x_1) , \quad \text{for all } u, v \in X_0 .$$

Since the left hand side of this inequality is independent of v and the right hand side is independent of u there exists a constant, $-c$ say, such that

$$f_0(u) - p(u+x_1) \le -c \le f_0(v) + p(-v-x_1) . \tag{$^{**}_{**}$}$$

If $\lambda > 0$ we set $u = x_0/\lambda$ in eqn ($^{**}_{**}$) and obtain

$$f_0(x_0/\lambda) - p(x_0/\lambda+x_1) \le -c ,$$

or, multiplying by λ and using condition (2),

$$f_1(x_0+\lambda x_1) = f_0(x_0) + \lambda c \le p(x_0+\lambda x_1) , \quad \text{for } x_0 \in X_0 \text{ and } \lambda > 0 . \tag{$^{***}_{**}$}$$

If $\lambda < 0$ we set $v = x_0/\lambda$, in eqn ($^{**}_{**}$) and obtain

$$-c \le f_0(x_0/\lambda) + p(-x_0/\lambda-x_1) ,$$

or, multiplying by $-\lambda$ and using condition (2),

$$f_1(x_0+\lambda x_1) = f_0(x_0) + \lambda c \le p(x_0+\lambda x_1) , \quad \text{for } x_0 \in X \text{ and } \lambda < 0 . \tag{$^{***}_{***}$}$$

It follows from eqns ($^{***}_{**}$) and ($^{***}_{***}$) that eqn ($^{**}_{*}$) holds. That is, we have extended f_0, defined on the subspace X_0, to f_1, defined on the subspace X_1 strictly containing X_0, in such a way that

$$f_1(x) \le p(x) \ , \quad \text{for} \quad x \in X_1 = D(f_1) \ ,$$

where $D(f_1) = X_1$ denotes the domain of f_1.

Beyond this point, the proof depends upon the assumptions made about f_0, p, and X. Two special cases are considered in Problem 7.9.

In the general proof given below it is necessary to have recourse to the axiom of set theory known as the *axiom of choice*. Some preliminary definitions are first needed. A partially ordered set is said to be *totally ordered* if for every $a_1, a_2 \in A$ either $a_1 \ge a_2$ or $a_2 \ge a_1$. A subset A_0 of a partially ordered set A is *bounded above* if there exists $\hat{a} \in A$ such that $\hat{a} \ge a_0$ for all $a_0 \in A_0$. We can now state the axiom of choice in one of its equivalent forms namely *Zorn's lemma: Let* A *be a partially ordered set with the property that every totally ordered subset* A_0 *of* A *is bounded above. Then there exists* $\tilde{a} \in A$ *such that if* $a \in A$ *and* $a > \tilde{a}$ *then* $a = \tilde{a}$. We call $\tilde{a}$ a *maximal element of* A.

Returning to the proof of the theorem, let A be the set of all linear extensions g of f_0 such that: (1) $D(g)$, the domain of g, is a linear subspace of X; and (2) $g(x) \le p(x)$, for $x \in D(g)$. If $g_1, g_2 \in A$, we write $g_1 \ge g_2$ if g_1 is an extension of g_2; this induces a partial ordering on A (Problem 7.10).

If A_0 is a totally ordered subset of A, let

$$\hat{X} = \bigcup_{g \in A_0} D(g) \ .$$

If $\hat{x} \in \hat{X}$ then $\hat{x} \in D(g)$ for some $g \in A_0$; set $\hat{g}(\hat{x}) = g(\hat{x})$. Then $\hat{g}$ is well-defined, since if $\hat{x} \in D(g_1)$ and $\hat{x} \in D(g_2)$, for $g_1, g_2 \in A_0$, we must have $g_1(\hat{x}) = g_2(\hat{x})$. Furthermore, $\hat{g} \in A$ and $\hat{g} \ge g$ for all $g \in A_0$ (Problem 7.11). Thus, A_0 is bounded above.

It follows from Zorn's lemma that A has a maximal element, $\tilde{g}$ say, with domain $\tilde{X}$. We assert that $\tilde{X} = X$ since otherwise, as shown in the first part of this proof, $\tilde{g}$ could be extended to a subspace $\tilde{X}_1$ strictly containing $\tilde{X}$, which would contradict the fact that $\tilde{g}$ is maximal.

To conclude the proof we set $f = \tilde{g}$. □

With slight modifications in the assumptions and conclusions, Theorem 7.4 holds for a complex vector space X (Yosida [1968], p. 105); we do not discuss this because it has few applications in numerical analysis.

We now state a series of corollaries which are consequences of the Hahn-Banach theorem. (Two further corollaries are given as Problems 7.12 and 7.13). Some of these corollaries will be used in the examples in this chapter, while others will be applied in later chapters.

COROLLARY 7.5. *Let* X *be a normed vector space,* X_0 *a linear subspace of* X, *and* f_0 *a continuous linear functional on* X_0. *Then there exists a continuous linear functional* f *defined on* X *which is an extension of* f_0 *and satisfies* $\|f\| = \|f_0\|$.

Proof. We prove this in the case when X is a real vector space using Theorem 7.4. We set $p(x) = \|f_0\| \, \|x\|$ Then $f_0(x) \le p(x)$ for $x \in X_0$. Thus there exists an extension $f(x)$ which satisfies

$$f(x) \le p(x) = \|f_0\| \|x\| , \quad x \in X .$$

Since f is linear,

$$-f(x) = f(-x) \le \|f_0\| \; \|-x\| = \|f_0\| \; \|x\| .$$

Hence, $\|f(x)\| = |f(x)| \le \|f_0\| \; \|x\|$, so that $\|f\| \le \|f_0\|$. Since f is an extension of f_0, $\|f\| \ge \|f_0\|$. Therefore, $\|f\| = \|f_0\|$. □

COROLLARY 7.6. *Let* M *be a linear subspace of a normed vector space* X . *Let* $x \in X$ *be such that*

$$\inf_{m \in M} \|x-m\| = d > 0 .$$

Then there exists a continuous linear functional f *on* X *such that:*

(1) f *vanishes on* M ;

(2) $f(x) = 1$;

(3) $\|f\| = 1/d$.

Proof. Let X_0 be the linear space spanned by M and x . Let f_0 be the functional: $f_0(z) = \alpha$ for $z \in X_0$, where α is the unique constant such that $z = \alpha x + m$, with $m \in M$. One verifies (Problem 7.12) that f_0 is a continuous linear functional and that $\|f_0\| = 1/d$, and then applies Corollary 7.5. □

THEOREM 7.7. *Let* X *be a normed vector space, let* M *be a linear subspace in* X , *and let* $\mathcal{M}$ *be the set of continuous linear functionals* f *on* X *such that:*

(1) f *vanishes on* M .

(2) $\|f\| \le 1$.

Then, for any $x \in X$,

$$d = \inf_{m \in M} \|x-m\| = \max_{f \in \mathcal{M}} |f(x)| . \qquad (*)$$

Proof. If M is dense in X , $f \equiv 0$ satisfies eqn (*).

If M is not dense in X it follows from Corollary 7.6 that $\mathcal{M}$ is not empty. Let $x \in X$ and $d = \inf_{m \in M} \|x-m\|$. For any $f \in \mathcal{M}$ we have

$$|f(x)| = |f(x-m)| \le \|f\| \|x-m\| \le \|x-m\| , \quad \text{for all } m \in M .$$

Consequently, choosing m such that $\|x-m\| \le d + \varepsilon$, it follows that $|f(x)| \le d + \varepsilon$ for any $\varepsilon > 0$. Hence,

$$D \equiv \sup_{f \in \mathcal{M}} |f(x)| \le d . \qquad (**)$$

To show that $D = d$, and that D is attained by some $f \in M$, two cases must be considered. (1) If $d = 0$ then $x \in \overline{M}$; hence $f(x) = 0$ for each $f \in M$ since f is continuous and vanishes on M. (2) If $d > 0$ let $f_1 = df$ where f is as in Corollary 7.6; then $f_1 \in M$ and $f(x) = d$. □

REMARK 7.5. Theorem 7.7 is typical of a whole class of 'duality results', in which a minimization problem is converted into a dual maximization problem. (The best known example of such duality results arises in the theory of linear programming.)

Given the problem of approximating $x \in X$ by $m \in M$, we can obtain an *upper* bound for $d = \inf\|m-x\|$ by picking any $m \in M$ and computing $\|x-m\|$. Theorem 7.7 allows us to compute *lower* bounds for d. □

EXAMPLE 7.1. Let $X = C([0,1] \times [0,1])$, and let M be the subspace of X consisting of functions of the form $m : (s,t) \to P(s) + Q(t)$ where P and Q are polynomials in one variable. Let it be required to approximate $x_0 \in X$ where $x_0 : (s,t) \to st$.

Let $m_0 : (s,t) \to \frac{1}{2}(s+t) - \frac{1}{4}$, so that $m_0 \in M$. It is found by direct computation (Problem 7.14) that

$$d = \inf_{m \in M} \|x_0 - m\| \le \|x_0 - m_0\| = \frac{1}{4}. \qquad (*)$$

On the other hand, let $f : X \to R^1$ be the continuous linear functional

$$f : x \to [x(0,0)+x(1,1)-x(1,0)-x(0,1)]/4 .$$

Then (Problem 7.14) $\|f\| = 1$, f vanishes on M, and

$$f(x_0) = \frac{1}{4}. \qquad (**)$$

Using Theorem 7.7 it follows from eqns (*) and (**) that m_0 is a best approximation to x_0 in M.

Since X is not strictly convex, we cannot apply Theorem 6.22 to conclude that m_0 is unique, and indeed another best approximation to x_0 is given by (Problem 7.14),

$$m_1 : (s,t) \to s + t - \tfrac{1}{2}(s^2+t^2) - \tfrac{1}{4} . \quad \square$$

EXAMPLE 7.2. *Claim:* Let $X = C[0,1]$ and let M be the subspace consisting of polynomials of degree at most n. Let $x_0 \in X$ be given. Then the polynomial $m_0 \in M$ is a best approximation to x_0 if m_0 has the *uniform oscillation property* ; that is, if there are $n + 2$ points $0 \le t_0 < \ldots < t_{n+1} \le 1$ such that

$$x_0(t_k) - m_0(t_k) = (-1)^k \, \|x-m_0\| , \quad 0 \le k \le n+1 . \qquad (*)$$

Proof. Let $\mathcal{M}$ be the set of continuous linear functionals f on X such that: (1) f vanishes on M ; (2) $\|f\| \le 1$.

The divided differences of $x \in X$ are defined recursively as follows:

$$x[t_0] = x(t_0) ,$$

$$x[t_0,t_1] = (x[t_0] - x[t_1])/(t_0-t_1) ,$$

$$x[t_0,t_1,\ldots,t_k] = (x[t_0,\ldots,t_{k-1}] - x[t_1,\ldots,t_k])/(t_0-t_k) ,$$

so that (Problem 7.15)

$$x[t_0,\ldots,t_{n+1}] = \sum_{k=0}^{n+1} [x(t_k)/\prod_{\substack{j=0\\ j\neq 0}}^{n+1} (t_k-t_j)] = \sum_{k=0}^{n+1} (-1)^{n+1-k}\alpha_k x(t_k) ,$$

where the α_k are positive constants. Set

$$\ell(x) = x[t_0,\ldots,t_{n+1}]/\sum_{k=0}^{n+1} \alpha_k .$$

It can be shown that $\ell \in \mathcal{M}$ (Problem 7.15).

Since m_0 satisfies the uniform oscillation property (*),

$$|\ell(x_0)| = |\ell(x_0-m_0)| ,$$

$$= \left|\sum_{k=0}^{n+1} (-1)^{n+1-k} \alpha_k (-1)^k \|x_0-m_0\| \Big/ \sum_{k=0}^{n+1} \alpha_k\right| ,$$

$$= \|x_0-m_0\| .$$

Thus, using Theorem 7.7,

$$\|x_0-m_0\| \geq \inf_{m\in M} \|x_0-m\| = \max_{f\in M} \|f(x_0)\| \geq |\ell(x_0)| = \|x_0-m_0\| .$$

□

7.4 *THE HAHN-BANACH THEOREM: SEPARATION OF CONVEX SETS*

We now turn to the second way of viewing the Hahn-Banach theorem. Let X be a real (complex) vector space, let M and N be subsets of X, and let f be a real (complex) linear functional on X. Then f *separates* M and N if there exists a real constant c such that

$$\mathcal{R}f(m) \leq c \leq \mathcal{R}f(n) , \quad \text{for all } m \in M \text{ and } n \in N ,$$

where $\mathcal{R}f(m)$ denotes the real part of $f(m)$. If there exist real constants c and ε, with $\varepsilon > 0$, such that

$$\mathcal{R}f(m) \leq c - \varepsilon < c \leq \mathcal{R}f(n) , \quad \text{for all } m \in M \text{ and } n \in N ,$$

we say that f *strictly separates* M and N.

It follows from the corollary to the Hahn-Banach theorem given in Corollary 7.6 that: If M is a closed linear subspace of a normed vector space X and $x \notin M$, then there exists a continuous linear functional f which strictly separates M and x. We consider here various related results. In all cases M and N are assumed to be convex, and the problem may be viewed as fitting a *hyperplane* $H = \{x : \mathcal{R}f(x) = c - \varepsilon/2\}$ between convex sets M and N (see Figure 7.1).

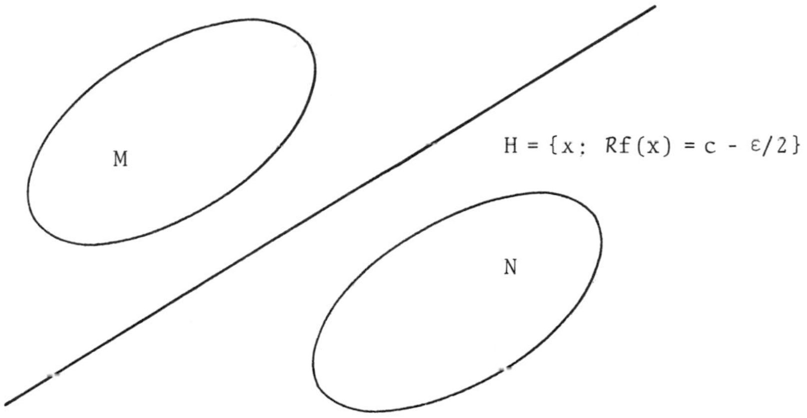

Figure 7.1. The hyperplane H *separating convex sets* M *and* N

In the analysis which follows we use the same approach as in Chapter 2 and consider a hierarchy of spaces. First we consider vector spaces and obtain the basic separation theorem (Theorem 7.8); by adding a topology we obtain continuity of the separating functional (Theorems 7.9 and 7.10); finally, in a locally convex topological vector space, we obtain strict separation (Theorem 7.11). This approach is only slightly longer than a direct proof of Theorem 7.11 would be, and brings out the contributions of the different assumptions.

We require two new concepts.

If K is a subset of a vector space X then a point $p \in K$ is an *internal point* of K if for each $x \in X$ there exists $\varepsilon > 0$ such that $p + \delta x \in K$ if $|\delta| \le \varepsilon$.

If K is a convex set in a real vector space X such that the origin is an internal point, the *support function*

F of K is defined by

$$F : x \in X \to \inf\{a \in R^1 : a>0 \text{ and } a^{-1}x \in K\} \in R^1 . \qquad (7.2)$$

Clearly, if $\alpha > 0$,

$$F(\alpha x) = \inf\{a: a>0 \text{ and } \frac{\alpha x}{a} \in K\} = \alpha F(x) . \qquad (*)$$

Also, for any $x, y \in X$ and $\varepsilon > 0$ let $a = F(x) + \varepsilon$ and $b = F(y) + \varepsilon$, so that $x/a \in K$ and $y/b \in K$. Since K is convex,

$$\frac{1}{a+b} [a(\frac{x}{a}) + b(\frac{y}{b})] = \frac{x+y}{a+b} \in K ,$$

so that

$$F(x+y) \le a + b = F(x) + F(y) + 2\varepsilon .$$

Letting $\varepsilon \to 0$ we obtain

$$F(x+y) \le F(x) + F(y) , \quad \text{for all } x,y \in X . \qquad (**)$$

From eqns (*) and (**) we conclude that every support function F satisfies conditions (1) and (2) of the Hahn-Banach theorem (Theorem 7.4). (Some further properties of support functions are given in Problem 7.16.)

THEOREM 7.8. *In a vector space, any two disjoint convex sets, one of which has an internal point, can be separated by a nonzero linear functional.*

Proof. We give the proof for a real vector space X .

Denote the disjoint convex sets by M and N . Assume that m_0 is an internal point of M and that n_0 is any point of N . We must show that there exists a nonzero linear functional f such that

$$f(m) \le f(n) , \quad \text{for } m \in M , n \in N . \qquad (*)$$

Since f is linear, this is equivalent to

$$f(m-n-m_0+n_0) \le f(-m_0+n_0) , \quad \text{for } m \in M , n \in N . \qquad (**)$$

Set $p = -m_0 + n_0$, and

$$K = \{m-n-m_0+n_0 : m \in M,\ n \in N\} = M - N + p .$$

Then eqn (**) can be written in the form

$$f(k) \le f(p) , \quad \text{for} \quad k \in K . \qquad (*{*}*)$$

It is readily checked that:

(1) $p \notin K$;

(2) K is a convex set (Problem 7.17) ;

(3) $\underline{0}$ is an internal point of K .

Let F be the support function of K, and set $f_0(\alpha p) = \alpha F(p)$ for $\alpha \in R^1$. Then f_0 is a linear functional defined on the one-dimensional subspace X_0 consisting of real multiples of p, and satisfies $f_0(x) \le F(x)$ for $x \in X_0$. Hence, by Theorem 7.4, we can extend f_0 to a linear functional on X satisfying $f(x) \le F(x)$ everywhere.

To conclude, we observe from eqn (7.2) that, since $p \notin K$,

$$f(p) = F(p) \ge 1 ,$$

while

$$f(k) \le F(k) \le 1 , \quad \text{for} \quad k \in K ,$$

so that eqns (*{*}*), (**), and (*) hold. □

We now consider topological vector spaces. Let K be a set in a topological vector space. A point $p \in K$ is an *interior point of* K if some neighbourhood of p belongs to K. An interior point of K is also an internal point (Problem 7.18).

THEOREM 7.9. *If a linear functional* f *on a real topological vector space separates two sets one of which has an interior point, then* f *is continuous.*

Proof. Denote the two sets by A and B and assume that p is an interior point of A. Then $\underline{0} + U \subset A - p$

for some neighbourhood U of $\underline{0}$. Thus, by Theorem 2.5 (part e), there is a balanced neighbourhood N of $\underline{0}$ such that $N \subset A - p$.

Let the linear functional f separate A and B . Since f is linear, f separates $A - p$ and $B - p$. Hence, replacing f by $-f$ if necessary,

$$f(n) \leq f(b-p) , \quad \text{for } n \in N \text{ and } b \in B .$$

That is the set $f(N)$ is bounded above. But $f(N) = f(-N)$ because N is balanced, and so $f(N)$ is bounded: $f(N) \subset [-a,+a]$, for some finite positive a . The theorem now follows with the help of Problem 7.19. □

Combining Theorems 7.8 and 7.9 we obtain

THEOREM 7.10. *In a topological vector space, any two disjoint convex sets one of which has an interior point can be separated by a non-zero continuous linear functional.* □

Finally, we consider locally convex topological vector spaces.

THEOREM 7.11. *In a locally convex topological vector space, any two disjoint closed convex sets one of which is compact can be strictly separated by a continuous linear functional.*

Proof. Denote the two convex sets by M and N , where M is compact. Set $A = M - N$.

A is convex and closed (see Theorem 6.24 (part 2)). Since $\underline{0} \notin A$, and X is locally convex there is a *convex* neighbourhood U of $\underline{0}$ such that U and A are disjoint. By Theorem 7.10, there exists a nonzero continuous linear functional f which separates A and U . Replacing f

by $-f$ if necessary, we have that

$$f(a) \geq f(u) \quad , \quad \text{for} \quad u \in U \quad \text{and} \quad a \in A \; . \tag{*}$$

Since f is nonzero there exists $x_1 \in X$ such that $f(x_1) = 1$. Choose $\varepsilon > 0$ for which $\varepsilon x_1 \in U$. Then, using eqn (*),

$$f(m) - f(n) = f(m-n) = f(a) \geq \sup_{u \in U} f(u) \geq f(\varepsilon x_1) = \varepsilon \; , \tag{**}$$

if $m \in M$ and $n \in N$. That is,

$$f(m) \geq c = \inf_{m \in M} |f(m)| \geq c - \varepsilon \geq f(n) \; , \quad \text{for } m \in M \text{ and } n \in N \; .$$

□

REMARK 7.6. The preceding theorems cannot be strengthened:

(1) Two disjoint convex sets cannot always be strictly separated as is shown by the example in R^1 with M the open positive real axis and N the open negative real axis.

(2) Two disjoint closed convex sets cannot always be strictly separated as is shown by the example in R^2 with

$$M = \{(x,y) : y \geq e^{-x}\} \; ,$$

and $N = -M$.

(3) One cannot eliminate the requirement, in Theorems 7.8 and 7.10, that one of the convex sets have an internal point. Tukey has given an example of two disjoint closed convex sets in a Hilbert space which cannot be separated. References to this, and other examples, are given by Dunford and Schwartz [1966, p. 461].

EXAMPLE 7.3: *LINEAR PROGRAMMING*

The following is a basic theorem in the theory of linear programming.

THEOREM 7.12. *(Farkas)*

Let A *be an* $m \times n$ *real matrix and* b *a row n-vector. Then precisely one of the following two possibilities arises:*

I. *There exists a non-negative row m-vector* x *such that* $xA = b$.

II. *There exists a column n-vector* y *such that* $by < 0$ *and* $Ay \geq 0$. *(Here* $Ay \geq 0$ *means that* $(Ay)_i \geq 0$ *for* $1 \leq i \leq m$.*)*

Proof. It is readily verified that I and II cannot be true simultaneously.

We therefore assume that I does not hold. The theorem will hold if we can show that II must hold.

We set

$$X = R^n \quad (\text{row } n \text{ vectors}),$$
$$M = \{zA\colon z \text{ any non-negative row m-vector}\},$$
$$N = \{b\}.$$

It is obvious that M and N are convex and that N is closed and compact, and it follows from Problem 6.30 that M is closed. Since possibility I does not hold, M and N are disjoint.

Applying Theorem 7.11 we conclude that there exists a linear functional f_1 which strictly separates M and N. But, every linear functional f on X is of the form

$$f\colon x \to xu = \sum_{i=1}^{n} x_i u_i ,$$

where u is a column n-vector. Thus, there exists a column n-vector y and constants ε and c such that

$$f(b) = by \leq c - \varepsilon , \text{ and } f(m) = my \geq c , \text{ for } m \in M .$$

Since $\underline{0} \in M$, we can conclude that $c \leq 0$. On the other hand, if $f(m) < 0$ for some $m \in M$ then, for sufficiently large positive α , $\alpha m \in M$ and $f(\alpha m) = \alpha f(m) < c$. Consequently, we may assume that $c = 0$. That is,

$$by < 0 , \text{ and } zAy \geq 0 \text{ for all non-negative m-vectors } z .$$

That is,

$$by < 0 , \text{ and } Ay \geq 0 . \quad \square$$

REMARK 7.7. There are difficulties in extending results on finite-dimensional linear programming to infinite dimensions. However, the theorem of Farkas can be generalized to infinite dimensions: see Kortanek [1977], and Craven and Koliha [1977]. □

EXAMPLE 7.4: *NON-NEGATIVE QUADRATURE FORMULAE*

The elegant theorem given below is due to Tchakaloff [1957].

In stating and proving the theorem it is convenient to use multi-indices. We recall (Section 4.10) that an n-index α is an n-tuple $(\alpha_1,\ldots,\alpha_n)$ of non-negative integers. If α is an n-index and $x = (x_j) \in R^n$, we set

$$|\alpha| = \sum_{j=1}^{n} \alpha_j \ ,$$

$$x^\alpha = \prod_{j=1}^{n} (x_j)^{\alpha_j}$$

If d is a positive integer, let

$$M_d = \{\alpha\colon \alpha \text{ an n-index, } |\alpha| \le d\} \ .$$

M_d has only a finite number of elements, N_d say. We assume that M_d is ordered,

$$M_d = \{\alpha^{(1)},\ldots,\alpha^{(N_d)}\} \ , \quad \text{say.}$$

For example, if $n = 2$ and $d = 2$ then $N_2 = 6$ and an appropriate ordering of M_2 is

$$\begin{aligned} M_2 &= \{\alpha^{(1)},\ldots,\alpha^{(6)}\} \ , \\ &= \{(0,0),(1,0),(0,1),(2,0),(1,1),(0,2)\} \ . \end{aligned}$$

Since there is a one-to-one correspondence between the integers 1 to N_d and the elements of M_d, we can use the multi-indices α as indices in vectors, arrays, and sums.

For example, if $x = (x_1, x_2) \in R^2$ and

$$a_{(0,0)} = 1 \ , \quad a_{(1,0)} = 2 \ , \quad a_{(0,1)} = 3 \ ,$$

then

$$\sum_{\substack{\alpha \in M_2 \\ |\alpha| \le 1}} a_\alpha x^\alpha = 1 + 2x_1 + 3x_2 \ .$$

THEOREM 7.13. *(Tchakaloff)*

Let D *be a closed bounded region in* R^n *with non-empty interior. Let* $\rho(x)$ *be non-negative and integrable on* D *. Let* d *be a fixed positive integer. Let* $M = M_d$ *be the ordered set of n-indices* α *satisfying* $|\alpha| \le d$ *. Let* M *have* N *elements.*

Then there exists a quadrature formula

$$I(f) = \int_D \rho(x) f(x) dx \approx I_M(f) = \sum_{i=1}^{N} w_i f(\xi_i) \ ,$$

such that:

(1) The formula is exact for polynomials of degree d *, that is, polynomials of the form*

$$p(x) = \sum_{\alpha \in M} a_\alpha x^\alpha \ ,$$

where the coefficients a_α *are real constants .*

(2) The quadrature nodes $\xi_i = (\xi_{i,1}, \dots, \xi_{i,n})$ *and quadrature weights* w_i *satisfy* $\xi_i \in D$ *and* $w_i \ge 0$ *.*

Proof. Since I and I_M are linear, condition (1) is equivalent to the N equations:

$$I_M(x^\alpha) = \sum_{i=1}^{N} w_i \xi_i^\alpha = I(x^\alpha) \ , \quad \alpha \in M \ .$$

That is,

$$wA = c \ , \tag{7.3}$$

where w is the row N-vector with components w_i , $c = (c_\alpha)$ is the row N-vector with components

$$c_\alpha = I(x^\alpha) = \int_D \rho(x)x^\alpha \, dx \ , \quad \alpha \in M \ ,$$

and $A = (a_{i\alpha})$ is the $N \times N$ matrix with components

$$a_{i\alpha} = \xi_i^\alpha = \prod_{j=1}^{n} (\xi_{i,j})^{\alpha_j} \ .$$

A particular example is given in Problem 7.20.

Since eqns (7.3) form a system of N equations in N unknowns there will, in general, be only one solution w , and the condition $w \geq 0$ may not be satisfied. We have to show that the points ξ_i may be chosen so that eqn (7.3) has a non-negative solution.

For $x \in D$ let $v(x)$ be the row N-vector $(x^\alpha; \alpha\in M)$. For example, for $n = d = 2$,

$$v(x) = (1, x_1, x_2, (x_1)^2, x_1x_2, (x_2)^2) \ .$$

With this notation the matrix A can be written as the row partitioned matrix $A = (v(\xi_i) \ ; \ 1 \leq i \leq N)$, and equations (7.3) take the form

$$\sum_{i=1}^{N} w_i v(\xi_i) = c \ . \tag{7.4}$$

Let $K \subset R^N$ be the set consisting of all non-negative linear combinations of N vectors v :

$$K = \{ \sum_{t=1}^{N} b_t v(x_t) : b_t \geq 0 \ , x_t \in D\} \ . \tag{7.5}$$

The following lemma about K will be proved later:

LEMMA

K *is a closed convex set in* R^N . □

We continue with the proof of the main theorem, and claim that $c \in K$. The proof is by contradiction. Assume that $c \notin K$. Then, by Theorem 7.11, there exists a hyperplane which strictly separates c and K . That is, there exists a column N-vector $a = (a_\alpha : \alpha \in M)$ such that

$$ca < 0 \; ; \quad \text{and} \quad ua \geq 0 \; , \quad \text{for} \quad u \in K \; . \qquad (*)$$

Let p be the polynomial of degree d with coefficients a_α . Then, for any $x \in D$, $v(x) \in K$ and so, from eqn (*),

$$p(x) = \sum_{\alpha \in M} a_\alpha x^\alpha = v(x)a \geq 0 \; .$$

Thus $p(x)$ is non-negative for $x \in D$. Recalling the definition of c ,

$$ca = \sum_{\alpha \in M} a_\alpha c_\alpha = \sum_{\alpha \in M} a_\alpha \int_D \rho(x) x^\alpha dx = \int_D \rho(x) p(x) dx \geq 0 \; ,$$

which contradicts eqn (*) .

Having proved that $c \in K$, we see from eqn (7.5) that c has a representation of the form

$$c = \sum_{t=1}^{N} b_t v(x_t) \; ,$$

where $b_t \geq 0$ and $x_t \in D$. That is, we have obtained non-negative weights $w_i = b_i$ and points $\xi_i = x_i$ such that eqn (7.4) is satisfied.

Proof of the lemma. Step 1: K *is a convex cone.* That is if $u_1, u_2 \in K$ and $\lambda_1, \lambda_2 \geq 0$ then $\lambda_1 u_1 + \lambda_2 u_2 \in K$.

Suppose that $u_1, u_2 \in K$. For any non-negative λ_1, λ_2 let $u = \lambda_1 u_1 + \lambda_2 u_2$. We assert that $u \in K$.

To see this we note that u_1 and u_2 are each the non-negative sum of N vectors v_s so that u is of the form

$$u = \sum_{s=1}^{p} b_s v_s \; , \quad \text{with} \quad b_s > 0 \; , \quad p \leq 2N \; . \tag{*}$$

Two cases arise: $p \leq N$ and $p > N$.
Case 1: If $p \leq N$ then, from eqns (*) and (7.5), $u \in K$.
Case 2: If $p > N$ then, from linear algebra we know that any p N-vectors are linearly dependent. Therefore, there exist coefficients γ_s, not all zero, such that

$$\sum_{s=1}^{p} \gamma_s v_s = 0 \; . \tag{**}$$

We may assume that at least one γ_s is positive, by, if necessary, changing the sign of every γ_s. Then

$$\mu = \max_s \, (\gamma_s / b_s) > 0 \; . \tag{$\overset{**}{*}$}$$

Subtracting $1/\mu$ times eqn (**) from eqn (*) we obtain,

$$u = \sum_{s=1}^{p} (b_s - \gamma_s/\mu) v_s = \sum_{s=1}^{p} b_s' v_s \; , \quad \text{say} \; .$$

It follows from eqn ($\overset{**}{*}$) that the b_s' are non-negative and at least one is zero. Thus, in eqn (*), p may be replaced by $p - 1$. Repeating the process we see that we may assume that $p \leq N$. But then, $u \in K$.

Step 2: K is closed. Let $\{u^{(r)}\}$ be a sequence of vectors in K which converges to $u = (u_i) \in R^N$,

$$u^{(r)} = (u_i^{(r)}) = \sum_{s=1}^{N} b_s^{(r)} v(x_s^{(r)}) \; ; \; b_s^{(r)} \geq 0 \; , \; x_s^{(r)} \in D \; , \; r = 1, 2, \ldots \; . \tag{$\overset{**}{**}$}$$

For any $x \in D$, the first coefficient of $v(x)$ is 1, so that, from eqn ($\overset{**}{**}$),

$$\sum_{s=1}^{N} b_s^{(r)} = u_1^{(r)} \to u_1 \; , \quad \text{as} \quad r \to \infty \; . \tag{$\overset{***}{**}$}$$

Since each $b_s^{(r)}$ is non-negative, we see from eqn (****) that the coefficients $b_s^{(r)}$ are bounded, $|b_s^{(r)}| \le b$ say. Hence, the 2N-vectors

$$(b_1^{(r)},\ldots,b_N^{(r)},\ x_1^{(r)},\ldots,x_N^{(r)})$$

belong to the set

$$S = [-b,+b]\times\ldots\times[-b,+b] \times D \ ,$$

in R^{2N}. Since S is bounded it is sequentially compact (Heine-Borel (Theorem 6.7)), so there exist convergent subsequences

$$\tilde{b}_s^{(r)} \to b_s \ , \quad \tilde{x}_s^{(r)} \to x_s \ , \quad \text{say.}$$

Remembering that $\{u^{(r)}\}$ is convergent, we conclude that

$$u^{(r)} \to \sum_{s=1}^{N} b_s v(x_s) \ ; \ b_s \ge 0 \ , \ x_s \in D \ , \quad \text{for} \quad 1 \le s \le N \ .$$

That is, $\{u^{(r)}\} \to u \in K$, so that K is closed. □

EXAMPLE 7.5: NON-NEGATIVE DIFFERENCE APPROXIMATIONS

Let $u \in C^\infty(-\infty,+\infty)$, and consider the following approximation to the second derivative of u at 0:

$$\ddot{u}(0) \approx -w_0 u(0) + \sum_{i=1}^{N-1} w_i u(x_i) \ , \tag{*}$$

where the weights w_i and nodes $x_i \ne 0$ are to be chosen so as to make the approximation as accurate as possible.

By expanding each of the terms on the right in a Taylor series, we see that the approximation will be of order m if

$$\sum_{i=1}^{N-1} w_i x_i^k = \begin{cases} w_0 \ , & \text{if} \quad k = 0 \ , \\ 2 \ , & \text{if} \quad k = 2 \ , \\ 0 \ , & \text{if} \quad k = 1 \quad \text{or} \quad 2 < k < m \ . \end{cases} \tag{**}$$

When using the approximation (*) to solve boundary value problems for elliptic differential equations, it is very desirable that the weights w_i be non-negative, because this implies that the maximum principle still holds.

We claim that, for a non-negative formula, the order m is at most 4. To see this, we rewrite eqns (**) in the matrix form

$$wA = b \,, \tag{$*_*^*$}$$

where b is the row m-vector,

$$b = (0,0,2,0,\ldots,0) \,,$$

w is the row N-vector,

$$w = (w_0, w_1, \ldots, w_{N-1}) \,,$$

and A is the $N \times m$ matrix,

$$A = \begin{pmatrix} -1 & 0 & 0 & \cdots & 0 \\ 1 & x_1 & x_1^2 & \cdots & x_1^{m-1} \\ 1 & x_2 & x_2^2 & \cdots & x_2^{m-1} \\ \cdot & \cdot & \cdot & \cdots & \\ 1 & x_{N-1} & x_{N-1}^2 & \cdots & x_{N-1}^{m-1} \end{pmatrix} .$$

If $m > 4$ let y be the column m-vector,

$$y^T = (0,0,-1,0,\alpha,0,\ldots,0) \,.$$

Then

$$by = -2 \,,$$

and

$$(Ay)_i = \begin{cases} 0 \,, & \text{if } i = 0 \,, \\ -x_i^2 + \alpha x_i^4 \,, & \text{if } 0 < i < m \,. \end{cases}$$

Picking $\alpha = \max_i(x_i^{-2})$, we have that $(Ay)_i \geq 0$ for all i. Consequently, from Farkas' theorem (Theorem 7.12), there is no non-negative solution of eqn ($*_*^*$) □.

PROBLEMS

7.1 Let A and B be distinct finite subsets of $[0,1]$. Show that there exists a polynomial $p(x)$ such that: (1) $p(x) = 0$ for $x \in A$; (2) $p(x) = 1$ for $x \in B$; and (3) $|p(x)| \leq 1$ for $x \in [0,1]$. [This is a hard problem].

7.2 If $A = \{t_0, t_1, \ldots, t_n\} \subset R^1$ and $f : A \to R^1$, then there exists a polynomial p of degree n, $p : R^1 \to R^1$, which is an extension of f. Show that this is not always true in two dimensions, by showing that there is no map $p : (x,y) \in R^2 \to a + bx + cy \in R^1$ which is an extension of $f : A \to R^1$ where

$$A = \{t_0, t_1, t_2\} = \{(0,0), (-1,-2), (+2,+4)\} \subset R^2 .$$

$$f(t_0) = 0 , \quad f(t_1) = -1 , \quad f(t_2) = +1 .$$

7.3 Let A and B be disjoint closed subsets of $[0,1]$. Show that there exists a real infinitely differentiable function f such that: (1) $f = 0$ on A; (2) $f = 1$ on B; (3) $|f| \leq 1$ on $[0,1]$. [Hint: First apply the Urysohn theorem to closed sets A_1 and B_1 strictly containing A and B, respectively, and then smooth using mollifiers (Section 4.11, page 96)].

7.4 Let $I_n(f) = \sum_{i=1}^{n} w_i f(x_i)$ be a quadrature formula for approximating the integral

$$I(f) = \int_B f(x)dx ,$$

where B is the unit ball in R^n. Use Urysohn's theorem to show that there exists a real non-negative continuous function $f : B \to R^1$ such that $I_n(f) = 0$ and $I(f) = 1$. {234}

7.5 Let A be the closed region in R^2 defined by:

$$A = \{(x_1,x_2): x_1^2+x_2^2 \le 1 \text{ and (either } x_1 \le 0 \text{ or } |x_2| \ge |x_1|^{3/2})\}$$

Let

$$f_c(x_1,x_2) = \begin{cases} 0 , & \text{if } x_1 \le 0 , \\ cx_1^2/(1+c^2x_1^2), & \text{if } x_1 > 0 \text{ and } x_2 > 0 , \\ -cx_1^2/(1+c^2x_1^2), & \text{if } x_1 > 0 \text{ and } x_2 < 0 , \end{cases}$$

where c is a positive constant.

Show that, on A, $|f_c| \le 1$, $|\partial f_c/\partial x_1| \le 9/(8\sqrt{3})$, and $|\partial f_c/\partial x_2| = 0$. Show also that if

$$p = \frac{1}{3^{1/2}c} , \quad \text{and} \quad q = \frac{1}{3^{3/4}c^{3/2}} = p^{3/2} ,$$

then

$$\frac{f_c(p,q)-f_c(p,-q)}{2q} = \frac{3^{3/4}}{4} c^{1/2} .$$

Hence show that if F_c is an extension of f_c to R^2 then

$$\sup_{x \in R^2} \left| \frac{\partial F_c}{\partial x_2} \right| \ge \frac{3^{3/4}}{4} c^{1/2} .$$

{234}

7.6 Let $f : t \in [0,1] \to -[t(1-t)]^{1/2} \in R^1$. Show that f is continuous and convex on $[0,1]$, but that there is no continuous convex extension of f to R^1. (Remark 7.2)

7.7 Let $f : A \to \overline{R}^1$, be convex, where A is a convex subset of a vector space X and $\overline{R}^1$ is the extended real line (Example 2.16).

Show that, with the convention that $0.\infty = 0$,

$$F : x \in X \to \begin{cases} f(x) , & \text{if } x \in A , \\ +\infty , & \text{otherwise} , \end{cases}$$

is a convex extension of f . (Remark 7.2, p. 235)

7.8 Develop the theory of the Riemann integral for functions in $C[a,b]$ by first defining the integral for stepfunctions and then using Theorem 7.3. {236}

7.9 Complete the proof of the Hahn-Banach theorem (Theorem 7.4) without using Zorn's Lemma in the following two special cases:

(a) X is finite dimensional.

(b) X is separable, and $p(x) = \|x\|$.

[Hint: In case (b), let $\{x_n\}$ be a denumerable set which is dense in X . Show that f_0 can be extended to the set

$$\tilde{X} = \{x_0 + \sum_{i \in I} \lambda_i x_i ;\ x_0 \in X_0 , I \text{ finite}\} ,$$

which is dense in X . Then apply Theorem 7.3.] {239}

7.10 Let A be the set of all linear extensions of a mapping. Let $g_1 \geq g_2$ if g_1 is an extension of g_2 . Prove that $\geq$ is an order relation with respect to which A is partially ordered. {239}

7.11 Prove that if $\hat{g}$ is defined as in the proof of Theorem 7.4 (Hahn-Banach), then $\hat{g} \in A$ and $\hat{g} \geq g$ for all $g \in A_0$. {239}

7.12 Show that the functional f_0 constructed in the proof of Corollary 7.6 satisfies conditions (1)-(3) of the corollary. {241}

7.13 Let X be a normed vector space. Let $x \subset X$. Prove that there exists a continuous linear functional f on X such that: (1) $\|f\| = 1$; (2) $f(x) = \|x\|$.

7.14 Complete Example 7.1 by showing:

(1) $\max_{0 \le s,t \le 1} |(st) - (\frac{1}{2}(s+t) - \frac{1}{4})| = \frac{1}{4}$.

(2) $\|f\| = 1$, $f : M \to \{0\}$, and $f(x_0) = \frac{1}{4}$.

(3) $\max_{0 \le s,t \le 1} |(st) - (s+t-\frac{1}{2}(s^2+t^2) - \frac{1}{4})| = \frac{1}{4}$. {242}

7.15 With the notation of Example 7.2, for $x \in C[0,1]$ let

$$\ell(x) = x[t_0,\ldots,t_{n+1}] / \sum_{k=0}^{n+1} \alpha_k ,$$

$$= \sum_{k=0}^{n+1} (-1)^{n+1-k} \alpha_k x(t_k) / \sum_{k=0}^{n+1} \alpha_k .$$

Show that:

(1) α_k is non-negative for all k .

(2) If x is a polynomial of degree r then $x[t_0,\ldots,t_n,t]$ is a polynomial of degree $\min(0,r-n-1)$. (Use induction on n) .

Hence deduce that ℓ annihilates polynomials of degree n , and that $\|\ell\| = 1$. {243}

7.16 Let $F : X \to R^1$ be the support function for a convex set K with internal point $\underline{0}$ in a real vector space X . Prove that

(1) $F(x) \le 1$, for $x \in K$,

(2) $F(x) < 1$ if x is an internal point of K .

(3) If $x \notin K$ then $F(x) \ge 1$. {246}

7.17 Let M and N be convex sets in a vector space. Show that $M+N$, $M-N = M+(-1)N$, and $M \cap N$ are convex. {247}

7.18 Prove that an interior point p of a set S in a topological vector space X is an internal point of S . {247}

7.19 Let f be a linear mapping, $f : X \to Y$, where X and Y are topological vector spaces. Assume that $f(N) \subset M$, where N is a neighbourhood of $\underline{0} \in X$ and M is topologically bounded. Show that f is continuous. {248}

7.20 Consider the following special case of Theorem 7.13: $n = 2$, $d = 2$, $D = [-1,1] \times [-1,1]$, $p \equiv 1$. Show that the eqns (7.3) take the form $wA = c$ where,

$$c = (4,0,0,4/3,0,4/3) ,$$

$$A = \begin{bmatrix} 1 & \xi_{1,1} & \xi_{1,2} & (\xi_{1,1})^2 & \xi_{1,1}\xi_{1,2} & (\xi_{1,2})^2 \\ 1 & \xi_{2,1} & \xi_{2,2} & (\xi_{2,1})^2 & \xi_{2,1}\xi_{2,2} & (\xi_{2,2})^2 \\ 1 & \xi_{3,1} & \xi_{3,2} & (\xi_{3,1})^2 & \xi_{3,1}\xi_{3,2} & (\xi_{3,2})^2 \\ 1 & \xi_{4,1} & \xi_{4,2} & (\xi_{4,1})^2 & \xi_{4,1}\xi_{4,2} & (\xi_{4,2})^2 \\ 1 & \xi_{5,1} & \xi_{5,2} & (\xi_{5,1})^2 & \xi_{5,1}\xi_{5,2} & (\xi_{5,2})^2 \\ 1 & \xi_{6,1} & \xi_{6,2} & (\xi_{6,1})^2 & \xi_{6,1}\xi_{6,2} & (\xi_{6,2})^2 \end{bmatrix} .$$

{253}

7.21 Prove that if the Laplacian operator $\nabla^2 u = u_{xx} + u_{yy}$ is approximated by a non-negative difference approximation at gridpoints,

$$u_{xx}(0,0) + u_{yy}(0,0) \approx -w_0 u(0,0) + \sum_{i=1}^{N} w_i u(n_i h, m_i h) ,$$

then the error is at least $0(h^4)$.

7.22 Show that the Laplace equation

$$u_{xx} + u_{yy} = 0 , \qquad (*)$$

is approximated with error $0(h^6)$ by the non-negative 'nine-point' difference approximation

$$\begin{aligned} 0 = -20u(0,0) &+ 4[u(h,0)+u(0,h)+u(-h,0)+u(0,-h)] + \\ &+ [u(h,h)+u(-h,-h)+u(h,-h)+u(-h,h)] . \end{aligned} \qquad (**)$$

Explain why this does not contradict Problem 7.21.

7.23 Let

$$Lu(x) = \frac{1}{2} \sum_{i,k=1}^{n} a_{ij}(x) \frac{\partial^2 u}{\partial x_i \partial x_j} + \sum_{j=1}^{n} b_j(x) \frac{\partial u}{\partial x_j} + c(x)\, u ,$$

be an elliptic operator in a domain $D \subset R^n$ with co-efficients which depend upon $x \in D$. Let there be a uniform grid with grid length h on D. Show that it is possible to choose a fixed finite set of points $m_s \in R^n$, with integer coefficients, such that, for small h, there is a difference approximation

$$L_h u(x) = \sum_s w_s(x,h) u(x+m_s h) = Lu(x) + O(h^2) , \quad x \in D ,$$

where $w_s(x,h) \geq 0$. [This is hard].

7.24 Use the Farkas lemma to prove the statement in Problem 6.30(a).

8
Bases and projections

Given a problem with solution u in an infinite dimensional space X we must seek approximate solutions in some n-dimensional space X_n. If $X_n \subset X$ we speak of an *internal approximation*; otherwise of an *external approximation*. Finite element approximations are usually internal approximations, while finite difference approximations are external approximations.

In general we do not know u, but we may have certain a-priori information about u; for example, if u is the solution of an integral or differential equation, then we often know that certain derivatives of u are bounded. One would like to choose X_n so that u can be adequately approximated by elements from X_n. In this chapter we give a general discussion of some of the criteria for choosing internal approximation spaces X_n.

In the last section of the chapter we prove the open mapping theorem and closed graph theorem; these results are needed in the present chapter and are also used in Chapter 9.

8.1 *SCHAUDER BASES*

A *basis* for a topological vector space X is a sequence $\{x_i\}$ in X such that to every $x \in X$ there corresponds a unique sequence of scalars $\lambda_i(x)$ for which

$$x = \lim_{n\to\infty} \sum_{i=1}^{n} \lambda_i(x)x_i . \tag{8.1}$$

Each element x_i in the basis has the convergent expansion

$$0x_1 + 0x_2 + \ldots + 0x_{i-1} + 1x_i + 0x_{i+1} + 0x_{i+2} \ldots .$$

Since the expansion (8.1) is unique, it follows that

$$\lambda_i(x_j) = \delta_{ij} = \begin{cases} 1 , & \text{if } i = j , \\ 0 , & \text{otherwise.} \end{cases} \tag{8.2}$$

It is a further consequence of the uniqueness of the expansion (8.1) that the coefficients $\lambda_i(x)$ depend linearly upon x (Problem 8.1), and the linear functionals λ_i are called the *coefficient functionals* of the basis $\{x_i\}$. If the coefficient functionals are continuous then the basis $\{x_i\}$ is called a *Schauder basis*.

THEOREM 8.1. *In a Banach space* X *every basis* $\{x_i\}$ *is a Schauder basis.*

Proof. Let $\{x_i\}$ be a basis in a Banach space X. Let Y be the space of scalar sequences $\{\alpha_i\}$ for which the limit

$$\lim_{n\to\infty} \sum_{i=1}^{n} \alpha_i x_i = \sum_{i-1}^{\infty} \alpha_i x_i \in X ,$$

exists. Y is a normed linear space with the norm

$$\|\{\alpha_i\};Y\| = \sup_n \|\sum_{i=1}^{n} \alpha_i x_i ;X\| . \tag{*}$$

Using the identity

$$\|a\| = \|a+b-b\| \le \|a+b\| + \|b\| ;$$

with $a = \alpha_n x_n$ and $b = \sum_{i=1}^{n-1} \alpha_i x_i$, it follows that

$$|\alpha_n| \|x_n\| \le \|\sum_{i=1}^{n} \alpha_i x_i ;X\| + \|\sum_{i=1}^{n-1} \alpha_i x_i ;X\| ,$$

$$\le 2 \sup_n \|\sum_{i=1}^{n} \alpha_i x_i ;X\| ,$$

$$= 2 \|\{\alpha_i\};Y\| . \tag{**}$$

If $\{y^{(p)}\}$ is a Cauchy sequence in Y with $y^{(p)} = \{\alpha_i^{(p)}\}$, then, applying eqn (**),

$$|\alpha_n^{(p)} - \alpha_n^{(q)}| \|x_n\| \le 2 \|y^{(p)} - y^{(q)} ;Y\| ,$$

so that, for each fixed n , $\{\alpha_n^{(p)}\}$ is a scalar Cauchy sequence converging to $\tilde{\alpha}_n$, say. An $\varepsilon\delta$-argument (Problem 8.2) shows that Y is complete, and hence Y is a Banach space.

The map $T : Y \to X$,

$$T : \{\alpha_i\} \in Y \to \sum_{i=1}^{\infty} \alpha_i x_i \in X ,$$

is continuous, linear, and one-to-one, and maps Y onto X. Thus, by the open mapping theorem, T^{-1} is continuous (see Theorem 8.7). Thus

$$\|\{\alpha_i\};Y\| \le \|T^{-1}\| \, \|\sum_{i=1}^{\infty} \alpha_i x_i;X\| . \tag{$*_*^*$}$$

Using eqns (8.1), (**), and ($*_*^*$), we obtain

$$|\lambda_n(x)| \, \|x_n\| \le 2\|\{\lambda_i(x)\};Y\| \le 2\|T^{-1}\| \, \|\sum_{i=1}^{\infty} \lambda_i(x)x_i\| = 2\|T^{-1}\| \, \|x\|.$$

□

With a basis $\{x_i\}$ in a vector space X we can associate the sequence of subspaces

$$X_n = \text{span}\{x_1,\dots,x_n\} \subset X , \tag{8.3}$$

and operators

$$P_n : x \in X \to \sum_{i=1}^{n} \lambda_i(x)x_i \in X_n \subset X . \tag{8.4}$$

Using eqn (8.2) we have

$$P_n : x_i \to x_i , \quad \text{for} \quad 1 \le i \le n . \tag{8.5}$$

Furthermore, since $P_n x \to x$ for all $x \in X$, the Banach-Steinhaus theorem (Theorem 5.4) implies that the P_n are uniformly bounded,

$$\|P_n\| \le M < \infty , \quad \text{for all} \quad n . \tag{8.6}$$

EXAMPLE 8.1.

Let $X = C[0,1]$,

$$x_1(t) = 1 - t , \quad 0 \le t \le 1 ,$$

$$x_2(t) = t , \quad 0 \le t \le 1 ,$$

$$h(t) = \begin{cases} 2t , & 0 \le t \le \frac{1}{2} , \\ 2 - 2t , & \frac{1}{2} \le t \le 1 , \\ 0 , & \text{otherwise} \end{cases}$$

$$x_{2^m+j+1}(t) = h(2^m t+1-j) , \quad j = 1,2,\ldots,2^m; \ m = 0,1,2,\ldots .$$

The coefficient functionals $\lambda_i(x)$ are given by,

$$\lambda_1(x) = x(t_1) ,$$

$$(*)$$

$$\lambda_i(x) = x(t_i) - \sum_{k=1}^{i-1} \lambda_k(x) x_k(t_i) , \quad \text{for } i > 1 ,$$

where

$$t_1 = 0 ,$$

$$t_2 = 1 ,$$

$$t_{2^m+j+1} = (2j-1)/2^{m+1} , \quad 1 \le j \le 2^m; \ m = 0,1,2,\ldots .$$

Behind this welter of formulae there lies a simple concept. The function $h(t)$ is the *hat function* which is piecewise linear, vanishes at $t = 0$ and $t = 1$, and is equal to 1 at $t = \frac{1}{2}$. The functions $x_3, \ldots,$ are hat functions with increasingly small supports — some of them are shown in Figure 8.1.

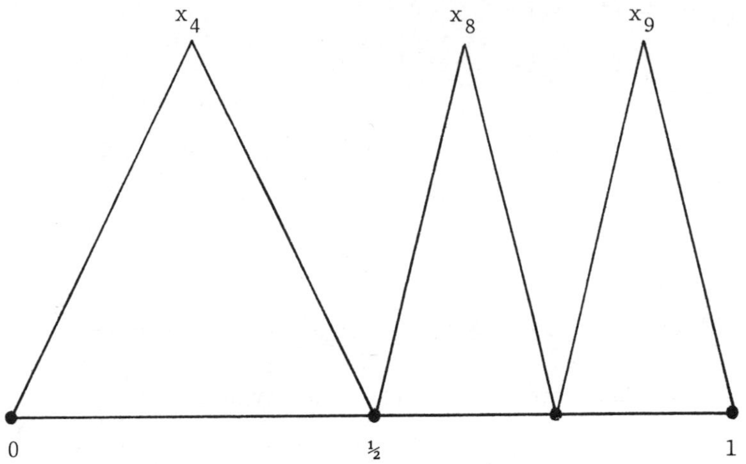

Figure 8.1. Some of the basis functions x_i *in Example 8.1.*

If P_n is defined by eqn (8.4) then the function $P_n x$ is the piecewise linear function which interpolates x at the points $t_1, t_2, \ldots, t_n$. (In particular, the functions $P_{2^m+1} x$ are the piecewise linear splines interpolating x at the equidistant knots $t_1, \ldots, t_{2^m+1}$.)

Using these facts it is readily shown (Problem 8.3) that

$$||P_n x - x|| \to 0 , \quad \text{as} \quad n \to \infty ,$$

so that $\{x_i\}$ is a basis. We see from eqn (*) that the coefficient functionals are continuous, so that $\{x_i\}$ is also a Schauder basis (this also follows from Theorem 8.1).

□

EXAMPLE 8.2. The functions $x_i : t \to t^{i-1}$ for $i = 1,2,\ldots,$ do not form a basis in $C[0,1]$.

To see this, we observe that if $\{x_i\}$ were a basis, then, by virtue of eqn (8.4), the associated projection P_n would map every polynomial of degree less than n into itself. That is, with the terminology of Section 5.5, P_n would be a polynomial operation of order $n-1$. It then follows from Theorem 5.7 that there exists $x \in C[0,1]$ such that $P_n x \not\to x$ as $n \to \infty$. □

For numerical purposes it is clearly useful if the underlying space X has a Schauder basis $\{x_i\}$, together with the associated spaces X_n (defined by eqn (8.3)) and mappings P_n (defined by eqn (8.4)). As Example 8.2 shows, the requirements for a Schauder basis can serve to eliminate undesirable approximations.

Bases are known for most of the common Banach spaces, but if $2 < p < \infty$ there is a subspace of ℓ^p which has no basis (Singer [1981, p. 2]). There is also a separable locally convex topological vector space with a basis which is not a Schauder basis (Marti [1969, p. 48]), and a Fréchet space which has no basis (Marti [1969, p. 127]). There is an extensive literature on bases: see Marti [1969] and Singer [1981].

8.2 *PROJECTIONS*

Let X be a vector space. A linear map $P : X \to X$ is a *projection in* X or a *projection of* X *onto* $P(X)$ if $P^2 = P$, that is, if $(y = Px) \Rightarrow (Py = y)$.

We have already encountered several examples of projections:

(1) $X = C[0,1]$; Px the polynomial of degree n which interpolates x at $n+1$ prescribed nodes.

(2) $X = C[0,1]$; Px the cubic spline which interpolates x at a prescribed set of knots (Section 5.6, page 144).

(3) X any topological vector space with a basis $\{x_i\}$; P_n the associated operator defined by eqn (8.4) mapping $X \to X_n = \text{span}\{x_1,\ldots,x_n\}$.

As the above examples show, projections occur naturally in numerical analysis. The theorem below gives an additional reason for interest in projections:

THEOREM 8.2. *Let* V *be a linear subspace in a normed vector space* X , *and let* P *be any projection of* X *onto* V . *Then*

$$||x-Px|| \le ||I-P|| \, d(x,V) \le (1+||P||) \, d(x,V) \, , \qquad (8.7)$$

where

$$d(x,V) = \inf_{y \in V} ||x-y|| \, .$$

Proof. For any $y \in V$ we have $y = Py$, and thus,

$$||x-Px|| = ||(x+y)-P(x+y)|| \le ||I-P|| \, ||x+y|| \le ||I-P|| \inf_{y \in V} ||x-y|| \, . \quad \square$$

Theorem 8.2 leads to the following comments:

(1) Even if V is a linear subspace, it is in general difficult to compute the proximity map (p. 212) which maps x onto a nearest point in V . (The best example is Chebyshev approximation, where the problem of computing the best approximation to $x \in C[0,1]$ in the subspace $V = \pi_n$ consisting of polynomials of degree n , is a difficult nonlinear problem.) However, as eqn (8.7) shows, if P is a projection of X onto V of small norm, $||P|| \le 4$ say, then the error $||x-Px||$ is at most 5 times the best possible error $d(x,V)$.

(2) Given X and V , this suggests the problem of determining the best lower bound, the *relative projection constant* denoted by $p(V,X)$, for the norm of any projection of X onto V . For example, it follows from eqn(5.30) (p. 142), that

$$p(\tilde{\pi}_n \, , \, \tilde{C}[0,2\pi]) \ge \frac{4}{\pi^2} \ln(\frac{n+1}{2}) \, .$$

Cheney and Price [1970] give an excellent survey of this problem.

(3) One must of course be careful not to be dogmatic and use only projections. Problem 8.7 gives an example of a case where the proximity map is easily computed.

We now derive some of the properties of projections.

Let P be a projection. Then $Q = I - P$ is also a projection since

$$Q^2 = Q.Q = (I-P)(I-P) = I - 2P + P^2 = I - P = Q .$$

Furthermore,

$$PQ = QP = P(I-P) = (I-P)P = P - P^2 = 0 ; \qquad (8.8)$$

that is, P and Q are *orthogonal*, $P \perp Q$.

Given a projection P , each $x \in X$ can be written as a sum

$$x = x_1 + x_2 ; \; x_1 \in M , \; x_2 \in N , \qquad (8.9)$$

where: (1) $x_1 = Px$, $x_2 = (I-P)x$; and (2) $M = R(P)$ and $N = R(I-P)$ are linear subspaces of X . ($R(P)$ is the range of P). The representation (8.9) is unique (Problem 8.8).

If M and N are linear subspaces of a vector space X such that a unique representation (8.9) holds for every $x \in X$ we write $X = M \oplus N$, and say that X is the *direct sum* of M and N .

There is a one-to-one relationship between projections in a space X and direct sums $X = M \oplus N$. On the one hand, as we have just seen, if P is a projection then $X = R(P) \oplus R(I-P)$. On the other hand, let $X = M \oplus N$, and define $P : X \to M$ by

$$P(x) = P(x_1+x_2) = x_1 ,$$

where x_1 and x_2 are as in eqn (8.9); then P is a projection of X onto M . (Problem 8.9)

Given a linear subspace M in a vector space X there always exists a projection of X onto M (Problem 8.10).

It is more interesting to consider the case when X is a topological vector space and ask when continuous projections exist. We have

THEOREM 8.3. *Let* $X = M \oplus N$ *be a Fréchet-space. Let* P *and* $I-P$ *be projections of* X *onto* M *and* N, *respectively.* *Then:* P *is continuous* $\Leftrightarrow$ M *and* N *are closed.*

Proof. $\Rightarrow$: To prove that M is closed it suffices to prove that M contains all its limit points (Theorem 3.2 (part 3)). Let $\{x_i\}$ be a sequence in M converging to $x \in X$. Since $x_i = Px_i$, and P is continuous,

$$x = \lim x_i = \lim Px_i = Px \in M .$$

Therefore, M is closed. The same argument shows that $N = R(I-P)$ is also closed.

$\Leftarrow$: Let $\{x_i\}$ be any sequence in X such that $x_i \to x \in X$ and $y_i = Px_i \to y \in X$. Consider the identity

$$\begin{aligned} x_i &= Px_i + (I-P)x_i , \\ &= y_i + z_i , \text{ say.} \end{aligned}$$

Since $x_i \to x$ and $y_i \to y$, the sequence $\{z_i\}$ must converge, $z_i \to z$ say, so that

$$x = y + z . \tag{*}$$

But the sequence $\{y_i\}$ lies in the closed set M, so $y \in M$. Similarly, $z \in N$. Therefore, from the uniqueness of representations of the form (*), $y = Px$.

In summary, if $x_i \to x$ and $Px_i \to y$, then $y = Px$. In the terminology of page 287, P is closed. In addition, P maps X into M which, as a closed linear subspace in a Fréchet space, is itself a Fréchet space. Applying Theorem 8.8 we conclude that P is continuous. □

REMARK 8.1. The condition in Theorem 8.2 that both M and N be closed, is necessary. The space ℓ^∞ is the Banach space of real sequences $\{x_n\}$ with $\|x\| = \sup|x_n| < \infty$ ℓ^∞ contains the closed linear subspace c_0 which consists of all sequences $\{x_n\}$ such that $x_n \to 0$ as $n \to \infty$. There is no continuous projection of ℓ^∞ onto c_0. (Marti [1969, p. 90]). □

8.3 *HILBERT SPACES*

As seen in the previous sections, a Banach space X may not have a continuous projection onto a given closed linear subspace, and may even not have a basis. As will be seen, here, Hilbert spaces (which, it will be recalled, are complete inner-product spaces) have these nice properties and many others.

Throughout this section, unless otherwise stated, all elements or points will be assumed to belong to a real or complex Hilbert space X with inner product (x,y) and norm $\|x\| = (x,x)^{\frac{1}{2}}$.

As shown in Section 2.8, the Schwarz inequality

$$|(x,y)| \leq \|x\|\,\|y\| ,$$

holds in an inner-product space. This implies that (x,y) is a continuous function of x and y.

Furthermore (Problem 2.20) the parallelogram identity holds:

$$\|x+y\|^2 + \|x-y\|^2 = 2[\|x\|^2 + \|y\|^2] .$$

Two points x and y are *orthogonal* if $(x,y) = 0$; this is also written, $x \perp y$. If

$$x = x_1 + x_2 + \ldots + x_n ,$$

where the x_i are mutually orthogonal ($x_i \perp x_j$ for $i \neq j$) then, as is easily proved by induction on n, the '*Pythagoras identity*' holds:

$$\|x\|^2 = \sum_{i=1}^{n} \|x_i\|^2 .$$

Two sets, E_1 and E_2, in X are said to be *orthogonal* if $(x_1,x_2) = 0$ for every $x_1 \in E_1$ and $x_2 \in E_2$; this is written as: $E_1 \perp E_2$. If E is a set in X then the set $E^\perp$ of all $x \in X$ orthogonal to E is called the *orthogonal complement of E*. It is readily shown that the orthogonal complement of a set is a closed linear subspace (Problem 8.12).

THEOREM 8.4. *Let* M *be a closed linear subspace in a Hilbert space* X. *Then:*

(1) *There exists a proximity map* P *which maps* $x \in X$ *onto the unique closest point* $Px \in M$.

(2) *For all* $x, z \in X$: $z = Px \Leftrightarrow z \in M$ *and* $(x-z) \perp M$.

(3) P *is linear, and hence a projection. (*P *is often called the orthogonal projection onto* M*).*

(4) $\|P\| = 1$.

(5) $X = M \oplus N$, *where* $N = M^\perp$ *is the orthogonal complement of* M. $M = R(P)$ *and* $N = R(I-P)$.

(6) *If* $M = \text{span}\{y_1,\ldots,y_n\}$, *where the* y_i *are non-zero and mutually orthogonal, then*

$$Px = \sum_{i=1}^{n} \frac{(x,y_i)}{(y_i,y_i)}\, y_i \,. \tag{8.10}$$

Proof. (1): Choose $x \in X$. We construct $z \in M$. If $x \in M$ set $z = x$. Otherwise, $d = \text{dist}(x,M) > 0$. As in Section 6.3.1, choose a minimizing sequence $\{z_n\}$ in M such that

$$\|z_n - x\|^2 \le d^2 + 1/n^2 \,, \quad \text{for } n \ge 1 \,. \tag{*}$$

Using the parallelogram identity and eqn (*) we find that

$$\begin{aligned} \|z_n - z_m\|^2 &= \|(z_n - x) - (z_m - x)\|^2 \,, \\ &= 2[\|z_n - x\|^2 + \|z_m - x\|^2] - \|(z_n - x) + (z_m - x)\|^2 \,, \\ &\le 4d^2 + 2/n^2 + 2/m^2 - \|(z_n - x) + (z_m - x)\|^2 \,. \end{aligned} \tag{**}$$

But, $(z_n+z_m)/2 \in M$ and so

$$||(z_n-x)+(z_m-x)||^2 = 4||(z_n+z_m)/2-x||^2 \geq 4d^2 .$$

Substituting into eqn (**), we see that

$$||z_n-z_m||^2 \leq 2/n^2 + 2/m^2 ,$$

so that $\{z_n\}$ is a Cauchy sequence in M. Remembering that M is closed, we conclude that $\{z_n\}$ converges to some element in M, z say.

The point z constructed above clearly satisfies

$$||z-x|| = d = \inf_{y \in M} ||x-y|| .$$

z is unique. [If $\tilde{z} \in M$ satisfies $||\tilde{z}-x|| = d$ then the parallelogram identity implies that

$$\begin{aligned} ||2x-(z+\tilde{z})||^2 &= 2[||x-\tilde{z}||^2 + ||x-z||^2] - ||z-\tilde{z}||^2 , \\ &= 4d^2 - ||z-\tilde{z}||^2 , \end{aligned}$$

which is only possible if $z = \tilde{z}$ since otherwise the mid-point $(z+\tilde{z})/2$ would be closer to x than z.].

Set $Px = z$.

(2) $\Rightarrow$: By construction, $z = Px \in M$. If $x - z$ is not orthogonal to M, then $(x-z,y) \neq 0$ for some $y \in M$. But then it is possible to choose a scalar α such that

$$\begin{aligned} ||x-z-\alpha y||^2 &= ||x-z||^2 - 2\mathrm{Re}[\bar{\alpha}(x-z,y)] + |\alpha|^2||y||^2 , \\ &< d^2 , \end{aligned}$$

contradicting the definition of d.

(2) $\Leftarrow$: For any $y \in M$, we have $y - z \in M$ and thus $(x-z) \perp (y-z)$. Using the Pythagoras identity,

$$\begin{aligned} ||x-y||^2 &= ||(x-z)-(y-z)||^2 , \\ &= ||(x-z)||^2 + ||y-z||^2 , \\ &\geq ||x-z||^2 , \end{aligned}$$

so that, from (1), $z = Px$.

(3) Let $x,y \in X$, and set $z = Px + Py$. Then $z \in M$ and

$$x + y - z = (x-Px) + (y-Py) ,$$

so that $(x+y-z) \perp M$. Thus, by (2), $P(x+y) = z = P(x) + P(y)$. Similarly, $P(\lambda x) = \lambda P(x)$, for any λ.

(4) For any $x \in X$,

$$x = Px + (I-P)x = x_1 + x_2 \text{ , say.}$$

Since $x_1 \in M$ and $x_2 = (x-x_1) \perp M$, the Pythagoras identity implies that

$$||x||^2 = ||x_1||^2 + ||x_2||^2 \geq ||x_1||^2 = ||Px||^2 ,$$

so that $||Px|| \leq ||x||$, with equality when $x \in M$.

(5) This is a restatement of (2).

(6) One verifies by direct computation that if Px is is given by eqn (8.10) then $Px \in M$ and $(x-Px) \perp M$. Now use (2). □

Theorem 8.4 permits a graphic illustration (Figure 8.2). The space X is the direct sum of the orthogonal complements M and N. Any point $x \in X$ can be written as the sum of two orthogonal elements x_1 and x_2, each of which is the unique closest point to x in its own subspace.

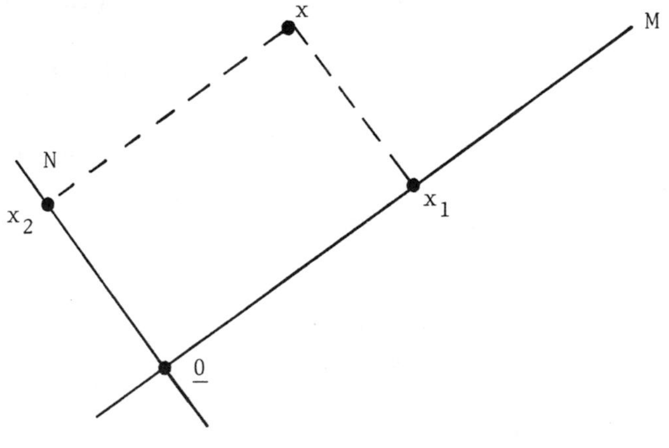

Figure 8.2. Graphic illustration of Theorem 8.4

REMARK 8.2. Theorem 8.4 may be contrasted with Theorem 6.22 where it was proved that there exists a proximity map onto a finite dimensional linear subspace of a Banach space. □

We now consider bases in Hilbert spaces. A system $\{x_\alpha;\alpha\in A\}$ of elements in a Hilbert-space X is *orthogonal* if every two distinct elements of the system are orthogonal; if, in addition, $\|x_\alpha\| = 1$ for every x_α then the system $\{x_\alpha\}$ is *orthonormal*. If every $x \in X$ can be approximated arbitrarily closely by finite linear combinations of the x_α then $\{x_\alpha\}$ is *fundamental*.

Let $\{x_i\}$ be a denumerable orthonormal system for a Hilbert-space X. For $x \in X$ the numbers $a_i = (x,x_i)$ are called the *Fourier coefficients* of x with respect to the system $\{x_i\}$ while $\sum_i a_i x_i$ is called the *Fourier series* of x with respect to the system $\{x_i\}$. The sums

$$s_n(x) = \sum_{i=1}^{n} a_i x_i = \sum_{i=1}^{n} (x,x_i) , \tag{8.11}$$

are called the *partial Fourier sums of* x. Comparing eqns (8.10) and (8.11) we see that $s_n(x)$ is the (proximity) projection of x onto $X_n = \text{span}\{x_1,\ldots,x_n\}$. Thus, from Theorem 8.4 (part 4),

$$\|s_n\| = 1 . \tag{8.12}$$

Furthermore, $s_n(x) \perp (x-s_n(x))$, so that, by Pythagoras,

$$\|x\|^2 = \|x-s_n(x)\|^2 + \|s_n(x)\|^2 . \tag{8.13}$$

From eqns (8.11) and (8.12),

$$\|s_n(x)\|^2 = \sum_{i=1}^{n} |a_i|^2 = \sum_{i=1}^{n} |(x,x_i)|^2 \le \|x\|^2 , \tag{8.14}$$

which is *Bessel's inequality*. It follows from Bessel's inequality that the real series of non-negative terms,

$$\sum_i |a_i|^2 = \sum_i |(x,x_i)|^2 \tag{*}$$

has partial sums which are bounded by $||x||^2$; therefore, the series (*) is convergent. Choose $\varepsilon > 0$. For any $m \geq n$,

$$||s_m(x)-s_n(x)||^2 = ||\sum_{i=n+1}^{m} (x,x_i)x_i||^2 = \sum_{i=n+1}^{m} |(x,x_i)|^2 = d_{m,n} \quad \text{say.} \tag{**}$$

Since the series (*) converges, there exists N such that $d_{m,n} \leq \varepsilon$ if $m,n \geq N$. Therefore, for any $x \in X$, the sequence $\{s_n(x)\}$ is a Cauchy sequence in X , and thus converges. Taking limits in eqn (8.13) we obtain,

$$\begin{aligned} ||x||^2 &= ||x-\lim s_n(x)||^2 + \lim||s_n(x)||^2 , \\ &= ||x-\lim s_n(x)||^2 + \sum_i |(x,x_i)|^2 . \end{aligned} \tag{8.15}$$

We conclude from eqn (8.15) that $\lim s_n(x) = x$ iff

$$||x||^2 = \sum_i |(x,x_i)|^2 . \tag{8.16}$$

Equation (8.16) is known as *Parseval's equality*, and an orthonormal system $\{x_i\}$ which satisfies Parseval's equality for all $x \in X$ is said to be *complete*. It is readily proved that a complete orthonormal system $\{x_i\}$ is a Schauder basis (Problem 8.13).

Let $\{z_i\}$ be a denumerable set in X . Then it is possible to construct an orthonormal system $\{x_i\}$ such that $\text{span}\{x_i : i \geq 1\} = \text{span}\{z_i : i \geq 1\}$. We begin by deleting any element z_i which is linearly dependent on its predecessors $z_1,\ldots,z_{i-1}$. The surviving elements will be denoted by $\{\tilde{z}_i\}$. We now construct $\{x_i\}$ by applying the *Gram-Schmidt algorithm:*

$$x_1 = \tilde{z}_1/||\tilde{z}_1|| .$$

$$\left.\begin{aligned} \tilde{x}_i &= \tilde{z}_i - \sum_{j=1}^{i-1} (\tilde{z}_i,x_j)x_j , \\ x_i &= \tilde{x}_i/||\tilde{x}_i|| . \end{aligned}\right\} \; i \geq 2 . \tag{8.17}$$

Each $\tilde{x}_i$ is nonzero, since otherwise $\tilde{z}_i$ would depend linearly upon $x_1,\ldots,x_{i-1}$, and hence upon $\tilde{z}_1,\ldots,\tilde{z}_{i-1}$, which is not possible. The construction of x_i is such that $x_i \perp x_j$ for $j < i$. Consequently, $\{x_i\}$ is a denumerable orthonormal system. It follows from the construction that each z_i is a linear combination of $x_1,\ldots,x_i$, so that $\text{span}\{z_i : i \geq 1\} = \text{span}\{x_i : i \geq 1\}$.

THEOREM 8.5. *Let* X *be a separable Hilbert space. Then:*

(1) X *has a fundamental denumerable orthonormal system* $\{x_i\}$.

(2) Let $\{x_i\}$ *be a denumerable orthonormal system in* X. *Then:* $\{x_i\}$ *is complete* $\Leftrightarrow$ $\{x_i\}$ *is fundamental.*

Outline of Proof. (1): Since X is separable, there exists a dense denumerable subset, $\{z_i\}$ say. Apply the Gram-Schmidt algorithm to $\{z_i\}$

(2) $\Rightarrow$*:* For any $x \in X$ and $\varepsilon > 0$, $\|s_n(x)-x\| < \varepsilon$ for large n.

(2) $\Leftarrow$*:* Given $\varepsilon > 0$, there exists $z_n \in X_n = \text{span}\{x_1,\ldots,x_n\}$ such that $\|z_n - x\| \leq \varepsilon$. But, for $m \geq n$, $s_m(x)$ is the proximity map of x onto $X_m \supset X_n$. Therefore,

$$\|s_m(x)-x\| \leq \varepsilon, \quad \text{for } m \geq n. \quad \square$$

We conclude this section with some remarks on the construction of complete orthonormal systems in Hilbert spaces:

REMARK 8.3. Under appropriate conditions, the eigenfunctions ϕ_i of certain eigenvalue problems for elliptic equations form a complete orthonormal system.

The archetypical example is the problem:

$$\ddot{\phi}(t) + \lambda\phi(t) = 0, \quad 0 \leq t \leq \pi,$$

$$\phi(0) = \phi(\pi) = 0,$$

with eigenfunctions

$$\phi_n(t) = \sin(nt), \quad n \geq 1.$$

These functions give rise, of course, to the classical Fourier sine series.

The study of such eigensystems is a sophisticated and specialized branch of analysis. See Courant and Hilbert [1953, Chapters 2 and 5] and Titchmarsh [1958]. □

REMARK 8.4. One can select a fundamental system $\{z_i\}$ and then apply the Gram-Schmidt algorithm (8.17) to construct a complete orthonormal system $\{x_i\}$. This approach is well-suited to obtaining orthogonal systems for spaces of functions with inner products of the form

$$(f,g) = \int_a^b w(t)f(t)g(t)dt ,$$

and is discussed in most texts on numerical analysis. A simple example is given in Problem 8.14. □

REMARK 8.5. To use the method described in Remark 8.4, it is of course necessary that $\{z_i\}$ is fundamental in X.

Consider the case $X = L^2(0,1)$. Then $C_0^\infty(0,1)$ is dense in X (Section 4.11, property (10), p. 95). But any $x \in C_0^\infty(0,1)$ can be approximated uniformly by polynomials (Weierstrass Theorem, Theorem 4.2). Thus, the system $\{(z_i: t \to t^{i-1}): i \geq 1\}$ is fundamental in $L^2(0,1)$. (The reader is asked to provide the details in Problem 8.15). This is generalized by a beautiful theorem due to Müntz which states that if $\{p_i\}$ is a sequence of real numbers such that $p_i > -\frac{1}{2}$, $p_i \neq 0$, and $p_i \to \infty$, then the system $\{t \to t^{p_i}\}$ is fundamental in $L^2(0,1)$ iff $\sum_{i=1}^{\infty} 1/p_i = \infty$. (see Davis [1963, p. 270].)

If X consists of functions defined on a non-compact set in R^n, then more care is needed. For example, let X consist of the measurable functions on $(0,\infty)$ with finite norm,

$$(x,x)^{\frac{1}{2}} = \|x\| = \left[\int_0^\infty w(t)|x(t)|^2 dt\right]^{\frac{1}{2}} < \infty ,$$

where the weight function $w(t)$ is prescribed.

If $w(t) = t^{-\ell n(t)}$, then it can be shown that every polynomial p belongs to X. However, if

$$f(t) = \sin(2\pi \ell n(t)) ,$$

then $||f-p|| \geq ||f||$ for every polynomial p. That is, the polynomials are not dense in X. (see Problem 8.16)

Fortunately, the polynomials *are* dense in X in the classical case $w(t) = e^{-t}$ which gives rise to the Laguerre polynomials. (Problem 8.17).

Natanson [1955, page 331] has a clear discussion of these questions. □

8.4 *CHOICE OF APPROXIMATING SUBSPACES, PROJECTIONS AND BASES*

In any space X there will be several possible choices: of approximating subspaces X_n, of projections P_n, and of bases $\{x_i\}$. For example, if $X = C[0,1]$ we may choose X_n to be the piecewise-linear splines $\$_n^1$ or the polynomials π_{n-1}. If we choose P_n to be defined by interpolation, we still must choose the points of interpolation. If $X_n = \pi_{n-1}$, then x_i can be the power t^{i-1} or the Legendre polynomial of degree $i - 1$.

We discuss each possible choice in turn:

CHOICE OF X_n

As regards the choice of the space X_n, the goal is to minimize the error dist(u, X_n) where u is the solution to the problem in hand. This often involves considerations which depend upon the specific problem and which are outside the scope of the present text: for example, if $u \in C[0,1]$ is known to behave like $t^{\frac{1}{2}}$ near $t = 0$, then it is helpful if at least one basis function x_i has a similar behaviour.

If it is only known that u belongs to a set $A \subset X$, then one possible strategy is to choose X_n so as to minimize the largest possible error. This idea is quantified by the concept of the *n-width* $d_n^X(A)$ of a set $A \subset X$,

$$d_n^X(A) = \inf_{X_n} \sup_{u \in A} \inf_{y \in X_n} ||u-y|| , \qquad (8.18)$$

where the outer infimum is taken over all n-dimensional subspaces X_n of X. Choosing an X_n for which $d_n^X(A)$ is attained, minimizes the maximum possible error in approximating u.

Bounds for the n-width of certain subsets of $C(\overline{\Omega})$ are known. For example, let A be the subset Λ_{01}^1 of $C[0,1]$ consisting of functions which are *Lipschitz continuous with constant 1*, that is,

$$|f(s)-f(t)| \le |s-t| , \quad \text{for} \quad s,t \in [0,1] .$$

It is known (Lorentz [1966, p. 135]) that

$$d_n^{C[0,1]}(\Lambda_{01}^1) = O(1/n) . \tag{8.19}$$

Also (Jackson theorem) $u \in \Lambda_{01}^1$ can be approximated with error $O(1/n)$ by some $p \in \pi_{n-1}$, and (obviously) by some spline in $\$_n^1$. Therefore, the spaces π_{n-1} and $\$_n^1$ are at worst 'almost optimal', and the theory of n-widths provides a rigorous justification for using either space to approximate elements in Λ_{01}^1.

In the applications to be given in the next chapter, we will often take X_n to be a piecewise linear polynomial space, since this choice makes it possible to illustrate concepts without becoming involved in a great deal of algebra. For future reference we note that if $X = C[0,1]$ and $\$_n^1$ is the space of piecewise linear polynomials on $[0,1]$ with uniform knots

$$t_i = \frac{i-1}{n-1} , \quad 1 \le i \le n ,$$

then (Problem 8.3)

$$\operatorname{dist}(u, \$_n^1) \le \omega(u, h) , \tag{8.20}$$

where $h = 1/(n-1)$ and $\omega(u,h)$ is the *modulus of continuity of* u,

$$\omega(u,h) = \max_{\substack{0 \le s,t \le 1 \\ |s-t| \le h}} |u(s)-u(t)| . \tag{8.21}$$

If u is differentiable, then

$$\omega(u,h) \leq \max_{0\leq s\leq 1} |u(s)| \ . \tag{8.22}$$

CHOICE OF PROJECTION P_n

If X is a Hilbert space, then one naturally chooses the projection $P_n\colon X \to X_n$ to be the proximity mapping. In other cases, it is desirable that $\|P_n\|$ should be small; see Theorem 8.2.

It is of course necessary that P_n should be readily computable, and this sometimes suggests taking P_n to be the projection for some 'nearby' Hilbert space. For example, one can often obtain a good approximation in $C[0,1]$ by using the projections associated with $L^2[0,1]$. Problems 8.19 to 8.21 pursue this idea.

CHOICE OF BASE ELEMENTS

For a given sequence of spaces X_n and projections P_n there may be several possible bases $\{x_i\}$. For theoretical purposes it often does not matter which basis is chosen. The choice of basis can, however, affect the numerical stability of a method.

In Chapter 9 we discuss several ways in which the solution $u \in X_n$ of the equation $Tu = g$ is approximated by the solution $u_n \in X_n$ of an equation

$$T_n u_n = g_n \ , \tag{*}$$

where $g_n \in X_n$, and where $T_n\colon X_n \to X_n$ has a bounded inverse T_n^{-1} . To solve eqn (*) numerically we convert it to a matrix problem by using the mapping

$$\phi_n\colon x = \sum_{i=1}^{n} \lambda_i x_i \in X_n \to \lambda = (\lambda_1,\ldots,\lambda_n) \in R^n \ . \tag{**}$$

Equation (*) then becomes the matrix problem,

$$A^{(n)} v^{(n)} = b^{(n)} \tag{***}$$

where $A^{(n)} = \phi_n T_n \phi_n^{-1}$, $v^{(n)} = \phi_n u_n$, and $b^{(n)} = \phi_n g_n$. From equation (2.3)(p. 43), the condition number for $A^{(n)}$ is $C_r = \|A^{(n)}\| \, \|A^{(n)})^{-1}\|$ which can be bounded by ,

$$\begin{aligned} C_r &= \|A^n\| \cdot \|(A^{(n)})^{-1}\| , \\ &= \|\phi_n T_n \phi_n^{-1}\| \cdot \|\phi_n T_n^{-1} \phi_n^{-1}\| , \\ &\le \|T_n\| \cdot \|T_n^{-1}\| \cdot [\|\phi_n\| \cdot \|\phi_n^{-1}\|]^2 . \end{aligned}$$

It is, therefore, desirable to choose the basis elements in such a way that the quantity $\|\phi_n\| \cdot \|\phi_n^{-1}\|$ is not large. Some examples follow:

EXAMPLE 8.3. If X is a Hilbert space, the natural choice is to pick $\{x_i\}$ as an orthonormal basis. In this case,

$$\|\phi_n x\|^2 = \sum_{i=1}^{n} |\lambda_i|^2 = \|x\|^2 ,$$

so that $\|\phi_n\| \cdot \|\phi_n^{-1}\| = 1$. □

EXAMPLE 8.4. If $X = C[-1,1]$ and $X_n = \pi_{n-1}$ then a poor choice is to pick $x_i : t \to t^{i-1}$. To see this, we note that the Chebyshev polynomial p of degree $n - 1$,

$$p: t \to \cos((n-1) \text{ arc cos } t) = \sum_{i=0}^{n-1} c_i t^i \text{ , say,}$$

is bounded by one on $[-1,1]$, so that

$$\|p;C[-1,+1]\| = 1 .$$

However, (Problem 8.22),

$$\|\phi_n(p);E^n\|^2 = \sum_{i=0}^{n-1} |c_i|^2 \ge 4^{n-2} ,$$

so that $\|\phi_n\|^2 \ge 4^{n-2}$. On the other hand, if $\lambda = (1,1,1,\ldots,1)$, then

$$(\phi_n^{-1}\lambda)(t) = \sum_{i=0}^{n-1} t^i ,$$

from which it is easily seen that $\|\phi_n^{-1}\| \ge \sqrt{n}$.

Thus,

$$\|\phi_n\| \cdot \|\phi_n^{-1}\| \geq [4^{n-2} \cdot n]^{\frac{1}{2}} ,$$

a rather large lower bound! □

EXAMPLE 8.5. If $X = C[-1,+1]$, $X_n = \pi_{n-1}$, and $x_i = L_{i-1}$, the Legendre polynomial of degree $i-1$, then it can be shown that (Problem 8.23)

$$\|\phi_n\| \|\phi_n^{-1}\| \leq 2n^2 . \quad □$$

The moral: Use orthogonal or 'nearly orthogonal' polynomials as basis elements in π_{n-1} when possible.

8.5 *THE INTERIOR MAPPING PRINCIPLE AND CLOSED GRAPH THEOREM*

THEOREM 8.6. *(Open mapping theorem)*

Let $T : X \to Y$ *be a continuous linear map of the Fréchet space* X *onto the Fréchet space* Y . *Then the image of every open set is open.*

Proof. The proof is given for the case when X and Y are Banach spaces. The open balls in X and Y with centre the origin and radius r will be denoted by $B_X(r)$ and $B_Y(r)$, respectively.

Since T maps X onto Y and X is the union of the balls $B_X(n)$, $n = 1,2,...,$ it follows that

$$Y = \bigcup_{n=1}^{\infty} T(B_X(n)) = \bigcup_{n=1}^{\infty} \overline{T(B_X(n))} .$$

By the Baire category theorem, one of the sets $\overline{T(B_X(n))}$ contains an open ball, $B_Y(y;R)$ say. Let $x \in (T^{-1}y) \cap \overline{B_X(n)}$. Then $\overline{T(B_X(n+\|x\|))} \supset B_Y(R)$. Thus there exists $r > 0$ such that

$$\overline{TB_X(1/2^k)} \supset B_Y(r/2^k) , \quad k = 0,1,2,... . \qquad (*)$$

Now let y be any point in $B_Y(r)$. From eqn (*) with $k = 0$, we know that for any $\varepsilon > 0$ there exists

$x \in B_X(1)$ such that $\|y-Tx\| \le \varepsilon$. In particular, taking $\varepsilon < r/2$, there exists $x_1 \in B_X(1)$ satisfying

$$y - Tx_1 \in B_Y(r/2) .$$

But then, from eqn (*) with $k = 1$, there exists $x_2 \in B_X(1/2)$ such that

$$(y-Tx_1) - Tx_2 \in B_Y(r/2^2) .$$

Repeating the process we obtain a sequence $x_1, x_2, \ldots,$ such that $x_i \in B_X(1/2^{i-1})$ and

$$y - Tz_i \in B_Y(r/2^i) , \qquad (**)$$

where $z_i = \sum_{j=1}^{i} x_j$. The sequence $\{z_i\}$ is a Cauchy sequence, since

$$\|z_m - z_n\| \le \sum_{j=n+1}^{m} \|x_j\| \le \sum_{j=n+1}^{m} 2/2^j \le 2^{1-n} , \quad \text{for } m > n . \qquad (***)$$

Therefore, $z_i \to z$ for some $z \in X$. Proceeding to the limit in eqn (**) we find that $y = Tz$. But, from eqn (***), $\|z\| \le \|z_1\| + 1 = 2$. In summary, given any $y \in B_Y(r)$ there exists $z \in \overline{B_X(2)}$ for which $y = Tz$. That is,

$$T(\overline{B_X(2)}) \supset B_Y(r) . \qquad (****)$$

Now let G be any open set in X, and let $y = Tx \in T(G)$. Since G is open there is a ball $B_X(\varepsilon)$ such that $x + \overline{B_X(\varepsilon)} \subset G$. But then, from eqn (****),

$$T(G) \supset T(x+\overline{B_X(\varepsilon)}) = y + T(\overline{B_X(\varepsilon)}) \supset y + B_Y(\varepsilon r/2) .$$

This means that for each $y \in T(G)$ there is a neighbourhood N such that $y + N \subset T(G)$. In other words, $T(G)$ is open. □

Problem 8.26 illustrates the open mapping theorem.

The next theorem is an immediate consequence of the open mapping theorem, and its proof is left as an exercise for the reader (Problem 8.24).

THEOREM 8.7. *Let* T *be a continuous linear one-to-one mapping of a Fréchet space* X *onto a Fréchet space* Y . *Then* T^{-1} *is a continuous linear mapping of* Y *onto* X . □

Let T be a map from $D(T) \subset X$ to Y , where X and Y are topological spaces. T is said to be *closed* if given any sequence $\{x_i\}$ in $D(T)$ such that $x_i \to x \in X$ and $Tx_i \to y \in Y$, it follows that $x \in D(T)$ and $y = Tx$.

EXAMPLE 8.6. Let $X = Y = C[0,1]$, and

$$(Tx)(t) = \frac{dx(t)}{dt} , \quad \text{for } x \in D(T) ,$$

where $D(T)$ consists of the functions in X which are continuously differentiable.

If $x_i \in D(T)$ and $Tx_i = y_i$ then

$$x_i(s) = x_i(0) + \int_0^s y_i(t)dt , \quad 0 \le s \le 1 . \qquad (*)$$

If furthermore $x_i \to x \in X$ and $Tx_i \to y \in Y$, then proceeding to the limit in eqn (*),

$$x(s) = x(0) + \int_0^s y(t)dt , \quad 0 \le s \le 1 ,$$

from which it follows that x is continuously differentiable, $x \in D(T)$, and $y = Tx$. Thus, T is closed. However, T is not continuous (Problem 8.25). □

THEOREM 8.8. *(Closed graph theorem)*

A closed linear map T *from a Fréchet space* X *to a Fréchet space* Y *is continuous.*

Proof. We consider the case when X and Y are Banach spaces.

We know (Problem 3.12) that the product space $Z = X \times Y$ is a Banach space with the norm,

$$\|(x,y);Z\| = \|x;X\| + \|y;Y\| \ . \qquad (*)$$

Since T is linear and closed, the set

$$Z_0 = \{(x,y) \in Z\colon\ x \in X\ ,\ \ y = Tx\}\ ,$$

is a closed linear subspace of Z. A closed subset of a Banach space is complete, and so Z_0 is a Banach space with the norm (*).

The maps

$$p_1\colon\ (x,y) = (x,Tx) \in Z_0 \to x \in X\ ,$$

$$p_2\colon\ (x,y) = (x,Tx) \in Z_0 \to y = Tx \in Y\ ,$$

are continuous. In addition, p_1 is one-to-one and has range X so that, by Theorem 8.7, the inverse map p_1^{-1} is continuous. But then, the composite map $p_2p_1^{-1}$ is continuous. That is:

$$p_2p_1^{-1} : x \in X \to p_2(p_1^{-1}x) = p_2((x,Tx)) = Tx \in Y\ ,$$

is continuous. □

PROBLEMS

8.1 Prove that if $\{x_i\}$ is a basis in a topological vector space, then the coefficients $\lambda_i(x)$ in eqn (8.1) depend linearly on x. {265}

8.2 Complete the proof of Theorem 8.1 by showing that the space Y of sequences $\{\alpha_i\}$ is complete. [Hint: To prove that $\{\tilde{\alpha}_i\} \in Y$, one must show that $\sum\tilde{\alpha}_i x_i$ is a Cauchy sequence in X. To do this, note that, for any $m > n$ and $q > p$,

$$\|\sum_{i=n}^{m} \tilde{\alpha}_i x_i\| \le \|\sum_{i=n}^{m} \alpha_i^{(p)} x_i\| + 2\sup_{k\le m}\|\sum_{i=1}^{k} (\alpha_i^{(q)} - \alpha_i^{(p)}) x_i\| + \qquad (*)$$

$$+\ 2\sup_{k\le m}\|\sum_{i=1}^{k} (\alpha_i^{(q)} - \tilde{\alpha}_i) x_i\|\ .]$$

{266}

8.3 Prove that if $P_n x$ is the piecewise linear approximation to $x \in C[0,1]$ defined in Example 8.1 then

$$\|P_n x - x\| \le \max_{\substack{0 \le s,t \le 1 \\ |s-t| \le \varepsilon_n}} |x(s) - x(t)| = \omega(x, \varepsilon_n) ,$$

where

$$\varepsilon_n = 1/2^{([\log_2 (n-1)])} , \quad n \ge 2 ,$$

and $[t]$ denotes the integer part of t. {268, 282}

8.4 Construct a Schauder basis for $C([0,1] \times [0,1]) = C(\overline{\Omega})$. [Hint: One possibility is as follows. Let

$$h(t) = \begin{cases} 1 - |t| , & \text{for } |t| \le 1 , \\ 0 , & \text{otherwise.} \end{cases}$$

so that $h(t)$ is a hat function with support $[-1,+1]$. For $n = 0,1,\ldots,$ let G_n be the gridpoints on $[0,1] \times [0,1]$ with gridsize 2^{-n}:

$$G_n = \{(i/2^n, j/2^n) : 0 \le i, j \le 2^n\} , \quad n \ge 0 .$$

Set

$$A_0 = G_0 ,$$

$$A_n = G_n \backslash G_{n-1} , \quad \text{for } n > 0 ,$$

$$x_{(u,v)}(s,t) = h(2^n(s-u))\, h(2^n(t-v)) , \quad \text{for } (s,t) \in \Omega \text{ and } (u,v) \in A_n .$$

Then, $x_{(u,v)}$ satisfies

$$x_{(u,v)}(s,t) = \begin{cases} 1 , & \text{if } s = u \text{ and } t = v , \\ 0 , & \text{if } (s,t) \in G_n \text{ and } s \ne u \text{ or } t \ne v . \end{cases}$$

The family of functions $x_{(u,v)}$ forms a Schauder basis for $C(\overline{\Omega})$.]

8.5 In Example 8.1 the Schauder basis in $C[0,1]$ is constructed using piecewise linear interpolation. Show that a Schauder basis can be constructed using piecewise cubic interpolation.

8.6 Let $X = C[0,1]$. Show that the following operators mapping X into itself are not projections:

(1) The Bernstein operator B_n which maps x into the Bernstein polynomial $B(n;x)$. (Section 4.8, p. 83).

(2) The operator P_n which maps x into the best approximation to x among the polynomials of degree n .

8.7 Let R_+^n be the non-negative orthant in R^n ,

$$R_+^n = \{(x_1,\ldots,x_n): x_i \geq 0 \ , \ 1 \leq i \leq n\} = \prod_{i=1}^{n} R_+^1 \ .$$

Let $(\cdot)_+ : R^1 \to R_+^1$ be defined by

$$(t)_+ = \max(0,t) \ .$$

Show that

$$p : (x_1,\ldots,x_n) \in R^n \to ((x_1)_+,\ldots,(x_n)_+) \in R_+^n$$

is a proximity map. {271}

8.8 Let P be a projection of a vector space X onto $M = R(P)$. Show that every $x \in X$ can be written uniquely as the sum of an element $x_1 \in M$ and an element $x_2 \in N = R(I-P)$. (eqn (8.9)) {271}

8.9 Let $X = M \oplus N$. Define $P : X \to M$ by $P(x_1+x_2) = x_1$, if $x_1 \in M$ and $x_2 \in N$. Prove that P is a projection of X onto M . {271}

8.10 Prove that if M is a linear subspace in a vector space X then there exists a projection of X onto M . [Hint: The proof is similar to parts of the proof of Theorem 7.4 (see p. 239). Let $X_0 = M$ and let P_0 be the identity map of X_0 onto M. If P_1 is a projection of X_1 onto M and $X_1 \neq X$, then P_1 may be extended: choose $x_2 \in X \backslash X_1$ and set $X_2 = \operatorname{span}\{X_1, x_2\}$; if $x \in X_2$ then $x = x_1 + \lambda x_2$; define $P_2 : X_2 \to M$ by $P_2(x_1 + \lambda x_2) = P_1(x_1)$. Use Zorn's lemma to show there exists a projection $P : X \to M$.] {271}

8.11 Let $M_1, M_2, \ldots, M_n$, be subspaces in a vector space X. X is said to be the *direct sum* of the M_i, $X = \oplus M_i$, if each x in X can be written uniquely as a sum,

$$x = \sum_{i=1}^{n} x_i \,, \quad \text{where } x_i \in M_i \,. \qquad (*)$$

Setting $P_i(x) = x_i$ if x is given by eqn (*), show that

(1) Each P_i is a projection.

(2) $P_i P_j = 0$ for $i \neq j$.

(3) $\sum_{i=1}^{n} P_i = I$, the identity map on X.

Conversely, given projections P_i in X satisfying (2) and (3), show that $X = \oplus M_i$ where M_i is the range of P_i.

8.12 Prove that the orthogonal complement of a set in a Hilbert space is a closed linear subspace. {274}

8.13 Prove that a denumerable orthonormal system $\{x_i\}$ in a Hilbert space is a Schauder basis iff it is complete. {278}

8.14 Let $Y = L^2(-1,+1)$ as defined in Definition 4.1 or 4.2.

(1) Show that Y is a real Hilbert space with inner-product

$$(x,y) = \int_{-1}^{+1} x(t)y(t)\,dt \,.$$

(2) Construct the first three Legendre polynomials by applying the Gram-Schmidt algorithm to the functions t^{i-1}, $i=1,2,\ldots$. (Remark 8.4, p.280). [These polynomials are used in Example 9.6, p.319].

8.15 Prove that the system $\{t^i : i=0,1,2,\ldots\}$ is fundamental in $L^2(0,1)$. {Remark 8.5, p. 280}

8.16 (1) Prove that

$$I_n = \int_0^\infty t^{-\ell n(t)}\, t^n \sin(2\pi\, \ell n(t))\,dt = 0 \ , \quad n = 0,1,2,\ldots$$

[Hint: Make the change of variables, $\ell n(t) = u + (n+1)/2$.]
{Remark 8.5, p. 281}

(2) Show that if $f \perp p$ in a Hilbert space then

$$\|f\| \leq \|f-p\| \ .$$ {Remark 8.5, p. 281}

8.17 Prove that the polynomials are dense in the Hilbert space X of real measurable functions on $(0,\infty)$ with inner product

$$(x,y) = \int_0^\infty e^{-t}x(t)y(t)\,dt = \int_0^\infty w(t)x(t)y(t)\,dt \ ,$$

by proving the following facts:

(1) $C_0^\infty(0,\infty)$ is dense in X.

(2) Parseval's identity (8.16) holds if $x(t) = y_r(t) = e^{-rt}$, $r = 0,1,2,\ldots$, and $x_i(t)$ is the normalized Laguerre polynomial

$$L_i(t) = \frac{e^t}{i!}\frac{d^i}{dt^i}(t^i e^{-t}) \ .$$

(3) If $f \in C_0^\infty(0,\infty)$, then

$$\int_0^\infty e^{-t}\left|f(t) - \sum_{r=0}^n a_r e^{-rt}\right|^2 dt = \int_0^1 \left|f(-\ell n(u)) - \sum_{r=0}^n a_r u^r\right|^2 du \ .$$

{Remark 8.5, p. 281}

8.18 Let $A = \{(x,y) \in E^2 : -\frac{1}{2} \leq x \leq \frac{1}{2} \ , \ -1 \leq y \leq 1\}$. Prove that the 1-width of A, $d_1^{E^2}(A)$, is equal to $\frac{1}{2}$.

8.19 If $X = C[-1,1]$, $X_n = \pi_{n-1}$, and $x \in X$, then the best approximation $\tilde{x} \in \pi_{n-1}$ to x is characterized by the uniform oscillation property (see Example 7.2) so that the error $x - \tilde{x}$ has at least n changes of sign on $[-1,+1]$. Let $p_1,\ldots,p_n$ be the first n orthonormal polynomials corresponding to the inner product

$$(f,g) = \int_{-1}^{+1} w(t)f(t)g(t)dt ,$$

where the weight function $w(t)$ is continuous and strictly positive, and p_i is of degree $i - 1$. Show that if

$$s_n x = \sum_{i=1}^{n} (x,p_i)p_i ,$$

then $x - s_n x$ also has at least n changes of sign in $[-1,+1]$. {283}

8.20 If, for $x \in C^{(2)}[-1,+1]$, $s_n x$ denotes the partial Fourier series expansion of x in terms of Legendre polynomials, show that

$$\|s_n x-x;\ C[-1,+1]\| = 0(n^{-\frac{1}{2}}) .$$

[This is hard to prove]. {283}

8.21 The Fourier sine series for the function $x(t) = (\pi-t)/2$, $0 \le t \le 2\pi$ is

$$\sum_{n=1}^{\infty} \frac{\sin(nt)}{n} .$$

Compute and plot the partial Fourier sine sums $s_n x$ for $n = 10, 100, 1000$. Verify the *Gibbs Phenomenon*, namely that, for large n,

$$\max_{0\le t\le 2\pi} |(s_n x)(t)| > 1.17 \quad \max_{0\le t\le 2\pi} |x(t)| = 1.17\ (\pi/2) .$$

{283}

8.22 Prove that if

$$p_n(t) = \cos((n-1)\text{arc}\cos t) = \sum_{i=0}^{n-1} c_i^{(n)} t^i , \text{ then}$$

$$\sum_{i=0}^{n-1} |c_i^{(n)}|^2 \ge 4^{n-2} , \quad \text{if } n \ge 2 . \qquad (*)$$

{Example 8.4, p. 284}

8.23 Let $X = C[-1,+1]$, $X_n = \pi_{n-1} \subset X$, and $x_i = P_{i-1}$, the Legendre polynomial of degree $i-1$. Show that if

$$\phi_n : \sum_{i=0}^{n-1} \lambda_i x_i \in X_n \to \lambda = (\lambda_1, \ldots, \lambda_n) \in E^n$$

then $\|\phi_n\| \le 2n^{3/2}$ and $\|\phi_n^{-1}\| \le \sqrt{n}$. [Hint: If $x = \sum_{i=0}^{n-1} \lambda_i x_i$, let $|\lambda_m| = \max |\lambda_i|$. Multiplying x by P_m, and integrating over $[-1,+1]$ we obtain

$$\left| \int_{-1}^{+1} x(t) P_m(t) dt \right| = |\lambda_m| \int_{-1}^{+1} \left| P_m(t) \right|^2 dt = \frac{2|\lambda_m|}{2n-1} . \qquad (*)$$

Since $|P_m(t)| \le 1$ in $[-1,+1]$, eqn (*) implies a lower bound for $\|x\|$ in terms of $|\lambda_m| = \|\lambda\|_\infty$.]

{Example 8.5, p. 285}

8.24 Use the open mapping theorem to prove that if T is a continuous linear one-to-one mapping of a Banach space X onto a Banach space Y, then T^{-1} is continuous. {Theorem 8.7, p. 287}.

8.25 Show that the operator

$$T : x \to dx/dt$$

of Example 8.6 is not continuous. {287}

8.26 Let $A = (a_{ij})$ be a real $m \times n$ matrix mapping R^n onto R^m:

$$A : x \in R^n \to Ax \in R^m .$$

Claim:

Let $Ax_0 = y_0$. Given a ball $B(y_0;\varepsilon)$ there is a ball $B(x_0;\delta)$ such that for every $y \in B(y_0;\varepsilon)$ there exists $x \in B(x_0;\varepsilon)$ for which $Ax = y$.

(1) Prove the claim using matrix theory.

(2) Prove the claim using the open mapping theorem.

(3) Show that the claim is false if A only maps R^n *into* R^m. {286}

8.27 Let L be a linear elliptic operator

$$Lu = \sum_{i,j=1}^{n} a_{ij}(x)\frac{\partial^2 u}{\partial x_i \partial x_j} + \sum_{j=1}^{n} b_i(x)\frac{\partial u}{\partial x_i} + c(x)u ,$$

with coefficients which are continuous on a bounded smooth domain $\Omega \subset R^n$. Let $X = C_0^2(\overline{\Omega})$, $Y = C(\overline{\Omega})$. Assume that for each $v \in Y$ there exists a unique $u \in X$ such that $Lu = v$. Use Theorem 8.7 or Theorem 8.8 to prove that u depends continuously upon v. That is, the boundary value problem $Lu = v$ has a solution which depends continuously upon the right hand side.

{93}

9
Approximate solution of linear operator equations

In this chapter we consider the solution of the linear operator equation

$$Tu = (I-K)u = g \ , \tag{9.1}$$

where K is a continuous linear mapping from a Banach space X into itself, $g \in X$ is given, and I is the identity mapping on X. Applications are made to integral equations

9.1 *PRELIMINARIES*

To approximate eqn (9.1) we introduce the approximating equation

$$T_n u_n = (I-K_n)u_n = g_n \ , \tag{9.2}$$

where $K_n : X \to X$ is bounded and linear, and $g_n \in X$. It is assumed that K_n is of *finite rank*, that is, the range $R(K_n)$ of K_n is finite dimensional; it will be arranged that $X_n = R(K_n)$ has dimension n. As will be seen shortly, the solution u_n of eqn (9.2) can, in general, be constructed numerically by solving a system of n linear algebraic equations. The choice of X_n, K_n, and g_n; the solution of eqn (9.2); and the relationship between u and u_n, form the content of this chapter.

REMARK 9.1. Since $u_n \in X$, u_n is an internal approximation to u. This substantially simplifies the analysis since the error $u - u_n$ can be bounded using the $\|\cdot;X\|$ norm.

In greater generality, we could approximate eqn (9.1) by

$$(\tilde{I}_n - \tilde{K}_n)\tilde{u}_n = \tilde{g}_n \ , \tag{9.3}$$

where $\tilde{K}_n$ is a mapping from a Banach space $\tilde{X}_n$ to another Banach space $\tilde{Y}_n$.

If $X = C(\Omega)$ for some $\Omega \subset R^m$, then some classical methods for solving eqn (9.1) yield approximations to u at a finite set of points in Ω; by suitably interpolating this data one can construct an approximation $u_n \in X$, and thereby bring such classical methods within the framework of the theory developed in the present chapter. □

To compute the solution u_n of eqn (9.2), let V_n be the space of real (complex) n-tuples with some norm (usually $\|\cdot\|_2$ or $\|\cdot\|_\infty$). By Theorem 6.8, there is an isomorphism ϕ_n between $R(K_n)$ and V_n. Two cases arise:

Case 1: If

$$g_n \in R(K_n) , \tag{9.4}$$

then (Problem 9.1) eqn (9.2) is equivalent to the matrix equations

$$\begin{aligned}
&\text{(a)} \quad & A^{(n)}v^{(n)} &= b^{(n)} , \\
&\text{(b)} \quad & v^{(n)} &= \phi_n u_n , \\
&\text{(c)} \quad & b^{(n)} &= \phi_n g_n , \\
&\text{(d)} \quad & A^{(n)} &= \phi_n (I-K_n)\phi_n^{-1} , \\
&\text{(e)} \quad & u_n &= \phi_n^{-1} v^{(n)} ,
\end{aligned} \tag{9.5}$$

where $A^{(n)}:V_n \to V_n$ is an $n \times n$ matrix, while $v^{(n)}$ and $b^{(n)}$ are n-vectors.

Case 2: If

$$g_n \notin R(K_n) \tag{9.6}$$

then we rewrite eqn (9.2) in the form

$$(I-K_n)(u_n-g_n) = K_n g_n , \tag{9.7}$$

which is equivalent (Problem 9.2) to the equations

$$
\begin{aligned}
&\text{(a)} \quad & A^{(n)}w^{(n)} &= c^{(n)} \ , \\
&\text{(b)} \quad & w^{(n)} &= \phi_n(u_n - g_n) \ , \\
&\text{(c)} \quad & c^{(n)} &= \phi_n K_n g_n \ , \\
&\text{(d)} \quad & A^{(n)} &= \phi_n(I-K_n)\phi_n^{-1} \\
&\text{(e)} \quad & u_n &= g_n + \phi_n^{-1}w^{(n)}
\end{aligned}
\tag{9.8}
$$

In analysing eqns (9.1) and (9.2) it is of course necessary to ensure that their solutions u and u_n do in fact exist. In Section 9.2 we discuss the general theory of the linear equations in Banach spaces, taking care to point out all the terrible things that can happen. In Section 9.3 we take a more optimistic view and give various conditions which ensure that eqns (9.1) and (9.2) can be solved. The remaining sections discuss specific numerical methods.

Our goal in this chapter is to give an introduction to the theory of the numerical solution of eqn (9.1), and to illustrate it by simple numerical examples. As the 'model equation' we use Love's integral equation,

$$((I-K)u)(s) \equiv u(s) - \frac{d}{\pi}\int_{-1}^{+1} \frac{u(t)}{d^2+(s-t)^2}\,dt = 1 \equiv g(s), \quad -1 \le s \le 1 \ , \tag{9.9}$$

which arises when considering the electric field between two parallel coaxial positively charged disks a distance D=-d apart (Sneddon [1966, p. 230]). Several other equations which arise in practice and to which the theory can be applied are given in the Problems. Section 9.7 discusses a nontrivial integral equation in some detail.

We conclude this section with a series of remarks.

REMARK 9.2. There are several texts in which the solution of eqn (9.1) is considered in great detail, for example Krasnoselskii, Vainikko et al [1972] and Kurpel [1976].

There are also several texts on the numerical solution of integral equations: Delves and Walsh [1974], Atkinson [1976], Ivanov [1976], Baker [1977], Jaswon and Symm [1977], and Golberg [1979]. □

REMARK 9.3. One aspect of the subject of integral equations is the tremendous variety among the equations which arise in practice. One encounters linear Fredholm integral equations of the second kind,

$$u(s) - \int_a^b k(s,t)u(t)dt = g(s) \ , \quad a \le s \le b \ , \qquad (9.10)$$

with a great variety of kernels $k(s,t)$. There is even more variety when linear equations of the first kind,

$$\int_a^b k(s,t)u(t)dt = g(s) \ , \quad a \le s \le b \ , \qquad (9.11)$$

and nonlinear equations are considered. (Zabreiko et al [1975]).

Because of the great variety among integral equations, it often happens that an equation does not fit neatly into known theoretical results. Furthermore, by manipulation of the equation, it is often possible to derive a-priori information about the solution which can be used to improve significantly the accuracy of numerical approximations; a simple example is given in Problem 9.3. □

REMARK 9.4. The operator K_n in eqn (9.2) maps bounded sets in X into bounded sets in $R(K_n)$. Since $R(K_n)$ is finite-dimensional, bounded sets in $R(K_n)$ are precompact (Theorems 6.3 and 6.9, p. 191 and 201). Therefore, K_n is a compact operator. In approximating eqn (9.1) by eqn (9.2) we are, therefore, approximating the operator K by a sequence of compact operators K_n . This leads to a number of comments:

(1) It follows from Theorem 6.25 (part 4) (p. 224) that if $||K_n - K|| \to 0$ then K must be compact. It is, therefore, no surprise that, for all the methods considered in this chapter, the convergence proofs require that K , or some related operator, be compact.

The general form of compact operators K on spaces such as $C[0,1]$ and $L^2(0,1)$ is known (Dunford and

Schwartz [1966, pages 496 and 507]): Kx is an integral,

$$Kx = \int x \, d\mu ,$$

where the measure μ must satisfy certain properties. This explains why the applications of the theory in this chapter are all to integral equations. It also explains why, when differential equations are treated, they are first reformulated as integral equations using Green's functions. (Problem 9.4).

Halmos and Sunder [1978] give a lively discussion of integral operators in L^2 spaces.

(2) If K_n is compact and $K_n x \to Kx$ for all $x \in X$, it does not necessarily follow that K is compact (Problem 9.5). Indeed, singular non-compact operator equations can be solved using finite-dimensional projections (Prössdorf [1978, chapter 11], Golberg [1979], Ivanov [1976]).

(3) Given two Banach spaces X and Y, and a compact operator $K: X \to Y$, it was long conjectured that there always exists a sequence of operators $K_n: X \to Y$ of finite rank such that $\|K_n - K\| \to 0$. This conjecture (which was called the *approximation problem*) was disproved by Enflo in 1973, and other counterexamples have been given by Szankowski. See Singer [1981].

(4) Petryshyn [1975] discusses the class of operators T such that the solution u of the equation $Tu = g$ can be obtained as the limit of the solutions u_n of equations $T_n u_n = g_n$, where each T_n is of finite rank. □

9.2 *LINEAR OPERATORS ON BANACH SPACES*

We recall (see p. 110) that the norm of a linear operator T mapping a normed linear space X into a normed linear space Y is defined to be

$$\|T\| = \sup_{\|x;X\| \le 1} \|Tx;Y\| .$$

If $T: X \to Y$ and $S: Y \to Z$, where X, Y, and Z are normed linear spaces, and S and T are continous linear mappings, then

$$ST \equiv S \circ T: X \to Z ,$$

and (Problem 9.6)

$$\|ST\| \leq \|S\| \, \|T\| . \tag{9.12}$$

We remind the reader that if $X = Y = Z$, then both ST and TS are defined but are, in general, not equal; simple examples involving 2×2 matrices are readily constructed (Problem 9.7).

The set of all continuous linear mappings from a normed linear space X to a normed linear space Y will be denoted by $B(X,Y)$, with the abbreviation $B(X) = B(X,X)$. (In $B(X,Y)$, 'B' may be thought of as an abbreviation for 'Bounded'.)

If $T \in B(X)$, then $T^n \in B(X)$ for $n = 1,2,\ldots,$ and eqn (9.12) becomes

$$\|T^n\| \leq \|T\|^n . \tag{9.13}$$

In Section 2.4 the inverse of a mapping $f: X \to Y$ was defined to be a multi-valued mapping with domain Y. For a linear mapping it is customary to use a slightly different definition which we will use unless otherwise indicated.

DEFINITION 9.1. The *linear mapping* $T: X \to Y$ has a *linear inverse* T^{-1} iff T is one-to-one (injective); T^{-1} is the unique operator such that $T^{-1}: R(T) \to X$ and

$$T^{-1}Tx = x , \quad \text{for } x \in D(T) = R(T^{-1}) , \tag{9.14}$$

$$TT^{-1}y = y , \quad \text{for } y \in D(T^{-1}) = R(T) . \tag{9.15}$$

□

The next theorem summarizes some of the elementary properties of inverses of linear operators. The proofs, which are left to the reader (Problem 9.8) use only the linearity of the spaces and operators, and depend in no way upon topological properties.

THEOREM 9.1. *Let* T *be a linear mapping of the vector space* X *into the vector space* Y, *possibly with linear inverse* T^{-1}.

(1) T^{-1} *exists iff* $(Tx = 0 \Leftrightarrow x = 0)$.

(2) *If* T^{-1} *exists then* T^{-1} *is linear and one-to-one.*

(3) *If* T^{-1} *exists then* T *is the inverse of* T^{-1}. *That is,* $(T^{-1})^{-1} = T$.

(4) *If* $S\colon Y \to X$ *and* $STx = x$ *for all* $x \in X$, *then* T^{-1} *exists and is equal to* $S|_{R(T)}$.

(5) *If* $S\colon Y \to X$ *and* $TSy = y$ *for all* $y \in Y$, *then* $R(T) = Y$.

(6) *If* X *has dimension* n *then* $R(T)$ *has dimension at most* n.

(7) *If* T^{-1} *exists and either* X *or* $R(T)$ *has dimension* n, *then both* X *and* $R(T)$ *have dimension* n.

(8) *If* $S\colon R(T) \to Z$ *is a linear mapping then* $(ST)^{-1}$ *exists iff both* S^{-1} *and* T^{-1} *exist, in which case* $(ST)^{-1} = T^{-1}S^{-1}$. □

Given $T \in B(X,Y)$, where X and Y are Banach spaces, the following possibilities arise concerning the range $R(T)$ of T and the existence of an inverse T^{-1}:

Properties of $R(T)$

(1) $R(T) = Y$. (9.16)

(2) $\overline{R(T)} = Y$ but $R(T) \neq Y$.

(3) $\overline{R(T)} \neq Y$.

Properties of T^{-1}

(a) T^{-1} exists and is continuous.

(b) T^{-1} exists but is not continuous. (9.17)

(c) T^{-1} does not exist.

Combining conditions (9.16) and (9.17) we obtain nine possible combinations.

If T^{-1} exists then T is one-to-one, and if $R(T) = Y$ then T is onto; therefore, it follows from the open mapping theorem (see Theorem 8.7) that case 1b cannot occur. It is left to the reader to prove that case 2a cannot occur (Problem 9.10). Of the remaining seven cases, four can occur if X and Y are finite dimensional spaces (Example 9.1) and all seven can occur in infinite dimensional spaces (Example 9.2).

EXAMPLE 9.1. The following examples illustrate those combinations of conditions (9.16) and (9.17) which can occur if X and Y are finite dimensional:

(1a) $X = Y = R^1$, $T: x \to x$.

(1c) $X = R^2$, $Y = R^1$, $T: \begin{pmatrix} x_1 \\ x_2 \end{pmatrix} \to x_1$.

(3a) $X = R^1$, $Y = R^2$, $T: (x) \to \begin{pmatrix} x \\ 0 \end{pmatrix}$.

(3c) $X = R^1$, $Y = R^1$, $T: x \to 0$.

The reader is asked to prove that cases 2b, 2c, and 3b cannot occur in finite dimensional spaces (Problem 9.11).

□

EXAMPLE 9.2. The following operators illustrate those combinations of conditions (9.16) and (9.17) which can occur in infinite-dimensional spaces. In all cases $X = Y = \ell^2$, $||T|| = 1$, and

$$x_k = (\overbrace{0,0,\ldots,0}^{k-1 \text{ terms}},1,0,\ldots) , \quad k \geq 1 ,$$

denotes the element with a 1 in position k and zeros elsewhere. Since T is linear, it suffices to define Tx_k , for all k:

(1a) $Tx_k = x_k$, for $k \geq 1$.

(1c) $Tx_1 = \underline{0}$, and $Tx_k = x_{k-1}$, for $k \geq 2$.

(2b) $Tx_k = 2^{1-k}x_k$, for $k \geq 1$.

(2c) $Tx_1 = 0$, and $Tx_k = 2^{1-k}x_{k-1}$ for $k \geq 2$.

(3a) $Tx_k = x_{k+1}$, $k \geq 1$.

(3b) $Tx_k = 2^{1-k}x_{k+1}$, $k \geq 1$.

(3c) $Tx_k = 0$.

The proof of these assertions is left to the reader (Problem 9.12). □

EXAMPLE 9.3. Let $X = L^2(0,\pi)$. The functions

$$x_1(t) = \frac{1}{\sqrt{\pi}} , \quad x_k(t) = \frac{\sqrt{2}}{\sqrt{\pi}} \cos(k-1)t , \quad k \geq 2 ,$$

form an orthonormal basis in X .

Set

$$k(s,t) = \sum_{k=1}^{\infty} \frac{x_k(s)x_k(t)}{2^{k-1}} ,$$

$$(Tx)(s) = \int_0^{\pi} k(s,t)x(t)dt .$$

Then $Tx_k = 2^{1-k}x_k$, so that the operator T is formally identical to the operator T in Example 9.2 (case 2b). In particular, the Fredholm integral equation of the first kind,

$$(Tu)(s) = \int_0^{\pi} k(s,t)u(t)dt = g(s) , \quad 0 \leq s \leq \pi ,$$

is ill-conditioned because T^{-1} is not continuous (see Section 2.10).

See Baker [1977, p. 635] for a discussion of numerical methods for such equations. □

REMARK 9.5. In our development of functional analysis so far, we have emphasized the similarities between finite dimensional and infinite dimensional spaces. However, as examples 9.1 and 9.2 show, operators on infinite-dimensional spaces can have a very complicated structure, and there are many difficult open problems, even in Hilbert spaces (Halmos [1967, 1970]). □

REMARK 9.6. Let A be an $m \times n$ real matrix, and let $g \in R^m$ be given. It sometimes happens in practical situations that the equation

$$Au = g \qquad (*)$$

does not have a unique solution, either because there is no solution $(g \notin R(A))$ or because A is not one-to-one.

The concept of a solution u of eqn (*) can be generalized as follows:

Let V be the set of points y which satisfy eqn (*) with minimal error:

$$V = \{y \in R^n : \|Ay-g\|_2 = \inf_{z \in R^n} \|Az-g\|_2\} . \qquad (9.18)$$

Then u is taken to be the point in V with minimum norm:

$$\|u\|_2 = \inf_{y \in V} \|y\|_2 , \quad u \in V . \qquad (9.19)$$

It is known that u is unique and that

$$u = A^+ g ,$$

where A^+, the *pseudo-inverse of A*, is the $n \times m$ real matrix which is uniquely characterized by the *conditions of Penrose*:

$$\begin{aligned} A^+AA^+ &= A^+ , \\ AA^+A &= A , \\ (A^+A)^T &= A^+A , \\ (AA^+)^T &= AA^+ . \end{aligned} \qquad (9.20)$$

(See Problem 9.14 for an example.)

This approach can be extended to operator equations and is a useful method of handling ill-conditioned problems such as the Fredholm equation of the first kind introduced in Example 9.3.

For detailed discussions of generalized inverses see: Bouillon and Odell [1971], Rao and Mitra [1971], Nashed [1976], and Campbell and Meyer [1979].

9.3 *CRITERIA FOR THE EXISTENCE OF INVERSES*

We begin with a theorem which is used so often to prove the existence of inverses, that it might well be called the 'fundamental theorem' of the subject.

THEOREM 9.2. *Let* X *be a Banach space and let* $A \in B(X)$ *satisfy* $||A|| < 1$. *Then* $(I-A)^{-1} \in B(X)$ *and*

$$||(I-A)^{-1}|| \le 1/(1-||A||) . \tag{9.21}$$

Proof. The mapping $(I-A)$ is one-to-one because if $(I-A)x_1 = (I-A)x_2$ then

$$||x_1-x_2|| = ||A(x_1-x_2)|| \le ||A|| \; ||x_1-x_2|| < ||x_1-x_2|| ,$$

which implies that $x_1 = x_2$.

For $n \ge 1$ define $T_n: X \to X$ by

$$T_n = (I+A+...+A^n) .$$

Then

$$||T_n|| \le 1+||A||+...+||A||^n \le 1/(1-||A||) , \tag{*}$$

so that the operators T_n are uniformly bounded. Also, for any $x \in X$, $\{T_n x\}$ is a Cauchy sequence in X since, for $m > n$,

$$||(T_n-T_m)x|| \le [||A||^{n+1}+...+||A||^m]||x|| ,$$

$$\le ||A||^{n+1}||x||/(1-||A||) .$$

By assumption, X is a Banach space and hence complete, so that $\{T_n x\}$ converges to some point in X. It follows from the Banach-Steinhaus theorem (Theorem 5.4, part 4, p. 121) that there exists a bounded linear operator $T: X \to X$ defined by

$$Tx = \lim T_n x \ , \quad x \in X \ .$$

By direct computation, for any $x \in X$,

$$(I-A)T_n x = T_n(I-A)x = (I-A^{n+1})x \ ,$$

so that,

$$\|[(I-A)T_n - I]x\| = \|[T_n(I-A)-I]x\| \le \|A\|^{n+1}\|x\| \ .$$

Proceeding to the limit,

$$(I-A)Tx = T(I-A)x = x \ , \quad \text{for} \quad x \in X \ .$$

We conclude from Theorem 9.1, parts 4 and 5, that $(I-A)^{-1}$ exists, $T = (I-A)^{-1}$ and $D([I-A]^{-1}) = X$.

The bound (9.21) follows from eqn (*). □

REMARK 9.7. When trying to establish the existence of the inverse of an operator $T: X \to X$ it may well happen that T is not of the form $T = I - A$ with $\|A\| < 1$, so that Theorem 9.2 cannot be applied. There are several tricks which can be tried:

(1) Write $T = B^{-1}(BT)$ for some operator B and apply Theorem 9.2 to BT. This is the approach used in Theorem 9.3 below.

(2) Find a new norm, $\|\cdot\|_\wedge$ say, so that $T = I - A$ with $\|A\|_\wedge < 1$. This approach is particulary effective in finite dimensional problems (Problem 9.15).

(3) Find an operator $S: X \to X$ such that $ST = I - A$ and $\|A\| < 1$. See Problems 9.16 and 9.17, and Section 9.7.

□

THEOREM 9.3. *Let* X *and* Y *be Banach spaces and let* A *and* B *be continuous linear mappings of* X *into* Y . *Let* A *map* X *onto* Y *and let* A *have the continuous inverse* A^{-1} . *If* $\|A-B\| < 1/\|A^{-1}\|$ *then* B *maps* X *onto* Y , B *has a continuous inverse* B^{-1} *and*

$$\|B^{-1}\| \le \frac{\|A^{-1}\|}{1-\|A^{-1}\|\,\|A-B\|} , \tag{9.22}$$

$$\|B^{-1}-A^{-1}\| \le \|A^{-1}\|\,\|A-B\|\,\|B^{-1}\| . \tag{9.23}$$

Proof. Since $A^{-1}: Y \to X$ exists, we may write

$$B = A - (A-B) = A(I-C) , \tag{*}$$

where $C: X \to X$ is given by

$$C = A^{-1}(A-B) .$$

Noting that

$$\|C\| = \|A^{-1}(A-B)\| \le \|A^{-1}\|\,\|A-B\| < 1 , \tag{**}$$

it follows from Theorem 9.2 that $(I-C)$ maps X onto X , and that $(I-C)^{-1}$ exists and satisfies

$$\|(I-C)^{-1}\| \le \frac{1}{1-\|C\|} \le \frac{1}{1-\|A\|^{-1}\|A-B\|} . \tag{***}$$

Thus, from eqn (*) B maps X onto Y . Furthermore, using Theorem 9.1(8) we conclude that B^{-1} exists and is given by

$$B^{-1} = (I-C)^{-1}A^{-1} . \tag{****}$$

Equation (9.22) is an immediate consequence of eqns (***) and (****), while eqn (9.23) follows from the identity,

$$B^{-1}-A^{-1} = B^{-1}(A-B)A^{-1} . \quad \square$$

For a generalization see Problem 9.18.

The next theorem can be paraphrased as follows: If the equation $Tx = y$ can always be solved approximately then it can be solved exactly.

THEOREM 9.4. *Let* T *be a continuous linear mapping of* X *into* Y *where* X *is a Banach space and* Y *is a normed space. Let* M *and* q , $0 < q < 1$, *be constants such that for each* $y \in Y$ *there exists* $x \in X$ *satisfying*

$$\|Tx-y\| \leq q\|y\| \quad \text{and} \quad \|x\| \leq M\|y\| .$$

Then, for each $y \in Y$ *there exists* $x \in X$ *such that* $Tx = y$, *and*

$$\|x\| \leq M\|y\|/(1-q) . \tag{*}$$

Outline of Proof. Let $y \in Y$. Set $y_0 = y$. Define $x_0, x_1, \ldots$ and $y_1, y_2, \ldots,$ recursively by means of the relations: $\|Tx_i - y_i\| \leq q\|y_i\|$, $\|x_i\| \leq M\|y_i\|$, and $y_{i+1} = y_i - Tx_i$. Then the series $\sum_{i=0}^{\infty} x_i$ converges to a point $x \in X$ which satisfies $Tx = y$ and eqn (*). The details of the proof are left to the reader (Problem 9.19). □

If A is an $n \times n$ matrix then A^{-1} exists iff for each n-vector b there is at most one n-vector x satisfying $Ax = b$. More briefly: existence is equivalent to uniqueness. This is generalized in the final theorem in this section which forms part of the Fredholm theory for operator equations (see for example Kantorovich and Akilov [1964, p. 494]).

THEOREM 9.5. *Let* X *be a Banach space and let* $K: X \to X$ *be a continuous compact linear operator. Consider the equation*

$$(I-K)u = g \tag{*}$$

Then the following are equivalent:

(1) *Equation* (*) *has a solution for every* $g \in X$; *that is,* $R(I-K) = X$.

(2) *For each* $g \in X$ *there is at most one* u *satisfying eqn* (*); *that is,* $(I-K)^{-1}$ *exists.* □

9.4 *APPROXIMATION IN NORM*

As an immediate application of Theorem 9.3 we obtain

THEOREM 9.6. *Let* $K: X \to X$ *be a continuous linear operator mapping a Banach space* X *into itself. Let* $(I-K)^{-1} \in B(X)$. *Let* $\{K_n\}$ *be a sequence of continuous linear operators mapping* X *into* X *such that* $\|K-K_n\| \to 0$.

Then:

(1) *There exist constants* M *and* N *such that* $(I-K_n)^{-1} \in B(X)$ *and* $\|(I-K_n)^{-1}\| \le M$ *if* $n \ge N$.

(2) *If, for* $n \ge N$,

$$(I-K)u = g \quad \textit{and} \quad (I-K_n)u_n = g_n \ , \tag{9.24}$$

then

$$\|u-u_n\| \le \|(I-K)^{-1}(K-K_n)u_n\| + \|(I-K)^{-1}\| \ \|g_n-g\| \ . \tag{9.25}$$

(3) *If* $\|g_n-g\| \to 0$ *as* $n \to \infty$ *then*

$$\|u-u_n\| \to 0 \quad \textit{as} \quad n \to \infty \ .$$

Proof. First apply Theorem 9.3 to establish (1). Elementary manipulation of eqns (9.24) then leads to eqn (9.25). (Problem 9.20). □

REMARK 9.8. There are several possible variations of Theorem 9.6:

(1) The roles of K and K_n can be interchanged, so that by verifying numerically that $(I-K_n)^{-1} \in B(X)$ it can be shown that $(I-K)^{-1} \in B(X)$.

(2) The bounds for the error $\|u-u_n\|$ can be expressed in several ways. (Problem 9.21). □

As an application of Theorem 9.6 we set $X = C[a,b]$ and approximate the equation

$$((I-K)u)(s) = u(s) - \int_a^b k(s,t)u(t)dt = g(s), \quad a \le s \le b, \tag{9.26}$$

by

$$((I-K_n)u_n)(s) = u_n(s) - \int_a^b k_n(s,t)u_n(t)dt = g(t), \quad a \le s \le b, \tag{9.27}$$

where $k_n(s,t)$ is a suitably chosen *degenerate kernel*,

$$k_n(s,t) = \sum_{i=1}^{n} x_i(s)y_i(t) . \tag{9.28}$$

The range $R(K_n)$ of tho mapping $K_n: X \to X$ dcfincd by eqn (9.27) is the n-dimensional subspace of $C[a,b]$ spanned by $x_1,\ldots,x_n$ (which we assume, without loss of generality, to be linearly independent).

Equation (9.26) is of the form (9.2) with $g_n = g$ and can be solved using eqns (9.8). Let

$$\phi_n: x = \sum_{i=1}^{n} \lambda_i x_i \in X_n \to \lambda = (\lambda_1,\ldots,\lambda_n) \in R^n .$$

Then, eqns (9.8) become,

(a) $A^{(n)}w^{(n)} = c^{(n)}$,

(b) $w^{(n)} = \phi_n(u_n - g)$,

(c) $c^{(n)} = (c_i^{(n)}) = (\int_a^b y_i(t)g(t)dt)$, (9.29)

(d) $A^{(n)} = (a_{ij}^{(n)}) = (\delta_{ij} - \int_a^b y_i(t)x_j(t)dt)$,

(3) $u_n(s) = g(s) + \sum_{i=1}^{n} w_i^{(n)} x_i(s)$, $a \le s \le b$.

EXAMPLE 9.4. We approximate Love's integral equation (9.9) with $d = -1$ by the degenerate equation (9.27) in the space $X = C[-1,+1]$.

To begin, we use the uniform oscillation property of polynomials of best approximation (Example 7.2) to construct the best linear approximation in the maximum norm to $1/(1+t)$ on $C[0,4]$. It is found (Problem 9.22) that the best linear approximation is

$$p(t) = \alpha - t/5; \ \alpha = \frac{2}{5} + \frac{1}{\sqrt{5}} \doteq .8472, \tag{*}$$

the maximum error being

$$\beta = 1 - \alpha = \frac{3}{5} - \frac{1}{\sqrt{5}} \doteq .1528 .$$

For $s,t \in [-1,+1]$, we have $(s-t)^2 \in [0,4]$. Thus, setting

$$k(s,t) = \frac{-1}{\pi} \frac{1}{1+(s-t)^2} ,$$

$$k_3(s,t) = \frac{-1}{\pi} p((s-t)^2) = \frac{-1}{\pi} [\alpha-(s-t)^2/5] , \tag{**}$$

we have that

$$|k(s,t) - k_3(s,t)| \leq \beta/\pi < .0487 .$$

From Example 5.5, and Problem 9.3 (part 1),

$$||K-K_3||_\infty = \max_{-1\leq s\leq 1} \int_{-1}^{+1} |k(s,t)-k_3(s,t)|dt \leq 2\beta/\pi < .0973 ,$$

$$||K||_\infty = \max_{-1\leq s\leq 1} \int_{-1}^{+1} |k(s,t)|dt = \tfrac{1}{2} .$$

Expanding eqn (**), equation (9.28) becomes

$$k_3(s,t) = \sum_{i=1}^{3} x_i(s)y_i(t) ,$$

with

$$\pi x_1(s) = 1 , \qquad y_1(t) = \frac{t^2}{5} - \frac{\alpha}{2}$$

$$\pi x_2(s) = -\frac{2}{5}s , \qquad y_2(t) = t ,$$

$$\pi x_3(s) = +\frac{s^2}{5} - \frac{\alpha}{2} , \qquad y_3(t) = 1 .$$

Equation (9.29a) becomes

$$A^{(3)}w^{(3)} = \begin{pmatrix} 1+\gamma & 0 & \delta \\ 0 & 1+\frac{4}{15\pi} & 0 \\ -2/\pi & 0 & 1+\gamma \end{pmatrix} \begin{pmatrix} w_1^{(3)} \\ w_2^{(3)} \\ w_3^{(3)} \end{pmatrix} = \begin{pmatrix} -\gamma\pi \\ 0 \\ 2 \end{pmatrix} = c^{(3)}$$

whore

$$\gamma = -\int_{-1}^{+1} y_1(t)x_1(t)dt = -\int_{-1}^{+1} y_3(t)x_3(t)dt ,$$

$$= (\alpha - 2/15)/\pi \doteq .2272$$

$$\delta = -\int_{-1}^{+1} y_1(t)x_3(t)dt = \frac{-1}{\pi}\left(\frac{2}{125} - \frac{2\alpha}{15} + \frac{\alpha^2}{2}\right) \doteq -.0834 .$$

The solution is:

$$w_2^{(3)} = 0 ,$$

$$\begin{pmatrix} w_1^{(3)} \\ w_3^{(3)} \end{pmatrix} = \frac{1}{\Delta}\begin{pmatrix} 1+\gamma & -\delta \\ 2/\pi & 1+\gamma \end{pmatrix} \begin{pmatrix} -\gamma\pi \\ 2 \end{pmatrix} ,$$

where

$$\Delta = (1+\gamma)^2 + 2\delta/\pi .$$

Thus,

$$(w^{(3)})^T \doteq (-.4882, 0, 1.3764) ,$$

corresponding to

$$u_3(s) = g(s) + \sum_{i=1}^{3} w_i^{(3)} x_i(s) ,$$

$$\doteq g(s) + \frac{1}{\pi} (-.4882 + 0 + 1.3764(\frac{s^2}{5} - \frac{\alpha}{2})) ,$$

$$\doteq .6590 + .0876s^2 . \qquad (9.30)$$

From Theorem 9.6 we obtain the rigorous error bound,

$$\|u-u_3\|_\infty \le \|(I-K)^{-1}(K-K_3)u_3\|_\infty ,$$

$$\le \frac{1}{1-\|K\|_\infty} \cdot \|K-K_3\|_\infty \cdot \|u_3\|_\infty ,$$

$$< (2) \cdot \left(\frac{2\beta}{\pi}\right) \cdot u_3(1) ,$$

$$< .1453 . \qquad (9.31)$$

A sharper bound for $\|u-u_3\|_\infty$ could be obtained by computing $\|(K-K_3)u_3\|_\infty$ instead of bounding this quantity by by $\|K-K_3\|_\infty \cdot \|u_3\|_\infty$, but this would be rather laborious. For more numerical results see Problems 9.23 and 9.24. □

REMARK 9.9. At first sight it might seem that it would be difficult to construct degenerate kernels k_n such that $\|K_n-K\| \to 0$, but in fact such kernels can be constructed by interpolation (Problem 9.25) or by Galerkin's method (Remark 9.10, p. 318)

The degenerate kernel method requires the evaluation of the $O(n^2)$ integrals in eqns (9.29c) and (9.29d), and this makes it an inefficient method when high accuracy is required. When low accuracy and rigorous error bounds are required, the method is very useful. □

9.5 *PROJECTION METHODS*

We recall (page 269) that a projection is a linear mapping P of a space into itself such that $P^2 = P$.

In a projection method, the equation

$$Tu = (I-K)u = g\ , \tag{9.1}$$

is approximated by

$$T_n u_n = (I-K_n)u = g_n\ , \tag{9.2}$$

where

$$K_n = P_n K: X \to X\ ,\ g_n = P_n g \in X_n\ , \tag{9.32}$$

and $P_n : X \to X$ is a projection of X onto the subspace X_n so that $R(P_n) = X_n$. We have not assumed that $u_n \in X_n$, but this follows since, from eqns (9.2) and (9.32).

$$u_n = P_n(Ku_n) + P_n g \in X_n\ .$$

P_n is a projection, so that $P_n u_n = u_n$. Thus, eqns (9.2) and (9.32) are equivalent to:

$$P_n(I-K_n)u_n = P_n T u_n = P_n g\ ,\quad u_n \in X_n\ . \tag{9.33}$$

LEMMA 9.7. *Let* $\{S_n\}$ *be a sequence of bounded linear operators mapping a Banach space* X *into a Banach space* Y. *Let the sequence* $\{S_n x\}$ *be convergent for every* $x \in X$, *and let* A *be a precompact set in* X.

Then $\{S_n x\}$ *converges uniformly on* A. *That is, given* $\varepsilon > 0$ *there exists* N *such that*

$$\|S_n x - \lim_{n\to\infty} S_n x\| \le \varepsilon\ ,\quad \text{if}\quad x \in A \quad \textit{and}\quad n \ge N\ .$$

Proof. Choose $\varepsilon > 0$. From Theorem 6.3 the set A is totally bounded so that there exists an ε-net for A, $x_1, x_2, \ldots, x_m$, say.

Let $Sx = \lim S_n x$. It follows from the Banach-Steinhaus theorem (Theorem 5.4) that $\|S\| \le M$ and $\|S_n\| \le M$, for some constant M and all n.

Choose N such that

$$\|Sx_k - S_n x_k\| \le \varepsilon\ ,\quad \text{for}\quad 1 \le k \le m \quad \text{and}\quad n \ge N\ .$$

Now let x be any point in A. By construction, $\|x - x_k\| \le \varepsilon$ for some x_k. Therefore, for $n \ge N$,

$$\begin{aligned}\|S_n x - Sx\| &\le \|S_n x_k - Sx_k\| + \|(S_n - S)(x - x_k)\|\ ,\\ &\le \varepsilon + 2M\varepsilon\ . \quad \square\end{aligned}$$

Applying Lemma 9.7 we obtain

THEOREM 9.8. *Let* $K: X \to X$ *be a continuous compact linear operator mapping a Banach space* X *into itself.*
Let $(I-K)^{-1} \in B(X)$. *Let* $\{P_n\}$ *be a sequence of projections on* X *such that* $P_n x \to x$ *as* $n \to \infty$, *for all* $x \in X$. *Let* $K_n = P_n K$.

Then $\|K-K_n\| \to 0$ *as* $n \to \infty$. *Consequently, there exist constants* M *and* N *such that* $(I-K_n)^{-1} \in B(X)$ *and satisfies* $\|(I-K_n)^{-1}\| \le M$ *if* $n \ge N$. *Furthermore, if, for* $n \ge N$,

$$(I-K)u = g \quad \textit{and} \quad (I-K_n)u_n = P_n g ,$$

then

$$\|u-u_n\| \le \|(I-K)^{-1}(K-K_n)u_n\| + \|(I-K)^{-1}\| \, \|P_n g-g\| , \qquad (9.34)$$

so that

$$\|u-u_n\| \to 0 \quad \textit{as} \quad n \to \infty . \quad \square$$

Proof. Let B be the unit ball in X. Since K is compact, $A = K(B)$ is precompact.

Applying Lemma 9.7 we can conclude that

$$\|K-P_n K\| = \sup_{x \in B} \|Kx-P_n(Kx)\| ,$$

$$= \sup_{y \in A} \|y-P_n y\| ,$$

$$\to 0 , \quad \text{as} \quad n \to \infty .$$

The theorem now follows from Theorem 9.6. $\square$

EXAMPLE 9.5: THE GALERKIN METHOD.

Let X be a separable Hilbert space with inner product $(\cdot,\cdot)$. Let X_n be the subspace of X,

$$X_n = \text{span}\{x_1,\dots,x_n\} ,$$

where the x_i are orthogonal.

In the *Galerkin method* of solving

$$Tu = (I-K)u = g , \tag{9.1}$$

one seeks $u_n \in X_n$,

$$u_n = \sum_{j=1}^{n} \lambda_j^{(n)} x_j , \tag{9.35}$$

such that the *residual*

$$r_n = Tu_n - g , \tag{9.36}$$

is orthogonal to X_n .

With the notation of Section 8.3 and the help of Theorem 8.4, the equation for u_n can be written in three equivalent ways (Problem 9.26):

(1) $$r_n \perp X_n \Longleftrightarrow (Tu_n - g) \perp X_n . \tag{9.37}$$

(2) $$(r_n, x_i) = (Tu_n - g, x_i) = 0 , \quad 1 \le i \le n . \tag{9.38}$$

(3) $$P_n r_n = P_n T u_n - P_n g = 0 , \tag{9.39}$$

where P_n is the orthogonal projection of X onto X_n .

Comparing eqns (9.2), (9.33), and (9.39), we see that the Galerkin method is a special form of projection method. In particular, given a separable Hilbert space X , we know from Theorem 8.5 that there exists at least one fundamental denumerable orthonormal system $\{x_i\}$. Such a system $\{x_i\}$ is complete and forms a Schauder basis, so that if $X_n = \text{span}\{x_1,\ldots,x_n\}$ and P_n is the orthogonal projection of X onto X_n , then $\|P_n\| = 1$ and $P_n x \to x$ as $n \to \infty$ for all $x \in X$. Thus, by Theorem 9.8, with this choice of P_n and X_n the Galerkin method yields a convergent sequence of approximations u_n in the case when K is compact.

To compute u_n , eqn (9.38) is expanded into matrix form using eqn (9.35):

$$u_n = \sum_{j=1}^{n} \lambda_j^{(n)} x_j ,$$

where $\lambda^{(n)} = (\lambda_1^{(n)}, \ldots, \lambda_n^{(n)})$, is the solution of the system of n linear algebraic equations:

$$\begin{aligned} &(a) \quad A^{(n)}\lambda^{(n)} = c^{(n)} , \\ &(b) \quad c^{(n)} = (c_i^{(n)}) = ((g,x_i)) , \quad (1 \le i \le n) , \qquad (9.40) \\ &(c) \quad A^{(n)} = (a_{ij}^{(n)}) = ((x_j,x_i) - (Kx_j,x_i)) . \quad (1 \le i,j \le n) . \end{aligned}$$

See also Ikebe [1972]. □

REMARK 9.10. When the original equation (9.1) is a Fredholm integral equation

$$(Tx)(s) = (I-K)x(s) = x(s) - \int_a^b k(s,t)x(t)dt ,$$

then the Galerkin method can also be considered as a degenerate kernel method.

To prove this we begin by noting that for any $x,y \in X$,

$$\begin{aligned} (x,Ky) &= \int_a^b x(s)ds \int_a^b k(s,t)y(t)dt , \\ &= \int_a^b y(t)dt \int_a^b x(s)k(s,t)ds , \\ &= \int_a^b y(s)ds \int_a^b k(t,s)x(t)dt , \\ &= (K^*x,y) , \end{aligned} \tag{9.41}$$

where

$$(K^*x)(s) = \int_a^b k(t,s)x(t)dt . \tag{9.42}$$

K^* is the *Hilbert space adjoint* of K.

Remembering that for a Galerkin method $K_n = P_nK$ where P_n is the orthogonal projection of X onto X_n, it follows from eqn (8.10) that, for any $x \in X$,

$$K_n x = \sum_{i=1}^{n} \frac{(Kx, x_i)}{(x_i, x_i)} x_i .$$

Applying eqns (9.41) and (9.42) we obtain,

$$\begin{aligned}(K_n x)(s) &= \sum_{i=1}^{n} \frac{(x, K^* x_i)}{(x_i, x_i)} x_i(s) , \\ &= (x, \sum_{i=1}^{n} \frac{K^* x_i}{(x_i, x_i)} x_i(s)) , \\ &= \int_a^b k_n(s,u)x(u)du ,\end{aligned}$$

whcre

$$k_n(s,u) = \sum_{i=1}^{n} \frac{x_i(s)}{(x_i, x_i)} \int_a^b k(t,u)x_i(t)dt . \qquad (9.43)$$

Thus the operator K_n corresponds to the degenerate kernel k_n defined by eqn (9.43).

By Theorem 9.8, $\|K_n - K\| \to 0$, as $n \to \infty$. □

EXAMPLE 9.6. We illustrate the Galerkin method by applying it to Love's integral equation (9.9) with $d = -1$.

Set $X = L^2(-1,+1)$ and introduce the orthogonal basis given by multiples of the Legendre polynomials (see Problem 8.14), so that

$$\begin{aligned}x_1(t) &= 1 , \\ x_2(t) &= t , \\ x_3(t) &= (3t^2-1)/2 .\end{aligned}$$

For any real continuous function F ,

$$\int_{-1}^{+1}\int_{-1}^{+1} \frac{F(s,t)}{1+(s-t)^2}\, ds dt = \int_{-1}^{+1}\int_{-1}^{+1} \frac{F(-s,-t)}{1+(s-t)^2}\, ds dt , \qquad (*)$$

and, setting $u = s + t$, $v = s - t$,

$$\int_{-1}^{+1}\int_{-1}^{+1} \frac{F(s,t)}{1+(s-t)^2}\, ds dt = \frac{1}{2}\int_{-2}^{+2} \frac{dv}{1+v^2} \int_{-(2-|v|)}^{2-|v|} F\left(\frac{u+v}{2}, \frac{u-v}{2}\right) du . \qquad (**)$$

Using eqns (9.9), (*), and (**), we obtain, after substantial effort (Problem 9.27) that

$$\pi((Kx_j, x_i)) = - \begin{pmatrix} 4\alpha-2\beta & 0 & 4-\beta-3\alpha \\ 0 & \frac{4}{3}\alpha-\frac{7}{3}\beta+\frac{2}{3} & 0 \\ 4-\beta-3\alpha & 0 & \frac{4}{5}\alpha-\frac{63}{20}\beta+\frac{17}{10} \end{pmatrix} , \qquad (9.44)$$

$$\doteq \begin{pmatrix} 2.81916 & 0 & -.12617 \\ 0 & .26518 & 0 \\ -.12617 & 0 & .05085 \end{pmatrix} ,$$

where

$$\alpha = \text{arc tan } 2 = \int_0^2 \frac{dv}{1+v^2} \doteq 1.107148 ,$$

$$\beta = \frac{1}{2} \ell n\ 5 = \int_0^2 \frac{v}{1+v^2}\, dv \doteq .804718 .$$

Furthermore,

$$c^{(3)} = ((g, x_i)) = \begin{pmatrix} 2 \\ 0 \\ 0 \end{pmatrix} ,$$

and

$$((x_j, x_i)) = \begin{bmatrix} 2 & 0 & 0 \\ 0 & 1 & 0 \\ 0 & 0 & 2/5 \end{bmatrix} .$$

Substituting into eqn (9.40), we obtain,

$$\begin{bmatrix} 2.8974 & 0 & -.04160 \\ 0 & 1.08441 & 0 \\ -.04160 & 0 & .41619 \end{bmatrix} \begin{bmatrix} \lambda_1^{(3)} \\ \lambda_2^{(3)} \\ \lambda_3^{(3)} \end{bmatrix} \doteq A^{(3)}\lambda^{(3)} = \begin{bmatrix} 2 \\ 0 \\ 0 \end{bmatrix} = c^{(3)} ,$$

with solution,

$$(\lambda^{(3)})^T \doteq (.69121,\ 0,\ .06670) .$$

This corresponds to

$$\begin{aligned} u^{(3)}(s) &= \sum_{j=1}^{3} \lambda_j^{(3)} x_j(s) , \\ &\doteq .69121 + .06670(3s^2-1)/2 , \\ &= .65786 + .10005\ s^2 . \end{aligned} \tag{9.45}$$

To estimate the error $\|u-u^{(3)}\|_2$ using Theorem 9.8, it is necessary to obtain bounds for $\|K\|_2$ and $\|K-K_3\|_2$.

It will be recalled (Example 5.2) that if A is a square matrix then

$$\|A\|_2 = \rho(A^T A)^{\frac{1}{2}} .$$

Similarly, $\|K\|_2$ can be expressed in terms of the eigenvalues of K^*K, but this is not useful in most practical work. Instead, we may use the bound (Problem 9.28),

$$||K||_2 \leq \left[\int_{-1}^{+1}\int_{-1}^{+1} [k(s,t)]^2 ds dt\right]^{\frac{1}{2}} ,$$

$$= \left[\frac{1}{\pi^2}\int_{-1}^{+1}\int_{-1}^{+1} [\frac{1}{1+(s-t)^2}]^2 ds dt\right]^{\frac{1}{2}} . \qquad (9.46)$$

Using numerical quadrature, we find that

$$||K||_2 \dot{\leq} .4737 .$$

It follows from eqn (9.43) that K_3 corresponds to the degenerate kernel

$$\pi k_3(s,t) = \sum_{i=1}^{3} \frac{x_i(s)}{(x_i,x_i)} \int_{-1}^{+1} \frac{x_i(u)}{1+(u-t)^2} du ,$$

$$= \int_{-1}^{+1} \frac{1}{1+(u-t)^2} \left[\frac{1}{2} + \frac{3}{2} su + \frac{5}{2} \cdot \frac{3s^2-1}{2} \cdot \frac{3u^2-1}{2}\right] du ,$$

$$= \frac{15}{4}(3s^2-1) + \frac{6s + 15t(3s^2-1)}{8} \ln\left[\frac{1+(1-t)^2}{1+(1+t)^2}\right] +$$

$$+ \frac{(15t^2-20)(3s^2-1) +12st +4}{8} [\arctan(1-t)$$

$$+ \arctan(1+t)] .$$

Using numerical quadrature, we find that

$$||K-K_3||_2 \leq \left[\int_{-1}^{+1}\int_{-1}^{+1} [k(s,t)-k_3(s,t)]^2 ds dt\right]^{\frac{1}{2}} \doteq .0280 .$$

Substituting these bounds into eqn (9.34), we obtain the error estimate

$$\|u-u_3\|_2 \le \frac{1}{1-\|K\|_2} \|K-K_3\|_2 \, \|u_3\|_2 ,$$

$$\dot{\le} \frac{1}{1-.4737} (.0512)(.5) < .027. \qquad (9.47)$$

For other numerical results see Problem 9.29. □

REMARK 9.11. A number of comments may be made regarding the computations in Example 9.6:

(1) The evaluation of the coefficients (x_i, Kx_j) requires the evaluation of two-dimensional integrals.

Even if it is possible to evaluate (x_i, Kx_j) in closed form, this is often a laborious task which is subject to human error. The computation of the coefficients in eqn (9.44) took several pages of algebraic manipulation; a number of errors were made which were detected by comparing the final results with those obtained by numerical quadrature. To handle values of n greater than 3 one would probably resort to programs for symbolic manipulation.

Another possibility is to compute the terms (x_i, Kx_j) using numerical quadrature. This introduces additional errors which must be analysed (Problem 9.30).

(2) In obtaining error estimates such as that in eqn (9.47) it is not usually necessary to obtain the best possible estimate. Judicious simplifications can lead to useful error estimates with much less effort. For example, the bound (9.46) for $\|K\|_2$ is usually adequate, and one might even use the yet cruder bound

$$\|K\|_2 \le \left[\frac{1}{\pi^2} \int_{-1}^{+1} \int_{-1}^{+1} \left[\frac{1}{1+(s-t)^2}\right]^2 ds\,dt\right]^{\frac{1}{2}} < \frac{2}{\pi} . \qquad \square$$

We conclude this section with the observation that projection methods form a very broad class of methods, and many numerical methods can be reformulated as projection methods by ingenious choice of X_n and P_n. One such method, the *collocation method*, is considered in Problems 9.31 and 9.32, while another method, the *Ritz-Galerkin method*, is briefly introduced in Problem 9.33.

9.6 *COLLECTIVELY COMPACT OPERATORS AND NYSTRÖM'S METHOD*

The theory of collectively compact operators was developed by Anselone [1971] to handle the Nyström method of solving integral equations The Nyström method can be analysed by other means (see e.g., Krasnoselskii et al [1972, p. 265]), but the theory of Anselone remains a very instructive generalization of the concept of a compact operator.

Nyström's method is as follows (Nyström [1930]). It is required to solve the Fredholm integral equation of the second kind in $X = C[a,b]$:

$$u(s) - \int_a^b k(s,t)u(t)dt = g(s) , \quad a \le s \le b . \tag{9.48}$$

where $k \in C([a,b] \times [a,b])$.

Assume that we are given a sequence of quadrature formulae Q_n of the form (5.15)(p. 124),

$$Q_n x = \sum_{j=1}^{n} w_j^{(n)} x(t_j^{(n)}) \doteq Qx = \int_a^b x(t)dt , \tag{9.49}$$

where the weights $w_j^{(n)}$ and nodes $t_j^{(n)}$ are known, such that $Q_n x \to Qx$ as $n \to \infty$ for all $x \in C[a,b]$.

We approximate eqn (9.48) by

$$u_n(s) - \sum_{j=1}^{n} w_j^{(n)} k(s,t_j^{(n)}) u_n(t_j^{(n)}) = g(s) , \quad a \le s \le b , \tag{9.50}$$

If this equation has a solution u_n, then u_n can be obtained using only linear algebra: the values $u_n(t_j^{(n)})$,

$j = 1,2,\ldots,n$ may be found by solving the system of n linear algebraic equations,

$$u_n(t_i^{(n)}) - \sum_{j=1}^{n} w_j^{(n)} k(t_i^{(n)}, t_j^{(n)}) u_n(t_j^{(n)}) = g(t_i^{(n)}) , \quad 1 \le i \le n , \tag{9.51}$$

and the values of $u_n(s)$ for $s \in [a,b]$ are then given by

$$u_n(s) = g(s) - \sum_{j=1}^{n} w_j^{(n)} k(s, t_j^{(n)}) u_n(t_j^{(n)}) , \quad a \le s \le b . \tag{9.52}$$

To analyse Nyström's method we begin by introducing operator notation. Let

$$X = C[a,b] . \tag{9.53}$$

Then eqns (9.48) and (9.50) may be written, respectively, in the form

$$(I-K)u = g , \tag{9.54}$$

and

$$(I-K_n)u_n = g , \tag{9.55}$$

where $K: X \to X$ and $K_n: X \to X$ are given by

$$(Kx)(s) = \int_a^b k(s,t)x(t)dt , \quad a \le s \le b , \tag{9.56}$$

and

$$(K_n x)(s) = \sum_{j=1}^{n} w_j^{(n)} k(s, t_j^{(n)}) x(t_j^{(n)}) , \quad a \le s \le b . \tag{9.57}$$

It is not true that $\|K_n - K\|_\infty \to 0$. Indeed,

$$\|K_n - K\|_\infty \ge \|K\|_\infty . \tag{9.58}$$

To prove this one observes that, given n and $\varepsilon > 0$, it is always possible (Problem 9.34) to construct $x \in X = C[a,b]$ such that $\|x\|_\infty = 1$ and

$$\begin{aligned} &(1)\quad x(t_j^{(n)}) = 0 \quad \text{for} \quad 1 \le j \le n\,, \quad \text{so that} \quad K_n x = 0\,. \\ &(2)\quad \|Kx\| > \|K\| - \varepsilon\,. \end{aligned} \tag{9.59}$$

It is a consequence of the inequality (9.58) that we cannot directly use Theorem 9.6 as was done in the analysis of the degenerate kernel method and the projection method. Instead, we use the concept of collective compactness.

It will be recalled (page 222) that $T: X \to Y$ is compact if T maps bounded sets in X into precompact sets in Y. If $\mathcal{T}$ is a family of mappings T from a topological space X into a topological space Y, then $\mathcal{T}$ is *collectively compact* if the set

$$\bigcup_{T \in \mathcal{T}} T(U)$$

is precompact whenever $U \subset X$ is bounded. In particular (Problem 9.35), if X is a Banach space, then $\mathcal{T} \subset B(X)$ is collectively compact iff the set

$$\{Tx:\ T \in \mathcal{T},\ \|x\| \le 1\} \tag{9.60}$$

is sequentially compact.

EXAMPLE 9.7. If K_n is defined by eqn (9.57) then the family $\mathcal{T} = \{K_n : n = 1,2,\ldots\}$ is collectively compact.

To prove this it must be shown that the set

$$S = \{K_n x:\ n = 1,2,\ldots, \text{ and } \|x\| \le 1\}$$

is sequentially compact.

Since, by assumption, $Q_n x \to Qx$ for all $x \in X$ (see eqn (9.49)), Theorem 5.5 (p. 124) guarantees the existence of a constant M such that

$$\sum_{j=1}^{n} |w_j^{(n)}| \le M\,, \quad \text{for all} \quad n\,.$$

Thus, for any $x \in X$,

$$\|K_n x\| \le \sum_{j=1}^{n} |w_j^{(n)}| \, \|k(\cdot,t_j^{(n)})\|_\infty \, |x(t_j^{(n)})| ,$$

$$\le M \|x\| \max_{a\le s,t\le b} |k(s,t)| ,$$

so that S is bounded.

Since, by assumption, $k(s,t)$ is continuous on $[a,b] \times [a,b]$, k is uniformly continuous. In particular, given $\varepsilon > 0$ there exists $\delta > 0$ such that

$$|k(s_1,t) - k(s_2,t)| \le \varepsilon , \quad \text{if} \quad |s_1 - s_2| \le \delta .$$

Thus, for any $x \in X$,

$$|(K_n x)(s_1)-(K_n x)(s_2)| \le \sum_{j=1}^{n} |w_j^{(n)}| \, |k(s_1,t_j^{(n)})-k(s_2,t_j^{(n)})| \, |x(t_j^{(n)})| ,$$

$$\le M \|x\|_\infty \varepsilon , \qquad (*)$$

if $|s_1 - s_2| \le \delta$, so that S is equicontinuous.

The Ascoli-Arzela theorem (Theorem 6.13) states that if $S \subset C[a,b]$ is bounded and equicontinuous then S is sequentially compact. □

We can now analyse the Nyström method:

LEMMA 9.9. *Let* X *be a Banach space. Let* $\{K_n\}$ *be a collectively compact sequence of operators, such that* $K_n \in B(X)$ *and* $K_n x \to Kx$ *for all* $x \in X$.

Then

$$\|(K-K_n)K_n\| \to 0 \quad \text{as} \quad n \to \infty .$$

Proof. Let B be the unit ball in X and let

$$A = \bigcup_{n=1}^{\infty} K_n(B) .$$

Then, by assumption, A is precompact.

Applying Lemma 9.7 we see that, given $\varepsilon > 0$, there exists N such that

$$\|K_n y - Ky\| \le \varepsilon , \quad \text{for} \quad y \in A \quad \text{and} \quad n \ge N .$$

That is,

$$\|(K_n - K)K_n x\| \le \varepsilon , \quad \text{for} \quad x \in B \quad \text{and} \quad n \ge N . \quad \square$$

THEOREM 9.10. *Consider the Nyström method as defined by equations (9.48) to (9.57).*

Assume that: $k(s,t)$ *is continuous;* $Q_n x \to Qx$ *for all* $x \in X = C[a,b]$; $(I-K)^{-1} \in B(X)$.

Then, for n *sufficiently large:*

(1) $\Delta \equiv \|(I-K)^{-1}(K_n - K)K_n\| < 1$,

(2) *Eqn (9.55) has a unique solution* u_n *for every* $g \in X$.

(3) $$\|u - u_n\| \le \frac{\|(I-K)^{-1}\| \, \|(K_n - K)g\| + \Delta\|(I-K)^{-1}g\|}{1 - \Delta} .$$

Consequently, $\|u - u_n\| \to 0$ *as* $n \to \infty$.

Proof. *(1):* Choose $x \in X$. Now choose $\varepsilon > 0$. Then there exists $\delta > 0$ such that

$$|k(s',t) - k(s'',t)| \le \varepsilon , \quad \text{for} \quad t \in [a,b] \text{ and } |s'-s''| \le \delta .$$

Consequently, from eqn (9.56),

$$|(Kx)(s') - (Kx)(s'')| \le (b-a)\|x\|_\infty \varepsilon , \quad \text{if} \quad |s'-s''| \le \delta . \qquad (*)$$

As shown in Example 9.6 (see eqn (*) on page 327),

$$|(K_n x)(s') - (K_n x)(s'')| \le M\|x\|_\infty \varepsilon ; \quad \text{if} \quad |s'-s''| \le \delta , \qquad (**)$$

for some constant M .

Now choose a δ-net $s_1, \ldots, s_m$ on $[a,b]$. For each s_i,

$$(K_n x)(s_i) = Q_n(k(s_i,\cdot)x(\cdot)) \to Q(k(s_i,\cdot)x(\cdot)) = (Kx)(s_i), \text{ as } n \to \infty ,$$

so that, for some N,

$$|(K_n x)(s_i) - (Kx)(s_i)| \le \varepsilon , \quad \text{for} \quad 1 \le i \le m \quad \text{and} \quad n \ge N . \qquad (\overset{*}{_*})$$

Combining eqns (*), (**), and ($\overset{*}{_*}$), we obtain:

$$\begin{aligned} \|K_n x - Kx\|_\infty &= \max_{s\in[a,b]} |(K_n x)(s) - (Kx)(s)| , \\ &\le [(b-a)+M]\|x\|_\infty \varepsilon + \max_{1\le i\le m} |(K_n x)(s_i) - (Kx)(s_i)| , \\ &\le [(b-a)+M]\|x\|_\infty \varepsilon + \varepsilon , \quad \text{if} \quad n \ge N . \end{aligned}$$

That is, for any $x \in X$,

$$K_n x \to Kx , \quad \text{as} \quad n \to \infty . \qquad (\overset{**}{_{**}})$$

(1) now follows from Example 9.7 and Lemma 9.9.

(2): Let n be such that $\Delta < 1$. Using Problem 9.17 we see that $(I-K_n)^{-1}$ exists and is bounded. Since K_n is compact, the Fredholm theory (see Theorem 9.5) allows us to conclude that the equation

$$(I-K_n)u_n = g$$

has a unique solution for every $g \in X$.

(3): Follows from Problem 9.17. □

EXAMPLE 9.8. Once again we consider Love's integral equation with $d = -1$.

The quadrature formula Q_n of eqn (9.49) is chosen to be the three-point Gaussian quadrature formula

$$\begin{aligned} Qx = \int_{-1}^{+1} x(t)dt \doteq Q_3 x &= \sum_{j=1}^{3} w_j^{(3)} x(t_j^{(3)}) , \\ &= \frac{5}{9}\, x\left(-\sqrt{\frac{3}{5}}\right) + \frac{8}{9}\, x(0) + \frac{5}{9}\, x\left(+\sqrt{\frac{3}{5}}\right) , \end{aligned} \qquad (*)$$

with error (Davis and Rabinowitz [1975, p. 75]),

$$|Qx - Q_3 x| \le \frac{2^7 (3!)^4}{7(6!)^3} \|x^{(6)}\|_\infty = \gamma_3 \|x^{(6)}\|_\infty , \quad \text{say,} \qquad (**)$$

where $x^{(\ell)}$ denotes the ℓ-th derivative of x .

Setting

$$v^{(3)} = \begin{bmatrix} u_3(t_1^{(3)}) \\ u_3(t_2^{(3)}) \\ u_3(t_3^{(3)}) \end{bmatrix} \quad \text{and} \quad c^{(3)} = \begin{bmatrix} g(t_1^{(3)}) \\ g(t_2^{(3)}) \\ g(t_3^{(3)}) \end{bmatrix} ,$$

equation (9.51) takes the form

$$A^{(3)}v^{(3)} = \begin{bmatrix} 1+\frac{5}{9\pi} & \frac{5}{9\pi} & \frac{25}{153\pi} \\ \frac{25}{72\pi} & 1+\frac{8}{9\pi} & \frac{25}{72\pi} \\ \frac{25}{153\pi} & \frac{5}{9\pi} & 1+\frac{5}{9\pi} \end{bmatrix} v^{(3)} = \begin{bmatrix} 1 \\ 1 \\ 1 \end{bmatrix} = c^{(3)} ,$$

with solution

$$(v^{(3)})^T \doteq (.71944,\ .65550,\ .71944),$$

corresponding to

$$u_3(s) = g(s) - \sum_{j=1}^{3} w_j^{(3)} k(s,t_j^{(3)})\, v_j^{(3)} ,$$

$$\doteq 1 \quad - \frac{1}{\pi}\,\frac{.39969}{1+(s-\sqrt{\frac{3}{5}})^2} - \frac{1}{\pi}\,\frac{.39969}{1+(s+\sqrt{\frac{3}{5}})^2} - \frac{1}{\pi}\,\frac{.58267}{1+s^2} . \tag{9.61}$$

To obtain an error estimate we must bound the quantities appearing in Theorem 9.10 (part 3).

As a preliminary step, we consider the function

$$\phi(z) = \frac{1}{1+z^2} = \frac{1}{2}\left(\frac{1}{z+i} + \frac{1}{z-i}\right) .$$

Differentiating ℓ times we obtain,

$$\max_{-\infty<z<\infty} |\phi^{(\ell)}(z)| \le M_\ell = \ell! \; . \qquad (*^{*}_{*})$$

Since $\|K\|_\infty \le \frac{1}{2}$ (see Problem 9.4) we have

$$\|(I-K)^{-1}\|_\infty \le \frac{1}{1-\|K\|_\infty} \le 2 \; . \qquad (^{**}_{**})$$

Next, using the bound (**) for the quadrature error together with the bound ($*^{*}_{*}$) we obtain

$$\|(K_3-K)g\|_\infty = \max_{-1\le s\le 1} |Q_3(k(s,\cdot)) - Q(k(s,\cdot))| \; ,$$

$$\le \gamma_3 \max_{-1\le s\le 1} \max_{-1\le t\le 1} \left|\frac{\partial^6 k(s,t)}{\partial t^6}\right| \; ,$$

$$\le \frac{1}{\pi} \gamma_3 M_6 \; . \qquad (^{***}_{**})$$

Similarly,

$$\|(K_3-K)K_3\|_\infty = \max_{\|x\|\le 1} \|(K_3-K)K_3 x\|_\infty \; ,$$

$$\le \max_{\|x\|\le 1} \|(K_3-K) \sum_{j=1}^{3} w_j^{(3)} k(\cdot,t_j^{(3)}) x(t_j^{(3)})\| \; ,$$

$$\le \sum_{j=1}^{3} |w_j^{(3)}| \; \|(K_3-K)k(\cdot,t_j^{(3)})\| \; ,$$

$$\le \gamma_3 \max_{-1\le s\le 1} \max_{-1\le t\le 1} \max_{1\le j\le 3} \left|\frac{\partial^6 (k(s,t)k(t,t_j^{(3)}))}{\partial t^6}\right| .$$

Using the formula of Leibnitz we obtain

$$\left|\frac{\partial^6 (k(s,t)k(t,t_j^{(3)}))}{\partial t^6}\right| \le \frac{1}{\pi^2} \sum_{\ell=0}^{6} \binom{6}{\ell} M_\ell M_{6-\ell} = \frac{1}{\pi^2} \sum_{\ell=0}^{6} \binom{6}{\ell} \ell!(6-\ell)! \; ,$$

$$= 7.6!/\pi^2 \; ,$$

so that

$$\|(K_3 - K)K_3\|_\infty \le 7 \frac{\gamma_3}{\pi^2} M_6 \ . \tag{$^{***}_{***}$}$$

Combining the bounds $(^{**}_{**})$, $(^{***}_{\ **})$ and $(^{***}_{***})$ and using Theorem 9.10 (part 3), we find that

$$\Delta = \|(I-K)^{-1}(K_3 - K)K_3\| \le 2 \cdot \frac{7\gamma_3 M_6}{\pi^2} = \frac{16}{25\pi^2} < .065 \ .$$

and

$$\|u - u_3\|_\infty \le \frac{\|(I-K)^{-1}\| \, \|(K_3 - K)g\| + \Delta \, \|(I-K)^{-1}g\|}{1 - \Delta} ,$$

$$< \frac{2 \dfrac{\gamma_3 M_6}{\pi} + 2\Delta}{1 - \Delta} < .17 \ . \tag{9.62}$$

For another numerical example see Problem 9.37. □

9.7 *THE INTEGRAL EQUATION OF RADON FOR THE DIRICHLET PROBLEM WITH CORNERS*

The methods discussed in the previous three sections can be applied to the Fredholm integral equation

$$u(s) - \int_a^b k(s,t)u(t)\,dt = g(s) \ , \quad a \le s \le b \ ,$$

with continuous kernel $k(s,t)$, and finite interval $[a,b]$. Interesting questions arise when these conditions are not satisfied, as happens when $[a,b]$ is unbounded and hence not compact (Remark 6.8, p. 201) or when $k(s,t)$ has singularities (Problems 9.55 and 9.56).

To conclude this chapter we consider in some detail the solution of yet another type of integral equation, namely the Riemann-Stieltjes integral equation associated with the Dirichlet problem for a domain with corners. The interesting feature of this equation is that the operator K is the sum of a bounded operator E and a compact operator F.

Let Γ be a rectifiable Jordan curve in the xy-plane which is the boundary of a simply-connected domain $\mathcal{D}$ (see Figure 9.1).

Let Γ have the parametric representation

$$x = x(s) , \quad y = y(s) , \quad 0 \le s \le S , \tag{9.63}$$

where S is the length of Γ and s denotes arc length measured in the anticlockwise direction from a fixed point $P(0)$. The point $(x(\sigma), y(\sigma)) \in \Gamma$ is denoted by $P(\sigma)$, and the length of the shorter of the two arcs joining $P(s)$ and $P(\sigma)$ will be denoted by $d(s,\sigma)$ so that

$$d(s,\sigma) = \min_{k=-1,0,1} |s - \sigma + k S| .$$

The distance between $P(s)$ and $P(\sigma)$ is denoted by $|P(s) - P(\sigma)|$.

Let $g = g(s)$ be defined on Γ and let $U = U(x,y)$ be the solution of the Dirichlet problem

$$U_{xx} + U_{yy} = 0 , \quad (x,y) \in \mathcal{D} , \tag{9.64}$$

$$U = g , \quad (x,y) \in \Gamma . \tag{9.65}$$

If Γ and g are smooth then it is well-known (Courant and Hilbert [1962, p. 298]) that U is equal to the potential corresponding to a *double-layer* of density $u = u(s)$ on Γ. That is, for $Q \in \mathcal{D}$

$$U(Q) = \int_\Gamma u(\sigma) \frac{\partial}{\partial n(\sigma)} \ln|Q - P(\sigma)| d\sigma = \frac{1}{\pi} \int_\Gamma u(\sigma) d\omega_Q(\sigma) , \tag{9.66}$$

where $n(\sigma)$ is the unit outward normal to Γ at $P(\sigma)$ and $|Q - P(\sigma)|$ is the distance from Q to $P(\sigma)$ (see Figure 9.1). The density u satisfies the Fredholm integral equation

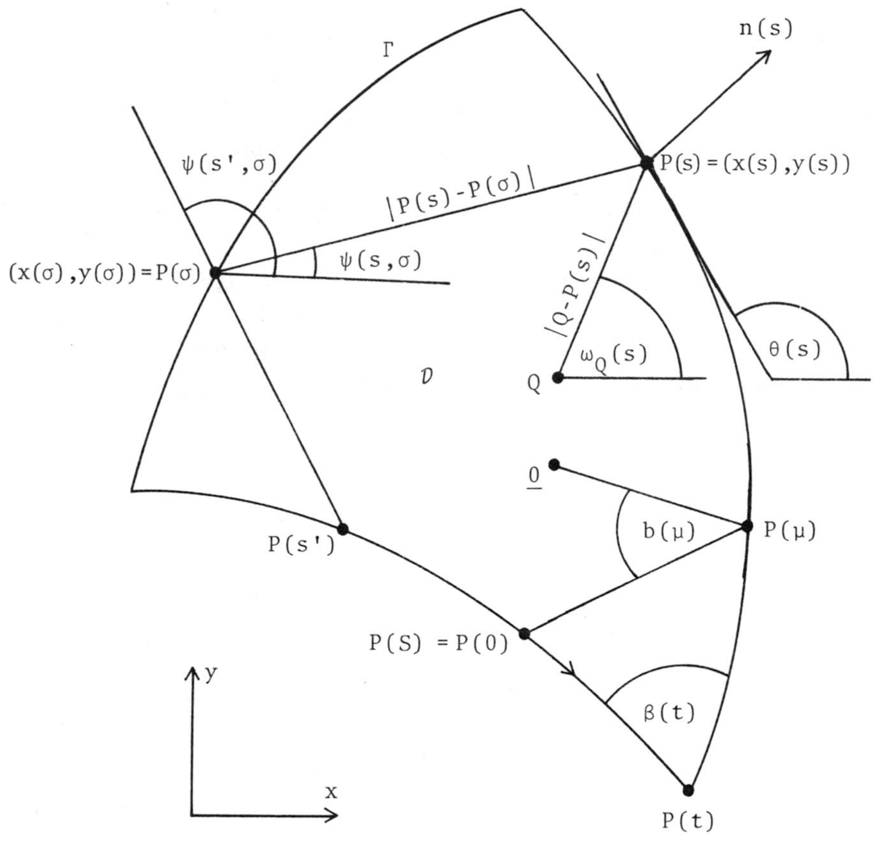

Figure 9.1: The curve Γ *and domain* $\mathcal{D}$

$$u(s) + \int_0^S u(\sigma)k(s,\sigma)d\sigma = g(s) \ , \quad 0 \leq s \leq S \ , \tag{9.67}$$

where $k(s,\sigma)$ is the known smooth function

$$k(s,\sigma) = \frac{1}{\pi}\frac{\partial}{\partial\sigma} \text{arc tan}\left[\frac{y(s) - y(\sigma)}{x(s) - x(\sigma)}\right] = \frac{1}{\pi}\frac{\partial\psi(s,\sigma)}{\partial\sigma} \ . \tag{9.68}$$

The integral equation (9.67) can be solved by any of the methods previously discussed in this chapter (Problem 9.38). In view of the close connection between conformal mapping and Dirichlet problem, it is not surprising that the kernel (9.68) (the *Neumann kernel*) also occurs in several of the integral equations in the theory of conformal mapping (see Gaier [1964] and Problem 9.39).

When Γ has corners the above theory is no longer adequate. A satisfactory theory was first developed by Radon [1919] using Stieltjes integrals, which we briefly review.

THE STIELTJES INTEGRAL

A real function $G(t)$ defined for $t \in [a,b]$ is of *bounded variation* provided that there exists a constant M such that

$$\sum_{j=1}^{n} |G(t_j) - G(t_{j-1})| \leq M \ , \tag{9.69}$$

for every partition $\pi = \{t_i\}$ of $[a,b]$,

$$a = t_0 < t_1 < \ldots < t_n = b \ . \tag{9.70}$$

If G is of bounded variation then the smallest constant M for which eqn (9.69) holds is called the *total variation* of G and is denoted by

$$\int_a^b |dG(t)| \ .$$

Given two real functions F and G defined on [a,b], and a partition $\pi = \{t_j\}$ of [a,b], we may form sums

$$\sum_{j=1}^{n} F(\xi_j)[G(t_j) - G(t_{j-1})] , \tag{9.71}$$

where the points ξ_j satisfy $\xi_j \in [t_{j-1}, t_j]$. If the sums (9.71) converge to a limit as $|\pi| \to 0$, where

$$|\pi| = \max_{1 \le j \le n} |t_j - t_{j-1}| , \tag{9.72}$$

irrespective of the choice of the ξ_j, then this limit is denoted by

$$\int_a^b F(t)dG(t) , \tag{9.73}$$

and is called the *Stieltjes integral* of F with respect to G over [a,b]. The Stieltjes integral is a natural extension of the Riemann integral, for which $G(t) = t$.

Basic properties of the Stieltjes integral include (Riesz and Nagy [1955, p. 105]):

(1) The Stieltjes integral (9.72) exists if: (a) F is continuous and G is of bounded variation; or (b) F is of bounded variation and G is continuous.

(2) The formula for integration by parts, namely,

$$\int_a^b F(t)dG(t) + \int_a^b G(t)dF(t) = F(b)G(b)-F(a)G(a) , \tag{9.74}$$

is valid whenever one of the integrals exists.

(3) If G is continuously differentiable, then

$$\int_a^b F(t)dG(t) = \int_a^b F(t) \frac{dG(t)}{dt} dt , \tag{9.75}$$

where the integral on the right is a Riemann integral.

(4) If $a \le c \le b$ then

$$\int_a^b F(t)dG(t) = \int_a^c F(t)dG(t) + \int_c^b F(t)dG(t) \; . \qquad (9.76)$$

We can now describe the theory of Radon for the Dirichlet problem (9.64), (9.65). It is assumed henceforth that the boundary Γ of $\mathcal{D}$ has the following properties:

(1) Γ is of *bounded rotation*. That is, there exists a function $\theta(s)$ which is of bounded variation and is such that Γ is the curve

$$\left.\begin{aligned} x(s) &= x(0) + \int_0^s \cos\theta(\sigma)d\sigma \; , \\ y(s) &= y(0) + \int_0^s \sin\theta(\sigma)d\sigma \; . \end{aligned}\right\} \quad 0 \le s \le S \; , \qquad (9.77)$$

The reader may find it helpful to note that if $P(s) = (x(s),y(s))$ is a smooth point on Γ then, modulo 2π , $\theta(s)$ is the angle between the x-axis and the tangent to Γ at $P(s)$ (see Figure 9.1).

(2) Γ has no cusps, so that

$$0 < \beta(t) < 2\pi \; , \qquad (9.78)$$

where $\beta(t)$ is the interior angle at $P(t)$ (see Figure 9.1).

We will need several of the properties of curves of bounded rotation. The theory of such curves, which is quite subtle, is given in the original paper of Radon (see also Cryer [1970]). In our numerical examples we consider only polygons for which the properties needed are almost obvious, and the reader may choose to think only of this case.

Let $\psi(s,\sigma)$ be as in Figure 9.1:

$$
\text{(a)} \quad \left.\begin{aligned} \cos\psi(s,\sigma) &= [x(s)-x(\sigma)]/|P(s)-P(\sigma)|, \\ \sin\psi(s,\sigma) &= [y(s)-y(\sigma)]/|P(s)-P(\sigma)|, \end{aligned}\right\} \quad \text{for} \quad s < \sigma ,
$$

$$
\text{(b)} \qquad \psi(s,\sigma) = \psi(\sigma, s) , \quad \text{for} \quad s \ge \sigma , \tag{9.79}
$$

$$
\text{(c)} \qquad \psi(s,s) = \psi(s,s+0) = \lim_{\sigma \downarrow s} \psi(s,\sigma) .
$$

It can be shown that ψ can be defined on $[0,S] \times [0,S]$ so that it satisfies eqns (9.79) and is continuous except at points (s,s) corresponding to corners $P(s) \in \Gamma$.

For example, for the square shown in Figure 9.2, if we normalize ψ by setting $\psi(0,0+0) = \pi$, then (Problem 9.41),

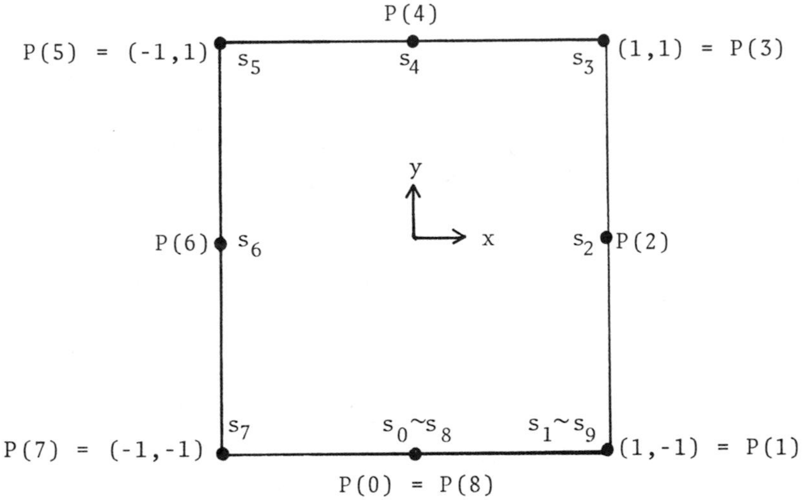

Figure 9.2: Γ and $\mathcal{D}$ in the case of a square.

$$\psi(0,3) = 3\pi/2 - \arctan(1/2) ,$$

$$\psi(0,4) = 3\pi/2 ,$$

$$\psi(3,\sigma) = \begin{cases} 3\pi/2 - \arctan \frac{1-\sigma}{2} , & \text{for } 0 \le \sigma \le 1 , \\ 3\pi/2 , \ \text{for } 1 \le \sigma < 3 , & \\ 2\pi , \ \text{for } 3 \le \sigma \le 5 , & \\ 2\pi + \arctan \frac{\sigma-5}{2} , & \text{for } 5 \le \sigma \le 7 \\ \frac{5\pi}{2} - \arctan \frac{9-\sigma}{2} , & \text{for } 7 \le \sigma \le 8 . \end{cases} \tag{9.80}$$

In the general case, it can be shown that ψ has the following properties:

(1) For fixed s, $\psi(s,\sigma)$ is of bounded variation in σ and

$$\int_0^S |d_\sigma \psi(s,\sigma)| \equiv \int_\Gamma |d_\sigma \psi(s,\sigma)| \le \int_\Gamma |d\theta(\sigma)| - \pi . \tag{9.81}$$

(2) If C is a closed subarc of Γ and $P(s) \in C$ then

$$\int_C |d_\sigma \psi(s,\sigma)| \le \int_C |d\theta(\sigma)| . \tag{9.82}$$

(3) $\psi(s,S) - \psi(s,0) = \pi$, for $s \in [0,S]$. (9.83)

(4) For fixed $\xi \in [0,S]$,

$$\overline{\psi}(s) = \int_0^\xi \psi(s,\sigma)d\sigma \tag{9.84}$$

is a continuous function on $[0,S]$.

(5) $\psi(s,s+0) - \psi(s,s-0) = \pi - \beta(s) = \theta(s+0) - \theta(s-0)$, (9.85)

where $\beta(s)$ is the interior angle at $P(s)$. (See Figure 9.1).

If Γ is a polygon, the above properties of ψ are readily proved from first principles (Problem 9.42). See Problem 9.43 for another example.

For $x \in X = \tilde{C}[0,S]$, the space of continuous periodic functions on $[0,S]$ with the maximum norm, let

$$(Kx)(s) = \frac{1}{\pi}\int_0^S x(\sigma)d_\sigma\psi(s,\sigma) \equiv \frac{1}{\pi}\int_\Gamma x(\sigma)d_\sigma\psi(s,\sigma) , \qquad (9.86)$$

the integral being the Stieltjes integral of x with respect to $\psi(s,\sigma)$, s being treated as a parameter. $(Kx)(s)$ is defined since $\psi(s,\cdot)$ is of bounded variation.

Radon proved that if g is continuous then the Dirichlet problem (9.64), (9.65) is solved by the function

$$U(Q) = \frac{1}{\pi}\int_\Gamma u(\sigma)\, d_\sigma\omega_Q(\sigma) , \qquad (9.87)$$

provided that $u \in X = \tilde{C}[0,S]$ satisfies

$$(Tu)(s) \equiv ((I+K)u)(s) \equiv u(s) + \frac{1}{\pi}\int_\Gamma u(\sigma)d_\sigma\psi(s,\sigma) = g(s), \qquad (9.88)$$
$$\text{for} \quad 0 \le s \le S .$$

Here, the function ω_Q is as shown in Figure 9.1. Since ω_Q is readily shown to be of bounded variation (Problem 9.44), the integral (9.87) is defined.

Equation (9.88), which we call *Radon's equation*, is the promised generalization of eqn (9.67). Equation (9.88) reduces to eqn (9.67) when Γ is smooth as is seen from eqns (9.68) and (9.75). In the remainder of this chapter we analyse the properties of Radon's equation and show how it can be approximated numerically.

We begin by proving that $K: X \to X$. To see this, choose $\tau \in [0,S]$ and let $\{\tau_n\}$ be a sequence of real points converging to s . We show that

$$U_n x \equiv (Kx)(\tau_n) \to Ux \equiv (Kx)(\tau) . \qquad (9.89)$$

To prove this we note that $\{U_n\}$ is a sequence of linear functionals on X. Also, from Problem 9.40 and eqn (9.81),

$$\|U_n\| \leq \sup_{\|x\|\leq 1} \left| \frac{1}{\pi} \int_0^S x(\sigma)\, d_\sigma \psi(\tau_n,\sigma) \right| ,$$

$$\leq \frac{1}{\pi} \int_0^S |d\theta(\sigma)| - 1 .$$

Therefore, by appealing to our old friend, the Banach-Steinhaus theorem (Theorem 5.4) it suffices to show that $U_n x \to Ux$ on a dense subset E of X. This can be proved by taking E to be the subspace of piecewise linear functions and using the properties (9.83) and (9.84) of ψ : the details are left to the reader (Problem 9.45).

Equation (9.89) shows that Kx is a continuous function. From eqn (9.83) we see that

$$\psi(S,\sigma) = \psi(0,\sigma) + \pi ,$$

from which it follows immediately that $x(S) = x(0)$. Thus, as asserted, $K: X \to X$.

However, if Γ has corners, then K is not a compact mapping of X into X. To illustrate this we consider the curve Γ shown in Figure 9.2, and set $\tau_0 = 1$ so that $P(\tau_0)$ is the lower right corner of Γ. For $n \geq 1$ let

$$\tau_n = 1 - 1/n . \qquad (*)$$

If $\psi(0,0) = \pi$ then, (see eqn (9.80)),

$$\psi(\tau_n,\sigma) = \begin{cases} \pi , & \text{for } \tau_n \leq \sigma \leq 1 , \\ \pi + \arctan\left(\dfrac{\sigma-1}{1-\tau_n}\right), & 1 \leq \sigma \leq 3 . \end{cases}$$

$$\psi(1,\sigma) = \begin{cases} \pi , & \text{for } 0 \leq \sigma < 1 , \\ \dfrac{3\pi}{2} , & \text{for } 1 \leq \sigma \leq 3 . \end{cases}$$

Finally, let

$$z_n(\sigma) = \begin{cases} 1 - |1-\sigma|n , & \text{if } |\sigma-1|n \le 1 , \\ 0 , & \text{otherwise.} \end{cases}$$

Then,

$$\begin{aligned} (Kz_n)(\tau_n) &= \frac{1}{\pi} \int_0^S z_n(\sigma) d_\sigma \psi(\tau_n,\sigma) , \\ &= \frac{1}{\pi} \int_1^{1+1/n} z_n(\sigma) d_\sigma \psi(\tau_n,\sigma) , \\ &= \frac{1}{\pi} \int_0^{1/n} (1-nt) \frac{d}{dt} [\pi + \text{arc tan}(nt)] dt , \\ &= \frac{1}{\pi} [\text{arc tan } 1 - \tfrac{1}{2} \ell n\ 2] , \\ &\doteq .14 . \end{aligned} \qquad (**)$$

On the other hand,

$$\begin{aligned} (Kz_n)(1) &= \frac{1}{\pi} \int_0^S z_n(\sigma) d_\sigma \psi(1,\sigma) , \\ &= \frac{1}{\pi} \left(\frac{\pi}{2} z_n(1)\right) , \\ &= \frac{1}{2} . \end{aligned} \qquad (\overset{**}{*})$$

From eqns (*), (**) and ($\overset{**}{*}$) we see that given $\varepsilon = .25$ there exists no δ such that, for all n ,

$$|(Kz_n)(s') - (Kz_n)(s'')| \le \varepsilon$$

whenever $|s'-s''| \leq \delta$. That is, the family of functions

$$\mathcal{T} = \{Kz_n : n = 1,2,\dots\} ,$$

is not equi-continuous. Using the Ascoli-Arzela theorem (Theorem 6.13) we conclude that $\mathcal{T}$ is not sequentially compact, and hence, as asserted, that K is not compact,

Since the difficult behaviour of $\psi(s,\sigma)$ occurs near the points $s = \sigma$, this suggests the following splitting of K . Let $C_{s,\delta}$ denote that part of Γ lying within the circle with centre $P(s)$ and radius δ , and set, with an obvious notation,

$$(Kx)(s) = \frac{1}{\pi} \int_{C_{s,\delta}} x(t)d_\sigma\psi(s,\sigma) + \frac{1}{\pi} \int_{\Gamma\backslash C_{s,\delta}} x(t)d_\sigma\psi(s,\sigma) , \tag{9.90}$$

$$= (H_\delta x)(s) + (J_\delta x)(s) , \quad \text{say.}$$

It is plausible, and was proved by Radon, that, for sufficiently small δ , J_δ is compact, H_δ is bounded, and

$$\lim_{\delta\to 0} \|H_\delta\| = \frac{1}{\pi} \sup_{s\in\Gamma} |\pi-\beta(s)| .$$

Thus, given $\varepsilon > 0$, there exists a splitting of K ,

$$K = E + F , \tag{9.91}$$

where F is compact and

$$\|E\| \leq \sup_{\Gamma} |\pi-\beta(s)| + \varepsilon < 1 . \tag{9.92}$$

Instead of considering the splitting (9.90) in detail, we consider another splitting which is less elegant but which leads to numerical approximations.

The function $\psi(s,\sigma)$ has discontinuities at some of the points $s = \sigma$. Subtracting these out we obtain a function $\tilde{\psi}(s,\sigma)$ which is continuous in σ for each s:

$$\tilde{\psi}(s,\sigma) = \begin{cases} \psi(s,\sigma), & 0 \le \sigma < s, \\ \psi(s,s-0), & \sigma = s, \\ \psi(s,\sigma) - [\psi(s,s+0) - \psi(s,s-0)], & s < \sigma \le S. \end{cases} \tag{9.93}$$

LEMMA 9.11: *For any* $\varepsilon > 0$ *there exists* $\bar{\delta}$ *such that*

$$\left| \int_\Gamma x(t)\, d_\sigma \psi(s',\sigma) - \int_\Gamma x(t)\, d_\sigma \tilde{\psi}(s'',\sigma) \right| \le \|x\| [\sup_\Gamma |\pi - \beta(s)| + \varepsilon], \tag{9.94}$$

if $d(s',s'') \le \bar{\delta}$.

Outline of proof: Let $s \in [0,S]$. For $\delta > 0$ set

$$C_{s,\delta} = \{P \in \Gamma: |P - P(s)| \le \delta\}.$$

Let δ_1 and δ_2 be two parameters which satisfy

$$0 < \delta_1 < \delta_2 < \varepsilon. \tag{*}$$

Let $P(s')$, $P(s'') \in C_{s,\delta_1}$. Then, for any $x \in X$ with with $\|x\|_\infty \le 1$,

$$\left| \int_\Gamma x(\sigma)\, d_\sigma \psi(s',\sigma) - \int_\Gamma x(\sigma)\, d_\sigma \tilde{\psi}(s'',\sigma) \right| \le I_1 + I_2 + I_3, \tag{**}$$

where

$$I_1 = \left| \int_{C_{s,\delta_2}} x(\sigma)\, d_\sigma \psi(s',\sigma) \right| ,$$

$$I_2 = \left| \int_{C_{s,\delta_2}} x(\sigma)\, d_\sigma \tilde{\psi}(s'',\sigma) \right| ,$$

and

$$I_3 = \left| \int_{\Gamma \backslash C_{s,\delta_2}} x(\sigma)\, d_\sigma [\psi(s',\sigma) - \tilde{\psi}(s'',\sigma)] \right| .$$

The three terms I_i can be bounded. Firstly, using eqn (9.82),

$$I_1 \leq \int_{C_{s,\delta_2}} |d\theta(\sigma)| .$$

It is obvious for polygons, and is true in general, that

$$\int_{C_{s,\delta_2}} |d\theta(\sigma)| \to |\pi - \beta(s)| , \quad \text{as} \quad \delta_2 \to 0 ,$$

so that,

$$I_1 \leq \sup_{\Gamma} |\pi - \beta(s)| + \varepsilon/3 , \text{ for small } \delta_2 . \qquad (*\!*\!*)$$

Since $\tilde{\psi}(s'',\sigma)$ is continuous and of bounded variation as a function of σ ,

$$I_2 \leq \varepsilon/3 , \text{ for small } \delta_2 . \qquad (*\!*\!*\!*)$$

Finally, it is readily shown using analytical geometry (see Figure 9.3) that

$$|[\psi(s',\sigma_2)-\psi(s',\sigma_1)] - [\tilde{\psi}(s'',\sigma_2)-\tilde{\psi}(s'',\sigma_1)]|$$

$$= |[\psi(s',\sigma_2)-\psi(s',\sigma_1)] - [\psi(s'',\sigma_2)-\psi(s'',\sigma_1)]| ,$$

$$\leq \frac{2|\sigma_2-\sigma_1|\delta_1}{(\delta_2-\delta_1)^2} ,$$

for all $P(\sigma_1)$, $P(\sigma_2) \in \Gamma \backslash C_{s,\delta_2}$.

Remembering the definition of a Stieltjes integral (see eqn (9.71)), this implies that

$$I_3 \leq \frac{S\ \delta_1}{\delta_2^2} . \qquad (*^{*}_{**})$$

Combining eqns (*), (**), ($*_*$), ($^{*}_{**}$) and ($*^{*}_{**}$) we conclude that for each $s \in [0,S]$ there exists $\delta_1 = \delta_1(s) > 0$ such that

$$\left| \int_\Gamma x(\sigma)\, d_\sigma \psi(s',\sigma) - \int_\Gamma x(\sigma)\, d_\sigma \tilde{\psi}(s'',\sigma) \right|$$

$$\leq \|x\| [\sup_\Gamma |\pi - \beta(s)| + \varepsilon] , \qquad (9.95)$$

provided that (see page 333 for the definition of $d(s,\sigma)$) ,

$$d(s',s) \leq \delta_1(s) \text{ and } d(s'',s) \leq \delta_1(s) \qquad (9.96)$$

Let $\delta(s)$ denote the supremum of the values of $\delta_1(s)$ for which inequality (9.95) holds. Then, $\delta(s)$ is lower semi-continuous on $[0,S]$ (Problem 9.46) and hence, by Theorem 6.20, attains its minimum $\bar{\delta}$ at some point $\bar{s}$.

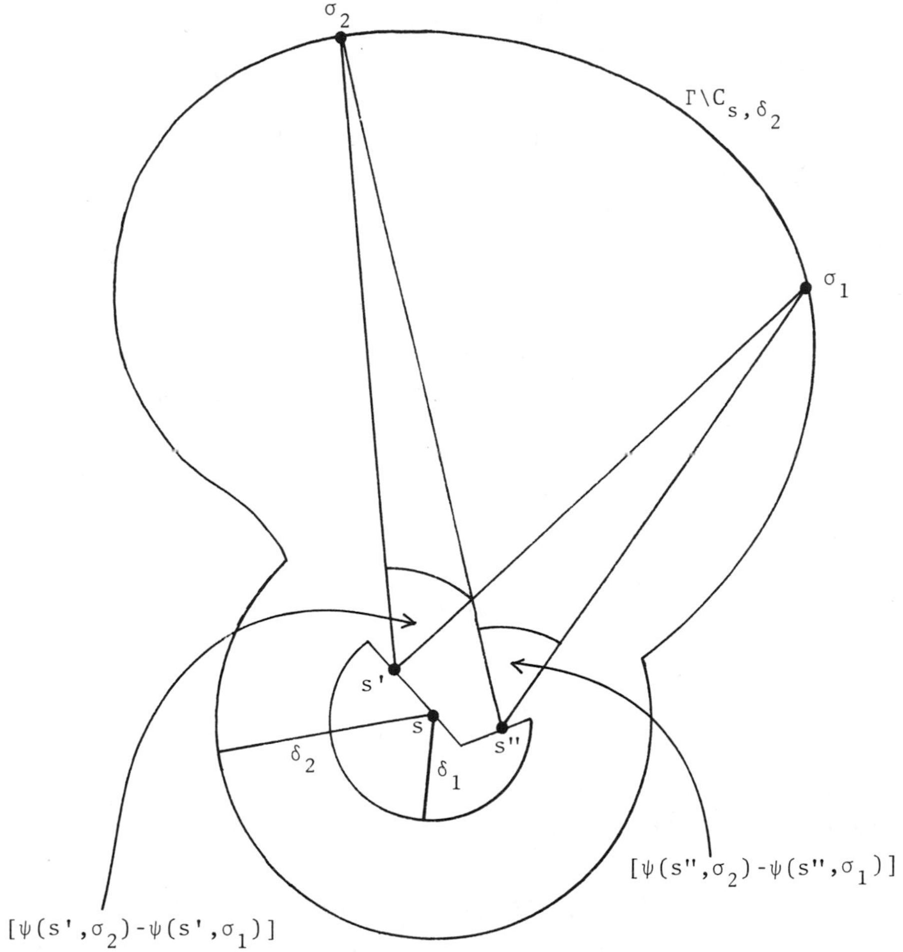

Figure 9.3: Geometry for proof of eqn (*****), page 346.

Thus,

$$\delta(s) \geq \inf_{\Gamma} \delta(s) \equiv \bar{\delta} = \delta(\bar{s}) > 0 .$$

□

We can now construct mappings E and F satisfying eqns (9.91) and (9.92).

Let $\tau = \{s_i\}$ be a partition of Γ:

$$s_n - S = s_0 < 0 = s_1 < s_2 < \ldots < s_n < S = s_{n+1}$$

consisting of n distinct points on Γ. The case $n = 8$ is illustrated in Figure 9.2. As usual $|\tau|$ denotes the largest spacing in τ,

$$|\tau| = \max_{1 \leq i \leq n} |s_i - s_{i+1}| .$$

Let P_n denote the projection of X onto the subspace X_n of periodic piecewise linear functions or linear splines with breakpoints s_i. That is,

$$P_n : y \in X \to \sum_{i=1}^{n} y(s_i) x_i(s) , \tag{9.97}$$

where the basis $\{x_1, \ldots, x_n\}$ consists of the 'hat-functions'

$$x_i(s) = \begin{cases} (s_{i+1} - s)/(s_{i+1} - s_i) , & \text{for } s \in [s_i, s_{i+1}] , \\ (s - s_{i-1})/(s_i - s_{i-1}) , & \text{for } s \in [s_{i-1}, s_i] , \\ 0 , & \text{otherwise.} \end{cases} \tag{9.98}$$

Set

$$(\tilde{K}x)(s) = \frac{1}{\pi} \int_{\Gamma} x(t) d_{\sigma} \tilde{\psi}(s, \sigma) , \tag{9.99}$$

and

$$J_n = P_n \tilde{K} , \tag{9.100}$$

so that

$$(J_n x)(s) = \frac{1}{\pi} \sum_{i=1}^{n} x_i(s) \int_\Gamma x(\sigma) d_\sigma \tilde{\psi}(s_i,\sigma) . \tag{9.101}$$

Finally, set

$$H_n = K - J_n . \tag{9.102}$$

It should be noted that $(\tilde{K}x)(s)$ is defined for $s \in [0,S]$ but that $\tilde{K}$ does not map X into X because

$$((K-\tilde{K})x)(s) = [\pi-\beta(s)]x(s) . \tag{9.103}$$

THEOREM 9.12. *Given* $\varepsilon > 0$ *let* $\bar{\delta}$ *be as in Lemma 9.11. Choose a partition* τ *of* $[0,S]$ *such that* $|\tau| < \bar{\delta}$.

Then J_n *as defined by eqn (9.101) is a bounded compact mapping of* X *into* X *such that*

$$\|H_n\| = \|J_n - K\| < \frac{1}{\pi} [\sup_\Gamma |\pi-\beta(s)| + \varepsilon] . \tag{9.104}$$

Proof. It is readily seen that $J_n: X \to X$ and that J_n is bounded and compact.

Let $x \in X$. Choose $s \in [0,S]$. Then $s \in [s_i, s_{i+1}]$ for some i . Since $x_i(s) + x_{i+1}(s) \equiv 1$ on $[s_i, s_{i+1}]$,

$$(Kx)(s) = \sum_{j=i}^{i+1} x_j(s) \int_\Gamma x(\sigma) d_\sigma \psi(s,\sigma) .$$

From the definition of J_n and the properties of the hat functions,

$$(J_n x)(s) = \sum_{j=i}^{i+1} x_j(s) \int_\Gamma x(\sigma) d_\sigma \tilde{\psi}(s_j,\sigma) .$$

Equation (9.104) is now an immediate consequence of eqn (9.94). □

Theorem 9.12 provides the key to the analysis of Radon's equation

$$Tu = (I+K)u = g . \qquad (9.88)$$

Since Γ has been assumed to have no cusps, it follows from Theorem 9.12 that, for large enough n, $\|H_n\| < 1$; set $E = H_n$ and $F = J_n$. By Theorem 9.2, $(I+E)^{-1} \in B(X)$. Rewriting Radon's equation in the form

$$(I+E+F)u = g , \qquad (9.105)$$

we see that it is equivalent to the equation

$$(I+(I+E)^{-1}F)u = (I+E)^{-1}g . \qquad (9.106)$$

But the product of a bounded operator and a compact operator is compact (Theorem 6.25, part 2). Hence, $(I+E)^{-1}F$ is compact, and the Fredholm theory (Theorem 9.5) is applicable.

Now, if Radon's equation has a solution $u \in X$, then eqn (9.87) defines the solution U of the Dirichlet problem (9.64), (9.65) with boundary data f. If Radon's equation has two distinct solutions u_1 and u_2 for the same right hand side g, then the difference $U_1 - U_2$ would be a solution of the Dirichlet problem with zero boundary data. From the theory of the Dirichlet problem we would have $U_1 = U_2$, which in turn implies that $u_1 = u_2$. (Problem (9.47)).

In summary, eqn (9.106) has at most one solution. The Fredholm theory now assures us that eqn (9.106) and eqn (9.88) have a unique solution for every $g \in X$. As discussed in section 9.2, the open mapping theorem implies that $(I+K)^{-1}$ is continuous. (see Theorem 8.7).

We have proved:

THEOREM 9.13. *Let* Γ *be of bounded rotation and without cusps.*

Then Radon's equation (9.88) has a unique continuous solution u *for every continuous* g.

Furthermore, $(I+K)^{-1} \in B(X)$, *where* K *is the Radon operator (9.86).* □

We now consider the numerical solution of Radon's equation. Several authors have contributed to this subject: Benveniste [1967], Bruhn and Wendland [1967], Cryer [1970] , Gaier [1964, p. 57], Petryshyn [1968], Miller [1979] , and Polsky [1962]. Here we use the projection method, as suggested by Bruhn and Wendland. We first discuss the theory and then conclude with a numerical example.

As in section 9.5, we replace Radon's equation

$$Tu = (I+K)u = (I+E+F)u = g ,$$

by

$$T_n u_n \equiv (I+P_n K)u_n \equiv (I+K_n)u_n = g_n \equiv P_n g . \qquad (9.107)$$

In our numerical example we will take the projection P_n to be defined by eqns (9.97) and (9.98), but this is not necessary. All that is required is that

$$P_n x \to x , \quad \text{as} \quad n \to \infty , \quad \text{for all} \quad x \in X \qquad (9.108)$$

and

$$\|P_n\| \cdot \|E\| < 1 - \mu , \qquad (9.109)$$

for some $\mu > 0$ and n sufficiently large. We will always assume that n is so large that eqn (9.109) holds.

Since K is not compact, Theorem 9.8 is not applicable. However, the proof of Theorem 9.13 suggests that we will be able to circumvent this by using eqn (9.106), and we now do this. We begin by setting

$$E_n = P_n E ,$$

and (9.110)

$$F_n = P_n F ,$$

so that eqn (9.107) becomes

$$T_n u_n \equiv (I+K_n)u_n \equiv (I+E_n+F_n)u_n = g_n \equiv P_n g \ . \qquad (9.111)$$

It follows from the assumptions (9.108) and (9.109) that

$$\|E_n\| \le \|P_n\| \, \|E\| < 1 - \mu \ ,$$

and

$$E_n x = P_n E x \to Ex \ , \quad \text{as} \quad n \to \infty \ .$$

We now state two lemmas, the simple proofs of which are left as problems for the reader (Problems 9.48 and 9.49).

LEMMA 9.14. *Let* X *be a Banach space. Let* $E \in B(X)$ *satisfy* $\|E\| < 1$. *Let* $\{E_n\}$ *be a sequence of operators in* $B(X)$ *such that: (1)* $\|E_n\| \le 1 - \mu$ *for some* $\mu > 0$; *and (2)* $E_n x \to Ex$ *for all* $x \in X$.

Then

$$(I+E_n)^{-1}x \to (I+E)^{-1}x \ , \quad \text{for all} \quad x \in X \ . \qquad \square$$

LEMMA 9.15. *Let* X *be a Banach space. Let* $\{G_n\}$ *be a sequence of operators in* $B(X)$ *such that* $G_n x \to Gx$ *for all* $x \in X$, *where* $G \in B(X)$. *Let* $F \in B(X)$ *be compact.*

Then

$$\|G_n F - GF\| \to 0 \ . \qquad \square$$

We now establish the existence of solutions u_n of the approximate eqn (9.107) and show that $u_n \to u$, thereby completing the basic theory for this numerical method of solving Radon's equation.

THEOREM 9.16. *For sufficiently large* n , *eqn (9.107) has a unique solution* u_n .

$u_n \to u$ *as* $n \to \infty$.

Proof. It follows from Lemma 9.14 that

$$(I+E_n)^{-1}x \to (I+E)^{-1}x\ , \quad \text{for} \quad x \in X\ . \tag{*}$$

Hence, using Lemma 9.15 twice, first with $G_n = P_n$ and $G = I$, and then with $G_n = (I+E_n)^{-1}$ and $G = (I+E)^{-1}$, we obtain

$$\|(I+E_n)^{-1}F_n-(I+E)^{-1}F\|$$

$$\le \|(I+E_n)^{-1}\|\ \|P_nF-IF\| + \|(I+E_n)^{-1}F-(I+E)^{-1}F\| \to 0\ . \tag{**}$$

In the course of proving Theorem 9.13 it was shown that $(I+(I+E)^{-1}F)^{-1} \in B(X)$. Combining this fact with eqn (**), it follows from Theorem 9.3 that $(I+(I+E_n)^{-1}F_n)^{-1} \in B(X)$. But $T_n = (I+E_n)(I+(I+E_n)^{-1}F_n)$, and so we can conclude that $T_n^{-1} \in B(X)$. This establishes the existence of the approximate solution u_n.

That $u_n \to 0$ follows from eqn (**) and Theorem 9.6 as applied to the equations

$$(I+(I+E)^{-1}F)u = (I+E)^{-1}g\ ,$$

$$(I+(I+E_n)^{-1}F_n)u_n = (I+E_n)^{-1}g_n\ . \quad \square$$

EXAMPLE 9.9. We consider the Dirichlet problem (9.64), (9.65) for a polygonal region, with the projection P_n defined by eqns (9.97) and (9.98).

Let $V_n = \ell_\infty^n$,

$$\phi_n\colon \sum_{i=1}^{n} \lambda_i x_i \in X_n \to (\lambda_1,\ldots,\lambda_n) \in V_n\ ;$$

$$v^{(n)} = \phi_n u_n = \begin{pmatrix} u_n(s_1) \\ \vdots \\ u_n(s_n) \end{pmatrix}, \quad \text{and} \quad b^{(n)} = \phi_n g_n = \begin{pmatrix} g(s_1) \\ \vdots \\ g(s_n) \end{pmatrix}. \tag{9.112}$$

Using eqns (9.5), (9.88), and (9.107) we find that

$$A^{(n)}v^{(n)} = b^{(n)} ,$$
$$A^{(n)} = (a_{ij}^{(n)}) = (\delta_{ij}+d_{ij}^{(n)}) . \tag{9.113}$$

where

$$d_{ij}^{(n)} = \frac{1}{\pi}\int_0^S x_j(\sigma)d_\sigma\psi(s_i,\sigma) ,$$
$$= \frac{1}{\pi}\delta_{ij}[\pi-\beta(s_i)] + \frac{1}{\pi}\int_{s_{j-1}}^{s_{j+1}} x_j(\sigma)d_\sigma\tilde{\psi}(s_i,\sigma) . \tag{9.114}$$

In general, the coefficients $d_{ij}^{(n)}$ must be evaluated numerically, but for simple curves Γ, and in particular for the case when Γ is a polygon, $d_{ij}^{(n)}$ can be expressed in closed analytic form.

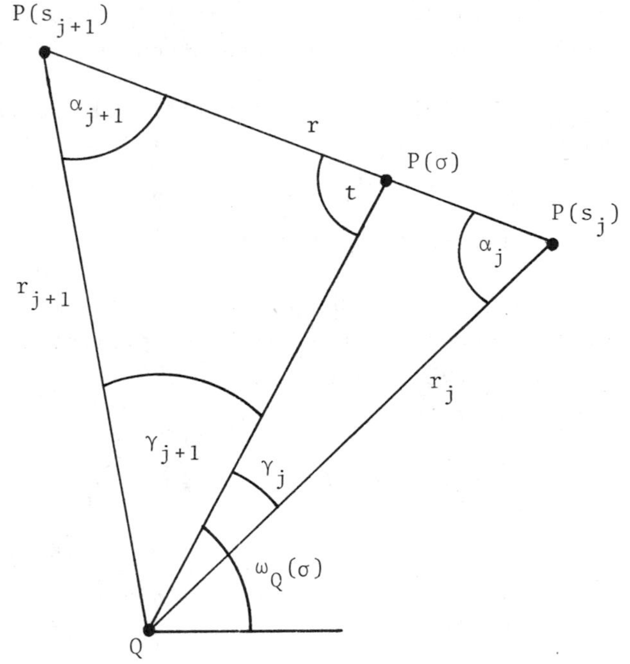

Figure 9.4: Computation of $d_{ij}^{(n)}$ *for Example 9.9.*

Consider the integral

$$\int_{s_j}^{s_{j+1}} z(\sigma)\, d_\sigma \omega_Q(\sigma) ,$$

where z is linear in σ. With the notation of Figure 9.4,

$$z(\sigma) = z(s_j) \frac{|P(s_{j+1})-P(\sigma)|}{|P(s_{j+1})-P(s_j)|} + z(s_{j+1}) \frac{|P(\sigma)-P(s_j)|}{|P(s_{j+1})-P(s_j)|} ,$$

$$= z(s_j) \frac{r_{j+1}}{r} \frac{\sin \gamma_{j+1}}{\sin t} + z(s_{j+1}) \frac{r_j}{r} \frac{\sin \gamma_j}{\sin t} .$$

Now

$$\int_{s_j}^{s_{j+1}} \frac{\sin \gamma_{j+1}}{\sin t}\, d\omega_Q(\sigma) ,$$

$$= \int_{\alpha_j}^{\pi-\alpha_{j+1}} \frac{\sin \gamma_{j+1}}{\sin t}\, dt ,$$

$$= \int_{\alpha_j}^{\pi-\alpha_{j+1}} \frac{\sin (t+\alpha_{j+1})}{\sin t}\, dt ,$$

$$= (\pi-\alpha_j-\alpha_{j+1}) \cos \alpha_{j+1} + \sin \alpha_{j+1}\, \ell n \left[\frac{\sin \alpha_{j+1}}{\sin \alpha_j}\right] ,$$

$$= \frac{\Delta\omega}{2r\, r_{j+1}} [r_{j+1}^2+r^2-r_j^2] + \frac{r_j}{r} \sin [\Delta\omega]\, \ell n\left(\frac{r_j}{r_{j+1}}\right) ,$$

where
$$r = |P(s_j)-P(s_{j+1})| ,$$

$$r_j = |Q-P(s_j)| ,$$

$$r_{j+1} = |Q-P(s_{j+1})| ,$$

and
$$\Delta\omega = \omega_Q(s_{j+1}) - \omega_Q(s_j) .$$

We thus find that

$$\int_{s_j}^{s_{j+1}} z(\sigma)d_\sigma\omega_Q(\sigma)$$

$$= \Delta\omega\ [z(s_{j+1}) + z(s_j)]/2 +$$

$$+ [z(s_{j+1}) - z(s_j)]\left[\Delta\omega \cdot \frac{r_j^2-r_{j+1}^2}{2r^2} + \frac{r_j r_{j+1}}{r^2} \sin(\Delta\omega)\,\ell n\left(\frac{r_{j+1}}{r_j}\right)\right] . \tag{9.115}$$

These formulae can also be used to evaluate the coefficients $d_{ij}^{(n)}$. Specifically, if $i \neq k$ and $i \neq k-1$ then

$$\int_{s_k}^{s_{k+1}} x_j(\sigma)d_\sigma\tilde{\psi}(s_i,\sigma) = \frac{1}{2}\Delta\tilde{\psi}\ [x_j(s_{k+1})-x_j(s_k)] +$$

$$+ [x_j(s_{k+1})-x_j(s_k)]\left[\Delta\tilde{\psi}\ \frac{r_k^2-r_{k+1}^2}{2r^2} + \frac{r_k r_{k+1}}{r^2} \sin(\Delta\tilde{\psi})\ \ell n\left(\frac{r_{k+1}}{r_k}\right)\right] \tag{9.116}$$

where

$$\Delta\tilde{\psi} = \tilde{\psi}(s_i,s_{k+1}) - \tilde{\psi}(s_i,s_k) ,$$
$$r = |P(s_k) - P(s_{k+1})| ,$$
$$r_k = |P(s_k) - P(s_i)| ,$$
$$r_{k+1} = |P(s_{k+1}) - P(s_i)| .$$

If $i = k$ or $i = k - 1$ then the integral is zero, since then $d_\sigma\tilde{\psi}(s_i,\sigma) = 0$ for $s_k \le \sigma \le s_{k+1}$.

We conclude with a few numerical results which were all obtained with a single programme. The programme accepted as input data: (1) the coordinates of the corners of a polygon Γ; (2) parameters to control the number of sub-divisions of each side of the polygon; and (3) a subroutine defining the function g on the right hand side of eqn (9.88). We give results for two problems. (Problem 9.50). For other numerical results see Problems 9.51 and 9.52.

Problem 1

Γ is the square shown in Figure 9.2, while

$$g(s) = 2(x(s))^2 + 2(y(s))^2 .$$

This problem arises when one considers the torsion of a cylindrical bar of cross-section Γ (Love [1944, p. 311]); it has been used as a model problem by several authors.

With the notation of eqns (9.112), (9.113), and (9.114), we find, for the case $n = 4$ taking $s_k = 2k-1$, for $1 \le k \le 4$, that

$$A^{(4)} \doteq \begin{pmatrix} 1.5000 & .1397 & .2206 & .1397 \\ .1397 & 1.5000 & .1397 & .2206 \\ .2206 & .1397 & 1.5000 & .1397 \\ .1397 & .2206 & .1397 & 1.5000 \end{pmatrix}, \quad b^{(4)} = \begin{pmatrix} 4 \\ 4 \\ 4 \\ 4 \end{pmatrix},$$

and

$$v^{(4)} = \begin{pmatrix} u_4(1) \\ u_4(3) \\ u_4(5) \\ u_4(7) \end{pmatrix} = \begin{pmatrix} 2 \\ 2 \\ 2 \\ 2 \end{pmatrix} .$$

For the case $n = 8$ with the points s_i as shown in Figure 9.2 we find that, $A^{(8)} \doteq$

$$\begin{pmatrix} 1.000 & .139 & .169 & .114 & .153 & .114 & .169 & .139 \\ .000 & 1.500 & .000 & .076 & .126 & .094 & .126 & .076 \\ .169 & .139 & 1.000 & .139 & .169 & .114 & .153 & .114 \\ .126 & .076 & .000 & 1.500 & .000 & .076 & .126 & .094 \\ .153 & .114 & .169 & .139 & 1.000 & .139 & .169 & .114 \\ .126 & .094 & .126 & .076 & .000 & 1.500 & .000 & .076 \\ .169 & .114 & .153 & .114 & .169 & .139 & 1.000 & .139 \\ .000 & .076 & .126 & .094 & .126 & .076 & .000 & 1.500 \end{pmatrix},$$

while

$$b^{(8)} = \begin{pmatrix} 2.000 \\ 4.000 \\ 2.000 \\ 4.000 \\ 2.000 \\ 4.000 \\ 2.000 \\ 4.000 \end{pmatrix}, \quad v^{(8)} \doteq \begin{pmatrix} .589 \\ 2.203 \\ .589 \\ 2.203 \\ .589 \\ 2.203 \\ .589 \\ 2.203 \end{pmatrix}.$$

The cases $n = 16$, 32, 64, and 128 were also solved, the points $s_k^{(n)}$ being given by

$$s_k^{(n)} = 1 + \frac{8(k-1)}{n}, \quad 1 \le k \le n .$$

The results for all values of n are given in Table 9.1. Because of the symmetry of the solution, only the values at $P(0)$, $P(\frac{1}{2})$, and $P(1)$ are given.

	N = 4	N = 8	N = 16	N = 32	N = 64	N = 128
P(0)	—	.58987	.68699	.70430	.70796	.70883
P(1/2)	—	—	1.13989	1.15536	1.15805	1.15862
P(1)	2	2.20376	2.23588	2.24196	2.24330	2.24363

Table 9.1: Selected values of P(s) *for Example 9.9, Problem 1.*

The successive differences in the values of P(1) are:

.20376, .03212, 00608, .00134, and .00033,

which strongly suggests that the error is $0(\frac{1}{n^2})$.

Problem 2

Γ is the curve shown in Figure 9.5 while g is given by

$$g(s) = (x(s))^3 - 3x(s)(y(s))^2 .$$

This problem, which illustrates re-entrant corners, is related to a problem solved by Benveniste [1967].

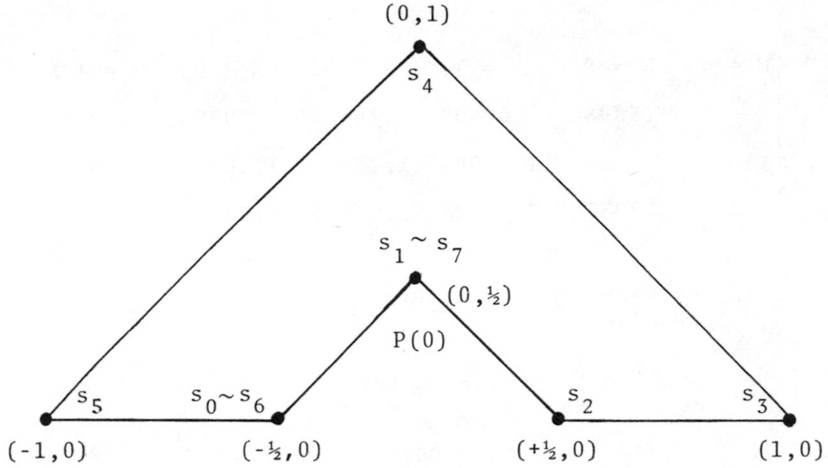

Figure 9.5: Problem 2 of Example 9.9

With the notation of Figure 9.5 and eqns (9.112), (9.113), and (9.114), we find, for the case $n = 6$ that

$$A^{(6)} \doteq \begin{pmatrix} .5000 & .0590 & .2693 & .8433 & .2693 & .0590 \\ -.1397 & 1.2500 & .4217 & .4201 & .1583 & -.1103 \\ -.0268 & .0933 & 1.7500 & .1397 & .1103 & -.0665 \\ .1866 & .1095 & .0472 & 1.5000 & .0472 & .1095 \\ -.0268 & -.0665 & .1103 & .1397 & 1.7500 & .0933 \\ -.1397 & -.1103 & .1583 & .4201 & .4217 & 1.2500 \end{pmatrix},$$

$$b^{(6)} = \begin{pmatrix} .00000 \\ .12500 \\ 1.00000 \\ .00000 \\ -1.00000 \\ -.12500 \end{pmatrix}, \text{ and } v^{(6)} = \begin{pmatrix} .00000 \\ -.02670 \\ .61248 \\ .00000 \\ -.61248 \\ .02670 \end{pmatrix}.$$

Combining eqns (9.66) and (9.112), each solution $v^{(n)}$ of eqn (9.113) determines an approximate solution

$$U^{(n)}(Q) = \frac{1}{\pi}\int_{\Gamma} (\phi_n^{-1}v^{(n)})(\sigma)\,d\omega_Q(\sigma) \;, \qquad (9.117)$$

of the Dirichlet problem

$$\nabla^2 U = 0 \;, \quad \text{in } \mathcal{D} \;,$$

$$U = g \;, \quad \text{on } \Gamma \;.$$

Using eqn (9.115), we see that the integral (9.117) can be evaluated analytically.

Since $x^3 - 3xy^2$ is a harmonic function, the solution of the Dirichlet problem in the present case is

$$U = x^3 - 3xy^2 \;.$$

The solution of Radon's equation was computed for several values of n, and then eqns (9.115) and (9.117) were used to compute $U^{(n)}(\frac{1}{4},\frac{1}{2})$. The results are summarized in Table 9.2. It is again observed that the convergence is quadratic.

n	$U^{(n)}(\frac{1}{4},\frac{1}{2})$	$U^{(n)}(\frac{1}{4},\frac{1}{2}) - U(\frac{1}{4},\frac{1}{2})$
6	.118794	.290669
12	-.084313	.087562
24	-.152471	.019404
48	-.167377	.004498
96	-.170762	.001113

Table 9.2: Values of $U^{(n)}(\frac{1}{4},\frac{1}{2})$ *for Example 9.9, Problem 2*

□

REMARK 9.12

In Chapter 1 we commented on the interaction between numerical functional analysis and classical numerical analysis. Example 9.9 illustrates these remarks.

1. The idea of reformulating the Dirichlet problem as an integral equation is due to Neumann [1870], who drew upon the physical concepts of potential theory. In particular, the solution $u(s)$ of the integral equation can be interpreted as the density of a double-layer on Γ. (A double-layer may be visualized as a curve or surface, one side of which is covered with a certain distribution of electrical charges, and the other side of which is covered with charges of equal magnitude but opposite sign.) Neumann considered not only smooth curves but curves with corners. *Thus, the original formulation of Radon's equation was based upon intuitive physical concepts.*

2. Neumann proved the existence of a solution to his integral equation for the case when Γ is convex (and thereby obtained the first rigorous existence proof for the Dirichlet problem). Although Neumann's work predated the birth of functional analysis by many years, Neumann's proof used concepts which were later to be generalized as part of the subject of functional analysis. *Thus, for Radon's equation, all existence proofs, and all convergence proofs for numerical methods, have depended upon functional analysis.*

3. There is a considerable literature on problems connected with the iterative solution of the linear algebraic equations (9.113), namely $A^{(n)}v^{(n)} = b^{(n)}$; see Todd and Warschawski [1955] and Gaier [1964, p. 52]. This work depends upon the rather special properties of the matrix $A^{(n)}$ (see Problem 9.53). Furthermore, there are at present no asymptotic error estimates for the difference $\|u-u_n\|$ between exact and approximate solutions of Radon's equation and it seems to the author that very detailed

classical analysis will be required to obtain such estimates. *Thus, Radon's equation also offers wide scope for classical numerical analysis.* □

PROBLEMS

9.1 Show that eqn (9.2) is equivalent to eqn (9.5) when $g_n \in R(K_n)$. {297}

9.2 Show that eqn (9.2) is equivalent to eqn (9.8). {297}

9.3 Consider Love's integral equation, eqn (9.9), with $d = -1$. Assume that there is a unique solution $u \in C[-1,1]$. Let m and M denote the minimum and maximum values of u, respectively. Prove:

(1) $\frac{1}{\pi} \arctan 2 \le \frac{1}{\pi} \int_{-1}^{+1} \frac{dt}{1+(s-t)^2} \le \frac{1}{2}$, for $-1 \le s \le 1$.

(2) $\frac{1}{2} \le m \le M \le 1$. (These bounds can be sharpened).

(3) u is even: $u(s) = u(-s)$, for $0 \le s \le 1$.
{Remark 9.3, p. 299}

9.4 Consider the two-point boundary value problem

$$\begin{aligned} &\ddot{u}(s) + F(s, u(s), \dot{u}(s)) = 0, \quad 0 \le s \le 1, \\ &u(0) = u(1) = 0, \end{aligned} \qquad (*)$$

where $F: R^3 \to R^1$ is a given continuous function.

Show that if $u \in C^2[0,1]$ satisfies eqn (*) then

$$u(s) = \int_0^1 K(s,t) F(t, u(t), \dot{u}(t))\,dt, \quad 0 \le s \le 1, \qquad (**)$$

where $K(s,t)$ is the Green's function

$$K(s,t) = \begin{cases} (1-s)t, & \text{if } t \le s, \\ (1-t)s, & \text{if } t \ge s. \end{cases}$$

{Remark (9.4)(1), p. 300}.

9.5 Let X be a Banach space with Schauder basis $\{x_i : 1 \le i < \infty\}$, and let P_n be the projection (see eqn (8.4)),

$$P_n : x = \sum_{i=1}^{\infty} \lambda_i(x) x_i \in X \to \sum_{i=1}^{n} \lambda_i(x) x_i \, .$$

Prove that:

(1) P_n is compact;

(2) $P_n x \to Ix$ for all x;

(3) I is not compact. {Remark 9.4 (2), p. 300}

9.6 Prove that if S and T are continuous linear mappings, $S: Y \to Z$ and $T: X \to Y$, then $\|ST\| \le \|S\| \, \|T\|$. Give an example where $\|ST\| \ne \|S\| \, \|T\|$. {eqn (9.12), p. 301}

9.7 Give two square matrices A and B such that $AB \ne BA$. {301}

9.8 Prove the elementary properties of linear inverses stated in Theorem 9.1.
[Hint: To prove part 4 we note that T is one-to-one since otherwise there would exist $x \ne \underline{0}$ such that $Tx = \underline{0}$; but then, $x = S(Tx) = S\underline{0} = S(T\underline{0}) = \underline{0} \ne x$. Furthermore, if $TSy \ne y$ for some $y \in R(T)$, then $TSTx \ne Tx$ where $y = Tx$; but $STx = x$, and so $Tx \ne Tx$.] {302}

9.9 Let $T: \begin{pmatrix} x_1 \\ x_2 \end{pmatrix} \in R^2 \to (x_1) \in R^1$,

$$S: (x_1) \in R^1 \to \begin{pmatrix} x_1 \\ 0 \end{pmatrix} \in R^2 \, .$$

Show that $TSy = y$ for all $y \in R^1$ but that T^{-1} does not exist. (Compare Theorem 9.1 (4), p. 302).

9.10 Prove that if $T \in B(X,Y)$ has a continuous inverse, and X is a Banach space, then $R(T)$ is closed, so that case 2a of eqns (9.16), (9.17) cannot occur. {303}

9.11 Prove that if X and Y are finite dimensional, then cases 2b, 2c, and 3b of conditions (9.16) and (9.17) cannot occur. {Example 9.1, p. 303}

9.12 Verify that the operators defined in Example 9.2 do in fact satisfy the stated combinations of conditions (9.16) and (9.17).
[Hint: In case (2b), the element $(1,\frac{1}{2},\frac{1}{4},\ldots,)$ does not belong to $R(T)$ since $(1,1,1,\ldots)$ does not belong to ℓ^2.]. {304}

9.13 Let $X = Y = C^1(0,1)$ and define $T: X \to Y$ by

$$(Tx)(s) = \dot{x}(s) \equiv \frac{dx}{ds}, \quad \text{for} \quad s \in (0,1).$$

Show that $R(T) = Y$, but that T^{-1} does not exist, so that T falls under case 1c of eqns (9.16), (9.17).

9.14 Let

$$A = \begin{pmatrix} 1 & 0 & 1 & 1 \\ 0 & 1 & -1 & 0 \\ 1 & 1 & 0 & 1 \end{pmatrix}, \quad A^+ = \frac{1}{15}\begin{pmatrix} 3 & 0 & 3 \\ -1 & 5 & 4 \\ 4 & -5 & -1 \\ 3 & 0 & 3 \end{pmatrix}.$$

Show that A and A^+ satisfy the conditions of Penrose, equations (9.20). {Remark 9.6, p. 305}.

9.15 Find the generalized inverse of

$$\begin{pmatrix} -1 & 0 & 1 & 2 \\ -1 & 1 & 0 & -1 \\ 0 & -1 & 1 & 3 \\ 0 & 1 & -1 & -3 \\ 1 & -1 & 0 & 1 \\ 1 & 0 & -1 & -2 \end{pmatrix}.$$

9.16 Prove: If $A, B, T \in B(X)$, $BT = I - A$, and $\|A\| < 1$, then T^{-1} exists. Furthermore, if $\tilde{B}$ denotes the restriction of B to $R(T)$, then:

(1) $T^{-1} = (I - A)^{-1}\tilde{B}$; (2) $T^{-1} - \tilde{B} = (I - A)^{-1} A\tilde{B}$,

(3) $\|T^{-1}\| \le \dfrac{\|B\|}{1 - \|A\|}$, (4) $\|T^{-1} - \tilde{B}\| \le \dfrac{\|A\| \cdot \|B\|}{1 - \|A\|}$.

[Hint: Use Theorem 9.1 (part 4).] {Remark 9.7, p. 307}

9.17 Prove: Let K, $L \in B(X)$. Assume that $(I - K)^{-1} \in B(X)$ and

$$\Delta = \|(I - K)^{-1}(L - K)L\| < 1. \qquad (*)$$

Then $(I - L)^{-1}$ exists and is bounded:

$$\|(I - L)^{-1}\| \le \frac{1 + \|(I-K)^{-1}\| \, \|L\|}{1 - \Delta} . \qquad (**)$$

Also,

$$\|(I-L)^{-1}y - (I-K)^{-1}y\| \le \frac{\|(I-K)^{-1}\| \, \|Ly - Ky\| + \Delta \, \|(I-K)^{-1}y\|}{1 - \Delta} , \qquad ({*}{*}{*})$$

for $y \in R(I-L)$.

[Hint: Use Problem 9.16 and the identity,

$$[I + (I-K)^{-1}L](I-L) = I - [(I-K)^{-1}(L-K)L] . \qquad ({*}{*}{*}{*})$$

(Brakhage [1960] and Anselone [1971, p. 8]).] {307, 329}

9.18 Show that Theorem 9.3 remains true if X and Y are normed spaces only one of which is a Banach space. [Hint: The proof of Theorem 9.3 given in the text requires only that X be a Banach space. If X is not a Banach space, set $B = (I-C)A$ where $C = (A-B)A^{-1}: Y \to Y$.] {308}

9.19 Complete the proof of Theorem 9.4 which states that 'an equation which can always be solved approximately can be solved exactly'. {309}

9.20 Prove Theorem 9.6. In particular, show that if $(I-K)u = g$, and $(I-K_n)u_n = g$ where K, K_n, $(I-K)^{-1}$, and $(I-K_n)^{-1}$ all belong to $B(X)$, then

$$\|u-u_n\| \le \|(I-K)^{-1}(K-K_n)u_n\| + \|(I-K)^{-1}\| \, \|g_n-g\| \, . \quad \{310\}$$

9.21 If $(I-K)u = g$ and $(I-K_n)u_n = g$, and K, K_n, $(I-K)^{-1}$ and $(I-K_n)^{-1}$ all belong to $B(X)$, show that

(a) $\|u-u_n\| \le \|(I-K_n)^{-1}(K_n-K)u\|$.

(b) $\|u-u_n\| \le \|(I-K)^{-1}(K-K_n)(I-K_n)^{-1}g\|$.

(c) $$\|u-u_n\| \le \frac{\|(I-K_n)^{-1}(K_n-K)u_n\|}{1-\|(I-K_n)^{-1}(K_n-K)\|},$$

if $\|(I-K_n)^{-1}(K_n-K)\| < 1$. {310}

9.22 Show that the optimum linear approximation in the maximum norm to $1/(1+t)$ on the interval $[0,4]$ is

$$p(t) = \frac{2}{5} + \frac{1}{\sqrt{5}} - \frac{t}{5} \, .$$

{Example 9.4, p. 312}.

9.23 Use the method of Example 9.4 with $n = 3$ to solve approximately Love's equation with $d = +1$. (Which arises when the coaxial disks have charges of opposite sign). {314}

9.24 Use the method of Example 9.4 with $n = 1$ to solve Love's equation with $d = -1$. {314}

9.25 Let

$$a = s_1 < s_2 < \ldots < s_n = b \text{ and } a = t_1 < t_2 < \ldots < t_n = b$$

be two partitions of $[a,b]$. Let $k_n(s,t)$ be the continuous piecewise bilinear function which interpolates $k(s,t)$ at the points (s_i,t_j), $1 \le i, j \le n$.

Show that $k_n(s,t)$ is a degenerate kernel of the form (9.28). Show also that if K and K_n are defined by eqns (9.26) and (9.27), respectively, then there are many choices of the points s_i and t_j such that $\|K-K_n\| \to 0$ as $n \to \infty$.
[Hint: See Problem 8.4] {Remark 9.9, p. 314}

9.26 Verify that eqns (9.37), (9.38), and (9.39) are equivalent. {317}

9.27 Verify the details of Example 9.6. {320}

9.28 Let $X = L^2(\Omega)$ for $\Omega \subset R^2$, and let $K: X \to X$ where

$$(Kx)(s) = \int_\Omega k(s,t)x(t)dt . \tag{*}$$

Show that

$$\|K\|_2 \le \left[\int_\Omega ds \int_\Omega |k(s,t)|^2 dt\right]^{\frac{1}{2}} . \quad \{321\}$$

9.29 Use the method of Example 9.6 with $n = 3$ to solve Love's integral equation with $d = +1$. {323}

9.30 If the projection method is used to solve a Fredholm integral equation and the coefficients $a_{ij}^{(n)}$ and $c_i^{(n)}$ are computed using numerical quadrature, then equation (9.40a), namely

$$A^{(n)}\lambda^{(n)} = c^{(n)} , \qquad (*)$$

is approximated by

$$\tilde{A}^{(n)}\tilde{\lambda}^{(n)} = \tilde{c}^{(n)} , \qquad (**)$$

where $\tilde{A}^{(n)}$ and $\tilde{c}^{(n)}$ denote the arrays of computed coefficients. When eqn (**) is solved numerically, one obtains an approximate solution, $\hat{\lambda}^{(n)}$ say.

Assuming that $|a_{ij}^{(n)} - \tilde{a}_{ij}^{(n)}| \le \varepsilon$, that $|c_i^{(n)} - \tilde{c}_i^{(n)}| \le \varepsilon$ and that $\tilde{C}^{(n)}$ is an approximate inverse of $\tilde{A}^{(n)}$, use Theorems 9.2 and 9.3 to obtain a computable bound for $\|\lambda^{(n)} - \hat{\lambda}^{(n)}\|$. {323}

9.31 Consider the integral equation

$$u(s) - \int_a^b k(s,t)u(t)dt = g(s) , \quad a \le s \le b , \qquad (*)$$

in the space $X = C[a,b]$.

The *collocation method* for eqn (*) consists of choosing a set of linearly independent functions $x_1,\ldots,x_n$ spanning $X_n \subset X$ and a set of points s_i ,

$$a \le s_1 < s_2 < \ldots < s_n \le b .$$

The approximate solution $u_n \in X_n$ of eqn (*) is constructed so that the corresponding residual r_n vanishes at the points s_i . Explicitly,

$$u_n = \sum_{j=1}^{n} \lambda_j^{(n)} x_j ,$$

where

$$u_n(s_i) - \int_a^b k(s_i,t)u_n(t)dt = g(s_i) , \quad \text{for} \quad 1 \le i \le n . \qquad (**)$$

Show that if the $n \times n$ matrix $(x_j(s_i))$ is non-singular, then the collocation method corresponds to a projection method with the projection $P_n : X \to X_n$ defined uniquely by

$$(P_n x)(s_i) = x(s_i) , \quad 1 \le i \le n . \qquad (***)$$

{324}

9.32 Apply the collocation method defined in Problem 9.31 to Love's integral equation with $d = \pm 1$. Take $n = 3$, x_i the Legendre polynomial of degree $i - 1$, and $(s_i) = (-1/\sqrt{3}, 0, +1/\sqrt{3})$.

{324}

9.33 Consider the two-point boundary value problem

$$\ddot{u}(s) + g(s) = 0 , \quad a \le s \le b , \qquad (*)$$

$$u(a) = u(b) = 0 . \qquad (**)$$

If v is any smooth function satisfying the boundary conditions (**) then, multiplying eqn (*) by v and integrating, we obtain,

$$\int_a^b \dot{u}(s)\dot{v}(s)\,ds = \int_a^b g(s)v(s)\,ds . \qquad (***)$$

Now choose an n-dimensional subspace X_n of $C[a,b]$ spanned by functions $v_1,\ldots,v_n$ satisfying eqn (**). The *Ritz-Galerkin method* finds $u_n \in X_n$ such that

$$\int_a^b \dot{u}_n(s)\dot{v}_i(s)\,ds = \int_a^b g(s)v_i(s)\,ds , \quad \text{for } 1 \le i \le n . \qquad (****)$$

As an example, choose X_n to be the subspace spanned by the piecewise linear functions with n knots uniformly distributed in (a,b) so that v_i is the hat function

$$v_i(s) = \begin{cases} (s-t_{i-1})/h\ , & \text{if} \quad s \quad [t_{i-1},t_i]\ , \\ (t_{i+1}-s)/h\ , & \text{if} \quad s \quad [t_i,t_{i+1}]\ , \\ 0\ , & \text{elsewhere,} \end{cases}$$

where $h = (b-a)/(n+1)$ and $t_i = a + ih$. If $u_n = \sum_{j=1}^{n} \lambda_j v_j$, show that eqn (**) becomes

$$Au_n = c\ ,$$

where $c = (c_i) = (\int_a^b g(s)v_i(s)ds)$, $1 \le i \le n$, and

$$A = \frac{1}{h}\begin{pmatrix} 2 & -1 & & & \\ -1 & 2 & \cdot & & \bigcirc \\ & \cdot & \cdot & \cdot & \\ & & \cdot & \cdot & \cdot \\ & \bigcirc & & \cdot & \cdot & \cdot \\ & & & & \cdot & -1 \\ & & & & -1 & 2 \end{pmatrix}.$$

We have deliberately avoided a rigorous justification of eqn (**) since this requires the use of Sobolev spaces. For results on the Ritz-Galerkin method see e.g. Schultz [1973] and Reddien [1980]. {324}

9.34 Show explicitly how to construct an $x \in C[a,b]$ for which $\|x\|_\infty = 1$, such that eqns (9.59) are satisfied. {325}

9.35 Show that if X and Y are normed spaces and T is a family of linear mappings of X into Y , then T is collectively compact iff

$$\{Tx\colon\ T \in \mathcal{T} \ \text{and}\ \|x\| \le 1\}$$

is a sequentially compact subset of Y . (See eqn (9.60), p. 326).

9.36 Let X be a Banach space. Let $\{K_n\}$ be a sequence of operators in $B(X)$ such that $K_n x \to Kx$ for all $x \in X$. Let $L \in B(X)$ be compact. Show that

$$\|(K_n - K)L\| \to 0 .$$

9.37 Apply Nyström's method to Love's integral equation with $d = +1$. Use three-point Gaussian quadrature.
{332}

9.38 A Dirichlet problem which is often used as a test problem is:

$$U_{xx} + U_{yy} = 0 , \quad \text{in } \mathcal{D} ,$$
$$U = x^2 - y^2 , \quad \text{on } \Gamma = \partial\mathcal{D} ,$$

where Γ is the ellipse

$$x = 5 \cos t , \; y = 3 \sin t , \; 0 \le t \le 2\pi .$$

Solve this problem numerically using the integral equation (9.67) and any of the numerical methods for integral equations described in sections 9.4 to 9.6.
{335}

9.39 With the notation of Figure 9.1, assume that Γ is smooth and that $\underline{0} \in \mathcal{D}$. Then there exists a unique conformal mapping, $w = f(z)$ say, which maps $\mathcal{D}$ onto the unit disk in the w-plane and satisfies the normalizing conditions: $f(\underline{0}) = \underline{0}$; $f(P(0)) = 1$; the point $P(s) \in \Gamma$ is mapped onto the point $f(P(s)) = e^{i\theta(s)}$, with $\theta(0) = 0$.

It can be shown (Gaier [1964, p. 9]) that θ satisfies the *integral equation of Gerschgorin*,

$$\theta(s) = \int_\Gamma k(s,t)\theta(t)dt - 2b(s) , \qquad (*)$$

where $b(\mu)$ is the angle subtended at $P(\mu)$ by $\underline{0}$

and $P(0)$ (see Figure 9.1), and where $k(s,t)$ is the Neumann kernel (9.68).

Solve eqn (*) numerically to obtain the mapping of the ellipse

$$x = a \cos t\ , \quad y = d \sin t\ , \quad 0 \le t \le 2\pi\ ,$$

onto the unit disk. Compare the answers with the exact solution

$$\theta(t) = \arc\tan\left[\frac{\frac{1}{2}(R^2-1)^2-\cos^2 t}{\frac{1}{2}(R^2+1)^2-\sin^2 t}\cdot\tan t\right] -$$

$$- 2\sum_{n=1}^{\infty}\frac{(-1)^n}{n}R^{-2n}\frac{\sin 2nt}{R^{4n}+1}\ ,$$

where $R = a + d$, and $a^2 - d^2 = 1$. Gaier [1964, p. 55] quotes many numerical experiments on this problem.

Gaier [1964] also derives the integral equations of Lichtenstein, Carrier, Banin, and Warschawski and Stiefel, all of which solve the problem of mapping $\mathcal{D}$ onto the unit disk using integral equations with the Neumann kernel. {335}

9.40 Let $F \in C[a,b]$ and let G be of bounded variation on $[a,b]$. Show that

$$\left|\int_a^b F(t)dG(t)\right| \le \max_{a\le t\le b}|F(t)|\int_a^b |dG(t)|\ . \qquad \{341\}$$

9.41 Verify the expression (9.80) for $\psi(3,\sigma)$ for the case of the square shown in Figure 9.2. {338}

9.42 Show from first principles that $\psi(s,\sigma)$ has the properties (9.81) to (9.85) when Γ is a convex polygon. {339}

9.43 Let $\mathcal{D}$ be the domain obtained by adjoining a right-angled cone with vertex $(\sqrt{2},0)$ to the unit circle.

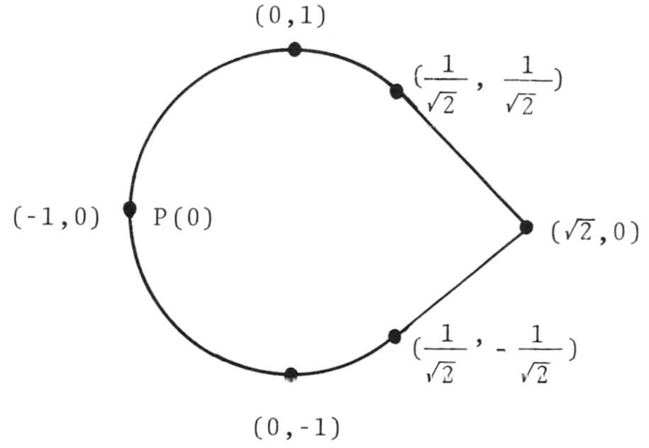

Show that if ψ is as defined in eqn (9.79) and $\psi(0,0) = \pi/2$ then,

$$\psi(S/2,0) = \pi\ ,$$

$$\psi(S/2,S/2-0) = 5\pi/4\ ,$$

$$\psi(S/2,S/2+0) = 7\pi/4\ ,$$

$$\psi(S/2,S) = 2\pi\ ,$$

where $S = 3\pi/2 + 2$.

Show that ψ satisfies the properties (9.81) to (9.85). {339}

9.44 Show that if Γ is of bounded rotation and ω_Q is as shown in Figure 9.1 then ω_Q is of bounded variation. {340}

9.45 Let $x \in \tilde{C}[0,S]$ be piecewise linear. Let $\{\tau_n\}$ be a sequence of points in $[0,S]$ converging to τ .

Show that

$$U_n x \equiv \int_\Gamma x(\sigma)\, d_\sigma \psi(\tau_n,\sigma) \to \int_\Gamma x(\sigma)\, d_\sigma \psi(\tau,\sigma) \equiv Ux \ ,$$

where $\psi(s,\sigma)$ is as in Figure 9.1 and Γ is of bounded rotation.
[Hint: Consider the case when x is the hat function

$$x(\sigma) = \begin{cases} 1-|\sigma-\xi| & \text{if } |\sigma-\xi| \le 1 \ , \\ 0 \ , & \text{otherwise.} \end{cases}$$

Then, using eqns (9.74) and (9.75), and the property (9.84),

$$U_n x = -\int_0^{\xi-1} \psi(\tau_n,\sigma)\, d\sigma + 2\int_0^{\xi} \psi(\tau_n,\sigma)\, d\sigma - \int_0^{\xi+1} \psi(\tau_n,\sigma)\, d\sigma \to Ux \ .$$

{341}

9.46 Let

$$\delta(s) = \sup_{P(s)\in\Gamma} \delta_1(s) \ ,$$

where $\delta_1(s)$ satisfies eqns (9.95) and (9.96). Show that $\delta(s)$ is lower semi-continuous. {209, 346}.

9.47 With the notation of section 9.7, let Γ be piecewise smooth and, for some $u \in \tilde{C}[0,S]$,

$$\int_\Gamma u(\sigma)\, d_\sigma \omega_Q(\sigma) = 0 \ , \qquad (*)$$

for all $Q \in \mathcal{D}$. Show that $u \equiv 0$.
[Hint: First show that if eqn(*) holds for a piecewise constant function u then $u \equiv 0$.] {350}.

9.48 Prove Lemma 9.14. {352}

9.49 Prove Lemma 9.15. {352}

9.50 Write a programme to implement the numerical algorithm in Example 9.9 for solving Radon's equation for a polygonal region. {357}

9.51 Solve approximately Radon's equation for the region shown below

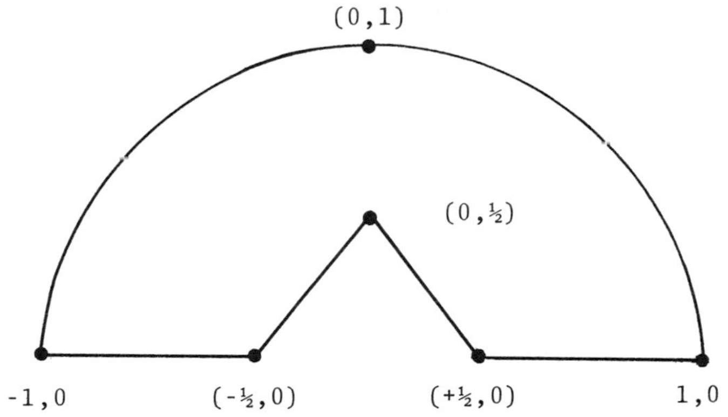

with $g(s) = (x(s))^3 - 3x(s)(y(s))^2$.

Use m subdivisions on each line segment and approximate the semi-circle by a polygon with 4m equal sides, for m = 1,2,3,6. Using the approach in Example 9.9, Problem 2 compute $U^{(n)}(.5,.5)$. (If $m = 6$ there are 48 points, and this problem was solved by Benveniste [1965].) {357}

9.52 Let $\mathcal{D}$ be the semi-circular region shown below

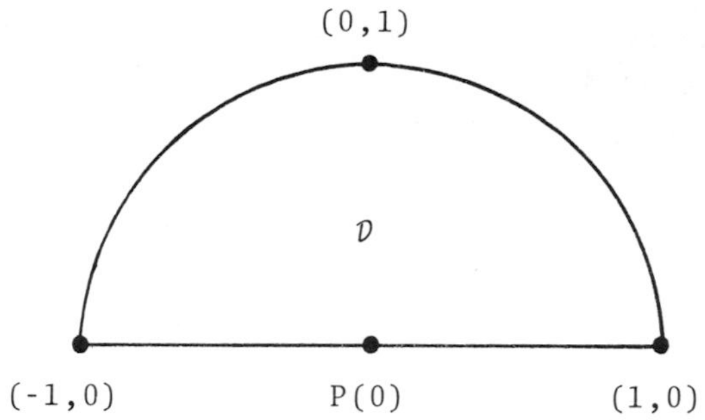

If $g(s)$ is defined by

$$g(s) = \begin{cases} \frac{3}{2}\,x(s), & \text{if } y(s) = 0\,, \\ \frac{3}{2}\,x(s) + \frac{y(s)}{2\pi}\,\ell n\left[\frac{1-x(s)}{1+x(s)}\right], & \text{if } y(s) > 0\,, \end{cases}$$

then, as stated by Miller [1979, p. 83], the exact solution of Radon's equation is

$$u(s) = x(s)\ .$$

With this data, solve Radon's equation numerically using m subdivisions on the diameter of $\mathcal{D}$ and $2m$ equal subdivisions on the semi-circle.

{357}

9.53 Inspection of the matrices $A^{(n)}$ given in Example 9.9 shows that these matrices have a rather special structure.

Show that if Γ is convex and the method of Example 9.9 is used to solve Radon's equation, then the resulting matrix $A^{(n)}$ is non-negative and irreducibly diagonally dominant. Hence show that $A^{(n)}$ is non-singular.

If Γ is not convex, can $A^{(n)}$ be singular?
{362}

9.54 Consider Radon's equation (9.88). Let

$\pi = \{s_n = s_0 < \ldots < s_{n+1} = s_1\}$ be a partition of Γ.

Let X_n be the space of periodic piecewise constant functions on $[0,S]$ with basis $\{x_1,\ldots,x_n\}$ where

$$x_i(s) = \begin{cases} 1, & \text{for } s \in (s_i, s_{i+1}), \\ \tfrac{1}{2}, & \text{for } s = s_i \text{ or } s = s_{i+1}, \\ 0, & \text{otherwise}, \end{cases}$$

and let $P_n : x \in X \to P_n x = \sum_{j=1}^{n} \tfrac{1}{2} x(s_j)[x_{j-1} + x_j] \in X_n$.

Let $\mathcal{D}$ denote the space of functions $\psi(s)$ defined on $[0,S]$ such that

(a) $\psi(s+0)$ exists, for $0 \le s < S$.
(b) $\psi(s-0)$ exists, for $0 < s \le S$.
(c) $\psi(s) = \tfrac{1}{2}[\psi(s+0) + \psi(s-0)]$, for $0 < s < S$.

Then $\mathcal{D}$ is a Banach space when equipped with the maximum norm. (Problem 4.17).

Replace Radon's eqn (9.88) by the approximate equation

$$(I + P_n K)u_n = P_n g. \qquad (*)$$

Does eqn (*) have a solution? Can one show that $u_n \to u$ where u is the solution of Radon's equation? (see Benveniste [1965] and Cryer [1970, p. 57]).

9.55 Consider the integral equation

$$x(s) + \int_a^b k(s,t)x(t)\,dt = g(s), \qquad (*)$$

where k is a *weakly singular* kernel of the form

$$k(s,t) = a(s,t)\,\ell n \frac{1}{|s-t|} + b(s,t),$$

where a and b are smooth.

Replace eqn (*) by

$$(1+A_i)\xi_i + h \sum_{\substack{j=0 \\ j\neq i}}^{n} k(t_i,t_j)(\xi_j-\xi_i) = \gamma_i \ , \quad 0 \le i \le n \ , \qquad (**)$$

where

$$A_i = \int_a^b k(t_i,t)dt \ , \quad h = \frac{b-a}{n} \ , \quad t_i = a+ih \ , \quad \gamma_i = g(t_i) \ ,$$

and ξ_i is an approximation to $x(t_i)$.

Assume that eqn(*) has a unique solution $x \in C[a,b]$. Show that the solution of eqn (**) exists and converges, in an appropriate sense, to that of eqn (*) for sufficiently large n .

Kussmaal and Werner (1968) analyse this problem and give an application to the exterior problem for the Helmholtz equation. {332}

9.56 The integral equation

$$\int_0^s \frac{a(s,t)}{(s^p-t^p)^\mu}\, x(t)dt = g(s) \ , \quad 0 \le s \le S \le \infty \ , \qquad (*)$$

on the interval $[0,S]$, where μ and p are constants satisfying $0 \le \mu \le 1$ and $p > 0$, is called a *first kind Abel-type integral equation*, since it is a generalization of the classical Abel equation

$$\int_0^s \frac{x(t)}{(s-t)^{\frac{1}{2}}}\, dt = g(s) \ .$$

There are numerous applications of eqn (*) to physical problems. See Anderssen and de Hoog [to appear]. {332}

Notation (In order of introduction)

Symbol	*Meaning*	*Defined*
$\dot{x}, \ddot{x}$, etc	derivatives of x	Chapter 1 page 4
R^1	real line	Section 2.1 page 6
$\in$	membership in a set	Section 2.2 page 6
$\{a,b,c\}$	set consisting of elements a,b, and c	Section 2.2 page 7
$\emptyset$	the empty set	Section 2.2 page 7
$\{x\in A: P(x)\}$	subset of A for which P(x) is true	Section 2.2 page 7
$\{A_i: i\in I\}$	family of sets A_i	Section 2.2 page 7
$A \cup B$	union of sets A and B	Section 2.2 page 7
$A \cap B$	intersection of sets A and B	Section 2.2 page 7
$\bigcup_{i\in I} A_i$	union of sets A_i	Section 2.2 page 7
$\bigcap_{i\in I} A_i$	intersection of sets A_i	Section 2.2 page 7
$A \supset B$	A contains B	Section 2.2 page 7
$A \setminus B$	complement of A with respect to B	Section 2.2 page 7
B^c	complement of B	Section 2.2 page 8
$\square$	end of a logical entity	Section 2.2 page 8
$P \Rightarrow Q$	P implies Q	Section 2.2 page 8
$P \Leftrightarrow Q$	P is equivalent to Q	Section 2.2 page 8
iff	if and only if	Section 2.2 page 8
AND	logical and	Section 2.2 page 8
OR	logical or	Section 2.2 page 8
NOT	logical not	Section 2.2 page 8
(a,b)	open interval in R^1	Section 2.3 page 8
$[a,b]$	closed interval in R^1	Section 2.3 page 8
$\{X,\tau\}$	topological space X with topology τ	Section 2.3 page 9
$\tau_1 \supset \tau_2$	topology τ_1 stronger than topology τ_2	Section 2.3 page 9

NOTATION (In order of introduction)

Symbol	*Meaning*	*Defined*	
N_x	neighbourhood of x	Definition 2.2	page 10
β_x	base of neighbourhoods of x	Definition 2.2	page 10
β	base of neighbourhoods for a topology	Definition 2.2	page 10
$\rho(x,y)$	metric	Definition 2.3	page 14
$B(x;\varepsilon)$	open ball	Example 2.7	page 14
$B[x;\varepsilon]$	closed ball	Example 2.7	page 14
s	space of real sequences	Example 2.8	page 14
$f : X \to Y$	f maps X into Y	Section 2.4	page 16
$D(f)$	domain of f	Section 2.4	page 16
$R(f)$	range of f	Section 2.4	page 16
$f(x)$	value of f at x	Section 2.4	page 16
I	identity mapping	Section 2.4	page 16
R^1_+	non-negative real numbers	Section 2.4	page 16
$f\|_Z$	restriction of f to Z	Section 2.4	page 17
$f \circ g$	composite mapping	Section 2.4	page 18
$f(U)$	{set-valued mapping, correspondence, multi-valued mapping, multi-function}	Section 2.4	page 19
f^{-1}	inverse mapping	Section 2.4	page 19
$H(x)$	Heaviside step function	Example 2.11	page 20
$X_1 \times X_2$	Cartesian product of two sets	Section 2.4	page 21
$f(x_1,\cdot)$	restriction of $f(x_1,x_2)$	Section 2.4	page 22
$\prod_{\alpha\in A} X_\alpha$	Cartesian product of arbitrarily many sets	Section 2.4	page 23
R^n	real n-dimensional space	Example 2.13 (Section 4.4	page 23 page 72)
$\mathbb{C}$	complex numbers	Section 2.5	page 25
$\underline{0}$	origin in a vector space	Definition 2.4	page 26

NOTATION (In order of introduction)

Symbol	*Meaning*	*Defined*	
$U \pm V$	algebraic operations on sets in vector spaces	Section 2.5	page 26
ΛU	algebraic operations on sets in vector spaces	Section 2.5	page 26
$u \pm V$	algebraic operations on sets in vector spaces	Section 2.5	page 26
λU	algebraic operations on sets in vector spaces	Section 2.5	page 26
$\overline{R}^1$	extended real line	Example 2.16	page 26
span (A)	linear subspace spanned by A	Section 2.5	page 28
ℓ^p, $p < 1$	real sequences with $\Vert\cdot\Vert_p$ norm	Example 2.18	page 32
$p(x)$	seminorm	Definition 2.6	page 32
$x^{(n)}$	n-th derivative of x	Example 2.20	page 33
$\Vert\cdot\Vert$, $\Vert x\Vert$, $\Vert\cdot\Vert_X$, $\Vert x;L^\infty\Vert$	different norms	Section 2.7	page 36
$\Vert x\Vert_1$	1-norm in R^n	Example 2.21	page 37
	norm in ℓ_1^n	Section 4.3	page 72
	norm in ℓ^1	Section 4.6	page 79
	norm in $L^1(\Omega)$	Section 4.11	page 93
$\Vert x\Vert_2$	Euclidean norm in E^n	Example 2.21	page 37
	norm in ℓ_2^n	Section 4.3	page 72
	norm in ℓ^2	Section 4.6	page 79
	norm in $L^2(\Omega)$	Section 4.11	page 93
$\Vert x\Vert_\infty$	maximum norm in R^n	Example 2.21	page 37
	norm in ℓ_∞^n	Section 4.3	page 72
	norm in ℓ^∞	Section 4.6	page 79
	norm in $C[a,b]$	Section 4.8	page 80
	norm in $C(\overline{\Omega})$	Section 4.9	page 86
	norm in $L^\infty(\Omega)$	Section 4.11	page 94

NOTATION (In order of introduction)

Symbol	*Meaning*	*Defined*	
(x,y)	inner product	Definition 2.9	page 39
$C_r(z)$	(relative) condition number	Section 2.10	page 42
τ_S	relative topology on a set S	Problem 2.2	page 43
$K(R^n)$	metric space for closed bounded sets in R^n	Problem 2.3	page 43
$x_1,x_2,\ldots,x_n,\ldots$; $\{x_n\}$; $\{x_n: n=1,2,\ldots\}$	sequences	Definition 3.1	page 49
$x_n \to x$; $\lim x_n = x$	limits of sequences	Definition 3.1	page 49
$\overline{E}$	closure of set E	Section 3.1	page 50
$\geq$	order relation	Definition 3.2	page 52
$\{x_\alpha\}(\alpha\in A)$	generalized sequence	Section 3.1	page 52
$x_\alpha \to x(\alpha\in A)$	generalized limit of a sequence	Section 3.1	page 52
$\Delta=\{t_0<t_1<\ldots<t_n\}$	partition of $[t_0,t_n]$	Example 3.3	page 53
$\{x_n\}\sim\{x_n'\}$	equivalent Cauchy sequences	Section 3.3	page 57
C_0	real functions with compact support	Problem 3.7	page 64
C_v	real functions vanishing at ∞	Problem 3.7	page 64
$BF[0,1]$	real bounded functions on $[0,1]$	Problem 3.8	page 64
E^n	n-dimensional Euclidean space	Section 4.3	page 68
x^Ty	inner product on E^n	Section 4.3	page 69
e^i	unit n-vector	Section 4.3	page 69

NOTATION (In order of introduction)

Symbol	*Meaning*	*Defined*	
δ_{ij}	Kronecker delta	Section 4.3	page 70
ℓ_p^n	n-tuples with $\|\cdot\|_p$ norm	Section 4.3	page 72
S^n	unit sphere in R^{n+1}	Section 4.4	page 76
$\mathbb{C}^n$	complex n-tuples	Section 4.5	page 77
$\prod_n$	bilaterally infinite complex sequences with period n	Section 4.5	page 78
F_n	discrete Fourier transform	Section 4.5	page 78
i	$\sqrt{-1}$	Section 4.5	page 78
ℓ^p, $1 \le p \le \infty$	real sequences with norm $\|\cdot\|_p$	Section 4.6	page 79
c	convergent sequences	Section 4.7	page 79
C[a,b]	continuous functions on [a,b] with maximum norm	Section 4.8	page 80
π[a,b]	polynomials on [a,b]	Section 4.8	page 81
$\binom{n}{k}$	binomial coefficient	Section 4.8	page 83
B(n;x)(t)	Bernstein polynomial	Section 4.8	page 83
$C(\bar{\Omega})$	continuous functions on $\bar{\Omega} \subset R^n$ with maximum norm	Section 4.9	page 86
$\partial\Omega$	boundary of set Ω	Section 4.10	page 88
$(\alpha_1,\ldots,\alpha_n)$	multi-index	Section 4.10	page 88
D^α, α a multi-index	partial derivative operator	Section 4.10	page 88
$\|\alpha\|$	'length' of a multi-index	Section 4.10	page 88
$C^m(\Omega)$	m-times continuously differentiable functions on $\Omega \subset R^n$	Section 4.10	page 89

NOTATION (In order of introduction)

Symbol	*Meaning*	*Defined*
$C^{\infty}(\Omega)$	infinitely differentiable functions	Section 4.10 page 89
$C_0^{\infty}(\Omega)$	infinitely differentiable functions with compact support	Section 4.10 page 89
$C^m(\overline{\Omega})$	m-times uniformly continuously differentiable functions	Section 4.10 page 89
$C^m(\overline{\Omega};R^{\ell})$	m-times continuously differentiable mappings into R^{ℓ}	Section 4.10 page 90
		Section 4.10 page 90
$L^p(\Omega)$	Lebesgue space with $\|\cdot\|_p$	Section 4.11 page 93
J_{ε}	mollifier	Section 4.11 page 96
$\mu(E)$	Lebesgue measure	Chapter 4, appendix page 98
$\mu^*(E)$	exterior Lebesgue measure	Chapter 4, appendix page 98
a.e.	almost everywhere	Chapter 4, appendix, page 101
$f^+(t)$	positive part of f	Chapter 4, appendix, page 101
ψ_A	characterisitc function of set A	Chapter 4, appendix, page 101
$f^-(t)$	negative part of f	Chapter 4, appendix, page 102
$\$^1_{n+1}[a,b]$	linear splines on [a,b] with n+1 breakpoints	Problem 4.10 page 107
$\mathcal{D}[a,b]$	piecewise continuous functions	Problem 4.21 page 109
$\|T\|$	norm of a linear operator	Section 5.1 page 110
$\|A\|_{\infty}$	maximum row-sum norm of matrix A	Example 5.1 page 111
$\|A\|_2$	Euclidean norm of matrix A	Example 5.2 page 112
$\rho(A)$	spectral radius of matrix A	Example 5.2 page 112

NOTATION (In order of introduction)

Symbol	*Meaning*	*Defined*	
$\|A\|_1$	maximum column-sum norm of matrix A	Example 5.3	page 113
$w_k^{(n)}$	quadrature weights	Section 5.3	page 124
$t_k^{(n)}$	quadrature nodes	Section 5.3	page 124
E	shift operator	Section 5.4.1	page 130
Δ	forward difference operator	Section 5.4.1	page 130
M	averaging operator	Section 5.4.1	page 133
$T_m^{(k)}$	coefficients in Romberg array	Section 5.4.2	page 136
$\pi_n[a,b]$	polynomials of degree $\le$ n on [a,b]	Section 5.5	page 139
$\tilde{C}[a,b]$	periodic continuous functions on [a,b]	Section 5.5	page 140
$\tilde{\pi}_n[a,b]$	trigonometric polynomials of degree $\le$ n on [a,b]	Section 5.5	page 140
$E(t)$	solution operator for an initial value problem	Section 5.7	page 150
$C(\Delta t)$	finite difference approximation for an initial value problem	Section 5.7	page 152
$\mathcal{F}$	Fourier transform	Section 5.7	page 159
$\mathcal{F}^{-1}$	inverse Fourier transform	Section 5.7	page 159
T^{β}	translation operator	Section 5.7	page 159
$\hat{C}(\Delta t)$	transformed finite difference approximation for an initial value problim	Section 5.7	page 161
$G(k,\Delta t)$	amplification matrix for a finite difference approximation	Section 5.7	page 162
$R(k,\Delta t)$	spectral radius of amplification matrix	Section 5.7	page 168

NOTATION (In order of introduction)

Symbol	*Meaning*	*Defined*	
$\hat{E}(\Delta t)$	transformed solution operator	Section 5.7	page 170
κ	open cover of a set	Section 6.1	page 186
$\lambda_i(x)$	coefficient functional	Section 6.2.1 (Section 8.1	page 196 page 264
epi f	epigraph of a function f	Example 6.2	page 210
$d(x,V) = dist(x,V)$	distance from point x to set V	Section 6.3.2	page 212
$prox(V,\cdot)$	proximity mapping onto a set V	Section 6.3.2	page 212
R^n_+	non-negative orthant in R^n	Problem 6.30	page 230
$\tilde{a}$	maximal element of a partially ordered set	Section 7.3	page 239
R_z	real part of complex number z	Section 7.4	page 244
F	support function of a set	Section 7.4	page 246
M_d	multi-indices of length at most d	Example 7.4	page 251
X_n	n-dimensional subspace of X	Chapter 8	page 264
$span\{x_1,\ldots,x_n\}$	subspace spanned by $x_1,\ldots,x_n$	Section 8.1	page 266
P_n	projection of X onto X_n	Section 8.1	page 266
P	a projection	Section 8.2	page 269
$p(V,X)$	relative projection constant	Section 8.2	page 270
$P \perp Q$	orthogonal projections	Section 8.2	page 271
$M \oplus N$	direct sum of spaces M and N	Section 8.2	page 271
$x \perp y$	orthogonal elements in a Hilbert space	Section 8.3	page 273
$E_1 \perp E_2$	orthogonal sets in a Hilbert space	Section 8.3	page 274

NOTATION (In order of introduction)

Symbol	*Meaning*	*Defined*	
$E^{\perp}$	orthogonal complement of E	Section 8.3	page 274
$s_n(x)$	partial Fourier sum of x	Section 8.3	page 277
$d_n^X(A)$	n-width of $A \subset X$	Section 8.4	page 281
$\Lambda_{01}^1[0,1]$	Lipschitz continuous functions on [0,1]	Section 8.4	page 282
$\omega(u,h)$	modulus of continuity	Section 8.4	page 282
ϕ_n	bijection of X_n onto R^n	Section 8.4	page 283
$\oplus M_i$	direct sum of spaces M_i	Problem 8.1	page 291
$Tu=(I-K)u=g$	a linear operator equation, $u \in X$	Chapter 9	page 296
$T_n u_n=(I-K_n)u_n=g_n$	an approximation to Tu=g with $u_n \in X_n$	Section 9.1	page 296
$A^{(n)}v^{(n)}=b^{(n)}$, $A^{(n)}w^{(n)}=c^{(n)}$	matrix representations of the approximate equations $T_n u_n=g_n$	Section 9.1	page 297
$B(X,Y)$	continuous linear mappings from X into Y	Section 9.2	page 301
$B(X)$	continuous linear mappings from X into X	Section 9.2	page 301
T^{-1}	linear inverse of T	Definition 9.1	page 301
A^+	pseudo-inverse of matrix A	Remark 9.6	page 305
K^*	Hilbert space adjoint of K	Section 9.5	page 319
$\Gamma,\mathcal{D},n$, $\rho(s),d(s,\sigma)$, $\theta(s),b(\mu)$, $\beta(t),\psi(s,\sigma)$, $\omega_Q(s)$	Various geometric quantities associated with a domain $\mathcal{D} \subset R^2$ with boundary Γ	Figure 9.1	page 334

NOTATION (In order of introduction)

Symbol	*Meaning*	*Defined*	
$\int_a^b \lvert dG(t)\rvert$	total variation of G	Section 9.7	page 335
$\lvert\tau\rvert$	maximum spacing of partition τ	Section 9.7	page 334
$\int_a^b F(t)dG(t)$	Stieltjes integral	Section 9.7	page 336
$\tilde{\psi}(s,\cdot)$	smoothed $\psi(s,\cdot)$	Section 9.7	page 344
$T=I+E+F$	splitting of Radon's equation; $\lVert E\rVert<1$, F compact	Section 9.7	page 350
$T_n=I+E_n+F_n$	approximation to Radon's equation	Section 9.7	page 351

Index

List of theorems, lemmas, corollaries

References

ADAMS, R.A. (1975). *Sobolev Spaces*. Academic Press, New York. {*95; 205; 234*}.

ANDERSSEN, R.S. and DE HOOG, F.R. (To appear). Application and numerical solution of Abel-type integral equations. In *Proceedings of the international symposium on ill-posed problems: theory and practice. (Delaware, Newark, 1979)*. (Nashed Z. editor). {Pr 9.56, *380*}

AHLFORS, L.V. (1966). *Complex analysis*. Mc-Graw Hill, New York. {Sol 2.26}

ANSELONE, P.M. (1971). *Collectively compact operator approximation theory*. Prentice-Hall, Englewood Cliffs, N.J. {*324; 367;* Sol 9.17}

ARONSZAJN, N. and SMITH K.T. (1956). Functional spaces and functional completion. *Ann. Inst. Fourier Grenoble*. 6, 125-185. {*61*}

ATKINSON, K.E. (1976). *A survey of numerical methods for the solution of Fredholm integral equations of the second kind*. Society for Industrial and Applied Mathematics, Philadelphia. {*86; 202; 298*}

BAKER, C.T.H. (1977). *The numerical treatment of integral equations*. Oxford University Press, Oxford. {*86; 202; 298; 304*}

BANACH, S. (1932). *Théorie des opérations linéaires*. Monografje Matematyczne, Warsaw. {*1*}

BARRODALE, I. and ROBERTS, F.D.K. (1978). An efficient algorithm for discrete ℓ_1 linear approximation with linear constraints. *SIAM J. Numer. Anal.* 15, *603-611*. {*75*}

BAUER, F., GARABEDIAN, P. and KORN, D. (1977). *Supercritical wing sections III*. Springer-Verlag, Berlin. {*78, 235*}

BAUER, F.L., RUTISHAUSER, H. and STIEFEL, E. (1963). New aspects in numerical quadrature. In *Proceedings of symposia in applied mathematics, Vol. XV (Experimental arithmetic, high speed computing and mathematics)*. American Mathematical Society, Providence, R.I. {*139*}

BENVENISTE, J.E. (1967). Projective solution of the Dirichlet problem for boundaries with angular points. *SIAM J. Appl. Math.* 15, *558-568*. {*351; 359;* Pr. 9.51, *377;* Pr. 9.54, *379*}

BOULLION, T.L. and ODELL, P.L. (1971). *Generalized inverse matrices*. Wiley-Interscience, New York. {*306*}

BRAKHAGE, H. (1960). Über die numerische Behandlung von Integralgleichungen nach der Quadraturformelmethode. *Numerische Math.* 2, *183-196*. {Sol 9.17}

BRAMBLE, J.H. and HUBBARD, B.E. (1964) New monotone type approximations for elliptic problems. *Math Computation*. 18, *349-367*. {Sol 7.23}

BREZINSKI, C. (1977). *Accélération de la convergence en analyse numerique*. Springer-Verlag, Berlin. {*80; 136*}

BRUHN, G. and WENDLAND, W. (1967). Über die näherungsweise Lösung von linearen Funktionalgleichungen. In *Funktionalanalysis, Approximations-theorie, numerische Mathematik*. (Collatz L., Meinardus, G. and Unger, H., editors). Birkhauser Verlag, Basel; {*351*}

CAMPBELL, S.L. and MEYER, C.D. Jr. (1979). *Generalized inverses of linear transformations*. Pitman, London. {*306*}

CELLINA, A. (1972). On the nonexistence of solutions of differential equations in nonreflexive spaces. *Bull. Amer. Math. Soc.* 78, *1069-1972*. {*219*}

CHENEY, E.W. (1966). *Introduction to approximation theory*. McGraw-Hill, New York. {*86; 214*}

CHENEY, E.W. and PRICE, K.H. (1970). Minimal projections. In *Approximation theory*. (Talbot, A., editor). Academic Press, New York. *261-289*. {*270*}

COURANT, R. and HILBERT, D. (1953). *Methods of mathematical physics, Vol. I*. Interscience, New York. {*280*}

COURANT, R. and HILBERT. D. (1962). *Methods of mathematical physics, Vol. II. Partial differential equations*. Interscience, New York. {*76; 150; 333*}

COBB, E.B. and HARRIS, B. (1966). The characterization of the solution sets for generalized reduced moment problems and its application to numerical integration. *SIAM Rev.* 8, *86-99*. {Sol 5.14}

CRAVEN, B.D. and KOLIHA, J.J. (1977). Generalizations of Farkas' theorem. *SIAM J. Math. Anal.* 8, *983-997*. {*251*}

CRYER, C.W. (1970). The solution of the Dirichlet problem for Laplace's equation when the boundary data is discontinuous and the domain has a boundary which is of bounded rotation by means of the Lebesgue-Stieltjes integral equation for the double layer potential, Technical Report No. 99, Computer Sciences Dept. University of Wisconsin, Madison, Wisc. {*337; 351;* Pr 9.54, *379;* Sol 4.21}

CRYER, C.W. and TAVERNINI, L. (1972). The numerical solution of Volterra functional differential equations by Euler's method. *SIAM J. Numer. Anal.* 9, *105-129*. {Sol. 6.26}

DAVIS, P.J. (1963). *Interpolation and approximation*. Blaisdell, New York. {*280*}

DAVIS, P.J. (1970). Non-negative interpolation formulas for harmonic and analytic functions. In *Approximation theory*, (Talbot, A., editor). Academic Press, New York. *83-100*. {Sol. 7.22}

DAVIS, P.J. and RABINOWITZ, P. (1975). *Methods of numerical integration*. Academic Press, New York. {*86; 329*}

DE BOOR, C. (1978). *A practical guide to splines.* Springer-Verlag, New York. {*144; 145*}

DEIMLING, K. (1977). *Ordinary differential equations in Banach spaces.* Springer-Verlag, Berlin. {*219*}

DEJON, B. and HENRICI, P. (editors). (1969). *Constructive aspects of the fundamental theorem of algebra.* Wiley-Interscience, London. {*77*}

DELVES, L.M. and WALSH, J. (editors). (1974). *Numerical solution of integral equations.* Clarendon Press, Oxford. {*86; 298*}

DENNIS, J. and SCHNABEL, R. (1979). *Quasi-Newton methods for unconstrained nonlinear problems.* Preprint. {*74*}

DE VORE, R.A. (1972). *The approximation of continuous functions by positive linear operators.* Springer-Verlag, Berlin. {*144*}

DUNFORD, N. and SCHWARTZ, J.T. (1966). *Linear operators. Part I: General theory.* Interscience, New York. {*41; 66; 98; 122; 123; 175; 193; 205; 234; 249; 299*}

DUNFORD, N. and SCHWARTZ, J.T. (1966). *Linear operators. Part II.* Interscience, New York. {*40*}

EGGLESTON, H.G. (1963). *Convexity.* Cambridge University Press, Cambridge. {*43; 206*}

EKELAND, I. and TEMAN, R. (1976). *Convex analysis and variational problems.* North-Holland, Amsterdam. {*211*}

EVANS, L.C. (1977). Application of nonlinear semigroup theory to certain partial differential equations. In *Nonlinear semigroups and applications.* (Crandall, M.G., editor) Wiley and Sons, New York. {*156*}

FADDEEVA, V.N. (1952). *Linear algebra.* Dover, New York. {*3; 74*}

FORSYTHE, G.E. and WASOW, W.R. (1960). *Finite-difference methods for partial differential equations.* John Wiley and Sons, New York. {Sol 7.23}

FREDHOLM, I. (1900). Sur une nouvelle méthode pour la résolution du probleme de Dirichlet. *Jrnl öfver. kongl. vet.-akad. för.* 57, *39-46.* {*2*}

FUGLEDE, B. (1957). Extremal length and functional completion. *Acta Math.* 98, *171-219*, {*61*}

GAIER, D. (1964). *Konstruktive Methoden der konformen Abbildung.* {*335; 351; 362;* Pr. 9.39, *373*}

GARABEDIAN, P.R. (1964). *Partial differential equations.* John Wiley and Sons, New York. {*78; 207; 235*}

GOLBERG, M.A. (editor). (1979). *Solution methods for integral equations.* Plenum, New York. {*298; 300*}

HAHN, H. (1922). Über Folgen linearer Operationen. *Monatsh. für Math. Phys.* 32. {Sol 4.21}

HALMOS, P.R. (1950). *Measure theory*. Van Nostrand, New York. {*98*}

HALMOS, P.R. (1960). *Naive set theory*. Van Nostrand, New York. {*6*}

HALMOS, P.R. (1967). *A Hilbert space problem book*. Van Nostrand, New York. {*40; 305*}

HALMOS, P.R. (1970). Ten problems in Hilbert space. *Bull. Amer. Math. Soc.* 76, *887-933*, {*305*}

HALMOS, P.R. and SUNDER, V.S. (1978). *Bounded integral operators on L^2 spaces*. Springer-Verlag, Berlin. {*299*}

HALPERIN, I. (1950). Discontinuous functions with the Darboux property. *Amer. Math. Monthly* 57, *539-540*. {*44*}

HARDY, G.H., LITTLEWOOD, J.E. and POLYA, G. (1967). *Inequalities*. Cambridge University Press, Cambridge. {*68; 175*}

HENRICI, P. (1962). *Discrete variable methods in ordinary differential equations*. John Wiley and Sons, New York. {*92; 155; 211*}

HENRICI, P. (1974). *Applied and computational complex analysis - Vol. 1*. Wiley-Interscience, New York. {*78*}

HENRICI, P. (1979). Fast Fourier methods in computational complex analysis. *SIAM Review* 21, *481-527*. {*78;* Sol. 4.6}

HILLE, E. (1962). *Analytic function theory*. Blaisdell, Waltham, Mass. {*235*}

HILTON, P.J. and WYLIE, S. (1965). *Homology theory*. Cambridge University Press, Cambridge. {*76*}

HOUSEHOLDER, A.S. (1958). The approximate solution of matrix problems. *J. Assoc. Comp. Mach.* 5, *205-253*. {*74*}

IKEBE, Y. (1972). The Galerkin method for the numerical solution of Fredholm integral equations of the second kind. *SIAM Review* 14, *465-491*. {*318*}

ISAACSON, E. and KELLER, H.B. (1966). *Analysis of numerical methods*. John Wiley and Sons, New York. {*137;* Sol 8.20}

IVANOV, V.V. (1976). *The theory of approximate methods and their application to the numerical solution of singular integral equations*. Noordhoff International, Leyden. {*298; 300*}

JASWON, M.A. and SYMM, M.A. (1977). *Integral equation methods in potential theory and elastostatics*. Academic Press, New York. {*86; 298*}

KANTOROVICH, L.V. and AKILOV, G.P. (1964). *Functional analysis in normed spaces*. MacMillan, New York. {*309*}

KELLER, H.B. (1968). *Numerical methods for two-point boundary-value problems.* Blaisdel, Waltham, Mass. {*93*}

KELLEY, J.L. and NAMIOKA, I. (1976). *Linear topological spaces.* Springer-Verlag, New York. {*41; 54; 56*}

KNOPP, K. (1954). *Theory and application of infinite series.* Blackie and Son, London. {*136*}

KOROVKIN, P.P. (1960). *Linear operators and approximation theory.* Hindustan Publishing Corp., Delhi. {*144*}

KORTANEK, K.O. (1977). Constructing a perfect duality in infinite programming. *Applied math. and optimization* 3, *357-372.* {*251*}

KRASNOSELSKII, M.A., VAINIKKO, G.M., ZABREIKO, P.O., RUTITSKII, YA, B. and STETSENKO, V. YA. (1972). *Approximate solution of operator equations.* Wolters-Noordhoff, Groningen. {*298; 324*}

KRYLOV, V.I. (1962). *Approximate calculation of integrals.* MacMillan, New York. {*86; 125;* Sol 5.14}

KURPEL, N.S. (1976). *Projection-iterative methods for solution of operator equations.* American Math. Soc., Providence, R.I. {*298*}

KULISCH, U.W. and MIRANKER, W.L. (1980). *Computer arithmetic in theory and practice.* Academic Press, New York. {*46*}

KUSSMAUL, R. and WERNER, P. (1968). Fehlerabschätzungen für ein numerisches Verfahren zur Auflosung linearer Integralgleichungen mit schwachsingulären Kernen. *Computing* 3, *22-46.* {Pr 9.56, *380*}

LADAS, G. and LAKSHMIKANTHAM, V. (1972). *Differential equations in abstract spaces.* Academic Press, New York. {*219*}

LAMBERT, J.D. (1973). *Computational methods in ordinary differential equations.* John Wiley and Sons, London. {*92*}

LORENTZ, G.G. (1966). *Approximation of functions.* Holt, Rinehart and Winston, New York. {*214; 282*}

LOVE, A.E.H. (1944). *A treatise on the mathematical theory of elasticity.* Dover, New York. {*357*}

MANDELBAUM, R. (1980). Four-dimensional topology: an introduction. *Bull. Amer. Math. Soc.* (New Series) 2, *1-159.* {*77*}

MARTI, J.T. (1976). *Nonlinear operators and differential equations in Banach spaces.* Wiley, New York. {*219*}

MILLER, C.W. (1979). *Numerical solution of two-dimensional potential theory problems using integral equation techniques.* University of Iowa, Ph.D. thesis. {*351*}

Modern computing methods. (1961). Her Majesty's Stationery Office, London. {*132*}

MOORE, R.E. (1966) *Interval analysis.* Prentice-Hall, Englewood Cliffs, N.J. {Pr 2.12, *45*}

MOORE, R.E. (1979) *Methods and applications of interval analysis.* Society for Industrial and Applied Mathematics, Philadelphia. {Pr 2.12, *45*}

MOTZKIN, T.S. and WASOW, W. (1953). On the approximation of linear elliptic differential equations by difference equations with positive coefficients. *J. Math. Phys.* 31, *253-259.* {Sol 7.23}

NASHED, M.Z. (editor). (1976). *Generalized inverses and applications.* Academic Press, New York. {*306*}

NATANSON, I.P. (1954). *Theorie der Funktionen einer reellen Veränderlichen.* Akademie-Verlag, Berlin. {*8; 99; 207*}

NATANSON, I.P. (1955). *Konstruktive Funktionentheorie.* Akademie-Verlag, Berlin. {*143, 144, 214, 281*}

NEUMANN, C. (1870). Zur Theorie des logarithmischen und des Newtonschen Potentials. Ber. Verh. König. Sächs. Ges. Wiss. Leipzig Math.-Phys. Cl. 22, *49-56.* {*362*}

NIVEN, T. and ZUCKERMAN, H.S. (1964). *An introduction to the theory of numbers.* John Wiley and Sons, New York. {Pr 3.9, *65*}

NŸSTROM, E.J. (1930). Über die praktische Auflösung von Integralgleichungen. *Soc. Scient. Fenn., Comm. Phys.-Math.* 5, 22 pages. {*324*}

ORTEGA, J.M. and RHEINBOLDT, W.C. (1970). *Iterative solution of nonlinear equations in several variables.* Academic Press, New York. {*74*}

PARLETT, B.N. (1980). *The symmetric eigenvalue problem.* Prentice-Hall, Englewood Cliffs, N.J. {*74*}

PETRYSHYN, W.V. (1968). On projection-solvability and the Fredholm alternative for equations involving linear A-proper operators. *Archive Rat. Mech. Anal.* 30, *270-284.* {*351*}

PETRYSHYN, W.V. (1975). On the approximation-solvability of equations involving A-proper and pseudo-A-proper mappings. *Bull. Amer. Math. Soc.* 81, *223-312.* {*300*}

POLSKY, N.I. (1962). Projective methods in applied mathematics. *Soviet Math.* 3, *488-492.* {*351*}

PRÖSSDORF, S. (1978). *Some classes of singular equations.* North-Holland, Amsterdam. {*300*}

RADON, J. (1919). Über die Randwertaufgaben beim logarithmischen Potential. *Sitz.-Ber. Akad. Wis. Wien.* 128 Abt. IIa, *1123-1167.* {*335*}

RAO, C.R. and MITRA, S.K. (1971). *Generalized inverse of matrices and its applications.* John Wiley and Sons, New York. {*306*}

REDDIEN, G.W. (1980). Projection methods for two-point boundary value problems. *SIAM J. Review* 22, *156-171.* {*372*}

RHEINBOLDT, W.C. (1974). *Methods for solving systems of nonlinear equations.* Society for Industrial and Applied Mathematics, Philadelphia. {*74*}

RICHTMYER, R.D. and MORTON, K.W. (1967). *Difference methods for initial value problems.* Interscience, New York. {*165*}

RIESZ, F. and SZ.-NAGY, B. (1955). *Functional analysis.* Frederick Ungar, New York. {*336*}

ROBERTSON, A.P. and ROBERTSON, W.J. (1964). *Topological vector spaces.* Cambridge University Press, Cambridge. {*122*}

ROGERS, C.A. (1964). *Packing and covering.* Cambridge University Press, Cambridge. {*196*}

RUDIN, W. (1964). *Principles of mathematical analysis.* McGraw-Hill, New York. {*87*}

RUDIN, W. (1966). *Real and Complex Analysis.* McGraw-Hill, New York. {*87; 98;* Sol 5.28; Sol 7.8}

SARD, A. (1963). *Linear approximation.* American Mathematical Society, Providence, R.I. {Sol 5.14}

SCHULZ, M.H. (1973). *Spline analysis.* Prentice-Hall, Englewood Cliffs, N.J. {Pr 9.33, *372*}

SINGER, I. (1981). *Bases in Banach spaces II.* Springer-Verlag, Berlin. {*269; 300*}

SMIRNOV, V.I. (1964). *A course of higher mathematics.* V. Pergamon, Oxford. {*90; 91; 98*}

SNEDDON, I.N. (1966). *Mixed boundary value problems in potential theory.* North-Holland, Amsterdam. {*298*}

STERBENZ, P.H. (1974). *Floating-point computation.* Prentice-Hall, Englewood Cliffs, N.J. {*46*}

STROUD, A.H. (1971). *Approximate calculation of multiple integrals.* Prentice-Hall, Englewood Cliffs, N.J. {*88*}

SULLIVAN, D. (1981) For $n>3$ there is only one finitely additive rotationally invariant measure on the n-sphere defined on all Lebesgue measurable subsets. *Bull. Amer. Math. Soc.* (New Series) 4, *121-123.* {*100*}

TCHAKALOFF, V. (1957). Formules de cubature mécaniques à coefficients non négatifs. *Bull. Sci. Math.* Series 2, 81, *123-134.* {*251*}

TITCHMARSH, E.C. (1958). *Eigenfunction expansions associated with second order differential equations. II.* Clarendon Press, Oxford. {*280*}

TODD, J. and WARSCHAWSKI, S.E. (1955). On the solution of the Lichtenstein-Gerschgorin integral equation in conformal mapping: II. Computational experiments. *Nat. Bur. Standards, Appl. Math. Ser.* 42, *31-44.* {*362*}

VARGA, R.S. (1962). *Matrix iterative analysis.* Prentice-Hall. Englewood Cliffs, N.J. {*74*}

WAIT, R. (1979). *The numerical solution of algebraic equations.* John Wiley and Sons, New York. {*74*}

WHITNEY, H. (1934). Analytic extensions of differentiable functions defined in closed sets. *Trans. Amer. Math. Soc.* 36, *63-89.* {*234*}

WHITNEY, H. (1944). On the extension of differentiable functions. *Bull. Amer. Math. Soc.* 50, *76-81.* {Sol 7.5}

WILKINSON, J.H. (1963). *Rounding errors in algebraic processes.* Prentice-Hall, Englewood Cliffs, N.J. {*77;* Sol 2.26}

WILKINSON, J.H. (1965). *The algebraic eigenvalue problem.* Clarendon Press, Oxford, {*74; 77*}

WILKINSON, J.H. and REINSCH, C. (1971). *Linear algebra.* Springer-Verlag, New York. {*77*}

YOSIDA, K. (1968). *Functional analysis.* Springer-Verlag, New York. {*123; 159; 193; 241*}

YOUNG, S.W. (1967). Piecewise monotone polynomial interpolation. *Bull. Amer. Math. Soc.* 73, *642-643.* {Sol 7.1}

ZABREIKO, P.O., KOSHELEV, A.I., KRASNOSELSKII, M.A., MIKHLIN, S.G., RAKOVSHCHIK, L. ZABREIKO, P.O. et al. (1975). *Integral equations - a reference text.* Noordhoff International Publishing, Leyden. {*86; 299*}

Solutions

CHAPTER 2: SOLUTIONS

2.1 $\Leftarrow$: Let $G \in \tau_2$. We want to show that $G \in \tau_1$. Let $x \in G$. By Theorem 2.1, there exists $N_x^{(2)} \in \beta_x^{(2)}$ such that $N_x^{(2)} \subset G$. By assumption, there exists $N_x^{(1)} \in \beta_x^{(1)}$ such that $N_x^{(1)} \subset N_x^{(2)}$. Hence, $N_x^{(1)} \subset G$. Thus, for every $x \in G$ there exists $N_x^{(1)} \in \beta_x^{(1)}$ such that $N_x^{(1)} \subset G$. By Theorem 2.1, $G \in \tau_1$.

$\Rightarrow$: Let $N_x^{(2)} \in \beta_x^{(2)}$. Then there exists $G \in \tau_2$ such that $x \in G \subset N_x^{(2)}$. By assumption, $\tau_2 \subset \tau_1$ so that $G \in \tau_1$. By Theorem 2.1, there exists $N_x^{(1)} \in \beta_x^{(1)}$ such that $N_x^{(1)} \subset G$. Therefore, $N_x^{(1)} \subset G \subset N_x^{(2)}$.

2.2 Assume that X has a topology τ and that $Y \subset X$. Define

$$\tau_Y = \{G \cap Y \colon G \in \tau\} .$$

We check the conditions of Definition 2.1:

(1) $\emptyset = Y \cap \emptyset \in \tau_Y$; $Y = Y \cap Y \in \tau_Y$.

(2) If $(G_\alpha \cap Y) \in \tau_Y$ for $\alpha \in A$, then $\cup G_\alpha \in \tau$ so that $\cup(G_\alpha \cap Y) = (\cup G_\alpha) \cap Y \in \tau_Y$.

(3) If $(G_1 \cap Y)$, $(G_2 \cap Y) \in \tau_Y$ then $G_1 \cap G_2 \in \tau$ and so

$$(G_1 \cap Y) \cap (G_2 \cap Y) = (G_1 \cap G_2) \cap Y \in \tau_Y .$$

2.3 We must check that ρ satisfies the conditions of Definition 2.3. Clearly ρ is symmetric, non-negative, and satisfies $\rho(S,S) = 0$.

If $\rho(S_1,S_2) = 0$ then $S_1 \supset S_2$. To see this, assume the contrary. Then there exists $x \in S_1^c \cap S_2$. Since, by assumption, S_1 is closed we know that S_1^c is open, so that for some $\mu > 0$

$$B(x;\mu) = \{y \in R^n \colon \|x-y\|_2 < \mu\} \subset S_1^c .$$

This implies that $\delta(S_1;S_2) \geq \mu > 0$ which contradicts the assumption that $\rho(S_1,S_2) = 0$. Similarly, $S_2 \supset S_1$, and so $S_1 = S_2$.

To prove the triangle inequality, let

$$U(S_1;\delta_1) \supset S_2;\ U(S_2;\delta_2) \supset S_3;$$
$$U(S_2;\delta_3) \supset S_1;\ U(S_3;\delta_4) \supset S_2 \ .$$

Then, from the definition of $U(S;\delta)$,

$$U(S_1;\delta_1 + \delta_2) \supset S_3 \quad \text{and} \quad U(S_3;\delta_3 + \delta_4) \supset S_1 \ ,$$

so that

$$\rho(S_1,S_3) \le (\delta_1 + \delta_2) + (\delta_3 + \delta_4) \ .$$

Taking the infimum of the right hand side, we obtain

$$\rho(S_1,S_3) \le \rho(S_1,S_2) + \rho(S_2,S_3) \ .$$

2.4 See a forthcoming paper by John Halton, in which it is shown how the number of checks can be reduced by attempting to construct a base for a topology.

2.5 To show that $\{\beta_x\}$ is a base of neighbourhoods for a topology we have to check the four conditions of Theorem 2.2.

Conditions (a) and (b) are trivial.

(c): If $N_1 = N_x(n_1)$, $N_2 = N_x(n_2)$, set $N = N_x(m)$ with $m \ge \max\{n_1,n_2\}$. Then $N \subset N_1 \cap N_2$.

(d): If $N_x = N_x(n)$ take $N'_x = N_x(2n)$ and, for $y \in N'_x$, take $N_y = N_y(2n)$.

Then, for all $z \in N_y$,

$$|z(t)-x(t)| \le |z(t)-y(t)| + |y(t)-z(t)| \le \frac{1}{2n} + \frac{1}{2n} = \frac{1}{n} \ ,$$

so that

$$z \in N_x \quad \text{and hence} \quad N_y \subset N_x \ .$$

Therefore, $\{\beta_x\}$ is a base of neighborhoods for a topology.

2.6 The conditions of Theorem 2.2 are satisfied, and can be checked in the manner of Problem 2.5.

2.7 The topology defined in Problem 2.5 is stronger. To see this we apply Problem 2.1.

Choose $U(G_1, G_2, \ldots, G_n) \in \beta_f$. We construct $N_f(n) \subset U_f$. Since G_i is open and $0 \in G_i$, there exists $r_i > 0$:

$$B_i(0;r_i) = \{y \in R^1 : |y| < r_i\} \subset G_i \ .$$

Let $r = \min_{1 \le i \le n} r_i$; then $U_f(B_1(0;r), \ldots, B_n(0;r)) \subset U_f(G_1, \ldots, G_n)$.

It suffices to prove that for every $r > 0$ there exists n such that

$$N_f(n) = \{g: |g(t)-f(t)| \le \frac{1}{n}, \text{ for } t \in [0,1]\},$$

$$\subset U_f(B_1(0;r), \ldots, B_n(0;r)) \ .$$

Take any $g \in N_f(n)$, $n > \frac{1}{r}$. Then:

$$|g(t)-f(t)| \le \frac{1}{n} < r, \text{ for all } t \in [0,1],$$

$$\Rightarrow |g(t_k)-f(t_k)| < r \text{ for all } k=1,\ldots n,$$

$$\Rightarrow g \in U_f(B_1(0;r), \ldots, B_n(0;r) \ .$$

Hence $N_f(n) \subset U_f(G_1, \ldots, G_n)$.

2.8 (a) Given $[a,b] \subset [0,1]$, a continuous mapping $f: R^1 \to R^1$, and β such that $f(a) \le \beta \le f(b)$, the bisection method (see page 75) can be used to construct $\alpha \in [0,b]$ such that $f(\alpha) = \beta$. Hence all continuous functions have property D.

(b) Let $f(x) = \sin \frac{1}{x}$ for $0 < x \le 1$, and $f(0) = 0$. This function is continuous at every point except $x = 0$ and so has property D for all $[a,b]$ with $0 < a \le b \le 1$. Take $a = 0$, b arbitrary in $(0,1]$, $f(a) = 0$, $f(b)$ given. In any interval $[b/(1+2\pi b), b]$, $f(x)$ is continuous and takes on every value in $[-1,+1]$. Thus there is an α in $[0,b]$ such that $f(\alpha) = \beta$, for any β in $[-1,+1]$, and hence for any β between $f(0)$ and $f(b)$. Therefore, f has property D; but f is not continuous at $x = 0$.

2.9 Let $f\colon X \to Y$ and $g\colon Y \to Z$ be continuous mappings between topological spaces and let $h = g \circ f$.

Let $W \subset Z$ be open, and set $V = g^{-1}(W)$. Then

$$\begin{aligned} h^{-1}(W) &= \{x \in X\colon g(f(x)) \in W\}, \\ &= \{x \in X\colon f(x) \in V\}, \\ &= f^{-1}(V). \end{aligned}$$

g is continuous, and so, by Theorem 2.3, $V = g^{-1}(W)$ is open. Similarly, $f^{-1}(V)$ is open. A final application of Theorem 2.3 shows that h is open.

2.10 Let $f\colon X \to Y$ and $g\colon W \to X$. Then $(f \circ g)\colon w \in W \to = f(g(w)) \in Y$. If $V \subset Y$, $f^{-1}(V) = \{x\colon f(x) \in V\}$; if $U \subset X$, $g^{-1}(U) = \{w\colon g(w) \in U\}$. Thus $(f \circ g)^{-1}(V) = \{w\colon (f \circ g)(w) \in V\} = \{w\colon f(g(w)) \in V\} = \{w\colon g(w) \in f^{-1}(V)\} = (g^{-1} \circ f^{-1})(V)$. This is true for all $V \subset Y$; so $(f \circ g)^{-1} = g^{-1} \circ f^{-1}$.

2.11 Let $f\colon x \to (x, x_0)$. Let p_1 and p_2 denote the projection maps from $X \times X$ to X. Then $p_1 \circ f\colon x \to x$ and $p_2 \circ f\colon x \to x_0$ are clearly continuous. Hence, as shown on page 24, f is continuous. If $g\colon x \to x + x_0$ then $g = a \circ f$ where $a\colon (x,y) \in X \times X \to x + y$. The map a is continuous since X is a topological vector space (see Definition 2.5). Since a and f are continuous, so is their composition g (Problem 2.9).

2.12 ρ satisfies the conditions for a metric (Definition 2.3). The first two conditions (non-negativity and symmetry) are obvious. The third condition (triangle inequality) follows since if $[a_1, a_2]$, $[b_1, b_2]$, and $[c_1, c_2]$ are interval numbers then,

$$\begin{aligned} |a_1 - c_1| \le |a_1 - b_1| + |b_1 - c_1| &\le \max(|a_1 - b_1|, |a_2 - b_2|) + \\ &\quad + \max(|b_1 - c_1|, |b_2 - c_2|), \\ &= \rho([a_1, a_2], [b_1, b_2]) + \rho([b_1, b_2], [c_1, c_2]). \end{aligned}$$

$|a_2 - c_2|$ may be bounded similarly, and the triangle inequality follows.

X is not a vector space since, for instance,

$$(1-1)[0,1] = 0[0,1] = 0 \neq 1[0,1] + (-1)[0,1] =$$
$$= [0,1] + [-1,0] = [-1,1] \ .$$

2.13 No. For example, $\infty + (-\infty) \neq 0$.

2.14 This question often leads to considerable discussion because the answer depends upon one's judgement. For example, if the hardware provides two representations of zero do we regard this as a violation of the condition that the zero be unique or do we agree to identify the two representations in the same way that one identifies the decimal numbers 1.0 and $0.9\dot{9}9$? The following answer assumes reasonable binary arithmetic.

Let X be the set of binary floating-point numbers, $\pm m2^e$ with $m = (\cdot\beta_1\beta_2\ldots\beta_r)$, $\beta_1 = 1$, $0 \le \beta_k \le 1$ $(2 \le k \le r)$, $0 \le |e| \le E$. We check the conditions of Definition 2.4 with $x = m_1 2^{e_1}$, $y = m_2 2^{e_2}$, $z = m_3 2^{e_3}$ and $r = 8$.

(1) $(x+y)+z = x+(y+z)$: Consider $x+y$: if $e_1 \ge e_2$, use $(m_1 + m_2 2^{e_2 - e_1})2^{e_1}$, possibly adjusted by a factor of 2^k or 2^{-k} , to make $\beta_1 = 1$, rounded. E.g. let $x = -0.1 \times 2$, $y = .10000001 \times 2$, $z = .1 \times 2^{-8}$. Then $x+y = .1 \times 2^{-6}$, $(x+y)+z = .101 \times 2^{-6}$; $(y+z) = .10000001 \times 2$, $x+(y+z) = .1 \times 2^{-6}$: FALSE.

(2) $x+y = y+x$: TRUE.

(3) $\underline{0}$ exists: [$\underline{0}$ has a special representation, $m = .000\ldots0$, $e = -E$.] TRUE.

(4) Each x has an inverse $-x$: TRUE [Except with two's complement notation, e.g., 00 = 0, 01 = 1, 10 = -2, 11 = -1; then -2 has no inverse] .

(5) $(\lambda\mu)x = \lambda(\mu x)$: Since λ and μ are real, this becomes the associative law $(xy)z = x(yz)$. Define

$xy = m_1 m_2 \beta^{e_1+e_2} = yx$, where $m_1 m_2$ may be rounded-off. FALSE. (because, e.g., $x = .1 \times 2$, $y = .1 \times 2^E$, $z = .1 \times 2^{-E}$ gives overflow for xy, not for $x(yz)$; or because of roundoff problems: e.g., with rounding, $(.10110000 \times .10110000) \times .10000001 = (2^{-1} \times .11110010) \times .10000001 = 2^{-2} \times .11110100$, but $.10110000 \times (.10110000 \times .10000001) = .10110000 \times (2^{-1} \times .10110001) = 2^{-2} \times .11110011$).

(6) $(\lambda+\mu)x = \lambda x + \mu x$: FALSE (same reasons as for (5)).

(7) $\lambda(x+y) = \lambda x + \lambda y$: FALSE (same reasons as for (5)).

(8) $1x = x$: TRUE.

(9) $\underline{0}$ is unique: TRUE. (except possibly for notation: $.00000000\beta^e = \underline{0}$ for all e).

(10) $-x$ is unique $= (-1)\,x$: TRUE if we normalize before rounding.

(11) $\lambda\underline{0} = \underline{0}$: TRUE.

2.15 With the usual arithmetic rules, 0 as zero, and $-x$ as additive inverse, X is a vector space.

To show that $\{\beta_x\}$ is a base of neighbourhoods we check the conditions of Theorem 2.2:

(a) $\beta_x \neq \emptyset$;
(b) $x \in N_x$;
(c) given $N_x = \{y: x \le y < x + \frac{1}{m}\}$ and $N'_x = \{y: x \le y < x + \frac{1}{n}\}$, their intersection is

$$N_x \cap N'_x = \{y: x \le y < x + \frac{1}{\max(m,n)}\} \in \beta_x .$$

(d) if $N_x = \{y: x \le y < x + \frac{1}{n}\}$ set $N'_x = N_x$. If $y \in N'_x$, then $N_y = \{z: y \le z < y + \frac{1}{k}\} \subset N_x$ if $y + \frac{1}{k} < x + \frac{1}{n}$, i.e. $\frac{1}{k} < x + \frac{1}{n} - y$. Since $y < x + \frac{1}{n}$, we can choose k to make this possible.

To show that X is not a topological vector space, it suffices to show that the local base β_0 does not satisfy all the conditions (a)-(d) of Theorem 2.5.

But (a) is true [see (c) above with $x = 0$]; (b) is true [let $U = [0,\frac{1}{n})$. Then $[0,\frac{1}{2n}) + [0,\frac{1}{2n}) \subset [0,\frac{1}{n})$]. (c) does not hold, though. For, if $U = [0,\frac{1}{n})$, $\lambda = -1$, and $V = [0,\frac{1}{m})$ then $-V = (-\frac{1}{m},0] \not\subset U$.

2.16 ⇒: Assume that X is a topological vector space and that $\beta = \beta_{\underline{0}}$ is a local base. We verify conditions (a)-(d) of Theorem 2.5.

(a): Follows from Theorem 2.2(c).

(b): From Definition 2.5(1), there exist neighbourhoods $N_{\underline{0}}^{(1)}$ and $N_{\underline{0}}^{(2)}$ such that $N_{\underline{0}}^{(1)} + N_{\underline{0}}^{(2)} \subset U$. Choose $V \in \beta$ such that $V \subset N_{\underline{0}}^{(1)} \cap N_{\underline{0}}^{(2)}$.

(c): From Definition 2.5(2) there exist an open V_1 and $N = B(0;\mu)$ such that $NV_1 \subset U$. For each nonzero $\alpha \in N$, αV_1 is an open neighbourhood of $\underline{0} \in X$ because it is the inverse image of V_1 under the continuous map $x \to x/\alpha$. Thus NV_1 is open. Choose $V \in \beta$, with $V \subset NV_1$.

(d): Since $0x = \underline{0} \in U$, we can use continuity of multiplication to obtain λ such that $x \in \lambda U$. □

⇐: Now assume that X is a vector space and that β is a nonempty family of subsets satisfying (a)-(d). Let τ be defined as in Theorem 2.5. We first note that, by (c), if $U \in \beta$ then, for some V, $0V = \underline{0} \in U$. For $x \in X$ set $\beta_x = \{x + U: U \in \beta\}$, so that $\beta_{\underline{0}} = \beta$. Thus $G \in \tau$ iff for each $x \in G$ there exists $N_x \in \beta_x$ with $N_x \subset G$.

To verify that τ is a topology, it suffices to check the conditions of Theorem 2.2, which we denote by (2a)-(2d). Conditions (2a) and (2b) are trivial; (2c) is easily satisfied by virtue of (a); and (2d) is met for $N_z = x + U$ by taking $N'_x = x + V$ and $N_y = y + V$, where, by (b), $V + V \subset U$.

It remains to show that addition and multiplication are continuous. <u>Addition</u>. If $x + y = z$ and $N_z = z + U$, take $N_x = x + V$ and $N_y = y + V$, where, by (b), $V + V \subset U$. <u>Multiplication</u>. Let $\lambda x = z$, and $N_z = z + U$. Using (b), we obtain U_1 such that $U_1 + U_1 \subset U$. We will determine $r \in R^1$ and $M \in \beta$ such that

$$(\lambda+B(0;r))(x+M) \subset z + U_1 + U_1 \ . \qquad (*)$$

By (c) there exists $V_1 \in \beta$ such that $B[0;1]V_1 \subset U_1$. By (d) there exists μ such that $\mu x \in V_1$. Set $r = |\mu|/2$. Then $\mu x \in V_1 \Rightarrow B[0;1]\mu x \subset U_1$. Hence,

$$B(0;r)x \subset U_1 \ . \qquad (**)$$

Now let n be a power of 2 which is greater than $|\lambda| + r$. By repeatedly applying (b), we obtain $M_1 \in \beta$ such that $\sum_{k=1}^{n} M_1 \subset U_1$. Using (c), we obtain $M \in \beta$ such that $B[0;1]M \subset M_1$. For any $y \in \lambda + B(0;r)$,

$$yM = \sum_{k=1}^{n} \frac{y}{n}M \subset \sum_{k=1}^{n} B[0;1]M \subset \sum_{k=1}^{n} M_1 \subset U_1 \ . \qquad (\overset{**}{*})$$

Combining eqns (**) and ($\overset{**}{*}$) we obtain the desired condition (*).

(e): Take V as in (c) and $W = \{\lambda v: v \in V\}$.

(f) $\Rightarrow$: Let X be a Hausdorff space. Choose $x \in X$, $x \neq \underline{0}$. Then there exists a neighbourhood N of $\underline{0}$ which does not contain x . Pick $U \in \beta$ such that $U \subset N$. Then x U .

(f) $\Leftarrow$: Assume $\bigcap_{\beta} U = \underline{0}$. Choose $x,y \in X$, $x \neq y$. Then, for some $U \in \beta$, $x - y \notin U$. Pick $V \in \beta$ such that $V + V = U$ and use (c) to obtain $V_1 \in \beta$ such that $B[0;1]V_1 \subset V$. Then $(x+V_1) \cap (y+V_1) = \emptyset$.

2.17 Let $0 < p < 1$. Set $f(x) = (1+x)^p - 1 - x^p$, for non-negative x . Then $f(0) = 1$ and $\dot{f}(x) < 0$ if $x > 0$. Hence $(1+x)^p \leq 1 + x^p$ if $x \geq 0$. Replacing x by b/a we obtain

$$(a+b)^p \leq a^p + b^p \ ,$$

for non-negative a and b . It now follows readily

that ρ is a metric, and that conditions (a)-(d) of Theorem 2.5 are satisfied.

2.18 (a) p_k is a seminorm because $|\cdot|$ defines a norm in R^1.

(b). If J is a finite set of integers and $\varepsilon > 0$,

$$N(J;\varepsilon) = \{x: p_k(x) < \varepsilon \text{ for all } k \in J\} .$$

We have to show that:

(1) Given J and ε, there exists n such that $B(\underline{0};\frac{1}{n}) = \{x: \rho(x,\underline{0}) < \frac{1}{n}\} \subset N(J;\varepsilon)$.

(2) Given n there exist J and ε such that $N(J;\varepsilon) \subset B(\underline{0};\frac{1}{n})$.

Note that:

$$\rho(x,y) = \sum_{k=1}^{\infty} \frac{1}{2^k} \frac{p_k(x-y)}{1+p_k(x-y)} ,$$

$$\rho(x,\underline{0}) = \sum_{k=1}^{\infty} \frac{1}{2^k} \frac{p_k(x)}{1+p_k(x)} .$$

(1): Let $m = \max_J k$ and take n such that $n > 2^{m+1}/\varepsilon$, where, without loss of generality, $0 < \varepsilon < 1$. Then

$$x \in B(\underline{0};\tfrac{1}{n}) \Rightarrow \sum_{k=1}^{\infty} \frac{1}{2^k}\frac{p_k(x)}{1+p_k(x)} \le \frac{1}{n} < \frac{\varepsilon}{2^{m+1}} ,$$

$$\Rightarrow \frac{1}{2^k}\frac{p_k(x)}{1+p_k(x)} < \frac{\varepsilon}{2^{m+1}} , \quad \text{for all } k ,$$

$$\Rightarrow \frac{p_k(x)}{1+p_k(x)} < \frac{\varepsilon}{2} , \quad \text{if } k \le m ,$$

$$\Rightarrow p_k(x) < \frac{\varepsilon}{2-\varepsilon} < \varepsilon , \quad \text{if } k \in J .$$

Therefore, $B(\underline{0};\frac{1}{n}) \subset N(J;\varepsilon)$.

(2): Let m be such that $\sum_{k=m+1}^{\infty} \frac{1}{2^k} < \frac{1}{2n}$.

Take $J = \{1,2,\ldots,m\}$ and $\varepsilon < \frac{1}{2n}$. Then, for all $x \in N(J;\varepsilon)$:

$$\rho(x,\underline{0}) = \sum_{k=1}^{\infty} \frac{1}{2^k}\frac{p_k(x)}{1+p_k(x)} \le \sum_{k=1}^{m} \frac{1}{2^k}p_k(x) + \sum_{k=m+1}^{\infty} \frac{1}{2^k} ,$$

$$\le \sum_{k=1}^{m} \frac{1}{2^k}p_k(x) + \frac{1}{2n} \le \varepsilon \sum_{k=1}^{m} \frac{1}{2^k} + \frac{1}{2n} ,$$

$$\le \varepsilon + \frac{1}{2n} < \frac{1}{2n} + \frac{1}{2n} = \frac{1}{n} .$$

Therefore, $N(J;\varepsilon) \subset B(\underline{0};\frac{1}{n})$.

2.19 (a): (1) Take $x = (1,1,1,0,\ldots,0)$, $y = (0,0,0,1,1,\ldots,1)$. Then $\|x+y\|_1 = \|(1,1,\ldots,1)\|_1 = n$, $\|x\|_1 = 3$, and $\|y\|_1 = (n-3)$; yet x,y are linearly independent; so $\|\cdot\|_1$ is <u>not strictly convex</u>.

(2) Take $x = (1,0,0,\ldots,0)$, $y = (0,1,0,\ldots,0)$; then $\|x\|_1 = 1 = \|y\|_1$ and $\|x-y\|_1 = 2$, while $\|x+y\|_1 = 2$; so $\|\cdot\|_1$ is <u>not uniformly convex</u>.

(b): (1) Take $x = (1,1,0,\ldots,0)$, $y = (2,0,\ldots,0)$; then $\|x\|_\infty = 1$, $\|y\|_\infty = 2$, $\|x+y\|_\infty = \|(3,1,0,\ldots,)\|_\infty = 3 = \|x\|_\infty + \|y\|_\infty$; but x,y are linearly independent, so $\|\cdot\|_\infty$ is <u>not strictly convex</u>.

(2) If $x = (1,1,0,\ldots,0)$, $y = (0,1,1,0,\ldots,0)$; then $\|x\|_\infty = \|y\|_\infty = 1$, $\|x-y\|_\infty = \|(1,0,-1,0,\ldots,0)\|_\infty = 1$; but $\|x+y\|_\infty = \|(1,2,1,0,\ldots,0)\|_\infty = 2$, so $\|\cdot\|_\infty$ is <u>not uniformly convex</u>.

(c): (1) Let $\|x+y\|_2 = \|x\|_2 + \|y\|_2$. Minkowski's inequality (page 68) gives equality iff (x_k) , (y_k) are proportional. Hence $\|\cdot\|_2$ is <u>strictly convex</u>.

(2) Let $\|x\|_2 = \|y\|_2 = 1$, and $\|x-y\|_2 \ge \mu$. By the parallelogram identity $\|x+y\|_2^2 + \|x-y\|_2^2 = 2\|x\|^2 + 2\|y\|^2$. So $\|x+y\|^2 \le 4 - \mu = \{2\sqrt{1-\mu/4}\}^2$. Since $0 < \mu \le 2$, we know that $0 < \frac{\mu}{4} \le \frac{1}{2}$, and $\|\cdot\|_2$ is uniformly convex.

2.20 $\|x+y\|^2 = (x+y,x+y) = (x,x) + (y,y) + (x,y) + (y,x)$,
$= \|x\|^2 + \|y\|^2 + 2\text{Re}(x,y)$.
$\|x-y\|^2 = (x-y,x-y) = (x,x) + (y,y) - (x,y) - (y,x)$,
$= \|x\|^2 + \|y\|^2 - 2\text{Re}(x,y)$.

Thus:

$$\|x+y\|^2 + \|x-y\|^2 = 2\{\|x\|^2 + \|y\|^2\} . \qquad (*)$$

To prove inverse of Theorem 2.8, assume that $\|x\|^2 = (x,x)$ for some inner product $(\cdot,\cdot)$. Then:

$$\|x+y\|^2 - \|x-y\|^2 = 4\text{Re}(x,y) ,$$

$$\|x+iy\|^2 - \|x-iy\|^2 = -4\text{Re}[i(x,y)] = +4\text{Imag}(x,y) ,$$

so that

$$(x,y) = \frac{1}{4}\{(\|x+y\|^2) - \{(\|x-y\|^2) + i(\|x+iy\|^2 - \|x-iy\|^2)\} . (**)$$

It remains to check that if $\|\cdot\|$ satisfies eqn (*) and $(\cdot,\cdot)$ is defined by eqn (**), then $(\cdot,\cdot)$ is an inner product.

2.21 It is readily verified that $\|\cdot\|_\infty$ and $\|\cdot\|_1$ exist for all continuous functions and satisfy the conditions for a norm given in Definition 2.7.

If $x \in X$ then

$$\|x\|_1 = \int_0^1 |x(t)|dt \le \max_{[0,1]} |x(t)| \int_0^1 dt = \|x\|_\infty .$$

However, there is no constant α such that $\|x\|_\infty \le \alpha\|x\|_1$ To see this, let $x(t) = 0$ except for $\frac{1}{2} - \delta < t < \frac{1}{2} + \delta$, where $x(t)=M(1-|t-\frac{1}{2}|/\delta)$, with

$\delta = \frac{\varepsilon}{M}$ (see the figure). Then

$$\|x\|_1 = \int_0^1 |x(t)|dt = \frac{1}{2} \cdot 2\delta \cdot M = \varepsilon \quad \text{(arbitrarily small)},$$

while $\|x\|_\infty = \max_{[0,1]} |x(t)| = M$ (arbitrarily large.)

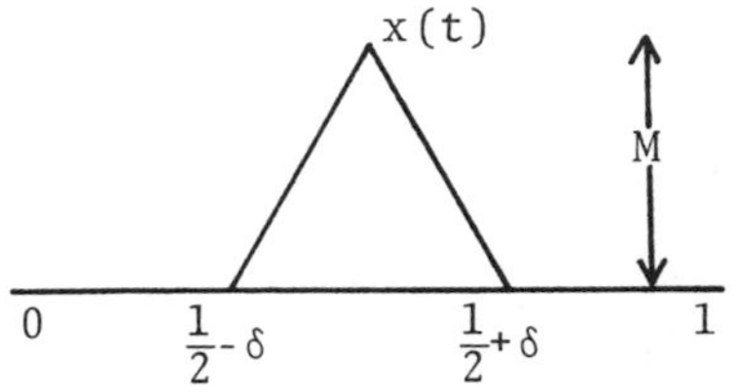

Figure: A 'spike' function.

2.22 Set $N(T;\varepsilon) = \{f: |f(t)| \le \varepsilon$ for $t \in$ finite set $T\}$. If X has a countable local base $\beta = \{B_1, B_2, \ldots\}$, then, since the $N(T;\varepsilon)$ form a local base, for every B_i there is an $N(T_i;\varepsilon_i) \subset B_i$. Each T_i is finite; so $\bigcup_{i=1}^{\infty} T_i = T$ is countable. Select $t \in [0,1]\backslash T$ ($[0,1]$ is uncountable, so such a t exists). Then consider $N(\{t\};\varepsilon)$. This neighbourhood of $\underline{0}$ contains no $N(T_i;\varepsilon_i)$ and hence no B_i. Thus β is not a local base and <u>X is not metrisable</u> (by Theorem 2.7).

2.23 We apply Theorem 2.7. If $x,y \in X$ and $x \neq y$, then $|x(s)-y(s)| > 2\varepsilon$ for some $\varepsilon > 0$ and some $s \in (-\infty,+\infty)$. If $n \ge s$ then $p_n(x-y) > 2\varepsilon$. Set $N = \{z: p_n(z) < \varepsilon\}$. Then $(x+N) \cap (y+N) = \emptyset$, and so X is <u>Hausdorff</u>.

Let

$$N(n;\varepsilon) = \{x \in X: |x(t)| \le \varepsilon \text{ for } |t| \le n\}.$$

It is readily checked that the topology on X has the countable local base

$$\beta = \{N(n;2^{-k}): n=1,2,\ldots \text{ and } k=1,2,\ldots\},$$

so that, by Theorem 2.7, X is metrisable.

For X to be normable, there must be a bounded convex neighbourhood of $\underline{0}$. Assume that such a neighbourhood exists, and denote it by C. Then, $N(n;\varepsilon) \subset C$ for some n and ε. Since C is bounded, for any $N(m;\delta)$ there is an α such that $C \subset \alpha N(m;\delta)$. But then $N(n;\varepsilon) \subset C \subset N(m;\alpha\delta)$. But now choose $m > n$. If $x \in N(n;\varepsilon)$, then $|x(t)|$ can be arbitrarily large for $t \in (n,m)$, and so $N(n;\varepsilon) \not\subset N(m;\alpha\delta)$. We have thus arrived at a contradiction and X is not normable.

2.24 Consider $x(t)$, $y(t)$ as shown below:

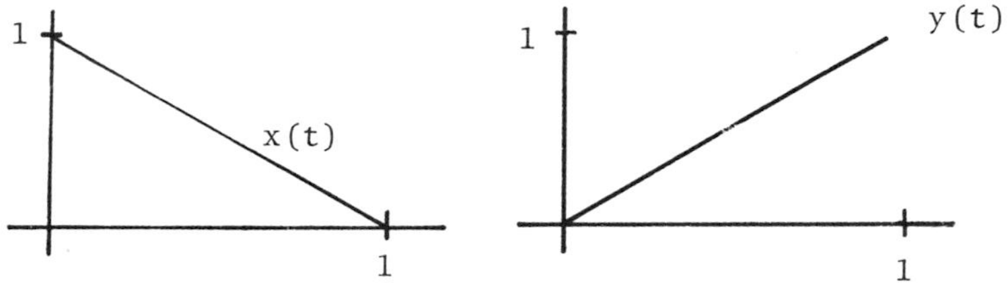

Then $\|x\|_\infty = \|y\|_\infty = 1$ and $\|x+y\|_\infty = 1$, $\|x-y\|_\infty = 1$ but $\|x+y\|^2 + \|x-y\|^2 = 2 \neq 4 = 2[\|x\|^2 + \|y\|^2]$. Since the parallelogram rule is not satisfied the space does not have an inner product satisfying $(x,x) = \|x\|_\infty^2$ (Theorem 2.8).

2.25 Proof of the Lemma.

$\Leftarrow$: Let N be any neighbourhood of $\underline{0}$. Then

$$B(\underline{0};r) \subset N \quad \text{for some } r > 0 .$$

Since E is bounded in norm there exists M such

that $\|x\| < M$ for $x \in E$. Hence,

$$E \subset B(\underline{0};M) = \frac{M}{r}B(\underline{0};r) \subset \frac{M}{r}N .$$

$\Rightarrow$: Take $N = B(\underline{0};1)$. Then, for some $\lambda > 0$, $E \subset \lambda B(\underline{0};1) = B(\underline{0};\lambda)$ so that $\sup_{x\in E} \|x\| \le \lambda < \infty$

2.26 If $f: \mathbb{C} \to \mathbb{C}$ and is differentiable,

$$C_r(z) = \limsup_{|h|\to 0} \frac{\frac{|f(z+h)-f(z)|}{|f(z)|}}{\frac{|h|}{|z|}}$$

$$= \limsup \frac{|z|}{|f(z)|} \left|\frac{f(z+h)-f(z)}{h}\right| ,$$

$$= \frac{|z|}{|f(z)|} \cdot |\dot{f}(z)| . \qquad (**)$$

$\frac{\partial p}{\partial t} = a_1 + 2a_2t + \ldots + kzt^{k-1} + \ldots + na_nt^{n-1}$ and

$\frac{\partial p}{\partial t}(x_0;z_0) \neq 0$ since, by assumption, x_0 is a simple root of $p(t;z_0)$.

By the implicit function theorem there exists a neighbourhood $N(z_0)$ of z_0 and $f: N(z_0) \to \mathbb{C}$ such that: $p(f(z);z) = 0$ for all $z \in N(z_0)$, $f(z_0) = x_0$, and $\dot{f}(z) = \frac{-\partial p/\partial z}{\partial p/\partial t}$.

Since $\frac{\partial p}{\partial z} = t^k$, $\dot{f}(z_0) = \dfrac{-(x_0)^k}{\frac{\partial p}{\partial t}(x_0;z_0)}$. Substituting into eqn (**) we obtain eqn (*).

In the case of the Wilkinson polynomial

$$\sum_{j=0}^{20} a_j t^j = (t-1)(t-2)\ldots(t-20) ,$$

we find that if $k = 19$, $x_0 = a_{19}$, and $z_0 = 15$ then $C_r(z_0) \doteq 1.5\ 10^{11}$.

2.27 For the topology defined using $\|\cdot\|_2$, a local base is given by $\{B_2(\underline{0};\frac{1}{m});\ m=1,2,\ldots\}$

$$B_2(\underline{0};\tfrac{1}{m}) = \{x\in R^n;\ [\sum x_i^2]^{\frac{1}{2}} < \tfrac{1}{m}\} .$$

For the topology defined using $\prod R^1$, it is readily verified that a local base is $\{B_\infty(\underline{0};\frac{1}{m}): m=1,2,...\}$,

$$B_\infty(\underline{0};\tfrac{1}{m}) = \{x \in R^n: \max |x_i| < \tfrac{1}{m}\} .$$

Now use Problem 2.1.

2.28 If $\|x+y\| = \|x\| + \|y\|$, then, by the parallelogram identity,

$$\|x-y\|^2 = \|x\|^2 + \|y\|^2 - 2\|x\|\cdot\|y\| ,$$

and hence

$$(x,y) + (y,x) = 2\|x\|\cdot\|y\| .$$

If $x \neq \underline{0}$ set $\lambda = \|y\|/\|x\|$. Direct computation shows that $\|y-\lambda x\|^2 = 0$.

2.29 Under the given hypotheses, eqn (2.3) follows almost immediately from eqn (2.2). See also Section 5.1, where the quantity $\sup\,[\|f(u)\|/\|u\|]$ is denoted by $\|f\|$.

CHAPTER 3: SOLUTIONS

3.1 Let $x_n \to x$ in τ_1. Let $V \in \tau_2$ be a neighbourhood of x. Then $V \in \tau_1$ since, by assumption, $\tau_1 \supset \tau_2$ Therefore, there exists N such that $x_n \in V$ if $n \geq N$, and we can conclude that $x_n \to x$ in τ_2.

3.2 (1) $V \cap \emptyset = \emptyset$ for every $V \subset X$. That is, $\overline{\emptyset} = \emptyset$.

(2) If $x \in E$ then, for every neighbourhood V of x, $x \in E \cap V$ and so $E \cap V \neq \emptyset$. That is, $E \subset \overline{E}$.

(3) $V \cap E_1 \neq \emptyset \Rightarrow V \cap E_2 \neq \emptyset$. Thus, $\overline{E}_1 \subset \overline{E}_2$.

(4) Consider $F = X \setminus \overline{E}$. If $x \in F$ then there exists a neighbourhood V of x such that $V \cap E = \emptyset$. There exists an open set G satisfying $x \in G \subset V$ and hence $G \cap E = \emptyset$. Let $y \in G$. Then G is also a neighbourhood of y. Since $G \cap E = \emptyset$, $y \notin \overline{E}$. Consequently $G \cap \overline{E} = \emptyset$, and hence $G \subset F$. From Theorem 2.1 we conclude that F is open, so that E is closed.

(5) Assume that $F = X \setminus E$ is open. For $x \in F$, there exists an open G satisfying $x \in G \subset F$, so that $G \cap E = \emptyset$. Thus $x \in F \Rightarrow x \notin \overline{E}$. That is, $x \notin E \Rightarrow x \notin \overline{E}$ and so $E \supset \overline{E}$. Now use (2).

(6) Using (3), $\overline{E}_i \subset \overline{E_1 \cup E_2}$ for $i = 1,2$ and so $\overline{E}_1 \cup \overline{E}_2 \subset \overline{E_1 \cup E_2}$. Using (2), we have $\overline{E}_1 \cup \overline{E}_2 \supset E_1 \cup E_2$. But $\overline{E}_1$ and $\overline{E}_2$ are closed, and thus $\overline{E}_1 \cup \overline{E}_2$ is also closed. Thus, by (3) and (5), $\overline{E_1 \cup E_2} \subset \overline{\overline{E}_1 \cup \overline{E}_2} = \overline{E}_1 \cup \overline{E}_2$.

(7) Let K be closed, $K \supset E$. By (2) and (5), $\overline{K} = K \supset \overline{E}$. Thus, $\overline{E}$ lies in every closed set K which contains E,

$$\overline{E} \subset \bigcap_{\substack{K \supset E \\ K \text{ closed}}} K .$$

On the other hand, $\overline{E}$ is closed, and so

$$\overline{E} \supset \bigcap_{\substack{K \supset E \\ K \text{ closed}}} K .$$

3.3 If E is dissected into a finite number m of closed triangles T_i, whose interiors are disjoint, the set of T_i,

$$\Delta = \{T_1, T_2, \ldots, T_m\},$$

is a *triangulation* of E. For example, in Figure 1 a triangulation $\Delta = \{T_1,T_2,T_3,T_4,T_5,T_6,T_7\}$ of an L-shaped region is shown.

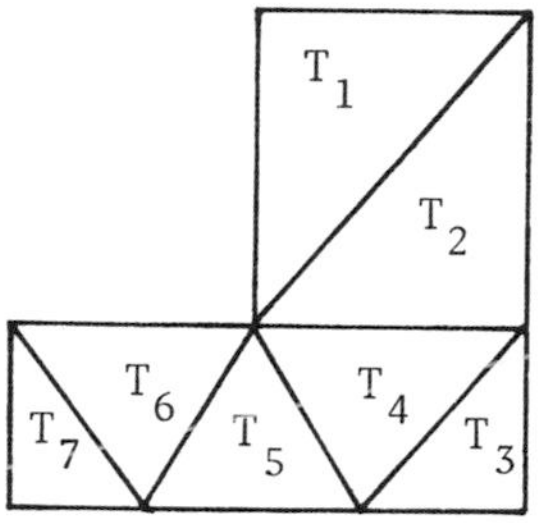

Figure 1: A typical triangulation.

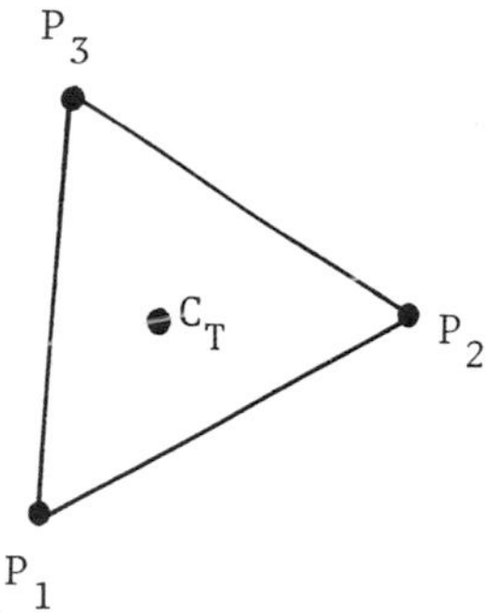

Figure 2: A typical triangle T.

We denote by A the set of all trangulations of E. If $\Delta_1, \Delta_2 \in A$, we say that $\Delta_1 \geq \Delta_2$ if for each $T^{(2)} \in \Delta_2$ there exists one or more $T^{(1)} \in \Delta_1$, $T_1^{(1)}, \ldots, T_m^{(1)}$, say, such that $T^{(2)} = \cup_j T_j^{(1)}$. Then A is a partially ordered set with respect to the ordering $\geq$. A is also well-ordered.

Given $\Delta_1, \Delta_2 \in A$, there are many ways to construct $\Delta \in A$ such that $\Delta_1 \leq \Delta$ and $\Delta_2 \leq \Delta$, and so A is also a directed set. [One such way is as follows. Let $S = \{P_1, \ldots, P_m\}$ denote the set of all points P such that P is a vertex of a triangle in $\Delta_1 \cup \Delta_2$. Construct the line segments which form the boundaries of triangles in $\Delta_1 \cup \Delta_2$. If two line segments intersect, adjoin the point of intersection to S. This provides a dissection of E into polygons each of which can be further subdivided into triangles. We have thus obtained a triangulation Δ such that $\Delta_1 \le \Delta$ and $\Delta_2 \le \Delta$.]

With each triangulation $\Delta \in A$ we associate the sum

$$x_\Delta = \sum_{T\in\Delta} f(C_T)|T| .$$

If x denotes the Riemann integral

$$x = \int_E f(t)dt ,$$

then

$$|x_\Delta - x| \le \left| \sum_{T\in\Delta} [\int_T f(t)dt - f(C_T)|T|] \right| ,$$

$$\le \sum_{T\in\Delta} \int_T |f(t) - f(C_T)|dt . \qquad (*)$$

Now let V be a neighbourhood of x in R^1, so that $B(x;\varepsilon) \subset V$ for some $\varepsilon > 0$. Since f is continuous, and E is bounded and closed, f is uniformly continuous on E (Theorem 6.12). Choose $\delta > 0$ so that

$$|f(x)-f(t)| \le \frac{\varepsilon}{\text{Area}(E)} , \text{ if } \|s-t\|_2 \le \delta \text{ and } s,t \in E . \qquad (**)$$

Choose a trangulation, Δ_V say, such that

$$\|s-t\|_2 \le \delta , \quad \text{if} \quad s,t \in T \in \Delta_V . \qquad (*{*\atop *})$$

Combining eqns (*), (**), and $(*{*\atop *})$, we see that $x_\Delta \in V$ if $\Delta \in A$ and $\Delta \ge \Delta_V$, so that $x_\Delta \to x(\Delta\in A)$.

3.4 Let $x_\alpha \to x(\alpha\in A)$. Let V be a neighbourhood of x. There exists $\beta \in A$ such that if $\alpha \in A$ and $\alpha \ge \beta$ then $x_\alpha \in V$. Since A_0 is cofinal in A, there exists $\beta_0 \in A_0$ such that $\beta_0 \ge \beta$. Thus, if $\alpha \in A_0$ and $\alpha \ge \beta_0$ then $x_\alpha \in V$. That is, $x_\alpha \to x(\alpha\in A_0)$.

3.5 $\Rightarrow$: If X and Y are isomorphic, there exists a linear bicontinuous map f of X onto Y. Since f is continuous, f maps some ball $B_X[\underline{0};\varepsilon]$ into $B_Y[\underline{0};1]$. That is, $\|x;X\| \le \varepsilon \Rightarrow \|f(x);Y\| \le 1$. Since f is linear, $\|f(x);Y\| \le \|x;X\|/\varepsilon$. Similarly, for some $\eta > 0$, $\|f^{-1}(y);X\| \le \|y;Y\|/\eta$, which is equivalent to $\|x;X\| \le \|f(x);Y\|/\eta$. Thus, eqn (*) holds with $\alpha_1 = \eta$ and $\alpha_2 = 1/\varepsilon$.

$\Leftarrow$: Let f be a linear map of X onto Y such that eqn (*) holds. Then

$$\|f(x+u) - f(x)\| = \|f(u)\| \le \alpha_2\|u\|,$$

so that f is continuous. Also, f is one-to-one since: $f(x_1) = f(x_2) \Rightarrow f(x_1 - x_2) = 0 \Rightarrow \alpha_1\|x_1 - x_2\| = 0 \Rightarrow x_1 = x_2$. Therefore, f^{-1} exists as a single-valued function. Applying eqn (*) again, we see that $\alpha_1\|f^{-1}(y)\| \le \|y\|$, which implies that f^{-1} is also continuous.

3.6 Let $\|x\| = \rho(x,\underline{0})$ where ρ is an invariant metric. Then

(a) $\|x\| = \rho(x,\underline{0}) \ge 0$. Also, $\|x\| = 0 \iff \rho(x,\underline{0}) = 0 \iff x = \underline{0}$.

(b) $\|-x\| = \rho(-x,\underline{0}) = \rho(-x+x,x) = \rho(\underline{0},x) = \rho(x,\underline{0}) = \|x\|$.

(c) $\|x+y\| = \rho(x+y,\underline{0}) = \rho(x+y-y,-y) = \rho(x,-y),$

$$\le \rho(x,\underline{0}) + \rho(\underline{0},-y) = \|x\| + \|y\|.$$

Conversely, let $\rho(x,y) = \|x-y\|$ where $\|\cdot\|$ is a quasinorm. Then,

(1) $\rho(x,y) = 0 \iff \|x-y\| = 0 \iff x - y = \underline{0} \iff x = y$.

(2) $\rho(x,y) = \|x-y\| = \|y-x\| = \rho(y,x)$.

(3) $\rho(x,y) = \|x-y\| = \|(x-z) - (z-y)\|,$

$$\le \|x-z\| + \|z-y\| = \rho(x,z) + \rho(z,y).$$

3.7 (a) <u>C_0 and C_v are normed spaces</u>

If $f,g \in C_0$, then $f(t) = 0$ outside $[-a_f,a_f]$, whence $\lambda f(t) = 0$ there also, and so $\lambda f \in C_0$; also $g(t) = 0$ outside $[-a_g,a_g]$, so $f(t) + g(t) = 0$ outside $[-a_m,+a_m]$, $a_m = \max\{a_f,a_g\}$ and so $f + g \in C_0$. Again, if $f,g \in C_v$ then $|f| \le \frac{\varepsilon}{2}$ outside

$[-a_{f,\varepsilon}, a_{f,\varepsilon}]$, and $|g| \le \frac{\varepsilon}{2}$ outside $[-a_{g,\varepsilon}, a_{g,\varepsilon}]$, so that $|f+g| \le \varepsilon$ outside $[-\max(a_{f,\varepsilon}, a_{g,\varepsilon})$, $+\max(a_{f,\varepsilon}, a_{g,\varepsilon})]$, and $f + g \in C_v$; and if $|f| \le \frac{\varepsilon}{|\lambda|}$ outside $[-a_{f,\varepsilon}, a_{f,\varepsilon}]$, then $|\lambda f| \le \varepsilon$, outside the same interval, so $\lambda f \in C_v$.

The above arguments have shown that the scalar product λf and vector sum $f + g$ are defined for all scalars λ and vectors f,g in both C_0 and C_v . To establish that C_0 and C_v are vector spaces we must show that the eleven conditions for a vector space (Definition 2.4) are satisfied. To establish further that C_0 and C_v are normed spaces, we must show that $\|\cdot\|$ is defined for all elements in the space, and that the three conditions for a norm (Definition 2.7) are satisfied. These elementary computations are omitted.

(b) Are C_0 and C_v Banach spaces?

Let $\{f_n\}$ be a Cauchy sequence in C_0 or C_v . Then, given $\varepsilon > 0$, there exists $N = N(\varepsilon)$ such that

$$\|f_p - f_q\| = \sup_{-\infty<t<\infty} |f_p(t) - f_q(t)| < \varepsilon , \text{ if } p,q \ge N . \tag{*}$$

In particular, $\{f_n(t)\}$ is a Cauchy sequence in R^1 for each fixed t . Since R^1 is complete, $\{f_n(t)\}$ has a limit, which we denote by $f(t)$, for each t .

For any fixed t ,

$$|f_p(t) - f(t)| \le |f_p(t) - f_q(t)| + |f_q(t) - f(t)| .$$

Let $p \ge N(\varepsilon)$. Choosing q sufficiently large, the first term in the above expression is less than ε while the second term is arbitrarily small. Thus,

$$|f_p(t) - f(t)| \le \varepsilon , \quad \text{if} \quad p \ge N(\varepsilon)$$

and hence

$$\sup_{-\infty<t<\infty} |f_p(t)-f(t)| \le \varepsilon \ , \ \text{if} \ \ p \ge N(\varepsilon) \ . \tag{**}$$

To shorten the proofs we make use of the space $C[a,b]$ (see Section 4.8.) If g is defined on $(-\infty,+\infty)$, we denote by $\tilde{g}$ the restriction of g to $[a,b]$. If g belongs to C_0 or C_v then $\tilde{g} \in C[a,b]$. $\{\tilde{f}_n\}$ is a Cauchy sequence in $C[a,b]$ and $C[a,b]$ is complete so that $\tilde{f}_n \to h$ for some $h \in C[a,b]$. Noting eqn (**), it follows that $h = \tilde{f}$ so that $\tilde{f}_n \to \tilde{f}$ in $C[a,b]$. In summary, the limit f of $\{f_n\}$ is continuous on every finite interval $[a,b]$, so that f is continuous on $(-\infty,+\infty)$.

For the remaining analysis, the spaces C_0 and C_v must be treated separately.

$\underline{C_v}$: Noting eqn (**), choose ε and p so that

$$\sup_{-\infty<t<\infty} |f_p(t)-f(t)| \le \varepsilon \ .$$

Since $f_p \in C_v$, there exists $a \in R^1$ such that $|f_p(t)| \le \varepsilon$ if $|t| \ge a$. Thus,

$$|f(t)| \le |f(t)-f_p(t)| + |f_p(t)| \le 2\varepsilon \ , \ \text{if} \ \ |t| \ge a \ .$$

We have already shown that f is continuous, so that $f \in C_v$. From eqn (*) $\|f_n-f\| \to 0$, and we conclude that <u>C_v is complete.</u>

$\underline{C_0}$: Define $f_n \in C_0$ as follows:

$$f_n(t) = \begin{cases} 1/(1+t), \ \text{if} \ 0 \le t \le n \ , \\ (1+2n-t)/(1+n)^2, \ \text{if} \ n \le t \le 2n+1 \ , \\ 0, \ \text{if} \ t > 2n+1 \ , \\ f(-t), \ \text{if} \ t < 0 \ . \end{cases}$$

Then

$$\|f_p-f_q\| \le 1/(1+\min\{p,q\}) \ ,$$

so that $\{f_n\}$ is a Cauchy sequence in C_0 . But $f_n(t) \to f(t)$, where $f(t) = 1/(1+|t|)$.

Since f does not vanish outside a finite interval, $f \notin C_0$, and so <u>C_0 is not complete.</u>

(c) <u>C_0 is dense in C_v</u>

If $f \in C_v$, choose $\varepsilon > 0$ and find a such that $|f(t)| \leq \varepsilon$ if $|t| \geq a$. Set

$$f_\varepsilon(t) = \begin{cases} f(-a)(t+a+1), \text{ if } t \in (-a-1,-a) , \\ f(t), \text{ if } |t| \leq a , \\ f(a)(-t+a+1), \text{ if } t \in (a,a+1) , \\ 0, \text{ if } |t| > a+1 . \end{cases}$$

Then $\|f-f_\varepsilon\| \leq 2\varepsilon$, and $f_\varepsilon \in C_0$ so C_0 is dense in C_v .

3.8 It is readily seen that the conditions of Definitions 2.4 and 2.7 are satisfied, so that $X = BF[0,1]$ is a normed space.

To prove that X is a <u>Banach space</u>, we need only show that the space is <u>complete</u>. Arguing as in Problem 3.7(b), if $\{f_n\}$ is a Cauchy sequence in X there is a function f to which it converges pointwise <u>and</u>

$$\sup_{[0,1]} |f_p(t)-f(t)| \leq \varepsilon , \text{ if } p \geq N(\varepsilon) . \qquad (*)$$

Choose p such that eqn (*) holds . Then

$$|f(t)| \leq |f_p(t)| + 1 \leq \|f_p\| + 1 , \quad \text{for } t \in [0,1] ,$$

so that f is bounded $\Rightarrow f \in X$. From eqn (*), $\|f_n-f\| \to 0$ so that X is <u>complete</u>.

To prove that X is not separable, we proceed by contradiction. If X is <u>separable</u>, there is a countable subset $F = \{f_i : i=0,1,\ldots\}$ which is dense in X . Divide $[0,1]$ into disjoint intervals $J_0 = [1,1]$, $J_1 = [0,2^{-1})$, $J_2 = [2^{-1},2^{-2}),\ldots, J_r = [2^{-r+1},2^{-r}),\ldots$. Then $[0,1] = \cup J_r$. Define g on $[0,1]$ as follows: If $t \in J_r$, then $g(t) = 1$ if $f_r(t) \leq 0$ and $g(t) = -1$ if $f_r(t) > 0$. Then $g \in X$. However, for every r,

$$\|g-f_r\| \geq \sup_{t \in J_r} |g(t)-f_r(t)| \geq 1 .$$

Thus, F is not dense in X, and, in consequence, X is <u>not separable</u>.

3.9 $x_n \to x = 0$. Using the theorem of Hurwitz, we see that there are infinitely many integers n such that $|k\pi - n| \leq \frac{1}{\sqrt{5}k}$, where k is an integer. For such (n,k), we have that $k > n/4$,

$$|x_n| = \frac{|\sin(n)|}{n^2} \leq \frac{\sin(\frac{1}{\sqrt{5}k})}{n^2} \leq \frac{\sin(\frac{4}{\sqrt{5}n})}{n^2} < \frac{4}{\sqrt{5}n^3} .$$

and

$$|x_{n+1}| = \frac{|\sin(n+1)|}{(n+1)^2} \geq \frac{\sin(1-\frac{1}{\sqrt{5}k})}{(n+1)^2} \geq \frac{\sin(\frac{1}{2})}{(n+1)^2} .$$

Hence, for such (n,k), $\left|\frac{x_{n+1}-x}{x_n-x}\right| > \frac{\sqrt{5}n}{4} \frac{\sin(\frac{1}{2})}{(1+\frac{1}{n})^2}$.

3.10 Assume that $x_n \not\to x$. Then there is a neighbourhood V of x and a subsequence $\{x_{n_i}\}$ such that $x_{n_i} \notin V$. By hypothesis, there is a sub-subsequence $\{x_{n_{i_j}}\}$ such that $x_{n_{i_j}} \to x$ as $j \to \infty$. This is a contradiction.

3.11 (a) Assume $\alpha \neq 0$, since otherwise $\overline{\alpha V} = \overline{V} = \underline{0}$.

Let $x \in \alpha\overline{V}$. Then $x = \alpha y$ with $y \in \overline{V}$. For each neighbourhood N_x of x, $N_y = \alpha^{-1}N_x$ is a neighbourhood of y. [N_x contains an open neighbourhood G_x of x. By continuity of multiplication, $G_y = \alpha^{-1}G_x$ is an open neighbourhood of y.] Since $y \in \overline{V}$, $N_y \cap V \neq \emptyset$, which implies that

$$N_x \cap \alpha V = \alpha N_y \cap \alpha V \neq \emptyset ,$$

so that $x \in \overline{\alpha V}$. Since x was an arbitrary point of $\alpha\overline{V}$, $\alpha\overline{V} \subset \overline{\alpha V}$.

On the other hand, $\overline{V}$ is closed and $f: x \to x/\alpha$ is continuous. Hence, $\alpha\overline{V} = f^{-1}(\overline{V})$ is the inverse image of a closed set under a continuous mapping, and so $\alpha\overline{V}$ is closed. By Problem 3.2(parts 2 and 5), $\overline{\alpha V} \subset \overline{\alpha\overline{V}} = \alpha\overline{V}$.

Combining these results we see that $\alpha\overline{V} \subset \overline{\alpha V} \subset \alpha\overline{V}$ so that $\alpha\overline{V} = \overline{\alpha V}$.

(b) The proof in (a) that $\alpha\overline{V} = \overline{\alpha V}$ depended upon the continuity of multiplication. Since addition is continuous, similar arguments show that, for all $x \in W$, $\overline{V} + x = \overline{V+x}$.

(c) Let $x = v + w \in \overline{V} + \overline{W}$. Let N_x be a neighbourhood of x . Then there exist N_v and N_w such that $N_v + N_w \subset N_x$. But $v \in \overline{V}$ and $w \in \overline{W}$, so $N_v \cap V \neq \emptyset$ and $N_w \cap W \neq \emptyset$. Thus

$$N_x \cap (V+W) \supset (N_v+N_w) \cap (V+W) \neq \emptyset .$$

Hence, $x = v + w \in \overline{V+W}$. Since x was an arbitrary point of $\overline{V} + \overline{W}$, $\overline{V} + \overline{W} \subset \overline{V+W}$.

3.12 Since each X_i is a vector space, so is $X = \prod X_i$.

Define $F: X = \prod_{i=1}^{\ell} X_i \to R^{\ell}$ by:

$$F(x) = (\|x_1;X_1\|, \ldots, \|x_\ell;X_\ell\|) .$$

Then:

$$\|x\| \equiv \|x;X\|_k = \|F(x);R^\ell\|_k , \quad k = 1,2,\infty .$$

To verify that X is a normed space we check the conditions of Definition 2.7:

(1) $\|x\|_k = \|F(x);R^\ell\|_k \geq 0$. Also, $\|x\|_k = 0$ $\Longleftrightarrow F(x) = 0 \Longleftrightarrow \|x_i;X_i\| = 0$ for $1 \leq i \leq \ell \Longleftrightarrow x = 0$.

(2) $\|\lambda x\|_k = \|F(\lambda x);R^\ell\|_k = \||\lambda| F(x);R^\ell\|_k = |\lambda| \|x\|_k$.

(3) Note that if $u, v \in R^{\ell}$ and $|u_i| \leq |v_i|$ for $1 \leq i \leq \ell$ then $\|u;R^{\ell}\|_k \leq \|v;R^{\ell}\|_k$ for $k = 1,2,\infty$. Thus,

$$\|x+y\|_k = \|F(x+y);R^{\ell}\|_k \leq \|F(x)+F(y);R^{\ell}\|_k ,$$

$$\leq \|F(x);R^{\ell}\|_k + \|F(y);R^{\ell}\|_k = \|x\|_k + \|y\|_k .$$

We now show that X is complete, and hence a Banach space. Let $\{x^n\}$ be a Cauchy sequence in X, where $x^n = (x_i^n) \in \prod X_i$. Since

$$\|x_i^p - x_i^q;X_i\| \leq \|x^p - x^q;X\|_k ,$$

$\{x_i^n\}$ is a Cauchy sequence in X_i for $1 \leq i \leq \ell$. X_i is complete, so, for $1 \leq i \leq \ell$, there exists $x_i \in X_i$ such that $\|x_i^n - x_i;X_i\| \to 0$ as $n \to \infty$. Set $x = (x_i) \in X$. Choose $\varepsilon > 0$, and determine N_i such that

$$\|x_i - x_i^n;X_i\| \leq \begin{cases} \varepsilon/(\ell^{\frac{1}{k}}) , & \text{if } k = 1,2 , \\ \varepsilon , & \text{if } k = \infty . \end{cases}$$

Let $N = \max_{1 \leq i \leq \ell} (N_i)$. Then

$$\|x - x^n;X\|_k \leq \begin{cases} [\sum_{i=1}^{\ell} (\|x_i - x_i^n;X_i\|)^k]^{\frac{1}{k}} , & \text{if } k = 1,2, \\ \max_i \|x_i - x_i^n;X_i\| , & \text{if } k = \infty , \end{cases}$$

$$\leq \varepsilon ,$$

if $n \geq N$. Thus, $\|x - x^n;X\| \to 0$ as $n \to \infty$, and X is complete.

CHAPTER 4: SOLUTIONS

4.1 Buniakowsky (Hardy, Littlewood, and Polya) and Bunyakovskii (Kantorovich and Akilov).

4.2 We have to prove that $\|\cdot\|_p$ is strictly convex and uniformly convex if $1 < p < \infty$, since the cases $p = 1$ and $p = \infty$ were considered in Problem 2.19.

Strict convexity. By Minkowski's inequality $(p > 1)$

$$\|x+y\|_p \le \|x\|_p + \|y\|_p ,$$

and equality holds iff x and y are proportional.

Uniform convexity. If $1 \le q \le 2$ and $0 \le s \le 1$ then

$$f(s) = (1+s)^q + (1-s)^q - 2(1+s^{\frac{q}{q-1}})^{q-1} \ge 0 , \qquad (*)$$

since

$$f(s) = 2\left[\sum_{k=1}^{\infty} [\binom{q}{2k}s^{2k} - \binom{q-1}{2k-1}s^{\frac{q(2k-1)}{q-1}} - \binom{q-1}{2k}s^{\frac{q(2k)}{q-1}}\right]$$

and each term in this series is non-negative.

Let $p \ge 2$ and set $q = p/(p-1)$. If $0 \le a < b$ set $s = (b-a)/(b+a)$ in eqn(*) and take the $(q-1)$th root of both sides. Using the elementary inequality

$$(\alpha+\beta)^r \le 2^{r-1}(\alpha^r+\beta^r) , \quad \text{if } \alpha,\beta \ge 0 \text{ and } r \ge 1 , \qquad (**)$$

we find that

$$|a+b|^p + |a-b|^p \le 2^{p-1}[|a|^p + |b|^p] \quad \text{if } a,b \in R^1 \text{ and } p \ge 2 . \qquad (*{*\atop *})$$

Thus, setting $a = x_i$, $b = y_i$, for $1 \le i \le n$, and summing, $\|x+y\|_p^p + \|x-y\|_p^p \le 2^{p-1}[\|x\|_p^p + \|y\|_p^p]$. If $\|x\|_p = \|y\|_p = 1$ and $\|x-y\|_p > \mu$ then $\|x+y\|_p^p < 2^p - \mu^p$, so that

$$\|x+y\|_p < (2^p-\mu^p)^{\frac{1}{p}} = 2(1 - (\mu/2)^p)^{\frac{1}{p}} .$$

For the case $1 < p < 2$ see Solution 4.7.

4.3 (a) Assume $1 \leq p < q < \infty$. Take $z_i = 1$ and $w_i = |x_i|^p$ for $i = 1,2,\ldots n$. Then by Hölder's inequalities (p. 67),

$$\sum_{i=1}^{n} z_i w_i = \sum_{i=1}^{n} |x_i|^p \leq \left(\sum_{i=1}^{n} 1\right)^{\frac{q-p}{q}} \left(\sum_{i=1}^{n} |x_i|^q\right)^{\frac{p}{q}} .$$

Taking the p-th root of this inequality we obtain

$$\|x\|_p \leq n^{(\frac{1}{p} - \frac{1}{q})} \|x\|_q . \qquad (*)$$

It is trivial to prove that $\|x\|_q \leq n^{\frac{1}{q}} \|x\|_\infty$, which, when combined with eqn (*), shows that

$$\|x\|_p \leq n^{(\frac{1}{p} - \frac{1}{q})} \|x\|_q , \quad \text{for} \quad 1 \leq p \leq q \leq \infty . \qquad (**)$$

If we take $x_i = 1$, $i = 1,\ldots n$ then

$$\|x\|_p = n^{\frac{1}{p}} , \quad \|x\|_q = n^{\frac{1}{q}} , \text{ and } \|x\|_p = n^{\frac{1}{p} - \frac{1}{q}} \|x\|_q ,$$

which shows that $n^{\frac{1}{p} - \frac{1}{q}}$ is the best constant for inequality (**).

(b) We now show that

$$\|x\|_q \leq \|x\|_p , \quad \text{if} \quad 1 \leq p \leq q \leq \infty . \qquad ({*}^{*}_{*})$$

To do so we first note that, by linearity, we may assume that $\|x\|_p = 1$. If $q < \infty$, $|x_i| \leq 1$ for all i , so that $|x_i|^q \leq |x_i|^p$, and hence

$$\|x\|_q = [\sum |x_i|^q]^{\frac{1}{q}} \leq [\sum |x_i|^p]^{\frac{1}{q}} = (\|x\|_p)^{\frac{p}{q}} = \|x\|_p .$$

If $q = \infty$ then eqn (${*}^{*}_{*}$) is trivial.

Equality is obtained in eqn (${*}^{*}_{*}$) if $x_1 = 1$ and $x_i = 0$ for $i > 1$.

4.4 The solution of a system of n linear algebraic equations by Gaussian elimination requires $n^3/3$ operations, where an operation is equal to one multiplication and one subtraction.

By assumption, one complex operation is equivalent to four real operations. Thus, the solution of eqn (*) requires $\frac{n^3}{3}$ complex operations $= \frac{4n^3}{3}$ real operations. The solution of eqn (**) requires $\frac{(2n)^3}{3}$ real operations $= \frac{8n^3}{3}$ real operations.

4.5 $f^{(1)}(0) = \frac{1}{2\pi}\int_{\theta=0}^{2\pi} f(e^{i\theta})e^{-i\theta}d\theta$.

The Trapezoidal Rule leads to:

$$f^{(1)}(0) \doteq \frac{1}{2\pi}\cdot\frac{\pi}{2}\cdot[\tfrac{1}{2}f(1) - if(i) - f(-1) + if(-1) + \tfrac{1}{2}f(1)],$$

$$= \frac{1}{4}[\tfrac{1}{2}\cdot 16 - i(1+i)^4 - 0 + i(1-i)^4 + \tfrac{1}{2}\cdot 16] ,$$

$$= 4. \quad \text{(The exact value !)}$$

4.6 See Henrici [1979, p. 482].

4.7 For the case $p \geq 2$ see Solution 4.2. If $1 < p \leq 2$ and $q = p/(p-1) \geq 2$, then it can be shown that if $x, y \in \ell^p$ then

$$(\|x+y\|_p)^q + (\|x-y\|_p)^q \qquad (*)$$
$$\leq 2((\|x\|_p)^p + (\|y\|_p)^p)^{q-1} ,$$

from which it follows that ℓ^p is uniformly convex if $1 < p \leq 2$. The proof of eqn(*) requires the use of several inequalities; see e.g. Adams [1975, p.34].

4.8 Let $\{x^k : k=1,2,\ldots\}$ be dense in ℓ^∞, where $x^k \in \ell^\infty$, $x^k = (x_i^k)$. Take $M > 1$ and define $x = (x_i) \in \ell_\infty$ by:

$$x_i = \begin{cases} x_i^i + 1 , & \text{if } |x_i^i| \le M , \\ 0 , & \text{if } |x_i^i| > M . \end{cases}$$

Then $\|x\|_\infty \le M + 1$. But

$$\|x - x^k\| \ge |x_k - x_k^k| \ge \min\{1, M\} = 1 ,$$

which contradicts the assumption that $\{x^k\}$ is dense in ℓ^∞.

4.9 Direct computation using elementary calculus.

4.10 Let Q be the set of all rational numbers.

If $f \in C[a,b]$ then f is uniformly continuous, so that given $\varepsilon > 0$ there exists $\delta > 0$ such that: $|t_1 - t_2| < \delta \Rightarrow |f(t_1) - f(t_2)| < \varepsilon/3$.

Let n be such that $\frac{b-a}{n} = \Delta t < \delta$, and set

$$t_0 = a, t_1 = a + \Delta t, \ldots, t_n = a + n\Delta t = b .$$

For each $i \in I = \{0,1,\ldots,n\}$ let $\phi_i \in Q$ be such that:

$$|f(t_i) - \phi_i| < \frac{\varepsilon}{3} .$$

Define $g \in C[a,b]$ piecewise linear by:

$$g(t) = \frac{(t-t_i)}{\Delta t}\phi_{i+1} + \frac{(t_{i+1}-t)}{\Delta t}\phi_i , \quad \text{if } t \in [t_i, t_{i+1}] .$$

Then an easy calculation shows that if $t \in [t_i, t_{i+1}]$

$$|f(t) - g(t)| \le |f(t) - f(t_i)| + |f(t_i) - \phi_i| + |\phi_i - g(t)| ,$$

$$\le \varepsilon .$$

Thus $\tilde{\$}^1$ is dense in $C[a,b]$. It is easily shown, using arguments similar to those on page 70, that $\tilde{\$}^1$ is denumerable.

It is an interesting exercise to prove that the points t_i^n can be required to be rational numbers instead of being of the form (*).

4.11 Let

$$u(t) = \begin{cases} \dfrac{2(t-a)}{b-a}, & \text{if } a \le t \le \dfrac{a+b}{2}, \\ 1, & \text{if } \dfrac{a+b}{2} \le t \le b, \end{cases}$$

and $v(t) = 1$. Then

$$\|u\|_\infty = \|v\|_\infty = 1; \quad \|u+v\| = 2 = \|u\| + \|v\|,$$

but $u \neq \lambda v$ for any real λ. Thus $C[a,b]$ is <u>not strictly convex</u>.

Also $\|u-v\| = 1$, so if $\mu \le 1$ there is no $\delta > 0$ such that:

$$\left.\begin{array}{l} \|u\| = \|v\| = 1 \\ \|u-v\| \ge \mu \end{array}\right\} \Rightarrow \|u+v\| \le 2(1-\delta).$$

Therefore, $C[a,b]$ is <u>not uniformly convex</u>.

4.12 By Theorem 2.6 the seminorms p_k induce a topology on $C_0^\infty(\Omega)$.

Since there are denumerably many p_k, the local base is denumerable and hence, by Theorem 2.7, the topology is metrisable.

Assume that there exists a topologically bounded convex neighbourhood of $\underline{0}$, T say. There must exist a base element, A say,

$$A = \{x \in X : |p_k(x)| \le \varepsilon \text{ for } 1 \le k \le n\},$$

such that $A \subset T$. Now,

$$U = \{x \in X: |p_{n+1}(x)| \leq 1\}$$

is a neighbourhood of $\underline{0}$. But, there exists no $\lambda \in R^1$ such that $A \subset \lambda U$. Thus, T is not topologically bounded, and so, by Theorem 2.7, the space is not normable.

4.13 We prove that if $f: \Omega \subset R^n \to R^1$ is uniformly continuous then there exists $g: \overline{\Omega} \to R^1$, continuous, such that:

$$g\Big|_{\Omega} \equiv f .$$

(a) For $x \in \Omega$ define $g(x) = \lim_{\substack{y \to x \\ y \in \Omega}} f(y)$. (*)

(b) We check whether eqn(*) defines $g(x)$ uniquely.

Let $\{y_n^1\}, \{y_n^2\}$ be sequences in Ω such that $y_n^1 \to x$ and $y_n^2 \to x$; we have to prove that $\lim_{n\to\infty} f(y_n^\ell)$, $\ell = 1,2$, exists and:

$$\lim_{n\to\infty} f(y_n^1) = \lim_{n\to\infty} f(y_n^2) .$$

Claim: $\{f(y_n^1)\}$ is a Cauchy sequence in R^1 .

Proof: Choose $\varepsilon > 0$. Since f is uniformly continuous, there exists $\delta > 0$ such that:

$$\|x_1 - x_2\| < \delta \Rightarrow |f(x_1) - f(x_2)| < \varepsilon .$$

Choose N such that $\|x - y_n^1\| < \frac{\delta}{2}$ if $n \geq N$. Then, it follows immediately that

$$|f(y_n^1) - f(y_n^2)| < \varepsilon \quad \text{if} \quad n,m > N . \quad \square$$

But R^1 is complete so that $f(y_n^1) \to \phi_1 \in R^1$.

Similarly $f(y_n^2) \to \phi_2 \in R^1$. We have to show that $\phi_1 = \phi_2$.

Take the sequence $\{t_n\}$ defined by:

$$t_{2n-1} = y_n^1, \quad t_{2n} = y_n^2, \quad n \geq 1 .$$

Then, since $t_n \to x$, the same argument as above shows that $\{f(t_n)\}$ is a Cauchy sequence, which implies that $\phi_1 = \phi_2$. Thus g is well-defined.

(c) We prove that g is uniformly continuous. Choose $\varepsilon > 0$. There exists $\tilde{\delta} > 0$ such that:

$$\|\tilde{x}_1 - \tilde{x}_2\| < \tilde{\delta} \Rightarrow |f(\tilde{x}_1) - f(\tilde{x}_2)| < \frac{\varepsilon}{3} .$$

Take $\delta = \frac{\tilde{\delta}}{3}$, and assume that $x_1, x_2 \in \overline{\Omega}$ satisfy $\|x_1 - x_2\| < \delta$. From the definition of g, there exist $\tilde{x}_1, \tilde{x}_2 \in \Omega$ such that:

$$|x_\ell - \tilde{x}_\ell| < \frac{\tilde{\delta}}{3} = \delta \quad \text{and} \quad |f(\tilde{x}_\ell) - g(x_\ell)| < \frac{\varepsilon}{3} ,$$

for $\ell = 1,2$. Thus,

$$\|\tilde{x}_1 - \tilde{x}_2\| \leq \|\tilde{x}_1 - x_1\| + \|x_1 - x_2\| + \|x_2 - \tilde{x}_2\| < \tilde{\delta} .$$

Hence

$$\begin{aligned} |g(x_1) - g(x_2)| &\leq |g(x_1) - f(\tilde{x}_1)| + |f(\tilde{x}_1) - f(\tilde{x}_2)| + \\ &\quad + |f(\tilde{x}_2) - g(x_2)| , \\ &\leq \frac{\varepsilon}{3} + \frac{\varepsilon}{3} + \frac{\varepsilon}{3} = \varepsilon . \end{aligned}$$

4.14 We show that $C^m(\overline{\Omega})$ is complete.

Let $\{u_k\}$ be a Cauchy sequence in $C^m(\overline{\Omega})$. Then there exist f_α such that

$$D^\alpha u_k \to f_\alpha \text{ in } C(\overline{\Omega}) , \quad \text{as } k \to \infty , \quad |\alpha| \leq m . \qquad (*)$$

[To prove this, use the fact that $C(\overline{\Omega})$ is a Banach space, for Ω bounded. (Section 4.8) Since the convergence in eqn (*) is with respect to the maximum norm, and since each function $D^\alpha u_k$ is uniformly continuous, f_α is also uniformly continuous.]

Set $f = f_0$. It remains to show that $D^\alpha f = f_\alpha$ This is done by induction. Let $x = (x_i) \in \Omega$. Let α, β and γ be n-indices satisfying $|\gamma| = 1$ and $\alpha = \beta + \gamma$. Then, for small $h \in R^1$,

$$D^\beta u_k(x+\gamma h) = D^\beta u_k(x) + \int_0^h D^\alpha u_k(x+\gamma s)ds .$$

Proceeding to the limit,

$$f_\beta(x+\gamma h) = f_\beta(x) + \int_0^h f_\alpha(x+\gamma s)ds ,$$

which implies that $f_\alpha = D^\gamma f_\beta$.

4.15 Let p_k satisfy eqn (*). Define $p_k^0 = p_k$ and

$$p_k^\ell(t) = x^{(k-\ell)}(m) + \int_m^t p_k^{\ell-1}(s)ds , \quad \text{for } \ell = 1,\ldots,m , \tag{$*_*^*$}$$

where $m = \frac{a+b}{2}$. Then

$$|x^{(k-\ell)}(t)-p_k^\ell(t)| \le \int_m^t |x^{(k-\ell+1)}(s)-p_k^{\ell-1}(s)|ds ,$$

$$\le \max_{a\le s\le b} |x^{(k-\ell+1)}(s)-p_k^{\ell-1}(s)| \cdot (\tfrac{b-a}{2}) .$$

It follows readily by induction that

$$\|x^{(k-\ell)}-p_k^\ell;C[a,b]\| \le \varepsilon(\tfrac{b-a}{2})^\ell , \quad \text{for } 0 \le \ell \le k . \tag{$^{**}_{**}$}$$

Set $p = p_k^k$. Combining eqns ($^{**}_{**}$) we obtain eqn (**).

We have thus shown that the polynomials are dense in $C^k[a,b]$. A straightforward generalization of the arguments on page 84 shows that each polynomial $p \in \pi[a,b]$ can be approximated by a polynomial with rational coefficients $\tilde{p} \in \tilde{\pi}[a,b]$ such that

$$\| p-\tilde{p};\ C^k[a,b] \| \le \varepsilon \ .$$

4.16 It is natural to use a generalization of the approach used in Problem 4.15.

Set $m = \frac{a+b}{2}$. Choose $\varepsilon > 0$, and $p_i(t)$, such that if α_i = (i-1 zeros, 1, n-i zeros) $\in R^n$,

$$\| D^{\alpha_i} x - p_i \| \le \varepsilon \ , \text{ for } 1 \le i \le n \ .$$

Define a polynomial $p: \Omega \to R^1$ by

$$p(m,m,m,\ldots,m) = x(m,m,\ldots,m) \ ,$$

$$p(t_1,t_2,\ldots,t_k,m,\ldots,m) = p(t_1,t_2,\ldots,t_{k-1},m,\ldots,m) +$$

$$+ \int_m^{t_k} p_k(t_1,\ldots,t_{k-1},t,m,\ldots m)dt \ ,$$

for $|t_k| < \frac{b-a}{2}$, $1 \le k \le n$. Then $p(t)$ approximates $x(t)$, but what about the derivatives?

4.17 (a) Assume $\{x_n\} \subset X$ is a Cauchy sequence. Then $\{x_n(t)\}$ is a Cauchy sequence in R^1 for all $t \in R^1$. We define:

$$x(t) = \lim_{n\to\infty} x_n(t) \ .$$

Using standard arguments we can establish that x is continuous and bounded, so that $x \in X$. Furthermore, $\| x-x_n \| \to 0$ so that X is complete.

(b) $x(t) = \sin(t^2)$ is bounded and continuous, and belongs to X . But $x \notin C(\overline{\Omega})$ because x is not uniformly continuous.

(c) If X were separable, a denumerable dense subset $\{x_n\}$ would exist. But we can easily construct $x \in X$ such that $\|x\| = 1$

and

$$\max_{n \le t \le n+1} |x(t) - x_n(t)| \ge 1 .$$

Thus, $\|x - x_n\| \ge 1$ for all n, which is a contradiction.

4.18 For any $t > a$ and $h > 0$,

$$u(t+h) = u(t) + \int_t^{t+h} \dot{u}(t)dt . \qquad (*)$$

By definition, u is continuous on $[a,b]$. From Problem 4.13, $\dot{u}$ can be extended to a continuous function on $[a,b]$ which we still denote by $\dot{u}$. We can thus let $t \to a$ in eqn (*) and obtain

$$\frac{u(a+h) - u(a)}{h} = \int_a^{a+h} \dot{u}(t)dt = \dot{u}(a+\theta h) , \quad \text{for some } \theta \in [0,1].$$

Hence,

$$\lim_{h\to 0} \frac{u(a+h) - u(a)}{h} = \dot{u}(a) = \lim_{t\to a} \dot{u}(t) .$$

4.19 Because u and its first derivatives are uniformly continuous, it suffices to check the continuity of v and its derivatives as the point $(-h, x_2)$ in the left half plane approaches the point $(0, x_2)$ on the x_2-axis:

$$\lim v(-h,x_2) = \lim \{3u(h,x_2)-2u(2h,x_2)\} = u(0,x_2) .$$

$$\lim \frac{\partial v}{\partial x_2}(-h,x_2) = \lim \{3 \frac{\partial u}{\partial x_2}(h,x_2) - 2 \frac{\partial u}{\partial x_2}(2h,x_2)\} ,$$

$$= \frac{\partial u}{\partial x_2}(0,x_2) .$$

$$\lim \frac{\partial v}{\partial x_1}(-h,x_2) = \lim \{-3\frac{\partial u}{\partial x_1}(h,x_2) + 4\frac{\partial u}{\partial x_1}(2h,x_2)\} ,$$

$$= \frac{\partial u}{\partial x_1}(0,x_2) .$$

4.20 It is readily checked that $f \in C^1(\bar{\Omega})$.

Any open set $\tilde{\Omega}$ containing $\bar{\Omega}$ must contain a neighbourhood of $(0,0)$. Let $x_1 = s$, $x_2 = s^4$,

$$\frac{\tilde{f}(x_1,x_2) - \tilde{f}(x_1,-x_2)}{x_2 - (-x_2)} = \frac{2s^2}{2s^4} = \frac{1}{s^2} .$$

This shows that $\frac{\partial \tilde{f}}{\partial x_2}$ is not bounded near $(0,0)$.

4.21 See Hahn [1922] and Cryer [1970, p. 118.]

CHAPTER 5: SOLUTIONS

5.1 Let $M_1 = \sup_{\|x\|_X \le 1} \|Tx\|_Y$, $M_2 = \sup_{\|x\|_X = 1} \|Tx\|_Y$, and

$M_3 = \inf\{\lambda : \|Tx\|_Y \le \lambda \|x\|_X, \text{ for all } x \in X\}$.

<u>$M_1 = M_2$:</u> (a) Obviously $M_1 \ge M_2$.

(b) For any $\varepsilon > 0$ there is an x_ε such that $\|x_\varepsilon\|_X \le 1$ and $\|Tx_\varepsilon\|_Y \ge M_1 - \varepsilon$.

Define $\tilde{x}_\varepsilon = \dfrac{x_\varepsilon}{\|x_\varepsilon\|_X}$. Then $\|\tilde{x}_\varepsilon\|_X = 1$ so that

$$M_2 \ge \|T\tilde{x}_\varepsilon\|_Y = \frac{1}{\|x_\varepsilon\|_X}\|Tx_\varepsilon\|_Y \ge \|Tx_\varepsilon\|_Y \ge M_1 - \varepsilon .$$

Since $\varepsilon > 0$ was arbitrary, $M_2 \ge M_1$.

<u>$M_2 = M_3$:</u> (a) Let $x \in X$ with $x \ne \underline{0}$. Then $\tilde{x} = \dfrac{x}{\|x\|_X}$ satisfies $\|\tilde{x}\|_X = 1$ and thus

$$\|T\tilde{x}\|_Y \le M_2 \iff \frac{1}{\|x\|_X}\|Tx\|_Y \le M_2 \iff \|Tx\|_Y \le M_2\|x\|_X .$$

If $x = \underline{0}$ then $\|x\|_X = 0$ and $\|Tx\|_Y = 0$. Therefore, $\|Tx\|_Y \le M_2\|x\|_X$ for all $x \in X$, and this implies that $M_2 \ge M_3$.

(b) For any $\varepsilon > 0$ there exists x_ε such that:

$$\|x_\varepsilon\| = 1 \text{ and } \|Tx_\varepsilon\|_Y \ge M_2 - \varepsilon ,$$

$$\Rightarrow M_3\|x_\varepsilon\|_X \ge \|Tx_\varepsilon\|_Y \ge (M_2 - \varepsilon)\|x_\varepsilon\|_X ,$$

$$\Rightarrow M_3 \ge M_2 - \varepsilon .$$

Therefore, $M_3 \ge M_2$.

5.2 Because $B = A^TA$ is symmetric, $B = U^TDU$ with U an orthogonal matrix, and so

$$\|Ax\|_2 = (x^TA^TAx)^{\frac{1}{2}} = (x^TBx)^{\frac{1}{2}} = (x^TU^TDUx)^{\frac{1}{2}} .$$

The linear map $U: x \in R^n \to y = Ux \in R^n$ is an isometric isomorphism: $\|x\|_2 = \|Ux\|_2 = \|y\|_2$. If $D = \mathrm{diag}(d_i)$ then

$$\|A\|_2 = \sup\{\|Ax\|_2 : \|x\|_2=1\} = \sup\{(x^TU^TDUx)^{\frac{1}{2}} : x^Tx=1\} ,$$
$$= \{(y^TDy)^{\frac{1}{2}} : y^Ty=1\} = \sup\{\sum d_iy_i^2 : \sum y_i^2=1\} ,$$
$$= \max_i |d_i|^{\frac{1}{2}} = [\rho(A^TA)]^{\frac{1}{2}} .$$

5.3 Let $M = \max_j \sum_{i=1}^n |a_{ij}| = \sum_{i=1}^n |a_{ij_0}|$, say.

(a) For any $x \in R^n$,

$$\|Ax\|_1 = \sum_{i=1}^n |(Ax)_i| = \sum_{i=1}^n |\sum_{j=1}^n a_{ij}x_j| ,$$
$$\le \sum_{i=1}^n \sum_{j=1}^n |a_{ij}||x_j| = \sum_{j=1}^n \left[|x_j| \sum_{i=1}^n |a_{ij}|\right] ,$$
$$\le \sum_{j=1}^n |x_j| M ,$$
$$= M\|x\|_1 .$$

Therefore $\|A\|_1 \le M$.

(b) Define $\tilde{x} = (\tilde{x}_j)$ by: $\tilde{x}_j = 0$ if $j \ne j_0$; $\tilde{x}_{j_0} = 1$. Then

$$\|Ax\|_1 = \sum_{i=1}^n |\sum_{j=1}^n a_{ij}\tilde{x}_j| = \sum_{i=1}^n |a_{ij_0}| = M\|\tilde{x}\|_1 .$$

Therefore, $\|A\|_1 \ge M$.

5.4 (a) $\|A_1\|_1 = \max_j \sum_{i=1}^n |a_{ij}| = \max\{4,6\} = 6$.

(b) $\|A_1\|_\infty = \max_i \sum_{j=1}^n |a_{ij}| = \max\{3,7\} = 7$.

(c) $A_1^TA_1 = \begin{bmatrix} 10 & -10 \\ -10 & 20 \end{bmatrix}$, so that

$$p(\lambda) = \det\begin{bmatrix}\lambda-10 & 10\\ 10 & \lambda-20\end{bmatrix} = \lambda^2 - 30\lambda + 100 \ .$$

The roots of $p(\lambda)$ are $\lambda_1 = 15 + 5\sqrt{5}$, and $\lambda_2 = 15 - 5\sqrt{5}$, so that $\rho(A_1^TA_1) = \max|\lambda_i| = 15 + 5\sqrt{5}$. $\|A_1\|_2 = [\rho(A_1^TA_1)]^{\frac{1}{2}} = \sqrt{5(3+\sqrt{5})}$.

(d) $\|A_2\|_1 = \max\limits_j \sum\limits_{i=1}^n |a_{ij}| = \max\{3,4\} = 4$.

(e) $\|A_2\|_\infty = \max\limits_i \sum\limits_{j=1}^n |a_{ij}| = \max\{3,4\} = 4$.

(f) If (λ,x) is an eigenpair of A_2 then, since A_2 is symmetric, (λ^2,x) is an eigenpair of $A_2^TA_2$. A_2 symmetric $\Rightarrow$ there are n real eigenvalues $\lambda_1,\ldots,\lambda_n$ of A_2 $\Rightarrow$ eigenvalues of $A_2^TA_2$ are $\lambda_1^2,\lambda_2^2,\ldots,\lambda_n^2$. Thus,

$$\|A_2\| = (\rho(A_2^TA_2))^{\frac{1}{2}} = (\max_{1\le i\le n} \lambda_i^2)^{\frac{1}{2}} = \max_{1\le i\le n}|\lambda_i| = \rho(A_2) \ .$$

If λ is an eigenvalue of A_2 then there exists $v \neq 0$ such that $A_2v = \lambda v$. That is:

$$-v_{i-1} + 2v_i - v_{i+1} = \lambda v_i, \quad 2 \le i \le n-1 ,$$
$$2v_1 - v_2 = \lambda v_1 ,$$
$$2v_n - v_{n-1} = \lambda v_n .$$

Defining $v_0 = v_{n+1} = 0$ we have

$$-v_{i-1} + 2v_i - v_{i+1} = \lambda v_i , \quad 1 \le i \le n . \qquad (*)$$

The characteristic polynomial for the difference equations (*) is

$$p(s) = s^2 + (\lambda-2)s + 1 = 0 ,$$

with roots $(2-\lambda \pm \sqrt{\lambda^2-4\lambda})/2$. From Gerschgorin's theorem we know that $0 < \lambda < 4$, so that the roots are complex and of modulus 1. We therefore try solutions of the form

$$v_i = \sin(\pi ik/(n+1)) \equiv \sin(\pi ikh) \ , \quad 0 \le i \le n+1 \ , \qquad (**)$$

where k is an integer between 1 and n . The conditions $v_0 = v_{n+1} = 0$ are satisfied. Substituting into eqn (*) from eqn (**) we find that:

$$(2-\lambda)\ \sin[\pi ikh] - \sin[\pi(i+1)kh] - \sin[\pi(i-1)kh] = 0 \ ,$$

so that v is an eigenvector of A_2 provided that

$$\sin(\pi ikh)[2-\lambda-2\cos(\pi kh)] = 0 \ . \qquad ({*}^{*}_{*})$$

The eigenvalues of A_2 are the roots of eqn $({*}^{*}_{*})$, namely

$$\lambda_k = 2 + 2\cos\left(\frac{\pi k}{n+1}\right) \ , \quad 1 \le k \le n \ ,$$

and $\|A\|_2 = \rho(A_2) = \max|\lambda_k| = 2 + 2\cos(\pi/(n+1))$.

5.5 Set $\sum_{i=1}^{n} |a_i| = M$.

(a) For any $x \in C[0,1]$,

$$\|Tx\| = \left|\sum_{i=1}^{n} x(t_i)a_i\right| \le \sum_{i=1}^{n} |x(t_i)|\,|a_i| \le \|x\|M \ .$$

Therefore, $\|T\| \le M$.

(b) Take $\tilde{x} \in C[0,1]$, piecewise linear, such that $\|\tilde{x}\| = 1$ and $\tilde{x}(t_i) = \mathrm{sign}(a_i)$, for $1 \le i \le n$. Then

$$\|T\tilde{x}\| = \left|\sum_{i=1}^{n} \tilde{x}(t_i)a_i\right| = \left|\sum_{i=1}^{n} |a_i|\right| = M\|\tilde{x}\| \ .$$

Therefore, $\|T\| \ge M$.

5.6 For any $x \in C[0,1]$,

$$\|Tx\| = \left|\int_0^1 (t-\tfrac{1}{2})x(t)dt\right| \le \|x\|_\infty \int_0^1 |t-\tfrac{1}{2}|dt = \tfrac{1}{4}\|x\|_\infty \ ,$$

so that $\|T\| \le \frac{1}{4}$.

For any $\varepsilon \in (0,\frac{1}{4})$ define x_ε by: $x_\varepsilon(t) = -1$ if $0 \le t \le \frac{1}{2} - \varepsilon$; $x_\varepsilon(t) = 1$ if $\frac{1}{2} + \varepsilon < t \le 1$; x_ε linear on $[\frac{1}{2} - \varepsilon, \frac{1}{2} + \varepsilon]$. Then, as is readily checked, $\|Tx_\varepsilon\| > \frac{1}{4} - \varepsilon^2$, so that $\|T\| \ge \frac{1}{4}$. Together with the previous result, this shows that $\|T\| = \frac{1}{4}$.

Now suppose that $x \in C[0,1]$ satisfies $\|x\| = 1$ and $\|Tx\| = \frac{1}{4}$. Then,

$$\|Tx\| = \left|\int_0^1 x(t)(t-\tfrac{1}{2})dt\right| = \frac{1}{4} = \int_0^1 |t-\tfrac{1}{2}|dt \ge \int_0^1 |x(t)(t-\tfrac{1}{2})|dt ,$$

which implies that

$$|x(t)(t-\tfrac{1}{2})| = |t-\tfrac{1}{2}| , \quad \text{for} \quad 0 \le t \le 1 .$$

This is not possible if $x \in C[0,1]$, and so there is no $x \ne 0$ such that $\|Tx\| = \|T\| \|x\|$.

5.7 (a) $iy_i = (i-1)y_{i-1} + x_i$, so that

$$2iy_i^2 = 2(i-1)y_iy_{i-1} + 2x_iy_i .$$

Rearranging, we obtain eqn (*).

(b) Consider the right hand side in eqn (*). We obtain

$$\begin{aligned}(1-2i)y_i^2 &+ 2(i-1)y_iy_{i-1} ,\\ &= -iy_i^2 + (i-1)y_{i-1}^2 - [y_i-y_{i-1}]^2(i-1) ,\\ &\le -iy_i^2 + (i-1)y_{i-1}^2 ,\end{aligned}$$

which leads to eqn (**).

(c) Summing eqn (**) we obtain:

$$\sum_{i=1}^n y_i^2 - 2\sum_{i=1}^n x_iy_i = -ny_n^2 ,$$

which implies eqn (***).

5.8 <u>(3) $\Leftrightarrow$ (4):</u> This follows from the definition of $\|T\|$ (see eqn (5.3)).

<u>(4) ⇒ (1)</u>: $\|Tx_n - Tx\| = \|T(x_n - x)\| \le M\|x_n - x\| \to 0$.
<u>(1) ⇒ (2)</u>: Obvious.

5.9 Let X be a complete metric space.

Assume that $X = \cup X_i$, where X_i is nowhere dense in X .

Choose a ball $B(x_0;r_0) \subset X$. Since X_1 is nowhere dense, there exists a ball $B(x_1;r_1)$ such that:

(1) $B(x_1;r_1) \cap X_1 = \emptyset$,
(2) $B(x_1;r_1) \subset B(x_0;r_0/3)$,
(3) $r_1 \le r_0/3$.

By induction, given $B(x_n;r_n)$ we can find $B(x_{n+1};r_{n+1})$ such that

(1) $B(x_{n+1};r_{n+1}) \cap X_{n+1} = \emptyset$,
(2) $B(x_{n+1};r_{n+1}) \subset B(x_n;r_n/3)$,
(3) $r_{n+1} \le r_n/3$.

Note that $B(x_{n+1};r_{n+1}) \cap X_k = \emptyset$ for $1 \le k \le n$.

If $m > n$ then

$$\rho(x_n,x_m) \le \rho(x_n,x_{n+1}) + \rho(x_{n+1},x_{n+2}) + \ldots + \rho(x_{m-1},x_m) , \qquad (*)$$
$$\le \frac{r_n}{3} + \frac{r_{n+1}}{3} + \ldots + \frac{r_{m-1}}{3} \le r_n \sum_{i=1}^{\infty} \left(\frac{1}{3}\right)^i = \frac{r_n}{2} .$$

Since $r_n \to 0$, it follows that $\{x_n\}$ is a Cauchy sequence, and since X is complete there exists $x_\infty \in X$ such that $x_n \to x_\infty$ and $\rho(x_n,x_\infty) \to 0$.

Now, using eqn (*),

$$\rho(x_n,x_\infty) \le \rho(x_n,x_m) + \rho(x_m,x_\infty) ,$$
$$\le r_n/2 + \rho(x_m,x_\infty) , \quad (\text{if } m \ge n)$$
$$< r_n \quad (\text{for } m \text{ large enough}) .$$

Therefore, $x_\infty \in B(x_n;r_n)$ for all $n \Rightarrow x_\infty \notin X_n$ for all $n \Rightarrow x_\infty \notin \cup X_n = X$. This is a contradiction and consequently X cannot be of the first category; that is, X is of the second category.

Note: Sometimes a set E is defined to be nowhere dense if $\bar{E}$ contains no nonempty open set (Definition 2). The definition of 'nowhere dense' given in Problem 5.9 will be called Definition 1.

These two definitions are equivalent:

Def(1) $\Rightarrow$ Def(2): Assume that E is nowhere dense in X (Def. 1). Let G be an open set in X. Then G contains an open ball, U say, which in turn, by Definition 1, contains a ball $B(x;r)$ such that $B(x;r) \cap E = \emptyset$. Then $B(x;r/2) \cap \bar{E} = \emptyset$, so that $G \not\subset \bar{E}$.

Def(2) $\Rightarrow$ Def(1): Assume that E is nowhere dense (Def. 2). Let G be a ball in X. Choose an open subset U of G. By Definition 2, $U \not\subset \bar{E}$, so that $U \cap \bar{E}^c \neq \emptyset$. But $U \cap \bar{E}^c$ is open, and thus contains an open ball, B say. Then $B \subset G$ and $B \cap E = \emptyset$.

5.10 Let F be a set in a Banach X and let E be the linear subspace spanned by F. Let T be a sequence of linear operators mapping X into a Banach space Y.

Assertion: $\lim_{n\to\infty} T_n x$ exists for each $x \in F \iff \lim_{n\to\infty} T_n x$ exists for each $x \in E$.

Proof:

$\Leftarrow$: Trivial.

$\Rightarrow$: Let $x \in E$. Then $x = \sum_{k=1}^{m} \alpha_k x_k$, where α_k is a scalar and $x_k \in F$, for $1 \le k \le m$. Hence.

$$\lim T_n x = \lim T_n[\sum_{k=1}^{m} \alpha_k x_k] = \lim \sum_{k=1}^{m} \alpha_k T_n(x_k) ,$$

$$= \sum_{k=1}^{m} \alpha_k \lim T_n(x_k) .$$

Therefore, $\lim T_n x$ exists for all $x \in E$.

5.11 Let $L: X \to Y$ be a bounded linear operator between Banach spaces X and Y. Let $T_n : X \to Y$ be a bounded linear operator for $n = 1,2,\ldots$. Then (1) $\iff$ (2) *AND* (3), where

(1) $\lim T_n x = Lx$ for all $x \in X$.

(2) $\lim_{n\to\infty} T_n x = Lx$ for all $x \in F$, with F fundamental in X .

(3) $\|T_n\| \le M$ for all n .

Proof:

(1) ⇒ (2) AND (3): Follows from the Banach-Steinhaus theorem.

(2) AND (3) ⇒ (1): Using (2), Problem 5.10 and the linearity of T_n and L , we see that

(2') $$\lim_{n\to\infty} T_n x = Lx\ , \quad \text{for} \quad x \in E\ ,$$

where E , the subspace spanned by F , is dense in X .

Using (2') and (3), and the Banach-Steinhaus theorem, we find that $Tx = \lim T_n x$ exists for all $x \in X$. From (2'), $Tx = Lx$ for $x \in E$, and it remains to show that $Tx = Lx$ for $x \in X$.

Choose $x \in X$. Since E is dense in X , there exists a sequence $\{x_k\}$ in E with limit x . Choose $\varepsilon > 0$ and find k such that $\|x-x_k\| < \varepsilon$. There exists N such that $\|T_n x_k - Lx_k\| < \varepsilon$ if $n \ge N$. Thus,

$$\begin{aligned}\|T_n x - Lx\| &\le \|T_n x - T_n x_k\| + \|T_n x_k - Lx_k\| + \|Lx_k - Lx\|\ , \\ &\le M\|x-x_k\| + \varepsilon + \|L\|\ \|x_k - x\|\ , \quad \text{if} \quad n \ge N\ , \\ &\le (1+M+\|L\|)\varepsilon\ , \quad \text{if} \quad n \ge N\ .\end{aligned}$$

Therefore, $Tx = \lim T_n x = Lx$.

5.12 The closed Newton-Cotes quadrature rule is obtained by integrating the polynomial which interpolates the integrand at equidistant points. Thus,

$$\int_{-1}^{+1} x(t)dt \doteq T_n(x) = \sum_{k=0}^{n} w_k^{(n)} x(t_k^{(n)})\ ,$$

where

$$t_k^{(n)} = kh - 1 \ , \quad h = 2/n \ , \quad \text{and}$$

$$w_k^{(n)} = h \int_0^n \prod_{i \neq k} \left(\frac{s-i}{k-i}\right) ds = A_{nk} \ , \ \text{say.}$$

For example,

$$A_{n0} = \frac{2}{n} \int_0^n \frac{(s-1)(s-2)\dots(s-n)}{(-1)^n n!} \, ds \ .$$

The coefficients A_{nk} are rational numbers, and it is convenient to express them in the form $A'_{nk} \cdot D_n$, where A'_{nk} is an integer. The computation of A_{nk} for small values of n is very easy on a computer. For large values of n , the coefficients A_{nk} grow rapidly in size and care must be taken to avoid round-off error.

Some of the coefficients A_{nk} are given below:

1	(1	1)								$\frac{2}{2}$
2	(1	4	1)							$\frac{2}{6}$
3	(1	3	3	1)						$\frac{2}{8}$
4	(7	32	12	32	7)					$\frac{2}{90}$
5	(19	75	50	50	75	19)				$\frac{2}{288}$
6	(41	216	27	272	27	216	41)			$\frac{2}{840}$
7	(751	3577	1323	2989	2989	1323	3577	751)		$\frac{2}{17280}$
8	(989	5888	-928	10496	-4540	10496	-928	5888	98)	$\frac{2}{28350}$

From Example 5.4,

$$\| T_n \| = \sum_{k=0}^{n} | w_k^{(n)} | = \sum_{k=0}^{n} | A_{nk} | \ .$$

5.13 For $x \in C[0,1]$ let

$$Tx = \int_0^1 x(t)dt \ , \quad T_n x = \sum_{r=1}^{n} \frac{1}{n} \sum_{k=0}^{m} w_k x\left(\frac{r-1+t_k}{n}\right) .$$

We wish to show that: $T_n x \to Tx$ for all $x \in X$ $\iff \sum_{k=0}^{m} w_k = 1$. We could of course use Steklov's theorem but it is easier to proceed directly.

$\Rightarrow$: Let $x(t) = 1$, for $0 \le t \le 1$. Then $Tx = 1$ and $T_n x = \sum_{k=0}^{m} w_k$. By assumption, $T_n x \to Tx$, and so $\sum_{k=0}^{n} w_k = 1$.

$\Leftarrow$: Let $x \in C[0,1]$. Choose $\varepsilon > 0$. Since x is uniformly continuous, there exists $\delta > 0$ such that $|x(s)-x(t)| \le \varepsilon$ if $|s-t| \le \delta$. Choose $N \ge 1/\delta$.

Let $n \ge N$, so that $x(t)$ varies by less than ε in each interval $[\frac{r-1}{n}, \frac{r}{n}]$. Then

$$Tx = \sum_{r=1}^{n} \int_{\frac{r-1}{n}}^{\frac{r}{n}} x(t)dt = \sum_{r=1}^{n} \frac{1}{n} x(\tfrac{r}{n}) + E_n^{(1)} ,$$

where

$$|E_n^{(1)}| \le \sum_{r=1}^{n} \frac{1}{n} \max_{\frac{r-1}{n} \le t \le \frac{r}{n}} |x(\tfrac{r}{n}) - x(t)| \le \varepsilon .$$

Also ,

$$T_n x = \sum_{r=1}^{n} \frac{1}{n} \sum_{k=0}^{m} w_k x(\tfrac{r}{n}) + E_n^{(2)} ,$$

$$= \sum_{r=1}^{n} \frac{1}{n} x(\tfrac{r}{n}) \sum_{k=0}^{m} w_k + E_n^{(2)} ,$$

$$= \sum_{r=1}^{n} \frac{1}{n} x(\tfrac{r}{n}) + E_n^{(2)} ,$$

where

$$|E_n^{(2)}| \le \sum_{r=1}^{n} \frac{1}{n} \sum_{k=0}^{m} |w_k| \max_{\frac{r-1}{n} \le t \le \frac{r}{n}} |x(\tfrac{r}{n}) - x(t)| ,$$

$$\le \varepsilon \sum_{k=0}^{m} |w_k| .$$

Combining the above, we see that

$$\|Tx - T_n x\| \le (1 + \sum_{k=0}^{m} |w_k|)\varepsilon , \quad \text{if} \quad n \ge N ,$$

so that $T_n x \to Tx$.

5.14 We begin by making three observations:

(a) The polynomials are dense in $C^1[a,b]$ (see Problem 4.15).

(b) If $T_n : C^1[a,b] \to R^1$ is defined by eqn (*) then, from eqn (5.14) (p. 116),

$$\frac{K_n}{2+b-a} \le \|T_n\| \le K_n , \qquad (**)$$

where

$$K_n = \sum_{k=1}^{n} (t_{k+1}^{(n)} - t_k^{(n)}) |\sum_{j=1}^{k} w_j^{(n)}| + |\sum_{j=1}^{n} w_j^{(n)}| , \qquad (***)$$

$$= R_n + |S_n| , \quad \text{say.}$$

(c) If $x(t) = 1$ for $a \le t \le b$, then

$$Tx = b - a \quad \text{and} \quad T_n x = \sum_{j=1}^{n} w_j^{(n)} = S_n .$$

We now prove the theorem.

⇒: (1) is trivial. From the Banach-Steinhaus theorem, we know that $\|T_n\| \le M$ for some M, and all n. Using eqns (**) and (*$\overset{*}{*}$) it follows that $R_n \le K_n \le (2+b-a)M$, so that (2) holds.

⇐: If $T_n p \to Tp$ for every polynomial p, then, by (c) above, as $n \to \infty$

$$S_n = \sum_{j=1}^{n} w_j^{(n)} \to b - a ,$$

so that $|S_n|$ is bounded by a constant, M_1 say. Together with (2) and eqns (**) and (*$\overset{*}{*}$), this implies that

$$\|T_n\| \le M + M_1 .$$

(1) now follows from the Banach-Steinhaus theorem. □

In this volume we discuss three approaches to quadrature:

(1) The general convergence theory a la Šteklov. (Theorem 5.5, p. 124).

(2) The Romberg transformation (p. 137).

(3) The existence of non-negative multi-dimensional formulae. (Theorem 7.13, p. 252).

There are many other interesting possibilities.

(1) The theorem proved above for $C^1[a,b]$ can be generalized to cover quadrature for functions in $C^m[a,b]$.

(2) A quadrature formula can be expressed as a Stieltjes integral (see p. 336),

$$T_n x = \sum_{k=0}^{n} w_k^{(n)} x(t_k^{(n)}) = \int_a^b x(t)d\mu_n(t) , \qquad (*)$$

where μ_n is piecewise constant with jumps of $w_k^{(n)}$ at the points $t_k^{(n)}$. This is the approach used by Krylov [1962, p. 243] in the analysis of quadrature formulae for functions which are analytic in a region of $\mathbb{C}^1$ which contains $[a,b]$.

(3) The construction of quadrature formulae which have minimum error is the source of many ingenious ideas. To mention but two, see Sard [1963] and Cobb and Harris [1966].

5.15 If $x = (x_n : 1 \le n < \infty) \in c$ there exists $x_\infty \in R^1$ such that $x_n \to x_\infty$. Thus, for all $\varepsilon > 0$, there exists N such that $|x_n - x_\infty| < \varepsilon$ if $n \ge N$.

Define $y^N = x_\infty u_0 + \sum_{n=1}^{N} (x_n - x_\infty) u_n$. y^N belongs to the linear subspace E spanned by $\{u_k\}$ and

$$\| x - y^N \| = \sup_{1 < n \le \infty} |x_n - y_n^N| = \sup_{n > N} |x_n - x_\infty| < \varepsilon .$$

Therefore E is dense in c and $\{u_k : 0 \le k \le \infty\}$ is a fundamental set in c.

5.16

n	p	n+p	$S_{n,p}$	$\tilde{u}_{n+p}$	$M\tilde{u}_{n+p-1}$	....		$M^p\tilde{u}_n$
0	0	0	$0.50000000 = \frac{1}{2}$	1				
1	0	1	$0.83333333 = \frac{5}{6}$	$-\frac{1}{3}$	$+\frac{1}{3}$			
1	1	2	$0.80000000 = \frac{4}{5}$	$\frac{1}{5}$	$-\frac{1}{15}$			
1	2	3	$0.79047619 = \frac{83}{105}$	$-\frac{1}{7}$	$+\frac{1}{35}$	$\frac{-2}{105}$		
2	2	4	$0.784126984 = \frac{247}{315}$	$-\frac{1}{9}$	$-\frac{1}{63}$	$\frac{2}{315}$	$\frac{-2}{315}$	
2	3	5	$0.784992784 = \frac{544}{693}$	$-\frac{1}{11}$	$\frac{1}{99}$	$\frac{-2}{693}$	$\frac{2}{1155}$	
2	4	6	$0.785259185 = \frac{35372}{45045}$	$\frac{1}{13}$	$-\frac{1}{143}$	$\frac{2}{1287}$	$\frac{-2}{3003}$	$\frac{8}{15015}$

5.17 With the notation of Section 5.4.1, let x_n denote the n-th partial sum of the original series, so that

$$x_0 = 0 \ , \ x_n = \sum_{s=0}^{n-1} \tilde{u}_s \ , \text{ and } \ \tilde{u}_n = x_{n+1} - x_n = (E-1)x_n \ .$$

Then

$$\begin{aligned} S_{n,p} &= \sum_{s=0}^{n-1} \tilde{u}_s + \frac{1}{2} \sum_{s=0}^{p} M^s \tilde{u}_n \ , \\ &= x_n + \frac{1}{2} \sum_{s=0}^{p} \left(\frac{1+E}{2}\right)^s \tilde{u}_n \ , \\ &= x_n + \sum_{s=0}^{p} \left(\frac{1+E}{2}\right)^s \left(\frac{E+1}{2} - 1\right) x_n \ , \\ &= x_n + \left[\left(\frac{1+E}{2}\right)^{p+1} - 1\right] x_n = \left(\frac{1+E}{2}\right)^{p+1} x_n \ , \\ &= \sum_{j=0}^{p+1} \frac{1}{2^{p+1}} \binom{p+1}{j} x_{n+j} \ . \end{aligned} \qquad (*)$$

The van Wijngaarden transformation transforms a sequence $x = (x_k)$ into the sequence $y = (y_\ell) = (S_{n_\ell, p_\ell})$. We know that $n_0 = 0$, $p_0 = 0$, and that $n_\ell + p_\ell = \ell$. Furthermore, both $\{n_\ell\}$ and $\{p_\ell\}$ are monotone sequences.

We now consider a fixed sequence $x = (x_k)$, which is transformed into a sequence (y_ℓ). Then the sequences $\{n_\ell\}$ and $\{p_\ell\}$ are also fixed.

From eqn (*),

$$y_\ell = S_{n_\ell, p_\ell} = \sum_{j=0}^{p_\ell+1} \frac{1}{2^{p_\ell+1}} \binom{p_\ell+1}{j} x_{n_\ell+j} = \sum_{k=n_\ell}^{\ell+1} \frac{1}{2^{p_\ell+1}} \binom{p_\ell+1}{k-n_\ell} x_k .$$

Set

$$a_{\ell k} = \begin{cases} 0 , & \text{if } k < n_\ell \text{ or } k > \ell + 1 , \\ \dfrac{1}{2^{p_\ell+1}} \dbinom{p_\ell+1}{k-n_\ell} , & \text{if } n_\ell \le k \le \ell + 1 , \end{cases} \qquad (**)$$

and let A be the infinite matrix $(a_{\ell k})$. Then the sequence (x_k) is transformed into the sequence $(y_\ell) = ((Ax)_\ell)$. We show below that A defines a regular method of summability, from which it follows that

$$\lim_{\ell\to\infty} y_\ell = \lim_{k\to\infty} x_k .$$

To show that A defines a regular method of summability we check the conditions of the theorem of Toeplitz:

(1) Let $N = \sup_{\ell \ge 0} n_\ell$. Two cases arise:

(a) If $N = \infty$, then, for every k there exists L such that $n_\ell > k$ for $\ell \ge L$. Hence, from eqn (**), $\lim_{\ell\to\infty} a_{\ell k} = 0$. Indeed $a_{\ell k} = 0$ for $\ell \ge L$.

(b) If $N < \infty$, then, since $\{n_\ell\}$ is a monotone sequence, $n_\ell = N$ for $\ell \ge L$. Then, for fixed k and large ℓ, $p_\ell = N - \ell$ and

$$a_{\ell k} = \begin{cases} 0 , & \text{if } k < N , \\ \dfrac{1}{2^{\ell+1-N}} \dbinom{\ell+1-N}{k-N} , & \text{if } k \ge N , \end{cases}$$

so that $a_{\ell k} \to 0$ as $\ell \to \infty$.

(2) $\sum_{k=0}^{\infty} a_{\ell k} = \sum_{k=n_\ell}^{\ell+1} \frac{1}{2^{p_\ell+1}} \binom{p_\ell+1}{k-n_\ell} = 1$.

(3) $\sum_{k=0}^{\infty} |a_{\ell k}| = 1$.

Thus, all the conditions of the theorem of Toeplitz are satisfied.

5.18 From eqn (5.22), $v_s = \frac{1}{2^{s+1}}(1-E)^s u_0$.

(1) $u_n = (\frac{1}{2})^n$. Then

$$v_s = \frac{1}{2^{s+1}} \sum_{j=0}^{s} \binom{s}{j}(-1)^j u_j = \frac{1}{2^{s+1}} \sum_{j=0}^{s} \binom{s}{j}(-1)^j (\frac{1}{2^j}) ,$$

$$= \frac{1}{2^{s+1}} \cdot (1 - \frac{1}{2})^s = \frac{1}{2}(\frac{1}{4})^s .$$

(2) $v_s = \frac{1}{2^{s+1}} (1 - \frac{1}{3})^s = \frac{1}{2^{s+1}}(\frac{2}{3})^s = \frac{1}{2}(\frac{1}{3})^s$.

(3) $v_s = \frac{1}{2^{s+1}} (1 - \frac{1}{4})^s = \frac{1}{2^{s+1}}(\frac{3}{4})^s = \frac{1}{2}(\frac{3}{8})^s$.

5.19 Forming the combination

$$\left(\frac{1}{\beta^v} - \frac{1}{\beta^{k+m+1}}\right)(*) + \left(\frac{1}{\beta^k} - \frac{1}{\beta^v}\right)(**)$$

we obtain eqn ($*_*^*$).

Eqn ($*_*^*$) holds for $k \le v \le k+m+1$, since, for example, the case when eqn (*) does not hold is $v = k+m+1$, but for this case eqn (*) contributes nothing to eqn ($*_*^*$) because the multiplying factor $(\frac{1}{\beta^v} - \frac{1}{\beta^{k+m+1}})$ is zero.

To compute $\alpha_0^{(k,m+1)}$ we must extract the first term in eqn ($*_*^*$). This leads to

$$\alpha_0^{(k,m+1)} = \frac{\frac{1}{\beta^k}\alpha_0^{(k+1,m)} - \frac{1}{\beta^{k+m+1}}\alpha_0^{(k,m)}}{\frac{1}{\beta^k} - \frac{1}{\beta^{k+m+1}}},$$

$$= \frac{\beta^{m+1}\alpha_0^{(k+1,m)} - \alpha_0^{(k,m)}}{\beta^{m+1} - 1}.$$

5.20 (1) *Claim:* There are constants C_s^m such that

$$T_m^{(k)} = \sum_{s=0}^{m} C_s^m T_0^{(k+s)}, \quad k = 1,2,\ldots .$$

Proof by induction: The case $m=0$ is trivially true with $C_0^0 = 1$.

Assume true for m. From eqn (5.24),

$$T_{m+1}^{(k)} = \frac{\beta^{m+1}T_m^{(k+1)} - T_m^{(k)}}{\beta^{m+1}-1},$$

$$= \frac{1}{\beta^{m+1}-1}\left\{\sum_{s=0}^{m} C_s^m \beta^{m+1} T_0^{(k+1+s)} - \sum_{s=0}^{m} C_s^m T_0^{(k+s)}\right\}, \qquad (**)$$

$$= \frac{1}{\beta^{m+1}-1}\left\{-C_0^m T_0^{(k)} + \sum_{s=1}^{m}(C_{s-1}^m\beta^{m+1} - C_s^m)T_0^{(k+s)} + \right.$$

$$\left. + C_m^m \beta^{m+1} T_0^{(k+m+1)}\right\},$$

$$= \sum_{s=0}^{m+1} C_s^{m+1} T_0^{(k+s)}.$$

(2) *Claim:* $\sum_{s=0}^{m} C_s^m = 1$, for $m \geq 0$.

Proof by induction: The hypothesis is true for $m = 0$. Assume true for m. From eqn (**),

$$\sum_{s=0}^{m+1} C_s^{m+1} = \frac{1}{\beta^{m+1}-1}\sum_{s=0}^{m} C_s^m(\beta^{m+1}-1) = \sum_{s=0}^{m} C_s^m = 1.$$

(3) We have established eqn (*) with $c_s = C^m_s$. Hence

$$\lim_{k\to\infty} T_m^{(k)} = \lim_{k\to\infty} \sum_{s=0}^{m} c_s T_0^{(k+s)} = \sum_{s=0}^{m} c_s \lim_{k\to\infty} T_0^{(k)} = \lim_{k\to\infty} T_0^{(k)} ,$$

since, from (2), $\sum_{s=0}^{m} c_s = 1$.

5.21 Continuing with the analysis of Problem 5.20 we introduce the polynomials

$$p_m(z) = \sum_{k=0}^{m} C^m_k z^k , \quad m \ge 0 .$$

Let E be the shift operator with respect to k so that $E(T_m^{(k)}) = T_m^{(k+1)}$. Then

$$T_m^{(k)} = \sum_{s=0}^{m} C^m_s T_0^{(k+s)} = \sum_{s=0}^{m} C^m_s E^s T_0^{(k)} = p_m(E) T_0^{(k)} ,$$

and, from eqn (**),

$$T_{m+1}^{(k)} = p_{m+1}(E) T_0^{(k)} = \frac{\beta^{m+1} T_m^{(k+1)} - T_m^{(k)}}{\beta^{m+1} - 1} ,$$

$$= \frac{\beta^{m+1} p_m(E) T_0^{(k+1)} - p_m(E) T_0^{(k)}}{\beta^{m+1} - 1} ,$$

$$= \left[\frac{\beta^{m+1} E p_m(E) - p_m(E)}{\beta^{m+1} - 1} \right] T_0^{(k)} .$$

This shows that

$$p_{m+1}(z) = \frac{\beta^{m+1} z - 1}{\beta^{m+1} - 1} \cdot p_m(z) . \qquad (\ast\ast\ast)$$

Let $t_m(z) = \sum_{k=0}^{m} c_{mk} z^k$, where the coefficients c_{mk} are as in the text (page 138). Then $c_{m,k} \equiv c_{mk} = C^m_{m-k}$ so that

$$t_m(z) = z^m p_m(\tfrac{1}{z}) . \qquad (\ast\ast\ast\ast)$$

We now consider the assertions in Problem 5.21:

(a) Using eqns ($*{*\atop *}$) and (${*\atop *}{*\atop *}$) we see that

$$t_{m+1}(z) = z^{m+1}p_{m+1}(\tfrac{1}{z}) = z^{m+1}\left(\frac{\beta^{m+1}\frac{1}{z}-1}{\beta^{m+1}-1}\right)p_m(\tfrac{1}{z}) ,$$

$$= \left(\frac{\beta^{m+1}-z}{\beta^{m+1}-1}\right)\cdot z^m p_m(\tfrac{1}{z}) ,$$

$$= \left(\frac{\beta^{m+1}-z}{\beta^{m+1}-1}\right)\cdot t_m(z) .$$

Since $t_0(z) \equiv 1$ it follows from this recurrence relation that

$$t_m(z) = \prod_{k=1}^{m}\left(\frac{\beta^k-z}{\beta^k-1}\right)\cdot t_0(z) = \prod_{k=1}^{m}\left(\frac{\beta^k-z}{\beta^k-1}\right) ,$$

$$(*{*\atop *}{*\atop *}) \quad = \prod_{k=1}^{m}\left(\frac{1-\beta^{-k}z}{1-\beta^{-k}}\right) .$$

(b) From eqn ($*{*\atop *}{*\atop *}$),

$$t_m(1) = \sum_{k=0}^{m} c_{mk} = \prod_{k=1}^{m}\left(\frac{1-\beta^{-k}\cdot 1}{1-\beta^{-k}}\right) = 1 .$$

(c) From eqn ($*{*\atop *}{*\atop *}$) ,

$$t_m(-z) = \prod_{k=0}^{m}\left(\frac{1+\beta^{-k}z}{1-\beta^{-k}}\right) = \sum_{k=0}^{m}(-1)^k c_{mk}z^k .$$

If $\beta > 1$, $t_m(-z)$ is the product of linear factors with strictly positive coefficients, so that $t_m(-z)$ has strictly positive coefficients. Hence

$$(-1)^k c_{mk} > 0 \quad \text{and} \quad t_m(-1) = \sum_{k=0}^{m}|c_{mk}| .$$

(d) If $0 < x < 1$, then

$$\ell n(\frac{1+x}{1-x}) = \ell n(1+x) - \ell n(1-x) ,$$

$$= 2\sum_{k=0}^{\infty} \frac{x^{2k+1}}{2k+1} < 2x(1 + \frac{1}{3} \sum_{k=1}^{\infty} x^{2k}) ,$$

$$= 2x(1 + \frac{1}{3} \frac{x^2}{1-x^2}) .$$

Hence, if $0 \le x \le \frac{1}{4}$, then

$$\ell n(\frac{1+x}{1-x}) < 2x(1 + \frac{1}{3} \frac{(\frac{1}{4})^2}{1-(\frac{1}{4})^2}) = \frac{92}{45} x .$$

Thus, using eqn (***) with $\beta = 4$,

$$\ell n\ t_m(-1) = \sum_{k=1}^{m} \ell n \left(\frac{1+4^{-k}}{1-4^{-k}}\right) < \frac{92}{45} \sum_{k=1}^{m} 4^{-k} ,$$

$$< \frac{92}{45} \cdot \frac{\frac{1}{4}}{1-\frac{1}{4}} = \frac{92}{45} \cdot \frac{1}{3} . .$$

Therefore,

$$t_m(-1) < \exp(\frac{92}{45} \cdot \frac{1}{3}) \doteq 1.9768 < 2 .$$

(e) From eqn (***),

$$t_m(-z) = \sum_{k=0}^{m} c_{mk}(-1)^k z^k = \prod_{k=1}^{m} \left(\frac{1+z4^{-k}}{1-4^{-k}}\right) .$$

By replacing the term $z4^{-k}$ by $z4^{-1}$ in the product we increase the size of the coefficients. Thus, we can conclude that $|c_{m,m-s}|$ is less than the coefficient of z^{m-s} in the polynomial

$$\prod_{k=1}^{m} \frac{1+\frac{z}{4}}{(1-4^{-k})} .$$

That is,

$$|c_{m,m-s}| \le \frac{\binom{m}{s}\left(\frac{1}{4}\right)^{m-s}}{\prod_{k=1}^{\infty}(1-4^{-k})} \longrightarrow 0 \ , \quad \text{as} \quad m \to \infty \ .$$

5.22 (1) Expanding in Taylor series about $t = 0$,

$$T_0^{(k)} = \frac{x(2^{-k})-x(-2^{-k})}{2\cdot 2^{-k}} = \sum_{s=0}^{\infty} \frac{x^{(2s+1)}(0)}{(2s+1)!}\left(\frac{1}{4^k}\right)^s .$$

We can apply Romberg's method with $\beta = 4$. For $x(t) = \sin t$:

$T_0^{(k)}$	$T_1^{(k)}$	$T_2^{(k)}$	$T_3^{(k)}$
0.841470985			
0.958851077	0.997977775		
0.989615837	0.999870757	0.999996956	
0.997397867	0.999991877	0.999999952	0.999999999

(2) Expanding in Taylor series about $t = 0$,

$$T_0^{(k)} = \frac{x(2^{-k})-x(0)}{2^{-k}} = \sum_{s=0}^{\infty} \frac{x^{(s+1)}(0)}{(s+1)!}\left(\frac{1}{2^k}\right)^s .$$

We can apply Romberg's method with $\beta = 2$. For $x(t) = \sin t$:

$T_0^{(k)}$	$T_1^{(k)}$	$T_2^{(k)}$	$T_3^{(k)}$
0.841470985			
0.958851077	1.076231170		
0.989615837	1.020380597	1.001763739	
0.997397867	1.005179897	1.000112997	0.999877177

5.23 Picking up where the hint stopped,

$$I_n > \sum_{k=1}^{n-1} \frac{2n+1}{(k+1)\pi} \int_{k\pi}^{(k+1)\pi} |\sin u|du ,$$

$$= \frac{2(2n+1)}{\pi} \sum_{k=1}^{n-1} \frac{1}{k+1} ,$$

$$> \frac{2(2n+1)}{\pi} \sum_{k=1}^{n-1} \int_{k+1}^{k+2} \frac{1}{t}\, dt ,$$

$$= \frac{2(2n+1)}{\pi} \int_{2}^{n+1} \frac{1}{t}\, dt ,$$

$$= \frac{2(2n+1)}{\pi} \ell n \ ((n+1)/2) .$$

5.24 We follow the steps outlined on page 142.

(1) Let $x \in \tilde{\pi}_n$. Then $T_n x = x$ so that

$$(T_n x^\tau)(s-\tau) = x^\tau(s-\tau) = x(s) .$$

Hence,

$$\frac{1}{2\pi} \int_0^{2\pi} (T_n x^\tau)(s-\tau)ds = x(s) .$$

Since $(S_n x)(s) = x(s)$, the identity (5.29) holds.

(2) Let $x \in \tilde{\pi} \backslash \tilde{\pi}_n$, so that x is a trigonometric polynomial, each of whose terms is of degree at least $n+1$. Then x is a linear sum of terms of the form sin mt or cos mt with $m > n$. Consider one such term, $x = \sin mt$ for example.

Then

$$x^{\tau}(t) = \sin m(t+\tau) = \sin mt \cos m\tau + \cos mt \sin m\tau ,$$

$$= x_1(t)y_1(\tau) + x_2(t)y_2(\tau) , \text{ say.}$$

Thus

$$(T_n x^{\tau})(s-\tau) = (T_n x_1)(s-\tau)\cdot y_1(\tau) + (T_n x_2)(s-\tau)\cdot y_2(\tau) .$$

$(T_n x_1)(s-\tau)$ and $(T_n x_2)(s-\tau)$ are trigonometric polynomials of degree n in $(s-\tau)$, and thus (by the usual trigonometric formulae) trigonometric polynomials of degree n in τ with coefficients which depend upon s . They are thus orthogonal to $y_1(\tau)$ and $y_2(\tau)$, which are trigonometric functions of degree $m > n$. Hence

$$\int_0^{2\pi} (T_n x^{\tau})(s-\tau)d\tau = 0 .$$

Since $S_n x = 0$, the identity (5.29) holds.

(3) We begin by observing that, for any $x \in \tilde{C}[0,2\pi]$, and $s,\tau \in [0,2\pi]$,

$$|(T_n x^{\tau+h})(s-\tau-h) - (T_n x^{\tau})(s-\tau)|$$

$$\le |(T_n x^{\tau+h})(s-\tau-h) - (T_n x^{\tau})(s-\tau-h)| +$$

$$+ |(T_n x^{\tau})(s-\tau-h) - (T_n x^{\tau})(s-\tau)| ,$$

$$\le \|T_n\| \, \|x^{\tau+h}-x^{\tau}\| + |(T_n x^{\tau})(s-\tau-h)-(T_n x^{\tau})(s-\tau)| ,$$

$$\to 0 \text{ as } h \to 0 ,$$

since:

(i) x is uniformly continuous, so that $\|x^{\tau+h}-x^{\tau}\| \to 0$ as $h \to 0$;

(ii) $T_n x^{\tau}$ is continuous.

We have thus shown that, for any $x \in \tilde{C}[0,2\pi]$, and $s \in [0,2\pi]$, $g(\tau) = T_n x^\tau (s-\tau)$ is a continuous function of τ, so that the (Riemann) integral of $g(\tau)$ exists,

$$V_n x = \frac{1}{2\pi} \int_0^{2\pi} (T_n x^\tau)(s-\tau)d\tau .$$

Choose $s \in [0,2\pi]$. Then: $V_n : \tilde{C}[0,2\pi] \to R^1$ is linear, and bounded:

$$\|V_n x\| \le \frac{1}{2\pi} \int_0^{2\pi} \|T_n x^\tau\| d\tau \le \|T_n\| \|x\| .$$

$U_n : x \to (S_n x)(s)$ is also a bounded linear operator from $\tilde{C}[0,2\pi]$ to R^1 (see eqn (5.28)). From (1) and (2) above, U_n and V_n coincide on the dense subset E of $\tilde{C}[0,2\pi]$ spanned by the trigonometric polynomials. For any $x \in \tilde{C}[0,2\pi]$, there exists a sequence $\{x_k\}$ in E converging to x. Thus,

$$(S_n x)(s) = U_n x = \lim_{k\to\infty} U_n x_k = \lim_{k\to\infty} V_n x_k = V_n x , \quad \text{for all } x \in \tilde{C}[0,2\pi] .$$

Since s was arbitrary, eqn (5.29) has been established for all $x \in \tilde{C}[0,2\pi]$.

5.25 (1) If $x(t) = t^2$ then

$$(B_n x)(t) = \sum_{k=0}^{n} \binom{j}{k} t^k (1-t)^{n-k} \left(\frac{k}{n}\right)^2 .$$

Using eqn (**) on page 82 of the text, and setting $\alpha = t$ and $\beta = 1-t$ therein, we obtain

$$(B_n x)(t) = [n(n-1)t^2 + nt]/n^2 .$$

(2) By definition, $H_n x$ has zero slope at the Chebyshev points $t_k^{(n)}$.

5.26 Define $h_i = t_{i+1}^{(n)} - t_i^{(n)}$, $0 \le i \le n-1$ and also:

$$h_m = \min_{0\le i\le n-1} h_i \; ; \; h_M = \max_{0\le i\le n-1} h_i \; ; \quad \alpha = h_M/h_m \; .$$

We modify eqns (5.32) through (5.39):

$$s(t) = \frac{1}{6h_i}[s_i''(t_{i+1}-t)^3 + s_{i+1}''(t-t_i)^3] +$$

$$+ \frac{1}{h_i}[x_i(t_{i+1}-t) + x_{i+1}(t-t_i)] -$$

$$- \frac{h_i}{6}[s_i''(t_{i+1}-t) + s_{i+1}''(t-t_i)] \text{ , on } I_i \; . \qquad (5.32')$$

Using eqn (5.32') we find that:

(1) $\dot{s}(0) = 0 \Rightarrow 2s_0'' + s_1'' = -\dfrac{6(x_1-x_0)}{h_0^2}$.

(2) $\dot{s}(1) = 0 \Rightarrow 2s_n'' + s_{n-1}'' = -\dfrac{6(x_n-x_{n-1})}{h_{n-1}^2}$.

(3) $\dot{s}(t_{i+0}) = \dot{s}(t_{i-0}) \Rightarrow$

$$\frac{-s_i'' h_i^2}{2h_i} + \frac{1}{h_i}[x_{i+1}-x_i] - \frac{h_i}{6}[s_{i+1}''-s_i''] =$$

$$= \frac{s_i'' h_{i-1}^2}{2h_{i-1}} + \frac{1}{h_{i-1}}[x_i-x_{i-1}] - \frac{h_{i-1}}{6}[s_i''-s_{i-1}''] \; .$$

Hence, eqn (5.33) becomes

$$Au = 6b \; , \qquad (5.33')$$

where

$$u = \begin{bmatrix} s''_0 \\ \vdots \\ s''_n \end{bmatrix}, \quad b = \begin{bmatrix} \frac{x_1 - x_0}{h_0} \\ \cdots\cdots \\ \frac{x_{i+1} - x_i}{h_i} - \frac{x_i - x_{i-1}}{h_{i-1}} \\ \cdots\cdots \\ - \frac{x_n - x_{n-1}}{h_{n-1}} \end{bmatrix} . \tag{5.34'}$$

$$A = (a_{ij}) = \begin{bmatrix} 2h_0 & h_0 & & & & \bigcirc \\ h_0 & 2(h_0+h_1) & h_1 & & & \\ & h_1 & 2(h_1+h_2) & h_2 & & \\ & & \cdots & \cdots & \cdots & \\ & \bigcirc & & h_{n-2} & 2(h_{n-2}+h_{n-1}) & h_{n-1} \\ & & & & h_{n-1} & 2h_{n-1} \end{bmatrix}$$

(5.35')

A is strictly diagonally dominant so that A is nonsingular. Therefore there exists a solution $s = T_n x$ of eqns (5.32') through (5.35').

Choose $v \in R^{n+1}$. Set $w = A^{-1}v \in R^{n+1}$. Then

$$\|w\|_\infty = \max |w_i| = |w_k| \text{ , for some } k .$$

Thus,

$$\begin{aligned} \|v\|_\infty &= \|Aw\|_\infty , \\ &\geq |(Aw)_k| , \\ &\geq |a_{kk}w_k| - \sum_{j \neq k} |a_{kj}w_j| , \\ &\geq \{|a_{kk}| - \sum_{j \neq k} |a_{kj}|\}|w_k| , \\ &\geq h_m \cdot |w_k| = h_m\|w\|_\infty = h_m\|A^{-1}v\|_\infty . \end{aligned}$$

Since $v \in R^{n+1}$ was arbitrary,

$$\|A^{-1}\|_\infty \leq \frac{1}{h_m} ,$$

and

$$\|u\|_\infty \le \|A^{-1}\|_\infty \cdot 6\|b\|_\infty \le \frac{6\|b\|_\infty}{h_m} . \qquad (5.36')$$

Also, from eqn (5.34'), $\|b\|_\infty \le \dfrac{4\|x\|_\infty}{h_m}$, so that

$$\max_i |s_i''| = \|u\|_\infty \le \frac{24\|x\|_\infty}{h_m^2} . \qquad (5.37')$$

From eqn (5.32'),

$$\|T_n x\| = \|s\|_\infty = \max_{a\le t\le b} |s(t)| ,$$

$$\le \frac{1}{6} \max_i |s_i''| \cdot \max_i \max_{0\le t\le h_i} \left[\frac{(h_i-t)^3+t^3}{h_i}\right] +$$

$$+ \max_i |x_i| \cdot \max_i \left[\max_{0\le t\le h_i} \left[\frac{(h_i-t)+t}{h_i}\right]\right] +$$

$$+ \frac{1}{6} \max_i |s_i''| \cdot \max_i \max_{0\le t\le h_i} [h_i\{(h_i-t)+t\}] ,$$

which implies that

$$\|T_n x\|_\infty \le \frac{2h_M^2}{6} \max_i |s_i''| + \max_i |x_i| ,$$

$$\le 8\|x\|_\infty \cdot \left(\frac{h_M}{h_m}\right)^2 + \|x\|_\infty \le \|x\|_\infty (8\alpha^2+1) .$$

Therefore,

$$\|T_n\| \le 8\alpha^2 + 1 . \qquad (5.38')$$

Now take $x \in C^1([a,b])$. Apply the Mean Value Theorem to eqn (5.34') to obtain

$$\|b\|_\infty \le 2\|\dot{x}\|_\infty .$$

Using eqn (5.36') it follows that

$$\max_i |s_i''| \le \frac{12\|\dot{x}\|_\infty}{h_m} .$$

Because of eqn (5.32'),

$$\|T_n x - x\|_\infty \le \frac{h_M^2}{3} \max |s_i''| + \|x-p\|_\infty ,$$

where

$$p(t) = \frac{x_i(t_{i+1}-t) + x_{i+1}(t-t_i)}{h_i} , \quad \text{for} \quad t \in I_i .$$

As in the text (page 149), $\|x-p\|_\infty \le h_M\|\dot{x}\|_\infty$, so that

$$\|T_n x - x\|_\infty \le \frac{4h_M^2}{h_m} \|\dot{x}\|_\infty + h_M\|\dot{x}\|_\infty ,$$

$$\le h_M\|\dot{x}\|_\infty \cdot (4\alpha+1) . \qquad (5.39')$$

It now follows, as in the text, that $T_n x \to x$ for every $x \in C[a,b]$.

5.27 One possibility is to replace condition (4) of eqn (5.31) by:

$$\dot{s}_n(0) = \dot{x}(0); \quad \dot{s}_n(1) = \dot{x}(1) . \qquad (4')$$

With this condition, s_n is called a *complete spline* and it can be shown (e.g. de Boor [1978, p. 63]) that

$$\|s_n - x\|_\infty = O(h^4)$$

if $x \in C^4[a,b]$.

5.28 Let

$$k(x,t) = \frac{1}{2\sqrt{\pi t}} e^{\frac{-x^2}{4t}} , \quad -\infty < x < \infty , \quad t > 0 .$$

Then

(1) $k \in C^{\infty}(R^1 \times R^1_+)$, $R^1_+ = \{t \in R^1 : t > 0\}$.

(2) $$\int_{-\infty}^{\infty} k(x,t)dx = \int_{-\infty}^{\infty} \frac{1}{2\sqrt{\pi t}} e^{-u^2} \cdot 2\sqrt{t}\, du = \frac{1}{\sqrt{\pi}} \int_{-\infty}^{\infty} e^{-u^2} du = 1 ,$$

for all $t > 0$.

$$\left(u = \frac{x}{2\sqrt{t}} , \; du = \frac{dx}{2\sqrt{t}}\right) .$$

(3) Direct computation shows that

$$\frac{\partial k}{\partial t} = \frac{\partial^2 k}{\partial x^2} = k(x,t) \frac{x^2 - 2t}{2t^2} , \text{ if } t > 0 .$$

(4) Remark: Let $f(t) = \int_a^b g(x,t)dx$. If $g(x,\cdot)$ is absolutely continuous for almost all x and $g_t(x,t)$ is Lebesgue measurable with

$$\int_a^b \int_0^T \left| \frac{\partial g}{\partial t}(x,s) \right| dxds < \infty , \text{ then for } 0 < t < T,$$

(i) $$\int_a^b g(x,t)dx = \int_a^b g(x,0)dx + \int_a^b \int_0^t g_t(x,s)dxds.$$

(ii) $$\int_a^b g(x,t)dx = f(t) \text{ is absolutely continuous.}$$

(iii) $$\frac{df}{dt}(t) = \frac{d}{dt} \int_a^b g(x,t)dx = \int_a^b \frac{\partial g}{\partial t}(x,t)dx . \qquad \square$$

To prove this we use Fubini's theorem (p.105) together with the fact that h is absolutely continuous iff h is the indefinite Lebesgue integral of its derivative (Rudin [1966, p.167]). From eqn (5.41),

$$u(x,t) = \int_{-\infty}^{\infty} k(x-z,t)u_0(z)dz .$$

Using the above remark we see that

$$\frac{\partial u}{\partial t}(x,t) = \int_{-\infty}^{\infty} \frac{\partial k}{\partial t}(x-z,t)\,u_0(z)dz \ ,$$

$$\frac{\partial^2 u}{\partial x^2}(x,t) = \int_{-\infty}^{\infty} \frac{\partial^2 k(x-z,t)}{\partial x^2}\,u_0(z)dt \ .$$

It now follows from (3) above that

$$\frac{\partial u}{\partial t} = \frac{\partial^2 u}{\partial x^2} \ .$$

(5) Finally, we show that $u(x,0) = u_0(x)$.

Choose $\varepsilon > 0$. By assumption, $u \in \overline{C(R^1)}$ so that u_0 is bounded and uniformly continuous. Thus, there exists $\delta > 0$ such that $|u_0(z)-u_0(x)| \le \varepsilon$ if $|x-z| \le \delta$. From eqn (5.41) and (2) above, we have, with an obvious notation, that, for any $h > 0$,

$$|u(x,h)-u_0(x)| \le \int_{-\infty}^{+\infty} k(x-z,h)\,|u_0(z)-u_0(x)|dz \ ,$$

$$\le \int_{|z-x|\ge\delta} \{\ \} + \int_{|z-x|<\delta} \{\ \} \ .$$

(a) $$\int_{|z-x|<\delta} \{\ \} \le \int_{|z-x|<\delta} k(x-z,h)\cdot\varepsilon dz \ ,$$

$$\le \varepsilon \int_{-\infty}^{\infty} k(x-z,h)dz = \varepsilon \int_{-\infty}^{\infty} k(z,h)dz = \varepsilon \ .$$

(b) $$\int_{|z-x|\geq\delta} \{\ \} \leq 2\|u_0\|_\infty \int_{|z-x|\geq\delta} k(x-z,h)dz\ ,$$

$$= \frac{4\|u_0\|_\infty}{\sqrt{\pi}} \int_{\delta/2\sqrt{h}}^{\infty} e^{-u^2} du.$$

But $\int_0^\infty e^{-u^2}du = \frac{\sqrt{\pi}}{2}$, so that $\lim_{h\to\infty} \int_{\delta/2\sqrt{h}}^{\infty} e^{-u^2}du = 0$.

□

Combining (a) and (b) above, we see that, for h small enough, $|u(x,h)-u_0(x)| \leq \varepsilon + \varepsilon = 2\varepsilon$.

5.29 5.44: In the hint it is proved that $\|E(t)u_0\| \leq \|u_0\|$. Hence,

$$E(t): X \to X \quad \text{for} \quad 0 \leq t < \infty\ .$$

5.45:

$$(E(t)(\lambda u_0+\mu v_0))(x) = \int_{-\infty}^{\infty} k(x-z,t)(\lambda u_0(z)+\mu v_0(z))dz\ ,$$

$$= \lambda\int_{-\infty}^{\infty} k(x-z,t)u_0(z)dz+\mu\int_{-\infty}^{\infty} k(x-z,t)v_0(z)dz\ ,$$

$$= (\lambda E(t)u_0)(x) + (\mu E(t)v_0)(x)\ ,$$

$$= (\lambda E(t)u_0 + \mu E(t)v_0)(x)\ .$$

5.46:

$$E(s)E(t)u_0(x) = \frac{1}{2\sqrt{\pi s}} \int_{-\infty}^{\infty} e^{\frac{-(x-y)^2}{4s}} \cdot \left\{ \frac{1}{2\sqrt{\pi t}} \int_{-\infty}^{\infty} e^{\frac{-(y-z)^2}{4t}} u_0(z)dz \right\} dy,$$

$$= \frac{1}{4\pi\sqrt{st}} \int_{-\infty}^{\infty} \left\{ \int_{-\infty}^{\infty} e^{-\left[y\frac{\sqrt{t+s}}{2\sqrt{st}}-\alpha\right]^2} dy \right\} e^{-\frac{(x-z)^2}{4(t+s)}} u_0(z)dz\ .$$

where

$$\alpha = \frac{2\sqrt{st}}{\sqrt{s+t}} \cdot \left(\frac{x}{4s} + \frac{z}{4t}\right) .$$

Now

$$\int_{-\infty}^{\infty} e^{-(ay+b)^2} dy = \int_{-\infty}^{\infty} e^{-u^2} \frac{du}{a} = \frac{\sqrt{\pi}}{a} .$$

Setting $a = \sqrt{t+s}/2\sqrt{st}$ and $b = -\alpha$, we obtain

$$(E(s)E(t)u_0)(x) = \frac{1}{4\pi\sqrt{st}} \cdot \sqrt{\pi} \cdot \frac{2\sqrt{st}}{\sqrt{t+s}} \cdot \int_{-\infty}^{\infty} e^{-\frac{(x-z)^2}{4(t+s)}} u_0(z)dz ,$$

$$= \frac{1}{2\sqrt{\pi(t+s)}} \cdot \int_{-\infty}^{\infty} e^{-\frac{(x-z)^2}{4(t+s)}} u_0(z)dz ,$$

$$= (E(s+t)u_0)(x) .$$

5.47: As in eqn (*) of the hint,

$$|(E(t)u_0(x))-u_0(x)|^2 \le \int_{-\infty}^{\infty} k(x-z,t)\{u_0(z)-u_0(x)\}^2 dz ,$$

so that

$$\|E(t)u_0-u_0\|^2 \le \int_{-\infty}^{\infty}\int_{-\infty}^{\infty} k(x-z,t)\{u_0(z)-u_0(x)\}^2 dzdx ,$$

$$= \int_{-\infty}^{\infty}\int_{-\infty}^{\infty} f(t,x,z)dzdx , \quad \text{say.} \qquad (**)$$

Using the inequality $(a+b)^2 \le 2(a^2+b^2)$, it follows that

$$\int_{-\infty}^{\infty}\int_{-\infty}^{\infty} f(t,x,z)dzdx \le 2\int_{-\infty}^{\infty}\int_{-\infty}^{\infty} k(x-z,t)\{u_0^2(z)+u_0^2(x)\}dzdx ,$$

$$= 2\{\|u_0\|^2 + \|u_0\|^2\} = 4\|u_0\|^2 .$$

Also,

$$k(u,t) = \frac{1}{2\sqrt{\pi t}} e^{-\frac{u^2}{4t}} \le k(u,t_0), \text{ for all } t \le t_0 ,$$

so that

$$f(t,x,z) \le f(t_0,x,z) , \quad \text{if} \quad t \le t_0 .$$

Then, by the Dominated Convergence theorem (Chapter 4, Appendix p. 104):

$$\lim_{t\to 0} \int_{-\infty}^{\infty}\int_{-\infty}^{\infty} f(t,x,z)dzdx = \int_{-\infty}^{\infty}\int_{-\infty}^{\infty} \lim_{t\to 0} f(t,x,z)dzdx = 0 ,$$

since, from elementary calculus,

$$\lim_{t\to 0} \frac{e^{-\frac{u^2}{4t}}}{2\sqrt{\pi t}} = 0 .$$

Therefore, for any $\varepsilon > 0$ we can find $\delta > 0$ such that

$$0 \le \int_{-\infty}^{\infty}\int_{-\infty}^{\infty} f(t,x,z)dzdx < \varepsilon^2 , \quad \text{if} \quad 0 < t < \delta .$$

Noting eqn (**),

$$\|E(t)u_0 - u_0\| < \varepsilon , \quad \text{if} \quad 0 < t < \delta .$$

5.48: By the hint:

$$\|E(t)u_0\|^2 \le \|u_0\|^2 \Rightarrow \|E(t)\| \le 1 .$$

5.30 Assume that $\frac{\partial u}{\partial t} = \frac{\partial^2 u}{\partial x^2}$ for $(x,t) \in R^1 \times (0,T)$, and that $\psi \in C_0^\infty[R^1 \times (0,T)]$. Then

$$\int_0^T \int_{-\infty}^{\infty} (\psi \frac{\partial u}{\partial t} - \psi \frac{\partial^2 u}{\partial x^2}) dxdt = 0 . \qquad (*)$$

Also

$$\int_0^T \int_{-\infty}^{\infty} \psi \frac{\partial u}{\partial t} dxdt = \int_{-\infty}^{\infty} \int_0^T \psi \frac{\partial u}{\partial t} dtdx ,$$

$$= \int_{-\infty}^{\infty} \left[\psi u \Big|_0^T - \int_0^T u \frac{\partial \psi}{\partial t} dt \right] dx ,$$

$$= -\int_{-\infty}^{\infty} \int_0^T u \frac{\partial \psi}{\partial t} dtdx , \qquad (**)$$

because ψ has compact support in $R^1 \times (0,T)$, so that $\psi(x,0) = \psi(x,T) = 0$ for all $x \in R^1$.

Similarly,

$$\int_0^T \int_{-\infty}^{\infty} \psi \frac{\partial^2 u}{\partial x^2} dxdt = \int_0^T \int_{-\infty}^{+\infty} \psi_{xx} u \, dxdt . \qquad (\overset{**}{*})$$

Combining eqns (*), (**), and ($\overset{**}{*}$), we see that u is a weak solution.

5.31 (1) It is readily verified that if $u_0(x) = \sin kx$ then the solution of the backwards heat equation is $u(x,t) = e^{+k^2 t} \sin kx$. With the same initial data, the solution of the heat equation is $e^{-k^2 t} \sin kx$. Thus, the former grows exponentially, while the latter decays exponentially.

(2) It suffices to replace λ by $-\lambda$ in eqn (5.81). The amplification matrix becomes

$$G(k,\Delta t) = \frac{1+(1-\theta)s}{1-\theta s} , \quad s = 4\lambda(\sin \frac{k\Delta x}{2})^2 ,$$

so that

$$\sup_k \| [G(k,\Delta t)]^n \| > |1+4\lambda|^n ,$$

and hence, from Theorem 5.10, the numerical method is unstable.

5.32 The DuFort-Frankel method is:

$$\frac{U(x,t+\Delta t)-U(x,t-\Delta t)}{2\Delta t} = \frac{U(x+\Delta x,t)+U(x-\Delta x,t)-U(x,t+\Delta t)-U(x,t-\Delta t)}{(\Delta x)^2} . \quad (*)$$

Define: $V(x,t) = \begin{pmatrix} U(x,t) \\ U(x,t-\Delta t) \end{pmatrix} = \begin{pmatrix} V_1(x,t) \\ V_2(x,t) \end{pmatrix}$;

then eqn (*) can be rewritten:

$$V_1(x,t+\Delta t) = V_2(x,t) + \mu\{V_1(x+\Delta x,t)+V_1(x-\Delta x,t)-V_1(x,t+\Delta t)-V_2(x,t)\} , \quad (**)$$

where $\mu = \dfrac{2\Delta t}{(\Delta x)^2}$.

But $V_2(x,t+\Delta t) = V_1(x,t)$, so that eqn (**) can be written as:

$$V_1(x,t+\Delta t) = V_2(x,t) + \mu\{V_2(x+\Delta x,t+\Delta t)+V_2(x-\Delta x,t+\Delta t)-V_1(x,t+\Delta t) - V_2(x,t)\} , \quad (***)$$

$$V_2(x,t+\Delta t) = V_1(x,t) ,$$

or in matrix form:

$$\begin{pmatrix} 1+\mu & 0 \\ 0 & 1 \end{pmatrix} V(x,t+\Delta t) + \begin{pmatrix} 0 & -\mu \\ 0 & 0 \end{pmatrix} [V(x+\Delta x,t+\Delta t)+V(x-\Delta x,t+\Delta t)] = \begin{pmatrix} 0 & 1-\mu \\ 1 & 0 \end{pmatrix} V(x,t) .$$

So, with the notation of eqn (5.77),

$$H_1 = \begin{pmatrix} 1+\mu & 0 \\ 0 & 1 \end{pmatrix} + \begin{pmatrix} 0 & -\mu \\ 0 & 0 \end{pmatrix} \cdot \{e^{ik\Delta x} + e^{-ik\Delta x}\} ,$$

$$= \begin{pmatrix} 1+\mu & -2\mu\cos(k\Delta x) \\ 0 & 1 \end{pmatrix} ,$$

$$H_0 = \begin{pmatrix} 0 & 1-\mu \\ 1 & 0 \end{pmatrix} ,$$

$$H_1^{-1} = \frac{1}{1+\mu} \begin{pmatrix} 1 & 2\mu\cos(k\Delta x) \\ 0 & 1+\mu \end{pmatrix} ,$$

$$G(k,\Delta t) = H_1^{-1} H_0 = \frac{1}{1+\mu} \begin{pmatrix} 2\mu\cos(k\Delta x) & 1-\mu \\ 1+\mu & 0 \end{pmatrix} .$$

5.33 $\Leftarrow$: Setting $w = (v,\underline{0})$ in eqn (5.71), we see that eqn (5.71) implies eqn (5.69).

$\Rightarrow$: Assume that $C: X \to X$ and $\|Cu\| \le K\|u\|$ for all $u \in X$. Let $w \in B$ iff $w = v_1 + iv_2$, $v_1, v_2 \in X$, $i = \sqrt{-1}$, and set $\|w\|^2 = \|v_1\|^2 + \|v_2\|^2$. Define $C: B \to B$ by $Cw = Cv_1 + iCv_2$. Then $\|Cw\|^2 = \|Cv_1\|^2 + \|Cv_2\|^2 \le K^2(\|v_1\|^2 + \|v_2\|^2) = K^2\|w\|^2$, so that $\|Cw\| \le K\|w\|$.

5.34 The explicit method of eqn (5.51) is:

$$U(x,t+\Delta t) = U(x,t) + \lambda\{U(x+\Delta x,t) - 2U(x,t) + U(x-\Delta x,t)\} ,$$

where $\lambda = \Delta t/(\Delta x)^2$. Using the translation operators $T^{(\beta)}$ we obtain,

$$C(\Delta t)U(t) = U(t) + \lambda\{T^{(1)}U(t) - 2T^{(0)}U(t) + T^{(-1)}U(t)\} .$$

Hence, as a special case of eqns (5.75) through (5.77),

$$H_1 = 1 \; ,$$

$$H_0 = 1 + \lambda\{e^{ik\Delta x} - 2 + e^{-ik\Delta x}\} \; ,$$

$$= 1 - 4\lambda \sin^2(\frac{k\Delta x}{2}) \; .$$

$$G(k,\Delta t) = H_1^{-1}H_0 = 1 - 4\lambda \sin^2(\frac{k}{2} \cdot (\frac{\Delta t}{\lambda})^{\frac{1}{2}}) \; .$$

5.35 Let $p = 1$. It was shown in the text (page 166) that the von Neumann condition is necessary for stability. Assume that the von Neumann condition holds. Then

$$|R(k,\Delta t)| \le 1 + K\Delta t \; ,$$

for some constant K . Since $G(k,\Delta t)$ is a 1×1 matrix,

$$\sup_k \| [G(k,\Delta t)]^n \| = \sup_k |R(k,\Delta t)|^n \; ,$$

$$\le [\exp(K\Delta t)]^n \; ,$$

$$\le \exp(KT) \; .$$

Applying Theorem 5.10, it follows that the method is stable.

5.36 We write eqn (5.88) in the form

$$T_1 = \tau + \lambda\theta T_2 + \lambda(1-\theta)T_3 \; . \qquad (*)$$

Then

$$T_1 = \sum_{k=1}^{3} \frac{(\Delta t)^k}{k!} \frac{\partial^k v(x,t)}{\partial t^k} + \frac{(\Delta t)^4}{4!} \bar{v}_{tttt} \; .$$

Expanding T_2 and T_3 in powers of Δt ,

$$T_2 = S_0(t) + \Delta t \; S_1(t) + \frac{(\Delta t)^2}{2!} S_2(t) + \frac{(\Delta t)^3}{3!} S_3(\bar{t}) \; ,$$

$$T_3 = S_0(t) \; ,$$

where

$$S_0(t) = v(x+\Delta x,t) - 2v(x,t) + v(x-\Delta x,t) ,$$

$$S_1(t) = v_t(x+\Delta x,t) - 2v_t(x,t) + v_t(x-\Delta x,t) ,$$

$$S_2(t) = v_{tt}(x+\Delta x,t) - 2v_{tt}(x,t) + v_{tt}(x-\Delta x,t) ,$$

$$S_3(t) = v_{ttt}(x+\Delta x,t) - 2v_{ttt}(x,t) + v_{ttt}(x-\Delta x,t) .$$

Expanding S_0 through S_3 in Taylor series, we obtain

$$S_j(t) = 2\sum_{k=1}^{3-j} \frac{(\Delta x)^{2k}}{(2k)!} \frac{\partial^{2k+j}v(x,t)}{\partial x^{2k}\partial t^j} + 2\,\frac{(\Delta x)^{8-2j}}{(8-2j)!}\,\frac{\partial^{8-j}v(\bar{x},t)}{\partial x^{8-2j}\partial t^j} ,$$

$$0 \le j \le 3 .$$

Substituting into eqn (*) we obtain

$$\tau = T_1 - \lambda S_0(t) - \lambda\theta\Delta t S_1(t) - \lambda\theta\,\frac{(\Delta t)^2}{2!}\,S_2(t) - \lambda\theta\,\frac{(\Delta t)^3}{3!}\,S_3(\bar{t}) ,$$

which is equivalent to eqn (5.89).

Equation (5.89) is of the form,

$$\tau = \frac{1}{2}\,v_{tt}(\Delta t)^2 - \frac{2\lambda}{4!}\,v_{xxxx}(\Delta x)^4 - \lambda\theta v_{xxt}(\Delta x)^2\Delta t + O((\Delta t)^3) .$$

Since $v_t = v_{xx}$, we obtain, by differentiation,

$$v_{tt} = v_{xxt} = v_{xxxx} . \qquad (**)$$

Using these relations, and setting $\Delta t = \lambda(\Delta x)^2$, we obtain

$$\tau = (\Delta x)^4[\frac{\lambda^2}{2} - \frac{\lambda}{12} - \lambda^2\theta]v_{tt} + O((\Delta x)^6) ,$$

so that $\tau = O((\Delta x)^6)$ provided that

$$\theta = \frac{1}{2} - \frac{1}{12\lambda} . \qquad (\overset{*}{_*})$$

Assuming that eqn ($\overset{*}{_*}$) holds, and using identities such as those in eqn (**), eqn (5.89) becomes

$$f(x) = \sum_{i=1}^{m} |b_i - \sum_{j=1}^{n} a_{ij} x_j| , \qquad (*)$$

where the coefficients b_i and a_{ij} are given. This problem is a discrete version of a continuous problem: Find $x \in R^n$ which minimizes

$$F(x) = \int_0^1 |b(t) - \sum_{j=1}^{n} x_j \phi_j(t)| dt , \qquad (**)$$

where the functions $b(t)$ and $\phi_j(t)$ are given. We obtain problem (*) from (**) by choosing m equidistant points $t_1,\ldots,t_m$ in $[0,1]$, replacing the integral by a sum over the points t_i, and setting $b_i = b(t_i)$, $a_{ij} = \phi_j(t_i)$.

For details about the computational solution of problem (*) see Barrodale and Roberts [1978].

REMARK 4.1. In the preceding discussion of R^n, the value of n did not play a rôle. It should, however, be realized that there are some subtle differences between the spaces R^n of different dimensions:

(1) Among all the spaces R^n, only R^1 is an *ordered topological vector space;* that is, only in R^1 can an order relation $\geq$ be introduced which is defined for all pairs of elements.

If $f : R^1 \to R^1$ is continuous then the equation $f(x) = 0$ can be solved by the *bisection method:*

(a) Find x_p, $x_n \in R^1$ such that $f(x\ ) \geq 0$ and $f(x_n) < 0$.

(b) Compute $x_m = (x_p+x_n)/2$. If $f(x_m) \geq 0$ replace x_p by x_m; otherwise replace x_n by x_m.

(c) Repeat step (b) until x_p and x_n are as close as desired. At each step there exists x lying between x_p and x_n such that $f(x) = 0$.

The bisection method can only be used in R^1 because it depends upon the fact that R^1 is ordered; in

(1) $\tau = \frac{1}{2} \Delta t \, \bar{v}_{tt} + \frac{(\Delta x)^2}{6} \bar{v}_{xxx}$.

(2) $\tau = \frac{1}{2} \Delta t \, \bar{v}_{tt} + \frac{\Delta x}{2} \bar{v}_{xx}$.

(3) $\tau = \frac{1}{2} \Delta t \, \bar{v}_{tt} + \frac{\Delta x}{2} \bar{v}_{xx}$.

(4) $\tau = \frac{1}{6} (\Delta t)^2 \bar{v}_{ttt} + \frac{1}{6} (\Delta x)^2 \bar{v}_{xxx}$.

(b) Computation of $G(k,\Delta t)$

The matrices H_0, H_1, and $G(k,\Delta t)$ of eqns (5.75) and (5.76) can be computed for each method:

(1) $H_1 = 1$,

$$H_0 = 1 - \frac{\lambda}{2} (e^{ik\Delta x} - e^{-ik\Delta x}) ,$$

$$= 1 - i\lambda \sin(k\Delta x) ,$$

$$G(k,\Delta t) = 1 - i\lambda \sin\left(\frac{k\Delta t}{\lambda}\right) .$$

(2) $H_1 = 1$,

$$H_0 = 1 - \lambda \cdot \{e^{ik\Delta x} - 1\} = 1 + \lambda - \lambda e^{ik\Delta x} ,$$

$$G = 1 + \lambda - \lambda \exp\left(\frac{ik\Delta t}{\lambda}\right) = 1 + \lambda - \lambda \cos\left(\frac{k\Delta t}{\lambda}\right) - i\lambda \sin\left(\frac{k\Delta t}{\lambda}\right) .$$

(3) $H_1 = 1$,

$$H_0 = 1 - \lambda(1 - e^{ik\Delta x}) ,$$

$$G(k,\Delta t) = 1 - \lambda + \lambda \exp\left(\frac{-ik\Delta t}{\lambda}\right) ,$$

$$= 1 - \lambda + \lambda \cos\left(\frac{k\Delta t}{\lambda}\right) - i\lambda \sin\left(\frac{k\Delta t}{\lambda}\right) .$$

(4) Set

$$V(x,t) = \begin{pmatrix} V_1(x,t) \\ V_2(x,t) \end{pmatrix} = \begin{pmatrix} U(x,t) \\ U(x,t-\Delta t) \end{pmatrix} .$$

We can write:

$$V_1(x,t+\Delta t) - V_2(x,t) + \lambda\{V_1(x+\Delta x,t) - V_1(x-\Delta x,t)\} = 0 ,$$
$$V_2(x,t+\Delta t) = V_1(x,t) ,$$

or

$$\begin{pmatrix} 1 & 0 \\ 0 & 1 \end{pmatrix} V(x,t+\Delta t) = \begin{pmatrix} -\lambda & 0 \\ 0 & 0 \end{pmatrix} \{V(x+\Delta x,t) - V(x-\Delta x,t)\} + $$
$$+ \begin{pmatrix} 0 & 1 \\ 1 & 0 \end{pmatrix} V(x,t) .$$

Thus,

$$H_1 = \begin{pmatrix} 1 & 0 \\ 0 & 1 \end{pmatrix} ,$$

$$H_0 = \begin{pmatrix} -\lambda & 0 \\ 0 & 0 \end{pmatrix} \cdot (e^{ik\Delta x} - e^{-ik\Delta x}) + \begin{pmatrix} 0 & 1 \\ 1 & 0 \end{pmatrix} ,$$

$$= \begin{pmatrix} -2i\lambda \sin(\frac{k\Delta t}{\lambda}) & 1 \\ 1 & 0 \end{pmatrix} .$$

$$G(k,\Delta t) = H_0 .$$

(c) <u>Consistency</u>

In the transformed domain, the original equation becomes

$$\frac{d\hat{w}}{dt}(k,t) = -ik\,\hat{w}(k,t) ,$$

so that

$$(\hat{E}(t)\hat{w}_0)(k) = e^{-ikt}\hat{w}_0(k) ,$$

and

$$(\hat{E}(\Delta t)v)(k) = [1 - ik\Delta t - \frac{k^2(\Delta t)^2}{2} + O(k\Delta t)^3)]v(k) . \qquad (**)$$

For methods (1) to (3) we find that

$$G(k,\Delta t) = 1 - ik\Delta t + O((k\Delta t)^2) ,$$

so that, by the arguments of Example 5.10 (page 170), all these methods are consistent.

To handle method (4) it is necessary to extend the technique used in the text. For a one-step method we test consistency by considering the error $[\hat{C}(\Delta t)-\hat{E}(\Delta t)]\hat{w}_0$ after one time step, starting with the correct values $\hat{w}_0$ at time $t = 0$. Similarly, for a two step method we must consider the error, W say, after one time step starting with the correct values at $t = 0$ and $t = \Delta t$. For method (4),

$$W = G(k,\Delta t)\begin{pmatrix}\hat{E}(\Delta t)\hat{w}_0\\ \\ \hat{w}_0\end{pmatrix} - \begin{pmatrix}\hat{E}(2\Delta t)\hat{w}_0\\ \\ \hat{E}(\Delta t)\hat{w}_0\end{pmatrix} .$$

By direct computation,

$$W = G(k,\Delta t)\begin{pmatrix}e^{-ik\Delta t}\\ \\ 1\end{pmatrix}\hat{w}_0 - \begin{pmatrix}e^{-2ik\Delta t}\\ \\ e^{-ik\Delta t}\end{pmatrix}\hat{w}_0 ,$$

$$= \begin{pmatrix}-2i\lambda \sin(\frac{k\Delta t}{\lambda})e^{-ik\Delta t} + 1 - e^{-2ik\Delta t}\\ \\ 0\end{pmatrix}\hat{w}_0 ,$$

$$= O((k\Delta t)^2) ,$$

so that method (4) is also consistent.

(d) Stability

For each method, we apply Theorem 5.10.

(1) $$\|G(k,\Delta t)\| = \sqrt{1+\lambda^2 \sin^2(\frac{k\Delta t}{\lambda})} ,$$

$$\sup_k \|[G(k,\Delta t)]^n\| = [\sup_k \|G(k,\Delta t)\|]^n = (1+\lambda^2)^{n/2} .$$

Therefore, there is no K such that

$$\|G(k,\Delta t)\|^n \le K , \text{ if } 0 \le n\Delta t \le T ,$$

and the method is <u>not stable</u>.

(2)

$$\|G(k,\Delta t)\| = \sqrt{(1+\lambda-\lambda \cos(\frac{k\Delta t}{\lambda}))^2 + (\lambda \sin(\frac{k\Delta t}{\lambda}))^2} ,$$

$$\sup_k \|[G(k,\Delta t)]^n\| = [\sup_k \|[G(k,\Delta t)\|]^n \ge (1+2\lambda)^n ,$$

so that the method is <u>not stable</u>.

(3)

$$\|G(k,\Delta t)\| \le |1-\lambda| + |\lambda e^{\frac{-ik\Delta t}{\lambda}}| = |1-\lambda| + |\lambda| .$$

Thus the method <u>is stable</u> if $\lambda \le 1$.

If $\lambda > 1$ then

$$\sup_k \|G(k,\Delta t)\| = \sup_k [(1-\lambda + \lambda (\cos \frac{k\Delta t}{\lambda}))^2 + (\lambda \sin(\frac{k\Delta t}{\lambda}))^2]^{\frac{1}{2}} ,$$

$$\ge |1-2\lambda| > 1 ,$$

and the method is <u>not stable</u>.

(4) $$G(k,\Delta t) = \begin{pmatrix} -2i\beta & 1 \\ 1 & 0 \end{pmatrix} ,$$

where $\beta = \lambda \sin(\frac{k\Delta t}{\lambda})$. The characteristic equation for G is

$$\mu^2 + 2i\beta\mu - 1 = 0 ,$$

so that the eigenvalues are

$$\mu = -i\beta \pm \sqrt{1-\beta^2} .$$

From the von Neumann condition, we see that for stability it is necessary that $|\beta| \le 1$, which implies that $|\lambda| \le 1$.

Now assume that $|\lambda| \le 1$, and let $\mu_1 = \mu_1(k,\Delta t)$ and $\mu_2 = \mu_2(k,\Delta t)$ be the eigenvalues of $G(k,\Delta t)$. By elementary matrix analysis,

$$G(k,\Delta t) = S\,D S^{-1} ,$$

where $D = \mathrm{diag}(\mu_1,\mu_2)$ and

$$S(k,\Delta t) = \begin{pmatrix} \mu_1 & \mu_2 \\ 1 & 1 \end{pmatrix} , \quad S^{-1}(k,\Delta t) = \frac{1}{\mu_1-\mu_2} \begin{pmatrix} 1 & -\mu_2 \\ -1 & \mu_1 \end{pmatrix} .$$

Hence,

$$[G(k,\Delta t)]^n = S\,D^n S^{-1} .$$

Since $|\lambda| < 1$, we have $|\mu_1-\mu_2| \ge 2\sqrt{1-\lambda^2} > 0$ and $|\mu_1| = |\mu_2| = 1$.
Thus, S and S^{-1} are bounded.
$D^n = \mathrm{diag}(\mu_1^n,\mu_2^n)$ is also bounded. Thus G^n is bounded, and hence the method is stable.

5.38 *Claim:* $y_n = \sum_{i=0}^{n-1} \binom{n-1}{i} (1-\alpha)^i \alpha^{n-1-i} x_{i+1}$.

Proof: By induction.

(1) $y_1 = x_1$, so true if $n = 1$.

(2) Assume true for n . Then

$$\begin{aligned} y_{n+1} &= \alpha y_n + (1-\alpha)Ty_n , \\ &= \alpha \sum_{i=0}^{n-1} \binom{n-1}{i} (1-\alpha)^i \alpha^{n-1-i} x_{i+1} + \\ &\qquad + (1-\alpha) \sum_{i=0}^{n-1} \binom{n-1}{i} (1-\alpha)^i \alpha^{n-1-i} x_{i+2} , \\ &= \alpha^n x_1 + \sum_{i=1}^{n-1} \left[\binom{n-1}{i} + \binom{n-1}{i-1} \right] (1-\alpha)^i \alpha^{n-i} x_{i+1} + \\ &\qquad + (1-\alpha)^n x_{n+1} , \\ &= \alpha^n x_1 + \sum_{i=1}^{n-1} \binom{n}{i} (1-\alpha)^i \alpha^{n-i} x_{i+1} + (1-\alpha)^n x_{n+1} , \\ &= \sum_{i=0}^{n} \binom{n}{i} (1-\alpha)^i \alpha^{n-i} x_{i+1} . \quad \square \end{aligned}$$

Thus, if

$$a_{ni} = \binom{n-1}{i} \cdot (1-\alpha)^i \alpha^{n-1-i} ,$$

then

$$y_n = \sum_{i=0}^{n-1} a_{ni} x_{i+1} .$$

If $n > N$,

$$\sum_{i=0}^{N-1} a_{ni} < N\alpha^{n-N} n^N \to 0 , \quad \text{as} \quad n \to \infty .$$

Let $\{x_n\}$ be a convergent sequence in X , $x_n \to \bar{x}$ say. Then $\|x_n\| \le M$ for some $M > 1$ and all n . Choose $\varepsilon > 0$. Now choose N such that

(1) $\|x_{i+1} - \bar{x}\| \le \varepsilon$ if $i \ge N$,

(2) $\displaystyle\sum_{i=0}^{N-1} a_{ni} < \varepsilon/(M+\|\bar{x}\|)$, if $n \ge N$.

Then, for $n > N$,

$$\begin{aligned} \|y_n - \bar{x}\| &\le \sum_{i=N}^{n-1} a_{ni} \|x_{i+1} - \bar{x}\| + \sum_{i=0}^{N-1} a_{ni} \|x_{i+1} - \bar{x}\| , \\ &\le \varepsilon \sum_{i=0}^{n-1} a_{ni} + (M+\|\bar{x}\|) \sum_{i=0}^{N-1} a_{ni} , \\ &\le \varepsilon + \varepsilon . \end{aligned}$$

5.39 We use the Banach-Steinhaus theorem.

(1) If p is a polynomial of degree n then $\Delta_h^j p(t) = 0$ for $j > n$. Consequently, the formal expansion (*) is exact for polynomials:

$$T_j p = \dot{p}(0) \,, \quad \text{for} \quad j > n \,.$$

(This could be proved directly by differentiating Newton's interpolation formula with forward differences:

$$f(t) \sim \sum_{n=0}^{\infty} \frac{1}{n!h^n} \prod_{j=0}^{n-1} (t-jh)\Delta_h^n f(0).)$$

The polynomials are dense in $X = C^1[0,1]$ (Problem 4.15), so that $T_n x \to \dot{x}(0)$ on a dense subset of X.

(2) Consider the function y which is piecewise linear and satisfies, for $h = 1/n$,

$$y(kh) = \begin{cases} 0, & k \text{ an even integer,} \\ h, & k \text{ an odd integer.} \end{cases} \qquad (**)$$

Then, it is readily seen that

$$\frac{1}{h} \Delta_h y_n(kh) = (-1)^k \,,$$

so that

$$\frac{1}{h} \Delta_h^j y_n(0) = 2^{j-1}(-1)^{j-1}, \quad j \geq 1 \,.$$

Consequently,

$$\frac{1}{h} \sum_{j=1}^{n} \frac{(-1)^j}{j} \Delta_h^j y(0) = \sum_{j=1}^{n} \frac{2^{j-1}}{j} \,. \qquad (\overset{*}{*}*)$$

The function y_n does not belong to X, but, as in Example 5.6, we can construct $x_\varepsilon \in X$ by rounding the corners of y_n. Bearing eqns (**) and (*$\dot{*}$) in mind, we see that such an x_ε may be chosen so that:

$$\|x_\varepsilon;C^1[0,1]\| = \max|x_\varepsilon(t)| + \max|\dot{x}_\varepsilon(t)| \le 2 ,$$

$$\|T_n x_\varepsilon\| = \sum_{j=1}^{n} \frac{2^{j-1}}{j} .$$

Thus, $\|T_n\| \to \infty$ as $n \to \infty$, and, by the Banach-Steinhaus theorem, there exists $x \in X$ such that $T_n x \not\to \dot{x}(0)$ as $n \to \infty$.

5.40 (1) There exists an eigenvalue λ and eigenvector u of A such that $|\lambda| = \sigma(A)$. But

$$\|A\| = \sup_{x\neq 0} \frac{\|Ax\|}{\|x\|} \ge \frac{\|Au\|}{\|u\|} = |\lambda| = \sigma(A) .$$

(2) It is a standard result in matrix theory, that every complex matrix is similar to an upper triangular matrix. Hence, $A = S^{-1}US$ where S is a non-singular $n\times n$ complex matrix and U is upper triangular.

For $\delta > 0$ let $D = \text{diag}\,(\delta^n,\delta^{n-1},\ldots,\delta^1)$, and set $V(\delta) = DUD^{-1}$. If $U = (u_{ij})$ then $V = (v_{ij})$ where

$$v_{ij} = \delta^{j-i}u_{ij} .$$

Consequently, $V(\delta) = \text{diag}(u_{ii}) + O(\delta)$, so that, for small δ, $V(\delta)$ is 'almost diagonal'.

Now define the norm $\|\cdot\|_\wedge$ by

$$\|x\|_\wedge = \|DSx\| ,$$

where

$$\|x\| = [\sum|x_i|^2]^{\frac{1}{2}} .$$

Then

$$\|A\|_{\wedge} = \sup_{\|x\|_{\wedge}=1} \|Ax\|_{\wedge} = \sup_{\|DSx\|=1} \|DSAx\| ,$$

$$= \sup_{\|y\|=1} \|DSAS^{-1}D^{-1}y\| , \quad (y = DSx)$$

$$= \sup_{\|y\|=1} \|DUD^{-1}y\| = \sup_{\|y\|=1} \|V(\delta)y\| ,$$

$$= \sup_{\|y\|=1} [\sum_{i=1}^{n} |u_{ii}|^2 |y_i|^2 + O(\delta)] .$$

Remembering that A and U are similar, each u_{ii} is equal to an eigenvalue of A, and so we can conclude from the preceding inequality that

$$\|A\|_{\wedge} \leq \sigma(A) + O(\delta) .$$

CHAPTER 6: SOLUTIONS

6.1 $\Rightarrow$: Let $\{H_\alpha : \alpha \in A\}$ be an open cover of S in S.

For each α there exists an open G_α in X such that:

$$H_\alpha = S \cap G_\alpha .$$

$S \subset \cup H_\alpha \subset \cup G_\alpha \Rightarrow \{G_\alpha\}$ is an open cover of S in X. Since S is compact in X, there exist $\alpha_1, \ldots, \alpha_m$ such that $S \subset \bigcup_{i=1}^{m} G_{\alpha_i}$. Hence,

$$S \subset S \cap \left(\bigcup_{i=1}^{m} G_{\alpha_i} \right) = \bigcup_{i=1}^{m} (G_{\alpha_i} \cap S) = \bigcup_{i=1}^{m} H_{\alpha_i}$$

Therefore we have extracted a finite subcover of S from the arbitrary open cover $\{H_\alpha\}$ so that S is compact in S.

$\Leftarrow$: If $\{G_\alpha\}$ is an open cover of S in X then $\{H_\alpha\}$ with $H_\alpha = G_\alpha \cap S$ is an open cover of S in S. There exists a finite subcover $\{H_\alpha : \alpha = \alpha_1, \alpha_2, \ldots, \alpha_m\}$. But

$$S \subset \bigcup_{i=1}^{m} H_{\alpha_i} \Rightarrow S \subset \bigcup_{i=1}^{m} G_{\alpha_i} ,$$

and thus S is compact in X.

6.2 $\Rightarrow$: Let N be a neighbourhood of x. Choose $B_1 = B(x;r_1) \subset N$, and $x_{n_1} \in B_1$. For $i = 2,3,\ldots$, we can construct $r_i = \frac{1}{2}\rho(x, x_{n_{i-1}})$, $B_i = B(x;r_i)$, and $x_{n_i} \in B_i$. Clearly, $n_i \neq n_j$ for $i \neq j$. All the points x_i belong to N.

$\Leftarrow$: Let $B_i = B(x;2^{-i})$. B_1 contains infinitely many x_n; choose one, x_{n_1}. For $i = 2,3,\ldots$ choose $x_{n_i} \in B_i$ such that $n_i > \max_{j<i} n_j$; this is possible because B_i contains infinitely many x_n. Then $\{x_{n_i}\}$ is a subsequence of $\{x_n\}$ which converges to x.

6.3 By Remark 6.1, Y_α is a compact space with respect to the topology induced on Y_α by X_α.

If $y_\alpha \in Y_\alpha$ then every neighbourhood $N(y_\alpha;Y_\alpha)$ of y_α in the induced topology is of the form $N(y_\alpha;Y_\alpha) = N(y_\alpha;X_\alpha) \cap Y_\alpha$, where $N(y_\alpha;X_\alpha)$ is a neighbourhood of y_α in X_α.

From Tychonoff's theorem (Theorem 6.4) Y is a compact space with respect to the product topology on Y, which is defined (see page 23) by the base of neighbourhoods $\beta = \{\beta_y\}$,

$$\beta_y = \prod_\alpha N(y_\alpha;Y_\alpha) = \prod_\alpha (N(y_\alpha;X_\alpha) \cap Y_\alpha),$$

$$= [\prod_\alpha N(y_\alpha;X_\alpha)] \cap Y .$$

We conclude that the product topology on Y is the same as the topology on Y induced by X. Thus, $Y \subset X$ is compact with respect to its relative topology, and hence, by Remark 6.1, Y is a compact subset of X.

6.4 Assume that S is sequentially compact but not bounded; then there exists a sequence $\{x_n\}$ such that: $x_n \in S$ and $|x_n| \geq n$, $n = 1,2,\ldots$. Since S is sequentially compact, there exists a subsequence $\{x_{n_i}\}$ such that:

$x_{n_i} \to x$ in R^1. Therefore, for some I,

$|x_{n_i} - x| < 1$, if $i \geq I$. This implies that

$|x_{n_i}| \leq |x| + |x_{n_i} - x| \leq |x| + 1$, if $i \geq I$,

which is a contradiction. Consequently, S is bounded.

6.5 The proof is sketched in the description of the problem.

6.6 The proof is contained in that of Theorem 6.4 (Tychonoff).

6.7 (a) If $G \subset E^2$ and $x \in [0,1]$ we denote by G_x^1 the one-dimensional 'cross-section' of G with the line segment $L_x = \{x\} \times [0,1]$. That is,

$$G_x^1 = \{y \in R^1 : (x,y) \in G\} .$$

We claim that if $G \subset E^2$ is open then $G_x^1 \subset R^1$ is open. To see this, let $y \in G_x^1$, so that $(x,y) \in G$. The topology on E^2 is the product topology, so that there exists a base element of the form $N_1 \times N_2$ such that N_1 and N_2 are open subsets of R^1 and $(x,y) \in N_1 \times N_2 \subset G$. Then, from the definition of G_x^1 , $y \in N_2 \subset G_x^1$. y was an arbitrary point of G_x^1 , and so, by Theorem 2.1, G_x^1 is open.

(b) Let κ be an open cover of $[0,1] \times [0,1]$, and let $x \in [0,1]$. Set

$$\kappa_x^1 = \{G_x^1 : G \in \kappa\} .$$

κ_x^1 is an open cover of $[0,1]$ (which is known to be compact). Extracting a finite subcover of κ_x^1 $\kappa_x^{1\prime}$ say, we obtain a finite subcover of κ ,

$$\kappa_x = \{G \in \kappa : G_x^1 \in \kappa_x^{1\prime}\} .$$

Then κ_x is a finite cover of L_x .

(c) Let κ be an open cover of $[0,1] \times [0,1]$. Noting Theorem 2.1, and the definition of the product topology on E^2 , each $G \in \kappa$ is the union of a possibly infinite number of open squares $S(\alpha;G)$,

$$G = \bigcup_{\alpha \in A(G)} S(\alpha;G) .$$

Set

$$\kappa^S = \{S : S = S(\alpha;G), \text{ for some } G \in \kappa \text{ and some } \alpha \in A(G)\} .$$

κ^S is an open cover of $[0,1] \times [0,1]$. As in (b) above, we may assert that, for each $x \in [0,1]$, there exists a finite open subcover κ_x^S of the line segment L_x . Remembering that each element of κ_x^S is a square, it follows from elementary geometrical arguments that there exists $\delta_x > 0$ such that κ_x^S is an open cover of the strip

$$S_x = \{(u,v) \in E^2 : u \in I_x \equiv (x-\delta_x, x+\delta_x) \cap [0,1],\ v \in [0,1]\} .$$

The intervals I_x form an open cover σ of the compact set $[0,1]$. We choose a finite subcover, σ' say.

Let

$$\kappa^{S'} = \{S \in \kappa^S : S \in \kappa_x^S \text{ and } I_x \in \sigma'\} .$$

Then $\kappa^{S'}$ is a finite subcover of $[0,1] \times [0,1]$. Each S in $\kappa^{S'}$ is contained in at least one $G \in \kappa$. Construct κ' by adjoining one $G \supset S$ to κ' for each $S \in \kappa^{S'}$. Then κ' is a finite subcover of $[0,1] \times [0,1]$.

(d) The reason that this proof was given was to illustrate the difference in difficulty between different methods of proof. For this problem, an indirect proof by contradiction using sequences is considerably simpler than the above 'direct' proof by 'construction'.

6.8 $$\lambda(x) = \begin{bmatrix} \lambda_1(x) \\ \vdots \\ \lambda_n(x) \end{bmatrix} \in E^n , \quad x = \sum_{i=1}^{n} \lambda_i(x) x_i \in X_0 .$$

In the arguments below we repeatedly use the fact that the representation of $x \in X_0$ in terms of the x_i is unique; that is, λ is one-to-one.

(1) <u>λ is linear</u>

$$(\alpha x)+(\beta y) = \sum_{i=1}^{n} \alpha\lambda_i(x)x_i + \sum_{i=1}^{n} \beta\lambda_i(x)x_i = \sum_{i=1}^{n} [\alpha\lambda_i(x)+\beta\lambda_i(y)]x_i ,$$

$$= (\alpha x+\beta y) ,$$

$$= \sum_{i=1}^{n} \lambda_i(\alpha x+\beta y)x_i ,$$

$$\Rightarrow \lambda_i(\alpha x+\beta y) = \alpha\lambda_i(x)+\beta\lambda_i(y) \Rightarrow \lambda \text{ is linear.}$$

(2) λ is onto:

Let z be any point in E^n. Set $x = \sum_{i=1}^{n} z_i x_i \in X_0$.
Then $\lambda(x) = z$.

(3) λ^{-1} is continuous:

For any $z \in E^n$,

$$\|\lambda^{-1}(z)\| \le \sum_{i=1}^{n} |z_i| \cdot \|x_i\| ,$$

$$= \|z\| \cdot \left[\sum_{i=1}^{n} \|x_i\|^2\right]^{\frac{1}{2}} .$$

$$\Rightarrow \quad \|\lambda^{-1}\| \le \left[\sum_{i=1}^{n} \|x_i\|^2\right]^{\frac{1}{2}} \quad \text{AND} \quad \lambda^{-1} \text{ is continuous.}$$

6.9 $\Leftarrow$: $\|y^k\| \le \sum_{i=1}^{n} |\lambda_i^{(k)}| \, \|x_i\| \le \max_{1 \le i \le n} |\lambda_i^{(k)}| \cdot \sum_{i=1}^{n} \|x_i\|$.

$\Rightarrow$: The proof is by contradiction. Assume that $\|y^k\| \le M$, $k = 1,2,\ldots,n$, but that, $|\lambda_{i_0}^{(k)}| \to \infty$ as $k \to \infty$.

Let $\mu_k = \max_i |\lambda_i^{(k)}|$. Then $\mu_k \to \infty$. For each k, there exists at least one i such that $|\lambda_i^{(k)}| = \mu_k$. Remembering that i can only take on the values 1 to n, we conclude that there exists a constant m, and a subsequence still denoted by $\{y^k\}$, such that $|\lambda_m^{(k)}| = \mu_k$, for all k.

Let $S = \prod_{i=1}^{n} [0,1]$, and

$$z^k = \frac{1}{\mu_k} y^k = \sum_{i=1}^{n} \frac{1}{\mu_i} \lambda_i^{(k)} x_i = \sum_{i=1}^{n} \alpha_i^{(k)} x_i , \text{ say.}$$

Then $\alpha^k = (\alpha_1^k, \ldots, \alpha_n^k)$ belongs to the compact set S for all k, and so there exists a subsequence, α^{k_j}, such that

$$\alpha^{k_j} \to \alpha = (\alpha_1, \ldots, \alpha_n) , \text{ say.}$$

This implies that

$$z^k \to \sum_{i=1}^{n} \alpha_i x_i = z \ , \quad \text{say.}$$

But $\|z^k\| \le \|y^k\|/\mu_k \le M/\mu_k \to 0$, so that $z = 0$. However, $\alpha_m = \lim_{k_j} \alpha_m^{\ j} = 1$, so that $\alpha \neq 0$. This is impossible because the x_i are linearly independent.

6.10 Since X is n-dimensional, there exists a basis $x_1,\dots,x_n$ such that every $x \in X$ has a unique representation

$$x = \sum_{j=1}^{n} \lambda_j(x) x_j \ ,$$

where $\lambda_j \in \mathbb{C}^1$. We can now use arguments similar to those of Theorem 6.8.

6.11 (a) Let X_0 be a real (complex) n-dimensional space. By Theorem 6.8 and Problem 6.10, there exists a bicontinuous linear mapping λ of X_0 onto $Y = R^n(\mathbb{C}^n)$.

Let $\{x_k\}$ be a Cauchy sequence in X_0 . Since

$$\|\lambda(x_k)-\lambda(x_\ell)\| \le \|\lambda\| \ \|x_k - x_\ell\| \ ,$$

$\{\lambda(x_k)\}$ is a Cauchy sequence in the complete space Y , and so $\lambda(x_k) \to \mu \in Y$. Then, $x_k \to \lambda^{-1}(\mu) \in X_0$. That is, X_0 <u>is complete</u>.

(b) Let $\tilde{X}_0$ be a copy of X_0 (with the same scalar multiplication, vector addition, and norm). Let λ be an isometric isomorphism of X_0 onto $\tilde{X}_0$.

If $\{x_n\}$ is a sequence in X_0 satisfying $x_n \to x \in X$, then $\{\lambda(x_n)\}$ is a Cauchy sequence in $\tilde{X}_0$. By (a), $\tilde{X}_0$ is complete, so that $\lambda(x_n) \to \tilde{z} \in \tilde{X}_0$. Then $x_n \to z = \lambda^{-1}(\tilde{z}) \in X_0$. That is, X_0 <u>is closed</u>.

6.12 The map $\lambda_i: x \in X \to \lambda_i(x) \in R^1$ is the composition of two mappings: $\lambda_i = p_i \circ \lambda$ where

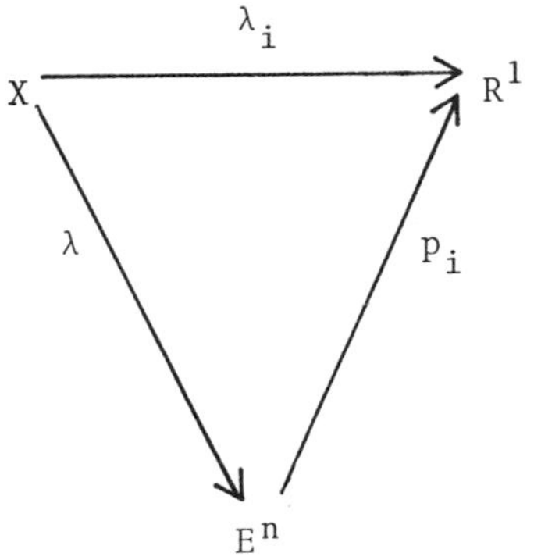

$\lambda: x \in X \to (\lambda_1(x),\ldots,\lambda_n(x)) \in E^n$, and p_i is the projection map, $p_i: (x_1,\ldots,x_n) \in E^n \to x_i \in R^1$.

In Theorem 6.8 it was shown that λ is continuous, while the definition of E^n implies that p_i is continuous (see page 24). Thus $\lambda_i = p_i \circ \lambda$ is continuous (Problem 2.9).

6.13 If $x \in S$, then

$$\|x\| = \sum_{i=0}^{m} \|x^{(i)}\|_\infty = \sum_{i=0}^{m} \|D^i x\|_\infty .$$

We introduce the notation,

$$S_i = \{x^{(i)}: x \in S\} , \quad i = 0,1,\ldots,m .$$

$\Leftarrow$: Since S is bounded, $\|x\| \le M$ for $x \in S$ and some constant M . Let $\{x_n\}$ be a sequence in S .

(a) *Claim:* S_i is a sequentially compact subset of $C[a,b]$ for $0 \le i \le m$.

Proof: S_m is equicontinuous by assumption.

If $i < m$, then, for any $x \in S$, it follows from the mean value theorem that:

$$|x^{(i)}(t_2)-x^{(i)}(t_1)| = |x^{(i+1)}(\xi)(t_2-t_1)| \ ,$$

$$\text{for some } \ \xi \in [t_2,t_1]$$

$$\le \|x^{(i+1)}\|_\infty \ |t_2-t_1| \ , \qquad (*)$$

$$\le M\cdot|t_2-t_1| \ .$$

Consequently, S_i is equicontinuous for $0 \le i \le m$.

Since S is bounded in $C^m[a,b]$, S_i is bounded in $C[a,b]$. Applying the Ascoli-Arzela theorem we conclude that S_i is a sequentially compact subset of $C[a,b]$.

(b) *Claim:* There is subsequence $\{x_{n,0}\}$ of $\{x_n\}$ such that

$$x_{n,0}^{(i)} \equiv D^i x_{n,0} \to F_i \in C[a,b] \quad \text{for} \quad 0 \le i \le m \ . \ (**)$$

Proof: Define $x_{n,m+1} = x_n$. Then use the following recursively for $i = m, m-1,\ldots,0$: Since $\{x_{n,i+1}^{(i)}\} \equiv \{D^i x_{n,i+1}\}$ is a sequence in S_i, which is a sequentially compact subset of $C[a,b]$, there is a subsequence of $\{x_{n,i+1}\}$, $\{x_{n,i}\}$ say, and an $F_i \in C[a,b]$ such that

$$x_{n,i}^{(i)} \to F_i \ , \quad \text{in} \quad C[a,b] \ .$$

(c) If $\{x_{n,0}\}$ satisfies eqn (**), then, for $t \in [a,b]$ and $0 \le i < m$,

$$F_i(t) = \lim_{n\to\infty} x_{n,0}^{(i)}(t) \ ,$$

$$= \lim_{n\to\infty} \left[x_{n,0}^{(i)}(a) + \int_a^t x_{n,0}^{(i+1)}(s)ds \right] \ ,$$

$$= F_i(a) + \int_a^t F_{i+1}(s)ds \ ,$$

from which we conclude that

$$D^i F_0(t) = F_i(t) \ , \quad 1 \le i \le m \ . \qquad (^{*}_{*})$$

(d) Combining eqns (**) and $(^{*}_{*})$, we see that the subsequence $\{x_{n,0}\}$ converges to $F_0 \in C^m[a,b]$. It follows that S is sequentially compact.

⇒: If either (1) or (2) does not hold, then, for some S_i , either S_i is not bounded or S_i is not equicontinuous. It follows from the Ascoli-Arzela theorem that there is a sequence $\{x_n\}$ such that $\{x_n^{(i)}\}$ does not contain a convergent subsequence in $C[a,b]$, so that $\{x_n\}$ cannot contain a convergent subsequence in $C^m[a,b]$.

6.14 The proof, which is omitted, is similar to that of Problem 6.13.

It should be noted that condition (2) cannot be replaced by the weaker condition that S_α be equicontinuous for $|\alpha| = m$, without making assumptions about Ω . The reason is that the inequality (*) of Problem 6.13 requires that the length of the path in Ω joining $t_1, t_2 \in \Omega$ should be bounded by a constant times $\|t_1 - t_2\|$. This is not true for the domain shown below (there are denumerably many 'cuts' of which only three are shown).

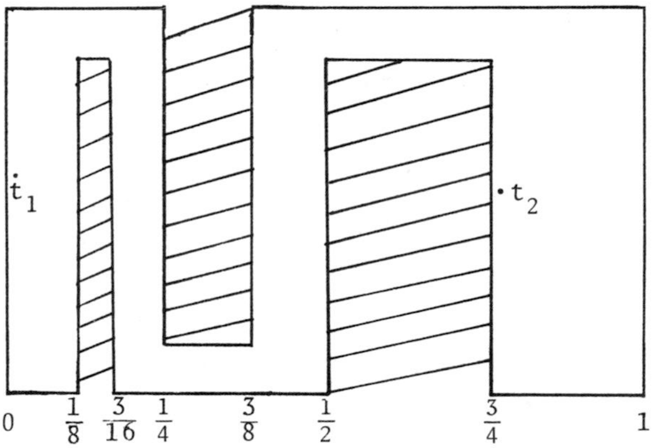

Figure: A bounded domain in R^2 *such that the path length between two points can be arbitrarily large.*

6.15 Given $\delta > 0$ choose an integer n such that $(1-\delta)^n < \frac{1}{2}$. Then,

$$|x_n(1)-x_n(1-\delta)| = 1 - (1-\delta)^n > \frac{1}{2} .$$

Consequently, S is not equicontinuous, and hence, by the Ascoli-Arzela theorem, S is not sequentially compact.

6.16 (a) $\|x_n\|_\infty = 1$ for all $n \Rightarrow S$ is bounded.

(b) x_n is a translation of x_0, so that it is enough to show that x_0 is uniformly continuous.

For any $\varepsilon > 0$ there exists $M > 0$ such that: $e^{-M^2} < \frac{\varepsilon}{2}$. Set $I = [-M,+M]$.

(1) x_0 is continuous on the compact set I, and so x_0 is uniformly continuous on I.

(2) If $t_1,t_2 \in R^1\backslash I$ then

$$|x(t_1)-x(t_2)| \leq \frac{\varepsilon}{2} + \frac{\varepsilon}{2} = \varepsilon .$$

It follows from (1) and (2) that x_0 is uniformly continuous on R^1.

(c) $\lim_{n\to\infty} x_n(t) = 0$, if $t \in R^1$, so that, if $x_{n_i} \to x$ we must have $x = \underline{0}$.

But $\|x_{n_i} - \underline{0}\| = \|x_{n_i}\| = 1$. Thus, $\{x_n\}$ contains no convergent subsequence, and S is not sequentially compact.

6.17 The proofs of the two theorems are very similar, and we only prove Theorem 6.15.

$\Rightarrow$: (1) Every sequentially compact set is bounded.

(2) S is totally bounded (Theorem 6.3). Choose an $(\varepsilon/3)$-net in c, $y^{(1)},\ldots,y^{(m)}$, say. Now choose N such that

$$|y_n^{(i)}-y_\infty^{(i)}| \leq \frac{\varepsilon}{3} , \quad \text{if} \quad n \geq N \quad \text{and} \quad 1 \leq i \leq m .$$

If $x \in S$ then

$$\|x-y^{(k)}\| \le \varepsilon/3 \quad \text{for some } k .$$

Thus

$$|x_\infty - y_\infty^{(k)}| = \lim_{n\to\infty} |x_n - y_n^{(k)}| \le \varepsilon/3 ,$$

and so

$$|x_n - x_\infty| \le |x_n - y_n^{(k)}| + |y_n^{(k)} - y_\infty^{(k)}| + |y_\infty^{(k)} - x_\infty| ,$$

$$\le \varepsilon , \text{ if } n \ge N .$$

$\Leftarrow$: Since S is bounded, the set $T = \{z \in R^1 : z = x_i$ for some $x = (x_j) \in S$ and some $i\}$ is a bounded subset of R^1. Choose an ε-net for T, $t_1,\ldots,t_m$ say.

By (2) there exists N such that

$$|x_i - x_\infty| \le \varepsilon , \quad \text{if } i \ge N \text{ and } x \in S .$$

Construct the $p = m^N$ elements of c of the form

$$y^{(s)} = (t_{k_1}^{(s)}, t_{k_2}^{(s)}, \ldots, t_{k_{N-1}}^{(s)}, t_{k_N}^{(s)}, t_{k_N}^{(s)}, t_{k_N}^{(s)}, \ldots) ,$$

where the first N terms of $(y_j^{(s)})$ are chosen from $t_1,\ldots,t_m$, and $y_j^{(s)} = y_N^{(s)}$ for $j > N$.

It is readily seen that $y^{(1)},\ldots,y^{(p)}$ are a 2ε-net for S, and so, by Theorem 6.3, S is sequentially compact.

6.18 Assume that $b > a$. A typical curve C is shown in the Figure.

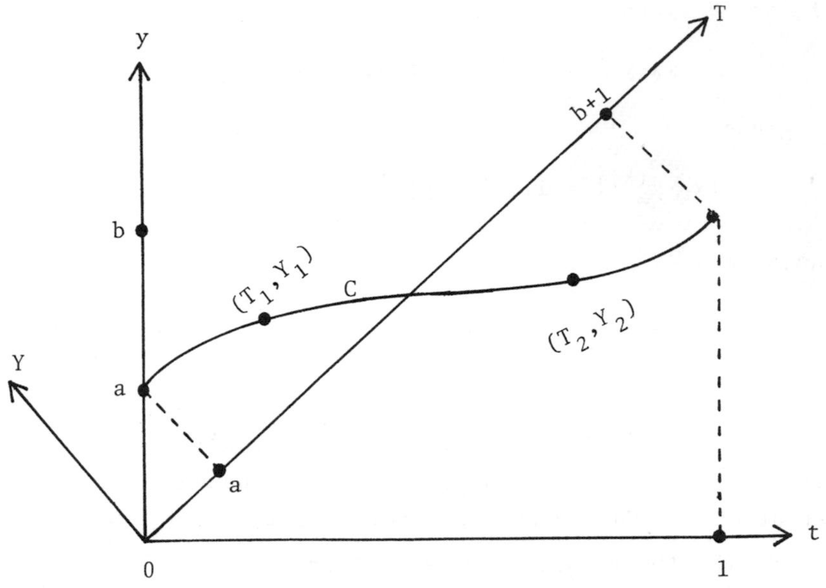

Figure: A monotonic curve C .

$$\begin{pmatrix} Y \\ T \end{pmatrix} = \begin{pmatrix} 1 & -1 \\ 1 & 1 \end{pmatrix} \begin{pmatrix} y \\ t \end{pmatrix} \; ; \; \begin{pmatrix} y \\ t \end{pmatrix} = \frac{1}{2} \begin{pmatrix} 1 & 1 \\ -1 & 1 \end{pmatrix} \begin{pmatrix} Y \\ T \end{pmatrix} .$$

(1) Let (Y_1, T_1) and (Y_2, T_2) be two points on C . Then

$$Y_1 - Y_2 = (y_1 - y_2) - (t_1 - t_2) ,$$

$$T_1 - T_2 = (y_1 - y_2) + (t_1 - t_2) .$$

By assumption, C is monotonic with respect to the yt coordinates, so that (y_1-y_2) and (t_1-t_2) have the same sign. Consequently,

$$|Y_1 - Y_2| \le |T_1 - T_2| .$$

Thus, on C , Y is a single-valued Lipschitz continuous function of T .

(2) The curves C_n may be represented by $Y = X_n(T)$, $a \le T \le b + 1$, $n = 1, 2, \ldots$. By (1), the functions

X_n are bounded and equicontinuous. Applying the Ascoli-Arzela theorem, we may extract a convergent subsequence X_{n_i} .

6.19 Let $\kappa \subset \tau_2$ be an open cover of S . Then $\kappa \subset \tau_1$. By assumption, S is compact (τ_1) , so there exists a finite subcover κ' . κ' is a finite subcover of the original cover κ , so S is compact (τ_2) .

6.20 If $f: V \to R^1$ is continuous and V is compact, then $f(V)$ is compact (Theorem 6.1, part 2). Thus $f(V) \subset R^1$ is bounded and closed. In R^1 , every closed and bounded set attains its supremum.

6.21 Let V be a compact subset of a topological space X , and $f: V \to R^1$ be lower semicontinuous.

(a) The sets

$$H_a = \{x \in V: f(x) > a\} , \quad a \in R^1 ,$$

form an open cover of V . Since V is compact, there is a finite subcover $H_{a_1},\ldots,H_{a_n}$, say. Thus

$$f(x) > \min a_j , \quad \text{for} \quad x \in V ,$$

so that $f(V)$ is bounded from below and

$$m = \inf_{x \in V} f(x) > -\infty .$$

(b) Let $\{x_n\}$ be a sequence in V , $x_n \to \bar{x} \in V$. For $\varepsilon > 0$ the set

$$N_\varepsilon = \{x \in V: f(x) > f(\bar{x}) - \varepsilon\}$$

is an open set containing $\bar{x}$, and is thus a neighbourhood of $\bar{x}$. Since $x_n \to \bar{x}$, there exists N such that $x_n \in N_\varepsilon$ if $n \geq N$. That is,

$$f(x_n) \ge f(\bar{x}) - \varepsilon, \quad \text{if} \quad n \ge N .$$

$$\Rightarrow \liminf_{n\to\infty} f(x_n) \ge f(\bar{x}) - \varepsilon ,$$

$$\Rightarrow \liminf_{n\to\infty} f(x_n) \ge f(\bar{x}) . \qquad (*)$$

(c) Choose a sequence $\{x_n\}$ in V such that $f(x_n) \to m$. Since, by Theorem 6.2, V is closed and sequentially compact there is a subsequence, also denoted by $\{x_n\}$ such that $x_n \to \bar{x} \in V$. Using eqn (*),

$$f(\bar{x}) \le \liminf_{n\to\infty} f(x_n) = \lim_{n\to\infty} f(x_n) = m ,$$

from which we conclude that $f(\bar{x}) = m$.

6.22 (a) The set

$$\{x\colon f(x) > t\} = \begin{cases} [0,1] = [0,1] \cap R^1 , & \text{if } t < 0 , \\ (t,1), & \text{if } 0 \le t < \frac{1}{2} , \\ (t,1] = [0,1] \cap (t,2) , & \text{if } \frac{1}{2} \le t \le 1, \\ \emptyset , & \text{if } t > 1 , \end{cases}$$

is open for all t.

(b) The set

$$\{x\colon f(x) < t\} = \begin{cases} \emptyset , & \text{if } t < 0 , \\ [0,t) = [0,1] \cap (-1,t), & \text{if } 0 \le t \le \frac{1}{2}, \\ [0,t) \cup \{1\} , & \text{if } \frac{1}{2} < t < 1 , \\ [0,1] , & \text{if } t \ge 1 . \end{cases}$$

is not open if $\frac{1}{2} < t < 1$.

(c) $\sup f(x) = 1$, which is not attained.

6.23 If S is not empty let

$$m = \inf_V f(x) = f(x_b) ,$$

so that

$$S = \{x \in V\colon f(x) = m\} .$$

(1) If $x_1, x_2 \in S$ and $\lambda \in [0,1]$ then

$$f(\lambda x_1 + (1-\lambda)x_2) \le \lambda f(x_1) + (1-\lambda)f(x_2) = m ,$$

so that $\lambda x_1 + (1-\lambda)x_2 \in S$.

(2) The proof is by contradiction. If $y \in S$ and $y \neq x$ then

$$f(\tfrac{1}{2}x + \tfrac{1}{2}y) < \tfrac{1}{2}f(x) + \tfrac{1}{2}f(y) = m ,$$

which is impossible.

(3) The proof is by contradiction. If $x \in S$ and $x \notin \partial V$, then $x \in V$ and $x \notin \overline{X\backslash V}$. That is, there exists a neighbourhood N of x such that $N \cap \overline{X\backslash V} = \emptyset$, and so, $N \subset V$.

Let $y \in V$, $y \neq x$. For $\lambda \in (0,1)$ let

$$x_\lambda = [x - \lambda y]/(1-\lambda) .$$

Since addition and multiplication are continuous, there exists $\varepsilon > 0$ such that $x_\varepsilon \in N$. But

$$m = f(x) = f((1-\varepsilon)x_\varepsilon + \varepsilon y) \ge (1-\varepsilon)f(x_\varepsilon) + \varepsilon f(y) \ge m .$$

Consequently, $f(y) = m$ for all $y \in V$, which is, by assumption, not possible.

(4) Let $x \in V$. For $\lambda \in (0,1)$ let $y_\lambda = (1-\lambda)y_b + \lambda x$. Then $y_\lambda \in V$ because V is convex, and $y_\lambda \in N$ for λ sufficiently small, by continuity of addition and multiplication. Thus, we may assume that $y_\lambda \in N \cap V$ and hence,

$$f(y_b) \le f(y_\lambda) \le (1-\lambda)f(y_b) + \lambda f(x) ,$$

which implies that $f(y_b) \le f(x)$ so that $y_b \in S$.

6.24 Continuing the arguments of Theorem 6.22, we have two possibilities:

(a) $u = (x - x_b)/2 = 0$. Then $x = x_b$, and $\|x - y_b\| = d = 0$, so that $y_b = x = x_b$.

(b) $u \neq 0$ and $v = \frac{x-y_b}{2} = \lambda u = \lambda\frac{x-x_b}{2}$.

(1) If $\lambda = 1$ then $x - y_b = x - x_b$ so that $x_b = y_b$.

(2) If $\lambda \neq 1$ then $x = (y_b - \lambda x_b)/(1-\lambda)$. Since $x_b , y_b \in X_0$, we conclude that $x \in X_0 \Rightarrow d = 0 \Rightarrow u = 0$. This is impossible.

In conclusion, in both cases $x_b = y_b$, and the approximation is unique.

6.25 The subspace X_0 consists of functions of the form $y_\alpha : t \to \alpha t$ for $\alpha \in R^1$. Clearly, $\|x-y_\alpha\| = \max\{1, |1-\alpha|\}$. The functions y_α , $\alpha \in [0,2]$, are all points of best approximation to x and satisfy $\|x-y_\alpha\| = 1$.

6.26 The arguments are very similar to those of Theorem 6.23 in the text, and we only indicate significant changes:

(1) x_n and f are vector-valued functions taking values in R^ℓ , and the appropriate changes must be made. For example,

$$|x_n(t)| \text{ becomes } \|x_n(t);R^\ell\|_\infty .$$

In Step 4, the Ascoli-Arzela theorem is applied successively to the components of x_n .

(2) In Step 2 construct $x_n \in C[\tau,\tau+\alpha]$ by:

(a) $t_{n,i} = \tau + \frac{i\alpha}{n}$, $0 \leq i \leq n$.

(b) $x_n(t_{n,0}) = x_n(\tau) = \xi$.

(c) $x_n(t_{n,i}) = \xi + \frac{\alpha}{n} \sum_{j=0}^{i-1} f(t_{n,i}, t_{n,j}, x_n(t_{n,j}))$.

(d) x_n is piecewise linear on $[\tau,\tau+\alpha]$.

(3) Choosing $\alpha < 1$, we are back to the problem considered in the text with

$$\dot{x}(t) = f(t,x(t),\phi(t-1)) .$$

For the more interesting problem with a variable delay $d(t)$,

$$\dot{x}(t) = f(t,x(t),x(t-d(t))) ,$$

see Cryer and Tavernini [1972].

6.27 Consider the case $n = 4m+1$.

$$x_n(3h) = x_n(2h) + \frac{3h}{2}[x_n(2h)]^{1/3} + hg(2h) ,$$

$$= \frac{1}{8}h^{3/2} + \frac{3h}{2}\cdot\frac{h^{\frac{1}{2}}}{2} + \frac{h}{8}(2h)^{\frac{1}{2}} \sin[m\pi+\pi/4] ,$$

$$\geq \frac{h^{3/2}}{8}(7-\sqrt{2}\sin(\pi/4)) > \frac{(3h)^{3/2}}{8} ,$$

so the induction hypothesis holds for $i = 3$. Assume that the hypothesis is true for i, so that $x_n(ih) \geq (ih)^{3/2}/8$. Then

$$x_n((i+h)h) = x_n(ih) + h[\tfrac{3}{2}(x_n(ih)^{1/3} + \frac{(ih)^{\frac{1}{2}}}{8}\cdot \sin(\frac{\pi}{2ih})] ,$$

$$\geq \frac{(ih)^{3/2}}{8} + \frac{3}{2}h\,\frac{(ih)^{\frac{1}{2}}}{2} - \frac{h(ih)^{\frac{1}{2}}}{8} ,$$

$$\geq \frac{h^{3/2}}{8}(i+1)^{3/2}$$

since, by the mean value theorem,

$$(i+1)^{3/2} - i^{3/2} \leq \frac{3}{2}(i+1)^{\frac{1}{2}} = \frac{3}{2}i^{\frac{1}{2}}\left(\frac{i+1}{i}\right)^{\frac{1}{2}} ,$$

$$< \frac{3}{2}i^{\frac{1}{2}}\cdot\left(\frac{3}{2}\right)^{\frac{1}{2}} < 5i^{\frac{1}{2}} .$$

6.28 Assume that X is a metric space.

(1) Let $\{\alpha x_n\}$ be a sequence in αV. V(pre) compact $\Rightarrow$ there is a subsequence $\{x_{n_i}\}$ in V such that $x_{n_i} \to x \in V\ (x \in X) \Rightarrow \alpha x_{n_i} \to \alpha x \in V(\in X)$.

(2) Let $\{v_n + w_n\}$ be a sequence in $V + W$ converging to $x \in X$. V compact $\Rightarrow$ there is a subsequence $\{v_{n_i}\}$ such that $v_{n_i} \to v \in V \Rightarrow w_{n_i} \to x - v$. W closed $\Rightarrow w_{n_i} \to w \in W \Rightarrow w = x - v \Rightarrow x = v + w \in V + W \Rightarrow V + W$ closed.

(3) Let $\{v_n+w_n\}$ be a sequence in $V + W$. V (pre) compact ⇒ there is a subsequence, still denoted by $\{v_n+w_n\}$, such that $v_n \to v \in V(\in X)$. W (pre) compact ⇒ there is a subsubsequence, still denoted by $\{v_n+w_n\}$ such that $w_n \to w \in W(\in X)$. Thus

$$v_n + w_n \to v + w \in V + W \ (\in X) ,$$

and $V + W$ is (pre) compact.

6.29 $T([0,1]) = \{0,1\}$, which is compact.

6.30 (a) Let

$$S = \{Ax:\ x \in R^n_+\}$$

We wish to show that S is closed.

Let $\{Ax^k\}$ be a sequence in S which converges to $z \in R^m$. We wish to show that $z \in S$.

Two possibilities arise:

Case 1: $\{x^k\}$ is bounded in R^n . Bounded sets in R^n are precompact, and so there exists a convergent subsequence, still denoted by $\{x^k\}$, such that $x^k \to x \in R^n$. $A:R^n \to R^m$ is continuous, and hence $Ax = z$. But $x^k \in R^n_+$, which implies that $x \in R^n_+$, and hence that $z \in S$.

Case 2: $\{x^k\}$ is not bounded in R^n , so that $\|x^k\|_\infty \to \infty$. Let $y^k = x^k/\|x^k\|_\infty$. The sequence $\{y^k\}$ is bounded; by compactness, we can select a subsequence still denoted by $\{y^k\}$ such that $y^k \to y \in R^n_+$, with $\|y\|_\infty = 1$.

$$Ay = \lim Ay^k = \lim \frac{Ax^k}{\|x^k\|_\infty} = 0 .$$

Let $u^k = x^k - \alpha_k y$ where the non-negative scalar α_k is chosen so that (i) $u^k \in R^n_+$ and (ii) at least one component of u^k is zero.

Since u^k has only n components, there is a constant i_1 and a subsequence, still denoted by $\{u^k\}$, such that $u_{i_1} = 0$ for all k.

In summary, if $\{x^k\}$ is unbounded, there exists a new sequence $\{u_k\}$ such that:
(i) $u^k \in R^n_+$; (ii) $u^k_{i_1} = 0$ for all k ; (iii) $Au^k \to z$. If $\{u^k\}$ is unbounded, the above arguments may be repeated to obtain a sequence $\{v^k\}$ say, such that $v^k \in R^n_+$, $Av^k \to z$ and v^k has two zero components. After at most n repetitions, we obtain a bounded sequence in R^n_+ converging to z , and may proceed as in Case 1.

(b) $f(R^1_+) = (0,1]$.

6.31 Let U be a bounded set in $C[a,b]$, so that

$$\|x\|_\infty \le M_1 , \quad \text{for} \quad x \in U .$$

A real continuous function attains its supremum on a compact set (Theorem 6.19), and hence M_2 and M_3 exist such that

$$M_2 = \max_{a \le s,t \le b} |K(s,t)| , \quad M_3 = \max_{-M_1 \le t \le M_1} |F(t)| .$$

If $x \in U$ then

$$|(Tx)(s)| \le \int_a^b |K(s,t)||F(x(t))|dt ,$$

$$\le \int_a^b M_2 M_3 dt = M_2 M_3 (b-a) , \quad \text{for} \quad s \in [a,b] ,$$

so that $T(U)$ is bounded.

Choose $\varepsilon > 0$. K is continuous on $[a,b] \times [a,b]$ and so, by Theorem 6.12, there exists $\delta > 0$ such that

$|K(s_1,t_1) - K(s_2,t_2)| \le \varepsilon$, if $|s_1-s_2|$, $|t_1-t_2| \le \delta$.

Let $x \in U$ and $|s_1-s_2| \le \delta$. Then

$$|(Tx)(s_1)-(Tx)(s_2)| \le \int_a^b |K(s_1,t)-K(s_2,t)|\cdot|F(x(t))|dt ,$$
$$\le M_3 \cdot (b-a) \cdot \varepsilon .$$

That is, $T(U)$ is equicontinuous.

It follows from the Ascoli-Arzela theorem (Theorem 6.13) that $T(U)$ is precompact, and that T is a compact operator.

6.32 Since T is linear, it suffices to show that T maps the unit ball $B = B[\underline{0};1] \subset L^p[a,b]$ into a precompact set. We apply Theorem 6.14 to the set $S = T(B)$.

Let $M = \max\limits_{a\le s,t\le b} |K(s,t)|$.

Choose $\varepsilon_1 > 0$. There exists $\delta > 0$ such that

$$|K(s_1,t_1)-K(s_2,t_2)| \le \varepsilon_1 , \qquad (*)$$

$$\text{if } |s_1-s_2|, |t_1-t_2| \le \delta .$$

(1) If $x \in B$ then by the Hölder inequalities (page 67 or Property (7) page 95), $x \in L^1[a,b]$ and

$$\int_a^b |x(t)|dt \le \alpha\|x\|_p \le \alpha \equiv (b-a)^{1-1/p} .$$

Thus, if $x \in B$ and $y = Tx$, then

$$|y(s)| \le \left|\int_a^b K(s,t)x(t)dt\right| \le M\alpha . \qquad (**)$$

In consequence

$$\|y\| = \|Tx\| = \left[\int_a^b |y(s)|^p ds\right]^{1/p} \le M\alpha(b-a)^{1/p} = M(b-a) , \qquad (***)$$

so that T is a bounded operator

$$\|T\| \le M(b-a) . \qquad (**)$$

(2) Set $\Omega = (a,b)$, and $G = (a+\delta, b-\delta)$, where $\delta > 0$ is as in eqn (*).

If $x \in B$ and $y = Tx$, then from eqn (**)

(i)

$$\int_{\Omega\setminus\overline{G}} |y(s)|^p ds = \int_a^{a+\delta} |y(s)|^p ds + \int_{b-\delta}^{b} |y(s)|^p ds ,$$

$$\le 2\delta (M\alpha)^p . \qquad (***)$$

(ii) For $|h| \le \delta$,

$$\int_\Omega |\tilde{y}(s)-\tilde{y}(s+h)|^p ds \le \int_{\overline{G}} + \int_{\Omega\setminus\overline{G}} = I_1 + I_2 , \quad \text{say.}$$

Since $|h| < \delta$, it follows from eqn (*) that

$$I_1 = \int_{\overline{G}} |y(s)-y(s+h)|^p ds ,$$

$$= \int_{\overline{G}} ds \left| \int_a^b [K(s,t) - K(s+h,t)]x(t)dt \right|^p ds ,$$

$$\le \int_{\overline{G}} (\varepsilon_1 \alpha)^p ds \le (b-a)\varepsilon_1^p \alpha^p .$$

On the other hand,

$$I_2 \le \int_{\Omega\setminus\overline{G}} (|\tilde{y}(s)|+|\tilde{y}(s+h)|)^p ds ,$$

$$\le 2\delta \, (2M\alpha)^p .$$

Combining the above inequalities,

$$\int_{\Omega} |\tilde{y}(s)-\tilde{y}(s+h)|^p ds \le (b-a)\varepsilon_1^p \alpha^p + 2\delta(2M\alpha)^p \ . \qquad (\overset{***}{***})$$

It follows from eqns $(*\overset{**}{**})$ and $(\overset{***}{***})$ that condition 2b of Theorem 6.14 is satisfied. We conclude that T is compact.

6.33 As in Problem 6.32, we let $S = T(B[0;1])$ and use Theorem 6.14 to show that S is compact.

Let $x \in B[0;1] \subset L^p(a,b)$ so that $\|x\|_p \le 1$. Then $y = Tx \in S$ is given by eqn (**).

(1) Using Hölder's inequality (p. 67) we see from eqn (**) that

$$|y(s)| \le M^{1/q} \ , \qquad (\overset{*}{*})$$

so that S is bounded.

(2) Set $\Omega = (a,b)$ and $G = (a+\delta, b-\delta)$ where $\delta > 0$ will be chosen later.

(a) From eqn $(\overset{*}{*})$,

$$\int_{\Omega\setminus\overline{G}} |y(s)|^p ds \le 2\delta M^{p/q} \ .$$

(b) From eqn (*) and Fubini's theorem it follows that $K \in L^q[\Omega\times\Omega]$. Thus, from the continuity of translation of functions in L^p (property 8, p. 95), given $\varepsilon > 0$ we can choose δ such that

$$\int_{\Omega} ds \int_{\Omega} |\Delta_h K(s,t)|^q dt \le \varepsilon \ , \quad \text{if} \quad |h| \le \delta \ , \qquad (\overset{**}{**})$$

where

$$\Delta_h K(s,t) = \tilde{K}(s+h, t+0) - K(s,t) \ .$$

Here, $\tilde{K}(u,v) = K(u,v)$ if $u,v \in \Omega$ and is zero otherwise.

Let

$$\Omega_\varepsilon = \{s\in\Omega: \int_\Omega |\Delta_h K(s,t)|^q dt > \varepsilon^{\frac{1}{2}}\} .$$

Then, from eqn ($\overset{*}{*}\overset{*}{*}$), $\mu(\Omega_\varepsilon) < \varepsilon^{\frac{1}{2}}$, if $|h| \le \delta$.

If $|h| < \delta$, then

$$\int_\Omega |\tilde{y}(s)-\tilde{y}(s+h)|^p ds = I_1 + I_2 ,$$

where

$$I_1 = \int_{\Omega\setminus\overline{G}} |y(s)-\tilde{y}(s+h)|^p ds \le 2\delta 2^p M^{p/q} ,$$

and

$$I_2 = \int_{\overline{G}} |y(s)-\tilde{y}(s+h)|^p ds .$$

Using Hölder's inequality, eqns (**) and ($\overset{*}{*}$), and the fact that $x \in B$, we obtain

$$I_2 \le \int_{\Omega_\varepsilon} |y(s)-\tilde{y}(s+h)|^p ds + \int_{\Omega\setminus\Omega_\varepsilon} ds \left[\int_\Omega |\Delta_h K(s,t)|^q dt\right]^{p/q} ,$$

$$\le \mu(\Omega_\varepsilon)\ 2^p M^{p/q} + \mu(\Omega)\varepsilon^{p/2q}$$

$$\le 2^p M^{p/q} \varepsilon^{\frac{1}{2}} + \mu(\Omega)\varepsilon^{p/2q} .$$

6.34 We note that

$$\int_0^a \ln t = a \ln a - a . \qquad (**)$$

(a) Using eqn (**) we find that

$$\|T\| = \max_{0\le s\le 1} \int_0^1 \Big|\ln|s-t|\Big| dt = 1 + \ln 2 ,$$

so that T is continuous.

(b) Let

$$(T_n x)(s) = \int_0^1 k_n(s,t)x(t)dt ,$$

where

$$k_n(s,t) = \begin{cases} \ln |s-t| , & \text{if } |s-t| \geq 1/n , \\ \ln 1/n , & \text{if } |s-t| \leq 1/n . \end{cases}$$

Then T_n is compact by Problem 6.31. Also,

$$\| T - T_n \| \leq \int_{-1/n}^{+1/n} |\ln |u| - \ln 1/n| du = 2/n .$$

It follows from Theorem 6.25, part 4 (p. 224) that T is compact.

6.35 Let $x \in \Omega$. There exists $a > 0$ such that the n-dimensional cube K with centre x , edges parallel to the axes, and edges of length $2a$, is contained in Ω . (see Figure).

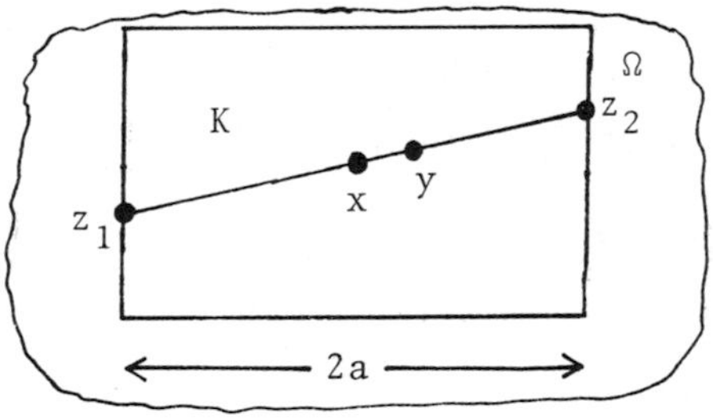

Figure: The n-dimensional cube K *in* Ω .

K has 2^n vertices; let $M_1 < \infty$ denote the maximum value attained by f on these vertices. Set $M = \max\{M_1, |f(x)|\}$. By repeated use of the convexity of f , it follows that $f \leq M$ in K .

Choose $\varepsilon > 0$, and let $y \in K$ satisfy $\|x-y\| \le \varepsilon a$. The line joining x and y may be extended to meet ∂K in points z_1 and z_2. (see Figure). Then

$$x = (1-\alpha)y + \alpha z_1 \quad \text{and} \quad y = (1-\beta)x + \beta z_2 ,$$

where $0 \le \alpha, \beta \le \varepsilon$. Thus,

$$f(x) \le (1-\alpha)f(y) + \alpha f(z_1) \le (1-\alpha)f(y) + \varepsilon M ,$$

$$f(y) \le (1-\beta)f(x) + \beta f(z_2) \le (1-\beta)f(x) + \varepsilon M ,$$

from which we obtain, by rearrangement,

$$(1-\alpha)(f(x)-f(y)) \le -\alpha f(x) + \varepsilon M \le 2\varepsilon M ,$$

$$(f(y)-f(x)) \le -\beta f(x) + \varepsilon M \le 2\varepsilon M ,$$

so that

$$|f(x)-f(y)| \le \frac{2\varepsilon}{1-\varepsilon} M .$$

6.36 Let U be the unit ball in X. Then $T(U)$ is bounded, so that

$$\|T\| = \sup_{x \in U} \|Tx\| < \infty .$$

6.37 Elementary manipulation.

6.38 A one-hundred and twenty-eighth note.

CHAPTER 7: SOLUTIONS

7.1 See Young [1967]. (This reference was drawn to our attention by Carl de Boor.)

7.2 The coefficients a,b, and c must satisfy the inconsistent system of equations,

$$\begin{aligned} a + 0b + 0c &= 0 \\ a - b - 2c &= -1 \\ a + 2b + 4c &= +1 \end{aligned}$$

7.3 See the hint in the text.

7.4 Take $A = \{x_1, x_2, \ldots, x_n\}$, and B a closed interval in $X = [0,1]$ such that B has positive measure $\mu(B)$ and is disjoint from A. Applying Urysohn's theorem, we see that there exists a continuous real non-negative function $g: X \to [0,1]$ such that $g(A) = \{0\}$ and $g(B) = \{1\}$. Then $I_n(g) = 0$ while $I(g) = \mu(B) > 0$. Take $f = g/\mu(B)$.

7.5 If $x_1, x_2 \geq 0$ then $f_c(x_1,x_2) = \frac{1}{c}\, g(cx_1)$, where $g(t) = t^2/(1+t^2)$. Therefore, $\frac{c}{\partial x_1} = \dot{g}(cx_1)$. Now

$$\dot{g}(t) = \frac{2t}{(1+t^2)^2}, \quad \ddot{g}(t) = \frac{2-6t^2}{(1+t^2)^3},$$

from which it follows that

$$\max_{0 \leq t \leq \infty} \dot{g}(t) = \dot{g}\left(\frac{1}{\sqrt{3}}\right) = 9/8\sqrt{3}.$$

The remainder of the problem requires only simple calculations and the use of Taylor's theorem. See Whitney [1944].

7.6 The function f is sketched below and it can be seen geometrically that f cannot be extended as a convex function to R^1 .

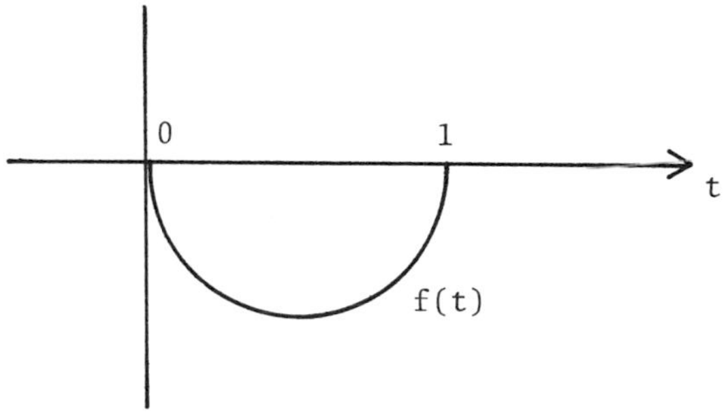

An analytical proof goes as follows. Let h be a convex extension of f to R^1 . Then

$$1 = (\frac{1-t}{2-t}) \cdot 2 + (1 - (\frac{1-t}{2-t})) \cdot t , \quad \text{for } t \in [0,1] ,$$

$$\Rightarrow h(1) = 0 \le (\frac{1-t}{2-t})h(2) + (1 - \frac{1-t}{2-t}) \cdot h(t) , \quad \text{for } t \in [0,1] ,$$

$$\Rightarrow 0 \le \frac{1-t}{2-t} h(2) - (\frac{1}{2-t}) \cdot [t(1-t)]^{\frac{1}{2}} , \quad \text{for } t \in [0,1] ,$$

$$\Rightarrow h(2) \ge \frac{1}{1-t} [t(1-t)]^{\frac{1}{2}} = \frac{t^{\frac{1}{2}}}{(1-t)^{\frac{1}{2}}} , \quad \text{for } t \in [0,1] ,$$

$$\Rightarrow h(2) = \infty .$$

7.7 Direct computation.

7.8 See, for example, Rudin [1966, p.19].

7.9 We continue the arguments given in the proof of Theorem 7.4.

(a) Let X have dimension n and X_0 have dimension $m < n$. As shown in the proof, f_0 can be extended to a functional f_1 defined on X_1 , where $X_1 \supset X_0$ has dimension $m + 1$. Repeating this process, we can extend f_0 to a functional f_{n-m} defined on a subspace X_{n-m} of dimension $m+(n-m) = n$. Thus, $X_{n-m} = X$.

(b) Let $\{x_n: n=1,2,...\}$ be a dense subset of X. Set $Y = X_0$, $f = f_0$. Now perform the following steps for $n = 1,2,...$:

Step n:

(1) If $x_n \in Y$, do nothing.

(2) If $x_n \notin Y$, let $\hat{Y}=\{\hat{y}=y+\lambda x_n: y\in Y \text{ and } \lambda\in R^1\}$, $\hat{f}: \hat{y} = y + \lambda x_n \in \hat{Y} \to \hat{f}(\hat{y}) = f(y) + \lambda c \in R^1$, where $c = \hat{f}(x_n)$ is chosen such that

$$\hat{f}(\hat{y}) \le p(\hat{y}), \quad \text{for all } \hat{y} \in \hat{Y}.$$

Then set $Y = \hat{Y}$, $f = \hat{f}$. □

Let $\tilde{X}$ be the space spanned by the x_i,

$$\tilde{X} = \{x_0 + \sum_{i\in I} \lambda_i x_i : I \text{ a finite subset of the integers.}\}.$$

If $\tilde{x} \in \tilde{X}$ then $\tilde{x} = x_0 + \sum_{i=1}^{n} \lambda_i x_i$ for some integer n and some λ_i. Thus, $f(\tilde{x})$ is defined. Furthermore,

$$f(\tilde{x}) \le p(\tilde{x}) = \|\tilde{x}\|, \quad \text{for all } \tilde{x} \in \tilde{X},$$

$$\Rightarrow -f(\tilde{x}) = f(-\tilde{x}) \le p(-\tilde{x}) = \|-\tilde{x}\| = \|\tilde{x}\|, \text{ for all } \tilde{x} \in \tilde{X},$$

$$\Rightarrow \quad |f(\tilde{x})| \le \|\tilde{x}\|, \quad \text{for all } \tilde{x} \in \tilde{X},$$

$\Rightarrow f$ is a bounded linear operator satisfying $\|f\| \le 1$.

By assumption, $\tilde{X}$ is dense in X, and we can now apply Theorem 7.3 to extend f to a bounded linear operator $\bar{f}$ defined on X and satisfying $\|\bar{f}\| = \|f\| = 1$. That is,

$$|\bar{f}(x)| = \|\bar{f}(x)\| \le \|\bar{f}\| \|x\| = p(x), \text{ for } x \in X.$$

7.10 Straightforward verification of the conditions of Definition 3.2, p. 52.

7.11 (1) If $x,y \in D(\hat{g})$, then $x \in D(g_1)$ and $y \in D(g_2)$ for some $g_1, g_2 \in A_0$. A_0 is totally ordered so $g_1 \geq g_2$ or $g_2 \geq g_1$. If $g_1 \geq g_2$, say, then $x,y \in D(g_1)$ and so $\lambda x + \mu y \in D(g_1) \subset D(\hat{g})$ for any scalars λ,μ. Therefore, $D(\hat{g})$ is a linear subspace. Also,

$$\hat{g}(\lambda x+\mu y) = g_1(\lambda x+\mu y) = \lambda g_1(x) + \mu g_1(y) = \lambda \hat{g}(x) + \mu \hat{g}(y) ,$$

and so $\hat{g}$ is linear.

(2) If $x \in D(\hat{g})$, then $x \in D(g_1)$ for some $g_1 \in A_0$. Thus $\hat{g}(x) = g_1(x) \leq p(x)$.

(3) If $x \in D(f_0)$ then $x \in D(g)$ for all $g \in A_0$, and so $x \in D(\hat{g})$. Clearly $\hat{g}$ is an extension of g for all $g \in A_0$.

7.12 If $z = \alpha x + m \in X_0$ and $\alpha \neq 0$ then

$$\|z\| = |\alpha| \; \|x-(-m/\alpha)\| \geq |\alpha| d , \tag{*}$$

since $(-m/\alpha) \in M$. Eqn (*) is trivially true if $\alpha = 0$. Thus, if $z = \alpha x + m \in X_0$ then

$$\|f_0(z)\| = |\alpha| \leq \|z\|/d ,$$

so that $\|f_0\| \leq 1/d$.

On the other hand, if $\varepsilon > 0$ there exists $m \in M$ such that $\|x-m\| < d + \varepsilon$. Thus,

$$\|f_0(x-m)\| = 1 > \|x-m\|/(d+\varepsilon) ,$$

so that $\|f_0\| > 1/(d+\varepsilon)$.

Combining these results it follows that $\|f_0\| = 1/d$.

7.13 Let

$$X_0 = \{\lambda x: \lambda \text{ a scalar}\} , \quad f_0: \lambda x \in X_0 \to \lambda\|x\| .$$

X_0 is a linear subspace of X, f_0 is a linear functional, $f(x) = \|x\|$, and

$$\|f_0(\lambda x)\| = |\lambda|\,\|x\| = \|\lambda x\| ,$$

so that $\|f_0\| = 1$. Now extend f to a linear functional f defined on X (Corollary 7.5, p. 240).

7.14 (1) $e(s,t) = st - \frac{1}{2}(s+t) + \frac{1}{4}$.

For fixed s , $e(s,t)$ is linear in t and attains its extrema at the points $t = 0$ and $t = 1$. Thus

$$\max_{0\le s,t\le 1} e(s,t) = \max_{0\le s\le 1} \max \{|e(s,0)|, |e(s,1)|\} ,$$

$$= \max_{0\le s\le 1} |\frac{s}{2} - \frac{1}{4}| = \frac{1}{4} .$$

(2) (a) Clearly $\|f\| \le 1$. On the other hand if $x_1 : (s,t) \to 1-2|s-t|$ then $\|x_1\| = 1$ and $\|f(x_1)\| = 1$. Thus $\|f\| = 1$.

(b) If $P: (s,t) \to P(s)$ then

$$f(P) = [P(0) + P(1) - P(1) - P(0)]/4 = 0 .$$

(3) Use the fact that

$$st - (s+t) + \frac{1}{2}(s^2+t^2) + \frac{1}{4} = \frac{1}{2}u^2 - u + \frac{1}{4} , \text{ where}$$

$$u = (s+t) .$$

7.15 (1) Let

$$x[t_0,t_1,\ldots,t_{n+1}] = \sum_{k=0}^{n+1} (-1)^{n+1-k}\alpha_k^{(n)} x(t_k) . \qquad (*)$$

We show, by induction on n , that $\alpha_k^{(n)} > 0$. Since

$$x[t_0,t_1] = \frac{x(t_1)-x(t_0)}{t_1-t_0} ,$$

$$= -\alpha_0^{(0)} x(t_0) + \alpha_1^{(0)} x(t_1) ,$$

with

$$\alpha_0^{(0)} = \alpha_1^{(0)} = 1/(t_1-t_0) > 0 ,$$

the induction hypothesis is true for $n = 0$. Assume true for n . Then

$$x[t_0,t_1,\ldots,t_{n+2}] = x[t_0,\ldots,t_{n+1}] - x[t_1,\ldots,t_{n+2}]/(t_0-t_{n+2}) ,$$

$$= [\sum_{k=0}^{n+1}(-1)^{n+1-k}\alpha_k^{(n)}x(t_k) -$$

$$- \sum_{k=1}^{n+2}(-1)^{n+1-(k-1)}\alpha_{k-1}^{(n)}x(t_k)]/(t_0-t_{n+2}) . \tag{**}$$

Comparing eqns (*) and (**) we obtain:

$$\alpha_0^{(n+1)} = \alpha_0^{(n)}/(t_{n+2}-t_0) > 0 ,$$

$$\alpha_{n+2}^{(n+1)} = \alpha_{n+1}^{(n)}/(t_{n+2}-t_0) > 0 ,$$

$$\alpha_k^{(n+1)} = (\alpha_k^{(n)} + \alpha_{k-1}^{(n)})/(t_{n+2}-t_0) > 0 , \quad 1 \le k \le n+1 ,$$

so that the hypothesis is satisfied for $n+1$.

(2) It is a standard result in numerical analysis, which will be found in many texts, that if x is a polynomial degree r then $x[t_0,t_1,\ldots,t_n,t]$ is a polynomial of degree $r-n-1$.

(3) Using the results of Example 5.4 (page 113) we find that

$$\|\ell\| = \sum_{k=0}^{n+1}|(-1)^{n+1-k}\alpha_k| \Big/ \sum_{k=0}^{n+1}\alpha_k = 1 .$$

7.16 (1) If $x \in K$, $\frac{x}{1} \in K$ and so $F(x) \le 1$.

(2) If x is an internal point of K then

$$\frac{x}{1-\delta} = x + (\frac{\delta}{1-\delta})\, x \in K$$

for small $\delta > 0$, so that $F(x) \le 1-\delta < 1$.

(3) If $F(x) < 1$ then, for some $\varepsilon > 0$, $x/(1-\varepsilon) \in K$. But K is convex, and so

$$x = (1-\varepsilon)(\frac{x}{1-\varepsilon}) + \varepsilon\, \underline{0} \in K .$$

7.17 Let S denote one of the sets $M + N$, $M - N$, and $M \cap N$. If $u, v \in S$ it is readily verified that $\lambda u + (1-\lambda)v \in S$ for $\lambda \in [0,1]$.

7.18 p an interior point $\Rightarrow N_p \subset S$ for some neighbourhood N_p of p. Let $x \in X$. Since X is a topological vector space, and $p + 0x = p$,

$$p + B(0;\delta)x \subset N_p$$

for some $\delta > 0$.

7.19 Let U be a neighbourhood of $\underline{0} \in Y$. M is bounded topologically $\Rightarrow M \subset \lambda U$ for some $\lambda > 0$. Then

$$f(\tfrac{1}{\lambda}N) = \tfrac{1}{\lambda}f(N) \subset \tfrac{1}{\lambda}M \subset U .$$

Therefore, $V = \frac{1}{\lambda}N$ is a neighbourhood of $\underline{0} \in X$ such that $f(V) \subset U$. That is, f is continuous at $\underline{0} \in X$. Since f is linear, f is continuous.

7.20 Direct computation.

7.21 We use the method of Example 7.5.

Assume that

$$u_{xx}(0,0) + u_{yy}(0,0) = -w_0 u(0,0) + \sum_{i=1}^{N} w_i u(n_i h, m_i h) + O(h^s) ,$$

where $s > 4$ and $w_i \geq 0$, $i = 0,\ldots,N$. Then:

$$wA = b \qquad (*)$$

where

$$w = (w_0, w_1, \ldots, w_N) ,$$
$$b = (0,0,0,2/h^2,0,2/h^2,0,0,0,0,0,0,0,0,0) ,$$

and

$$A = \begin{pmatrix} -1 & 0 & 0 & 0 & 0 & 0 & \ldots & 0 & \ldots & 0 \\ 1 & n_1 & m_1 & n_1^2 & n_1 m_1 & m_1^2 & \ldots & n_1^4 & \ldots & m_1^4 \\ \vdots & \vdots & \vdots & \vdots & \vdots & & & & & \\ 1 & n_N & m_N & n_N^2 & n_N m_N & m_N^2 & \ldots & n_N^4 & \ldots & m_N^4 \end{pmatrix} .$$

Take $y^T = (0,0,0,-1,0,0,0,0,0,0,+\alpha,0,0,0,0)$.
Then $by = -2/h^2 < 0$ and

$$Ay = \begin{pmatrix} 0 \\ -n_1^2 + \alpha n_1^4 \\ \cdots \\ -n_N^2 + \alpha n_N^4 \end{pmatrix} \geq 0 ,$$

if $\alpha = \max_i (1/n_i^2)$. By the Farkas lemma, there is no non-negative solution w of eqn (*).

7.22 It is a standard result in the theory of the numerical solution of elliptic equations that the nine-point approximation is $0(h^6)$. See also Davis [1970].

There is no contradiction with Problem 7.21, because the additional accuracy of the nine-point formula depends upon the fact that Laplace's equation is being solved. The formula is only an $0(h^4)$ approximation to $u_{xx} + u_{yy}$ for general u .

7.23 See Motzkin and Wasow [1953], Forsythe and Wasow [1960], Bramble and Hubbard [1964].

7.24 Assume that $b \in A(R_+^n)^c$ so that there is no $x \geq 0$ such that $b = Ax$. Then, by the Farkas lemma, there is a vector y such that $by < 0$ and $yA \geq 0$. By continuity there is a neighbourhood N_b of b such that if $z \in N_b$ then $zy < 0$ and $yA \geq 0$. By the Farkas lemma again, there is no $x \geq 0$ such that $z = Ax$. That is, $N_b \cap A(R_+^n) = \emptyset$.

In summary, if $b \in (A(R_+^n))^c$ then $b + N_b \subset (A(R_+^n))^c$ for some neighbourhood N_b of b . Thus, $(A(R_+^n))^c$ is open $\Rightarrow A(R_+^n)$ is closed.

CHAPTER 8: SOLUTIONS

8.1 If $x = \lim \sum \lambda_i(x)x_i$ and $y = \lim \sum \lambda_i(y)x_i$ then

$$\alpha x + \beta y = \lim \sum (\alpha\lambda_i(x) + \beta\lambda_i(y))x_i ,$$

so that $\lambda_i(\alpha x + \beta y) = \alpha\lambda_i(x) + \beta\lambda_i(y)$.

8.2 With the notation of Theorem 8.1, let $\{y^p\}$ be a Cauchy sequence in Y, where $y^{(p)} = \{\alpha_i^{(p)}\}$. As shown in the text, $\alpha_i^{(p)} \to \tilde{\alpha}_i$ as $p \to \infty$. We wish to show that $y = \{\tilde{\alpha}_i\} \in Y$ and that $y^{(p)} \to y$.

If $m > n$ and $q > p$ then

$$\|\sum_{i=n}^{m} \tilde{\alpha}_i x_i\| = \|\sum_{i=n}^{m} \alpha_i^{(p)} x_i + \sum_{i=1}^{m} [(\tilde{\alpha}_i - \alpha_i^{(q)}) + (\alpha_i^{(q)} - \alpha_i^{(p)})]x_i - \sum_{i=1}^{n-1} [(\tilde{\alpha}_i - \alpha_i^{(q)}) + (\alpha_i^{(q)} - \alpha_i^{(p)})]x_i\| ,$$

from which we obtain eqn (*) of the hint. Choose $\varepsilon > 0$. $\{y^{(p)}\}$ is a Cauchy sequence, so we may pick p such that if $q > p$ then

$$\sup_k \|\sum_{i=1}^{k} (\alpha_i^{(q)} - \alpha_i^{(p)})x_i\| \le \|y^{(q)} - y^{(q)}\| < \varepsilon/6 .$$

$y^{(p)} \in Y$ so we may pick N such that if $m \ge n \ge N$ then

$$\|\sum_{i=n}^{m} \alpha_i^{(p)} x_i\| < \varepsilon/6 .$$

Finally, $\alpha_i^{(q)} \to \tilde{\alpha}_i$, so that, given m we can choose $q > p$ such that

$$\|(\alpha_i^{(q)} - \tilde{\alpha}_i)x_i\| < \varepsilon/6m .$$

Substituting the above inequalities into eqn (*), we obtain that

$$\| \sum_{i=n}^{m} \tilde{\alpha}_i x_i \| < \varepsilon \ , \quad \text{if} \quad m,n \geq N \ .$$

That is, $y = \{\tilde{\alpha}_i\} \in Y$.

Given ε there exists P such that

$$\sup_n \| \sum_{i=1}^{n} (\alpha_i^{(p)} - \alpha_i^{(q)}) x_i \| = \| y^{(p)} - y^{(q)} \| \leq \varepsilon \ , \quad \text{if} \quad q \geq p \geq P \ ,$$

which implies that, for any n ,

$$\| \sum_{i=1}^{n} (\alpha_i^{(p)} - \alpha_i^{(q)}) x_i \| \leq \varepsilon \ , \quad \text{if} \quad q \geq p \geq P \ .$$

Letting $q \to \infty$, we have $\alpha_i^{(q)} \to \tilde{\alpha}_i$, and, for any n ,

$$\| \sum_{i=1}^{n} (\alpha_i^{(p)} - \tilde{\alpha}_i) x_i \| \leq \varepsilon \ , \quad \text{if} \quad p \geq P \ ,$$

so that

$$\| y^{(p)} - y \| = \sup_n \| \sum_{i=1}^{n} (\alpha_i^{(p)} - \tilde{\alpha}_i) x_i \| \leq \varepsilon \ , \quad \text{if} \quad p \geq P \ .$$

That is, $y^{(p)} \to y$.

8.3 Let $x \in C[0,1]$, and $n \geq 2^m + 1$. If the points $t_1, \ldots, t_n$ are ordered, $0 = s_1 < \ldots < s_n = 1$ say, then $|s_k - s_{k+1}| \leq 2^{-m} = \varepsilon_n$. Thus, if $s \in [s_k, s_{k+1}]$,

$$|(P_n x)(s) - x(s)| \leq \max_{s_k \leq s \leq s_{k+1}} \left| \frac{(s-s_k)(x(s_{k+1})-x(s)) + (s_{k+1}-s)(x(s_k)-x(s))}{s_{k+1}-s_k} \right| ,$$

$$\leq \max_{s_k \leq s \leq s_{k+1}} \{ |x(s_{k+1}) - x(s)| \ , \ |x(s_k) - x(s)| \} \ ,$$

$$\leq \omega(x, \varepsilon_n) \ .$$

8.4 Let $x \in C(\overline{\Omega})$. Following the suggestions in the hint one obtains a sequence of approximations to x , namely

$$y_n(s,t) = \sum_{k=0}^{n} \sum_{(u,v)\in A_k} \lambda_{(u,v)}(x)x_{(u,v)}(s,t) , \qquad (*)$$

where the coefficients $\lambda_{(u,v)} \in R^1$ are chosen so that

$$y_n(u,v) = x(u,v) , \quad \text{for} \quad (u,v) \in G_n . \qquad (**)$$

This is achieved by setting

$$\lambda_{(s,t)}(x) = x(s,t) - \sum_{k=0}^{n-1} \sum_{(u,v)\in A_k} \lambda_{(u,v)}(x)x_{(u,v)}(s,t) ,$$

$$\text{for} \quad (s,t) \in A_n .$$

That is, y_n is a piecewise bilinear approximation to x which interpolates x at the points in G_n .

Let

$$\omega(x;\varepsilon) = \sup_{\|(s_1,t_1)-(s_2,t_2);\ell_\infty^2\| \le \varepsilon} |x(s_1,t_1)-x(s_2,t_2)| .$$

Each point $P = (s,t) \in \overline{\Omega}$ lies in one of the squares of the grid G_n , S say. $y_n(s,t)$ is a weighted average of the values of x at the four corners, P_i say, of S . Thus, if $(s,t) = P \in \overline{\Omega}$

$$\begin{aligned} |y_n(s,t)-x(s,t)| &= |\sum_{i=1}^{4} \alpha_i [y_n(P_i)-x(P)]| , \\ &\le \sum_{i=1}^{4} \alpha_i |y_n(P_i)-x(P)| , \\ &\le \omega(x;2^{-n}) . \end{aligned}$$

To bring the representation (*) into the form (8.1) it is only necessary to linearly order the functions $x_{(u,v)}$. We omit this.

8.5 We give two possible solutions:

(1) Let

$$h(t) = \begin{cases} 4(|t-\frac{1}{2}|-\frac{1}{2})^2(4|t-\frac{1}{2}|+1), & \text{if } 0 \le t \le 1 \\ 0, & \text{otherwise} \end{cases} \qquad (*)$$

Then $h(t)$ is a hat function in $C^1[0,1]$ (see Figure 1).

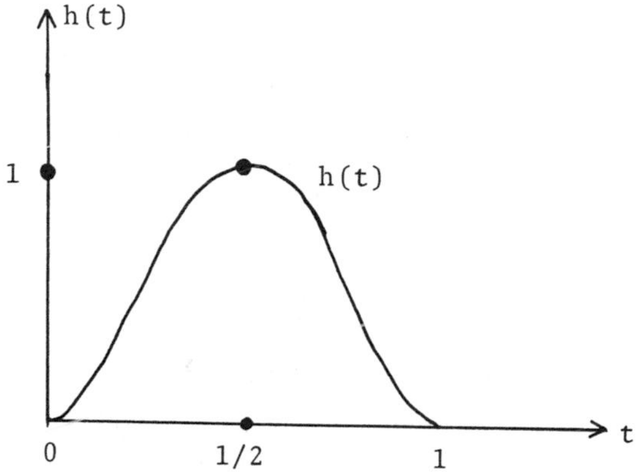

Figure 1: A smooth hat function

Set

$$x_1(t) = h(\tfrac{t+1}{2})\Big|_{[0,1]},$$

$$x_2(t) = h(\tfrac{t}{2})\Big|_{[0,1]},$$

$$x_{2^m+j+1}(t) = h(2^m t+1-j);\ j=1,2,\ldots,2^m;\ m=0,1,2,\ldots .$$

Let $x \in C[0,1]$ and let the coefficients $\lambda_i(x)$ be as in eqn(*) on page 267. $P_n x$ is a piecewise cubic approximation to x, which interpolates x at the points $t_1,\ldots,t_n$. It is plausible that, for all $x \in C[0,1]$,

$$P_n x = \sum_{i=1}^{n} \lambda_i(x)x_i \to x \ , \quad \text{as} \quad n \to \infty \ ,$$

but we do not know whether this is true.

(2) Let t_i , $i \geq 1$, be as in Example 8.1. For $i \geq 1$ let x_i be the cubic spline at the knots t_j , $1 \leq j \leq \max(2,i)$, which satisfies (see Section 5.6, p. 144)

$$\dot{x}(0) = 0 \ , \quad \dot{x}(1) = 0 \ ,$$

$$x_i(t_j) = \delta_{ij} \ , \quad 1 \leq j \leq \max(2,i) \ .$$

Let $x \in C[0,1]$ and let the coefficients $\lambda_i(x)$ be as in eqn(*) on page 267. Then

$$P_n x = \sum_{i=1}^{n} \lambda_i(x)x_i$$

is the cubic spline approximation to x at the knots $t_1,\ldots,t_n$. As shown in Section 5.6, $P_n x \to x$ as $n \to \infty$, so that $\{x_i\}$ is a Schaudér basis.

8.6 (1) Let $x_k \in C[0,1]$ where $x_k : t \to t^k$. As shown in Problem 5.25,

$$B_n x_2 = (1-\tfrac{1}{n})x_2 + \tfrac{1}{n}x_1 \ .$$

Thus

$$\begin{aligned}(B_n)^2 x_2 &= (1-\tfrac{1}{n})B_n x_2 + \frac{1}{n} B_n x_1 \ , \\ &= (1-\tfrac{1}{n})^2 x_2 + (\tfrac{2}{n} - (\tfrac{1}{n})^2)x_1 \ , \\ &\neq B_n x_2 \ .\end{aligned}$$

Hence, $(B_n)^2 \neq B_n$ so that B_n is not a projection.

(2) $P_n^2 = P_n$ but P_n is not linear. For example, take $n=0$. Clearly,

$$(P_0x)(t) = \frac{m+M}{2}, \quad \text{for } t \in [0,1],$$

where $m = \min x(t)$ and $M = \max x(t)$.

Now take $x_1(t) = t$, and $x_2(t) = \cos(2\pi t)$. $P_0x_1 \equiv \tfrac{1}{2}$ and $P_0x_2 \equiv 0$ but

$$P_0(x_1+x_2) = \frac{M+m}{2} \geq \frac{2+(\frac{1}{4}-1)}{2} = \frac{5}{8} \neq P_0x_1 + P_0x_2 .$$

8.7 It is readily checked that if $t \in R^1$ and $u \in R_+^1$ then $|t-u| \geq |t-(t)_+|$. Thus, if $x \in R^n$ and $y \in R_+^n$ then

$$(\|x-y\|_2)^2 = \sum_{i=1}^{n} (x_i-y_i)^2 \geq \sum_{i=1}^{n} (x_i-(x_i)_+)^2 = (\|x-p(x)\|_2)^2 .$$

{271}

8.8 Let

$$x_1 + x_2 = \bar{x}_1 + \bar{x}_2 , \qquad (*)$$

where $x_1, \bar{x}_1 \in M = R(P)$ and $x_2, \bar{x}_2 \in N = R(I-P)$. We wish to prove that $x_1 = \bar{x}_1$ and $x_2 = \bar{x}_2$.

By rearranging eqn(*) we see that

$$x_1 - \bar{x}_1 = \bar{x}_2 - x_2 = y , \quad \text{say}. \qquad (**)$$

Since $y = x_1 - \bar{x}_1 \in M$, there exists $z_1 \in X$ such that $y = Pz_1 \Rightarrow Py = P^2z_1 = y$. Similarly there exists $z_2 \in X$ such that $(I-P)z_2 = y \Rightarrow Py = Pz_2 - P^2z_2 = 0$. Combining these facts we conclude that $y = Py = 0$, so that, from eqn(*), $x_1 = \bar{x}_1$ and $x_2 = \bar{x}_2$.

8.9 (1) If $x = x_1 + x_2$, $x_1 \in M$, $x_2 \in N$,

$y = y_1 + y_2$, $y_1 \in M$, $y_2 \in N$,

then
$$\begin{aligned} P(\alpha x+\beta y) &= P((\alpha x_1+\beta y_1) + (\alpha x_2+\beta y_2)) , \\ &= \alpha x_1 + \beta y_1 = \alpha P(x) + \beta P(y) , \end{aligned}$$

because $\alpha x_1 + \beta y_1 \in M$ and $\alpha x_2 + \beta y_2 \in N$. Thus, P is linear.

(2) If $x \in M$ then

$$Px = P(x+\underline{0}) = x .$$

Thus, P maps X onto M .

(3) If $x = x_1 + x_2$, $x_1 \in M$ and $x_2 \in N$, then

$$P^2x = P(P(x_1+x_2)) = P(x_1) = P(x_1+\underline{0}) = x_1 = Px .$$

Thus, $P^2 = P$. We conclude that P fulfils all the requirements for a projection.

8.10 Introduce the set A of all linear extensions g of P_0 such that: (1) D(g), the domain of g , is a linear subspace of X; and (2) g is a projection of D(g) onto M . If $g_1, g_2 \in A$ we write $g_1 \geq g_2$ if g_1 is an extension of g_2; this induces a partial ordering on A .

Arguing as in the proof of Theorem 7.4 we conclude that A has a maximal element, P say. Then R(P) = X since otherwise, as shown in the hint, P could be extended to a subspace strictly containing R(P).

8.11 (1) $P_i(x) = x_i = \underline{0}+\underline{0}+\ldots+\underline{0}+x_i+\underline{0}+\ldots+\underline{0}$, so that $P_i(P_i(x)) = P_i(\underline{0}+\underline{0}+\ldots+\underline{0}+x_i+\ldots+\underline{0}) = x_i \Rightarrow P_i^2 = P_i$. The linearity of P_i follows readily from the uniqueness of the representation (*).

(2) $P_i(P_j(x)) = P_i(\underline{0}+\ldots+\underline{0}+x_j+\underline{0}\ldots+\underline{0}) = \underline{0}$, if $i \neq j$.

(3) $Ix = x = \sum_{i=1}^{n} x_i = \sum_{i=1}^{n} P_i(x)$ so that $I = \sum_{i=1}^{n} P_i$.

Conversely, if $P_1,\ldots,P_n$ satisfy (1), (2), and (3), then, for any $x \in X$

$$x = \sum_{i=1}^{n} P_i(x) = \sum_{i=1}^{n} x_i \ , \quad x_i \in M_i = R(P_i) \ ,$$
$$1 \le i \le n \ . \qquad (**)$$

We want to show that this representation is unique. Assume $x = \sum_{i=1}^{n} y_i$, $y_i \in M_i$, $1 \le i \le n$. Then $\sum_{i=1}^{n} (x_i - y_i) = 0$ so that , for any k ,

$y_k - x_k = \sum_{i \neq k} (x_i - y_i)$. It follows from the assumptions (1) and (2) that

$$y_k - x_k = P_k(y_k - x_k) = \sum_{i \neq k} P_k(x_i - y_i) \ ,$$

$$= \sum_{i \neq k} P_k P_i (x_i - y_i) = 0 \ ,$$

so that (**) is indeed unique.

8.12 Let M be a subset of a Hilbert space X , and let N be its orthogonal complement,

$$N = M^{\perp} = \{x \in X: \ (x,y) = 0 \ \text{ for all } \ y \in M\} \ .$$

For $y \in X$ introduce the mapping f_y ,

$$f_y: x \in X \to (x,y) \in R^1 (\text{or } \mathbb{C}^1) \ .$$

Then

$$N = \bigcap_{y \in X} f_y^{-1}(0) \ . \qquad (*)$$

f_y is linear, because (x,y) is linear (Definition 2.9, p. 39). f_y is bounded, since from the Schwarz inequality (p. 39)

$$\| f_y(x) \| = |(x,y)| \le \|y\| \ \|x\| \ .$$

As a bounded linear operator, f_y is continuous (Theorem 5.1, p. 118). $\{0\}$ is a closed set in X and therefore the inverse image $f_y^{-1}(0)$ is closed (Theorem 2.3, part 3, p. 20). The union of arbitrarily many open sets is open, from which it follows readily that the intersection of arbitrarily many closed sets is closed. Since N is the intersection of the closed sets $f_y^{-1}(0)$, N is closed. Each set $f_y^{-1}(0)$ is a linear subspace of X, from which it follows that N is also a linear subspace of X.

8.13 (1) Schauder $\Rightarrow$ complete:

If $\{x_i\}$ is a Schauder basis in X then there exist continuous linear functionals λ_i such that, for each $x \in X$,

$$S_n(x) = \sum_{i=1}^{n} \lambda_i(x)x_i \to x , \quad \text{as} \quad n \to \infty .$$

Since inner products are continuous, we conclude that, for every integer j,
$(S_n(x),x_j) = (\lambda_j(x)x_j, x_j) \to (x,x_j)$, as $n \to \infty$, so that $\lambda_j(x) = (x,x_j)$ and hence $S_n(x)$ is equal to the partial Fourier sum of x which we have denoted by $s_n(x)$. Thus, $s_n(x) \to x$ for all $x \in X$. As shown in the text (p. 278) it follows that $\{x_i\}$ is complete.

(2) Complete $\Rightarrow$ Schauder:

If $\{x_i\}$ is complete then, for all $x \in X$,

$$s_n(x) = \sum_{i=1}^{n} \lambda_i(x)x_i \to x \quad \text{as} \quad n \to \infty ,$$

where $\lambda_i(x) = (x,x_i)$. The λ_i are continuous linear functionals If $\sum_{i=1}^{n} \alpha_i x_i \to x$ then it is readily shown, using the orthogonality of the x_i that $\alpha_i = \lambda_i(x)$, so that the expansion in terms of the x_i is unique. We conclude that $\{x_i\}$ is a Schauder basis.

8.14 (1) Two possible approaches are sketched below.

(a) With the notation of Definition 4.1, p. 93, we have $Y = \hat{X}$. Let $x, y \in \hat{Y}$. Then there exist sequences $\{\tilde{x}_n\}$, $\{\tilde{y}_n\}$ in $\tilde{X} \subset \hat{X}$ such that

$$\tilde{x}_n \to x \quad \text{and} \quad \tilde{y}_n \to y\,, \quad \text{in } \hat{X}\,,$$

and corresponding sequences $\{x_n\}$ and $\{y_n\}$ in X. Since $\tilde{X}$ and X are isometrically isomorphic,

$$\|\tilde{x}_n - \tilde{y}_n\|^2 + \|\tilde{x}_n + \tilde{y}_n\|^2 =$$

$$\int_{-1}^{+1} ([x_n(t) - y_n(t)]^2 + [x_n(t) + y_n(t)]^2)\,dt\,.$$

It follows by elementary computation, and proceeding to the limit, that x and y satisfy the parallelogram law

$$\|x-y\|^2 + \|x+y\|^2 = 2[\|x\|^2 + \|y\|^2]\,, \qquad (*)$$

so that, by Theorem 2.8, Y is an inner-product space with an inner-product $(\cdot,\cdot)$ such that $\|x\|^2 = (x,x)$.

If $(\cdot,\cdot)$ denotes this inner-product on Y then,

$$\begin{aligned} 2(x,y) &= (x,x) + (y,y) - (x-y,\, x-y)\,, \\ &= \lim_{n\to\infty} \int_{-1}^{+1} ([x_n(t)]^2 + [y_n(t)]^2 - \\ &\qquad - [x_n(t) - y_n(t)]^2)\,dt\,, \\ &= 2 \lim_{n\to\infty} \int_{-1}^{+1} x_n(t) y_n(t)\,dt\,. \end{aligned}$$

Since the inner-product exists, the limit also exists.

(b) Starting from Definition 4.2, p. 94, we have the entire theory of Lebesgue integration at our disposal. If $x, y \in Y$ then, from Hölder's inequalities (p. 67), we know that the integral

$$(x,y) = \int_{-1}^{+1} x(t)y(t)dt$$

exists. It is readily checked that $(x,x)^{\frac{1}{2}} = \|x\|$ and that $(\cdot,\cdot)$ satisfies the conditions of Definition 2.9 (p. 39) for a real inner product.

(2) We note that

$$\alpha_n = \int_{-1}^{1} t^n dt = \begin{cases} \frac{2}{n+1}, & \text{if } n \text{ is even}, \\ 0, & \text{if } n \text{ is odd}. \end{cases}$$

With the notation of eqn (8.17) (p. 278),

$$\tilde{z}_i : t \to t^{i-1} \text{ and } \|\tilde{z}_i\| = \sqrt{\alpha_{2i-2}}, \text{ for } i \geq 1 .$$

(a) $x_1 = \tilde{z}_1/\|\tilde{z}_1\| = \tilde{z}_1/\sqrt{\alpha_0} = \tilde{z}_1/\sqrt{2}$.

(b) $\tilde{x}_2 = \tilde{z}_2 - (\tilde{z}_2,x_1)x_1 = \tilde{z}_2$,

$x_2 = \tilde{x}_2/\|\tilde{x}_2\| = \tilde{z}_2/\sqrt{\alpha_2} = \sqrt{\frac{3}{2}}\, \tilde{z}_2$.

(c) $\tilde{x}_3 = \tilde{z}_3 - (\tilde{z}_3,x_2)x_2 - (\tilde{z}_3,x_1)x_1$,

$$= \tilde{z}_3 - \frac{\alpha_2}{\sqrt{\alpha_0}} x_1 = \tilde{z}_3 - \frac{\sqrt{2}}{3} x_1 = \tilde{z}_3 - \frac{\alpha_2}{\alpha_0} \tilde{z}_1 .$$

Since $\tilde{x}_3 \perp x_1$, we have by Pythagoras,

$$\|\tilde{x}_3\|^2 = \|\tilde{z}_3\|^2 - \frac{\alpha_2^2}{\alpha_0} \|x_1\|^2 = \alpha_4 - \frac{\alpha_2^2}{\alpha_0} = \frac{8}{45} ,$$

so that

$$x_3 = \tilde{x}_3/\|\tilde{x}_3\| = [\tilde{z}_3 - \frac{\alpha_2}{\alpha_0}\tilde{z}_1]\sqrt{\frac{45}{8}} ,$$

$$= \frac{3}{2}\sqrt{\frac{5}{2}}\,[\tilde{z}_3 - \frac{1}{3}\tilde{z}_1] .$$

8.15 Let $f \in L^2(0,1)$. Choose $\varepsilon > 0$. Since $C_0^\infty(0,1)$ is dense in $L^2(0,1)$ (property (10), p. 95) there exists $g \in C_0^\infty(0,1)$ such that:

$$\|f-g\|_2^2 = (f-g,f-g) = \int_0^1 (f-g)^2 dt < (\tfrac{\varepsilon}{2})^2 .$$

By Weierstrass, there exists a polynomial

$p_n(t) = \sum_{i=0}^{n} a_i t^i$ such that:

$$\|g-p_n\|_\infty = \max_{0 \le t \le 1} |g(t)-p_n(t)| < \frac{\varepsilon}{2} ,$$

so that

$$\|g-p_n\|_2^2 = \int_0^1 (g-p_n)^2 dt \le \int_0^1 (\tfrac{\varepsilon}{2})^2 dt = (\tfrac{\varepsilon}{2})^2 .$$

Therefore:

$$\|f-p_n\|_2 \le \|f-g\|_2 + \|g-p_n\|_2 \le \frac{\varepsilon}{2} + \frac{\varepsilon}{2} = \varepsilon ,$$

and $\{t^i : 0 \le i < \infty\}$ is fundamental in $L^2(0,1)$.

8.16 (1) Making the suggested change of variables:

$$\ell n(t) = u + \frac{n+1}{2} , \quad du = \frac{1}{t} dt , \quad t = e^{u+\frac{n+1}{2}}$$

we obtain

$$I(n) = \int_{-\infty}^{\infty} e^{-[u+\frac{n+1}{2}]^2} \cdot e^{n[u+\frac{n+1}{2}]} \cdot$$

$$\cdot \sin[2\pi u + (n+1)\pi] \cdot e^{[u+\frac{n+1}{2}]} \cdot du ,$$

$$= \int_{-\infty}^{\infty} \exp\{-(u+\tfrac{n+1}{2}) \cdot [(u+\tfrac{n+1}{2})-(n+1)]\} \cdot$$

$$\cdot \sin[2\pi u+(n+1)\pi]du ,$$

$$= \int_{-\infty}^{\infty} \exp\{-u^2+(\tfrac{n+1}{2})^2\} \sin(2\pi u)\cdot(-1)^{n+1}du ,$$

$$= (-1)^{n+1}\exp\{(\tfrac{n+1}{2})^2\} \int_{-\infty}^{\infty} e^{-u^2}\cdot\sin(2\pi u)du = 0,$$

because the integrand is an odd function.

(2) By Pythagoras,

$$\|f-p\|^2 = \|f\|^2 + \|-p\|^2 \geq \|f\|^2 . \quad \square$$

Let the space X be as in Remark 8.5. We have shown in (1) that $f(t) = \sin(2\pi \ell n(t))$ is orthogonal to each polynomial p with respect to the inner product

$$(f,g) = \int_0^{\infty} w(t)f(t)g(t)dt; \; w(t) = t^{-\ell n(t)} .$$

By (2), $\|f-p\| \geq \|f\|$ for every polynomial p and so the polynomials are not dense in X , for this choice of weight function w .

8.17 (1) We begin by noting that $x \in X$ iff x is measurable and

$$\int_0^{\infty} w(t)x^2(t)dt = \int_0^{\infty} e^{-t}x^2(t)dt = \int_0^{\infty} (x(t)e^{-\frac{1}{2}t})^2 dt < \infty ,$$

so that $x \in X$ iff $w^{\frac{1}{2}}x \in L^2(0,\infty)$.

Now choose $x \in X$ and $\varepsilon > 0$. Then $f = w^{\frac{1}{2}}x \in L^2(0,\infty)$. Since $C_0^\infty(0,\infty)$ is dense in $L^2(0,\infty)$ (see p. 95), there exists $g \in C_0^\infty(0,\infty)$ such that

$$\int_0^\infty [f(t)-g(t)]^2 dt = \int_0^\infty w(t)[x(t)-h(t)]^2 dt \le \varepsilon^2 ,$$

where $h(t) = [w(t)]^{-\frac{1}{2}}g(t)$. We conclude that $h \in C_0^\infty(0,\infty)$ and that $\|x-h\| \le \varepsilon$, so that $C_0^\infty(0,\infty)$ is indeed dense in X.

(2) Let $y_r(t) = e^{-rt}$, where r is a non-negative integer. It is known from the theory of orthogonal polynomials that

$$\int_0^\infty e^{-t}L_i(t)L_j(t)dt = \delta_{ij} , \qquad (*)$$

$$L_i(t) = \sum_{k=0}^{i} \binom{i}{k} \frac{1}{k!}(-1)^k t^k . \qquad (**)$$

By direct computation,

$$\|y_r\|^2 = \int_0^\infty e^{-t}e^{-2rt}dt = \int_0^\infty e^{-(2r+1)t}dt = \frac{1}{2r+1} . \qquad (***)$$

Furthermore,

$$\begin{aligned}
(y_r,L_i) &= \int_0^\infty e^{-t} \sum_{k=0}^{i} \binom{i}{k} \frac{1}{k!}(-1)^k t^k \cdot e^{-rt}dt ,\\
&= \sum_{k=0}^{i} \binom{i}{k} \frac{1}{k!} (-1)^k \int_0^\infty e^{-(r+1)t}t^k dt ,\\
&= \sum_{k=0}^{i} \binom{i}{k} \frac{1}{k!} (-1)^k \cdot \frac{k!}{(r+1)^{k+1}} ,\\
&= \frac{1}{(r+1)} \cdot \sum_{k=0}^{i} \binom{i}{k} 1^{i-k} \cdot \left(\frac{-1}{r+1}\right)^k ,\\
&= \frac{1}{r+1} \cdot \left(1 - \frac{1}{r+1}\right)^i = \frac{r^i}{(r+1)^{i+1}} .
\end{aligned}$$

Thus,

$$\sum_{i=0}^{\infty} |(y_r, L_i)|^2 = \frac{1}{(r+1)^2} \cdot \sum_{i=0}^{\infty} \left(\frac{r}{r+1}\right)^{2i} = \frac{1}{(2r+1)} . \qquad (\substack{**\\**})$$

Comparing eqns ($*\substack{*}$) and ($\substack{**\\**}$) we see that Parseval's equality holds for $y_r(t) = e^{-rt}$.

(3) Make the substitution $u = e^{-t}$.

(4) Now let $x \in X$. Choose $\varepsilon > 0$. By (1) there exists $f \in C_0^{\infty}(0,\infty)$ such that $\|x-f\| \le \varepsilon$. The function g defined by $g(u) = f(-\ell n(u))$ belongs to $C[0,1]$ and so, by the Weierstrass theorem, there exists a polynomial

$$p(u) = \sum_{r=0}^{n} a_r u^r , \quad \text{say,}$$

such that $\|g-p;\ C[0,1]\| \le \varepsilon$. Let

$$h(t) = \sum_{r=0}^{n} a_r e^{-rt} = \sum_{r=0}^{n} a_r y_r(t) = p(e^{-t}) .$$

Using (3) we conclude that

$$\|x-h\| \le \|x-f\| + \|f-h\| = \varepsilon + \|g-p;\ L^2(0,1)\| \le 2\varepsilon .$$

In (2) it was shown that Parseval's equality holds for $e^{-rt} = y_r(t)$, $r = 0,1,2,\ldots$, with respect to the orthonormal system $\{L_i : i \ge 0\}$. That is (p. 278), the partial Fourier sums

$$s_m(y_r) = \sum_{i=0}^{m} (y_r, L_i) L_i$$

converge in X to y_r as $m \to \infty$. Thus there exists an integer M such that

$$\|s_M(y_r) - y_r\| \le \varepsilon/(n+1) \quad \text{for} \quad 0 \le r \le n .$$

Combining the above results we see that

$$\|x - \sum_{r=0}^{n} a_r \sum_{i=0}^{M} (y_r, L_i) L_i\| \le 3\varepsilon \ ,$$

so that the polynomials are dense in X .

8.18 Each linear subspace X_α in E^2 is a straight line passing through the origin with slope α (see Figure).

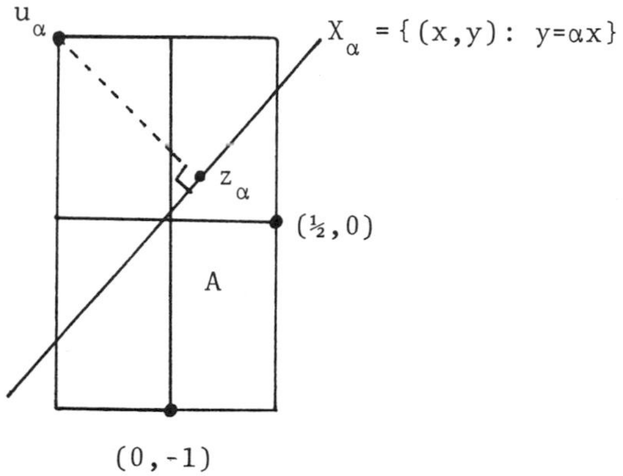

Figure: The set A *and subspace* X_α

Let

$$\sup_{u\in A} \inf_{z\in X_\alpha} \|z-u\| = \|z_\alpha - u_\alpha\| \ , \quad \text{say.}$$

Geometrically it is easily seen that u_α may be taken to be either the upper left or upper right corner of A (depending on the sign of α) . The corresponding closest point $z_\alpha \in X_\alpha$ is the projection of u_α on X_α . Geometrically it is now easily seen that $\|z_\alpha - u_\alpha\|$ is minimized when X_α is the vertical y-axis, in which case $u_\alpha = (\tfrac{1}{2},1)$, $z_\alpha = (0,1)$, and

$$\|u_\alpha - z_\alpha\| = \tfrac{1}{2} \ .$$

8.19 The proof is by contradiction. Let $e = x - s_n x \in X$. Assume that e has fewer than n changes of sign on $[-1,+1]$, and denote the points where these occur by $t_1,\ldots,t_m$. Define $p \in \pi_{n-1}$ by

$$p(t) = \prod_{j=1}^{m} (t-t_j) .$$

By construction, $p(t)e(t)$ does not change sign on $[-1,+1]$ so that $(p,e) \neq 0$.

On the other hand, $s_n x$ is the orthogonal projection of x onto π_{n-1} and so, by Theorem 8.4 (part 2), $(s_n x - x) \perp \pi_{n-1}$. But $e = -(s_n x - x)$ and $p \in \pi_{n-1}$, from which we conclude that $(p,e) = 0$. This is the desired contradiction.

8.20 This is surprisingly hard to prove. -- see Isaacson and Keller [1966, p. 206].

8.21 The following results were obtained

n	t	$2((s_n x)(t))/\pi$
10	.2800	1.0863
100	.0320	1.1682
1000	.0032	1.1774

8.22 By direct computation,

$$p_1(t) = 1 , \quad p_2(t) = t , \quad p_3(t) = 2t^2-1 ,$$

so that the hypothesis (*) is satisfied if $n = 2$ and $n = 3$.

In the general case we proceed by induction. It is readily verified that p_n satisfies the recurrence relation

$$p_n(t) = 2tp_{n-1}(t) - p_{n-2}(t) . \qquad (*)$$

It is also readily seen that the nonzero coefficients $c_{n-1}^{(n)}, c_{n-3}^{(n)}, c_{n-5}^{(n)} \ldots$ alternate in sign. Thus, from eqn (*), $|c_i^{(n)}| \geq 2|c_{i-1}^{(n-1)}|$ for $1 \leq i \leq n-1$. Thus

$$\sum_{i=0}^{n-1} |c_i^{(n)}|^2 \geq \sum_{i=1}^{n-1} |c_i^{(n)}|^2 \geq 4 \sum_{i=0}^{n-2} |c_i^{(n-1)}|^2 ,$$

which implies that

$$\sum_{i=0}^{n-1} |c_i^{(n)}|^2 \geq 4^{n-2} .$$

8.23 The inequality (*) implies that

$$2\|x\| \geq \left| \int_{-1}^{+1} x(t)P_n(t)dt \right| \geq \frac{\|\lambda\|_\infty}{n} \geq \frac{\|\lambda\|_2}{n^{3/2}} = \frac{\|\phi_n x\|}{n^{3/2}} ,$$

so that $\|\phi_n\| \leq 2n^{3/2}$.

On the other hand,

$$\|\phi_n^{-1}\lambda\| = \|\sum \lambda_i x_i\|_\infty \leq \sum |\lambda_i| = \|\lambda\|_1 \leq n^{\frac{1}{2}}\|\lambda\|_2 ,$$

so that $\|\phi_n^{-1}\| \leq n^{\frac{1}{2}}$.

8.24 Let $f = T^{-1}$ so that $f: Y \to X$. If G is any open set in X then $f^{-1}(G) = T(G)$, which is open by the open mapping theorem (Theorem 8.6). Applying Theorem 2.3 (part 2) (p. 20) we conclude that $f = T^{-1}$ is continuous.

8.25 For $n = 1,2,\ldots,$ define $x_n \in X = C[0,1]$ by $x_n(t) = t^n$. Then $x_n \in D(T)$. But

$$\|Tx_n\| = \|\dot{x}_n\| = n = n\|x_n\| .$$

Thus T is not a bounded operator, and hence, by

Theorem 5.1 (p. 118), T is not continuous.

Theorem 8.8 is not applicable because while $D(T)$ is indeed a normed linear space, it is not complete.

Consider for example the functions f_n,

$$f_n(t) = \begin{cases} t\,, & 0 \le t \le \frac{1}{2} - \frac{1}{n}\,, \\ 1-t, & \frac{1}{2} + \frac{1}{n} \le t \le 1\,, \\ (\frac{1}{2}-\frac{1}{n}) + \frac{2}{n\pi}\cos(\frac{n\pi}{2}(t-\frac{1}{2}))\,, & \frac{1}{2} - \frac{1}{n} \le t \le \frac{1}{2} + \frac{1}{n}\,. \end{cases}$$

Then $f_n \in D(T)$ and $f_n \to f$ in X, where

$$f(t) = \begin{cases} t\,, & 0 \le t \le \frac{1}{2}\,, \\ 1-t\,, & \frac{1}{2} \le t \le 1\,. \end{cases}$$

But $f(\frac{1}{2})$ does not exist, so that $f \notin D(T)$.

8.26 (1) Denote the columns of A by $v_1, \ldots, v_n$, where each v_i is an m-vector. Since A maps R^n onto R^m, A has rank m, and so at least m of the vectors v_i must be linearly independent. By renumbering if necessary we may assume that $v_1, \ldots, v_m$ are linearly independent. Let E be the $m \times m$ matrix with columns $v_1, \ldots, v_m$. E is nonsingular.

If $Ax_0 = y_0$ and $y \in B(y_0; \varepsilon)$ let $u = E^{-1}(y-y_0)$, and $x = x_0 + v$ where $v \in R^n$, $v_i = u_i$ for $1 \le i \le m$, $v_i = 0$ for $m+1 \le i \le n$. Then $Ax = y$ and $x \in B(x_0; \varepsilon\|E^{-1}\|)$.

(2) We apply the open mapping theorem (Theorem 8.6) with $X = R^n$, $Y = R^m$, and $T = A$. We can conclude that A maps the open ball $B(x_0; 1)$ onto an open set G in R^m containing y_0. Thus, for some $r > 0$,

$$A(B(x_0; 1)) \supset B(y_0; r)\,.$$

Take $\delta = \varepsilon/r$.

(3) Take $m = n = 1$ and $A = (0)$.

8.27 By assumption, L is a linear one-to-one mapping of X onto Y . L is continuous because

$$\| Lu; C(\overline{\Omega})\| \le M\| u; C_0^2(\overline{\Omega})\| ,$$

where

$$M = (n^2+n+1) \max_{\substack{1\le i,j\le n \\ x\in\overline{\Omega}}} \{|a_{ij}|, |b_i|, c\} .$$

Thus, by Theorem 8.7, L^{-1} is continuous.

CHAPTER 9: SOLUTIONS

9.1 Since $g_n \in R(K_n)$, the left and right hand sides of eqn (9.2) belong to the domain of ϕ_n. Also $u_n = g_n + K_n u_n \in R(K_n)$. Applying ϕ_n to eqn (9.2),

$$\phi_n(I-K_n)\phi_n^{-1}\phi_n u_n = \phi_n g_n . \qquad (*)$$

ϕ_n is an isomorphism, and so eqn (9.2) can be recovered from eqn(*) by applying ϕ_n^{-1} to both sides of eqn (*). That is, eqns (9.2) and (*) are equivalent.

Introducing the terminology of eqns (9.5b), (9.5c), (9.5d), and (9.5e), eqn (*) becomes eqn (9.5a).

9.2 Since $K_n g_n \in R(K_n)$, the arguments of Case 1 are applicable to eqn (9.7). Replacing g_n by $K_n g_n$ and u_n by $(u_n - g_n)$ in eqns (9.5), we obtain eqns (9.8).

9.3 Let

$$F(s) = \int_{-1}^{+1} \frac{dt}{1+(s-t)^2} = \arctan(1+s) + \arctan(1-s) .$$

(1) F is symmetric about $s = 0$ and

$$\dot{F}(s) = \frac{1}{1+(1+s)^2} - \frac{1}{1+(1-s)^2} < 0 , \text{ if } 0 \le s \le 1 .$$

Therefore,

$$2 \arctan 1 = \frac{\pi}{2} = F(0) \ge F(s) \ge F(1) = \arctan 2,$$

$$\text{for } -1 \le s \le 1 .$$

(2) If u satisfies eqn (9.9) with $d = -1$ then

$$((I-K)u)(s) = u(s) + \frac{1}{\pi}\int_{-1}^{+1} \frac{u(t)}{1+(s-t)^2}\, dt = 1 , \qquad (*)$$

$$-1 \le s \le 1 .$$

Let $m = u(s_1)$, $M = u(s_2)$. From eqn (*) with $s = s_1$,

$$1 = u(s_1) + \frac{1}{\pi}\int_{-1}^{+1} \frac{u(t)}{1+(s_1-t)^2}\,dt \le m + \frac{M}{\pi} F(s_1) \le m + \frac{1}{2} M . \qquad (**)$$

Similarly, setting $s = s_2$,

$$1 = u(s_2) + \frac{1}{\pi}\int_{-1}^{+1} \frac{u(t)}{1+(s_2-t)^2}\,dt \ge M + \frac{m}{\pi} F(s_2) ,$$

$$\ge M + \frac{m}{\pi} \text{arc tan } 2 . \qquad (***)$$

Finally, from (1) and eqn (5.11) (p. 113), $\|K\| = \frac{1}{2}$. Thus, using Theorem 9.2 (p. 306),

$$\max\{|m|, |M|\} = \|u\| = \|(I-K)^{-1}1\| \le \frac{1}{1-\|K\|} \cdot 1 = 2 . \qquad (****)$$

It follows easily from inequalities (**), (***), and (****), that $\frac{1}{2} \le m \le M \le 1$.

(3) Assume that u is a solution of Love's integral equation (*). Let $v(s) = u(-s)$ for $-1 \le s \le 1$. Then

$$v(s) = u(-s) = 1 - \frac{1}{\pi}\int_{-1}^{+1} \frac{u(t)}{1+(-s-t)^2}\,dt .$$

Making the substitution $r = -t$, we find that

$$v(s) = 1 - \frac{1}{\pi}\int_{-1}^{1} \frac{u(-r)}{1+(-s+r)^2}\,dr = 1 - \frac{1}{\pi}\int_{-1}^{+1} \frac{v(r)}{1+(s-r)^2}\,dr ,$$

so that v is also a solution of eqn (*). The assumption that eqn (*) has a unique solution then implies that

$$u(s) = v(s) = u(-s) , \quad -1 \le s \le 1 .$$

9.4 Replace the variable s in eqn (*) by t, multiply by $K(s,t)$, and integrate with respect to t from 0 to 1. We obtain eqn (**) provided that

$$\int_0^1 \ddot{u}(t)K(s,t)dt = -u(s) . \qquad (*\underset{*}{*})$$

To prove that eqn $(*\underset{*}{*})$ holds, we may use repeated integration by parts:

$$\int_0^1 \ddot{u}(t)K(s,t)dt = \int_0^s \ddot{u}(t)(1-s)t\,dt + \int_s^1 \ddot{u}(t)(1-t)s\,dt ,$$

$$= \Big[\dot{u}(t)(1-s)t\Big]_{t=0}^{t=s} - \int_0^s \dot{u}(t)(1-s)\,dt +$$

$$+ \Big[\dot{u}(t)(1-t)s\Big]_{t=s}^{t=1} + \int_s^1 \dot{u}(t)s\,dt ,$$

$$= \Big[-u(t)(1-s)\Big]_{t=0}^{t=s} + \Big[u(t)s\Big]_{t=s}^{t=1}$$

$$= -u(s) .$$

9.5 (1) Since $\{x_i\}$ is a Schauder basis, each λ_i is a bounded linear functional. Thus, P_n maps bounded sets in X into bounded sets in the finite-dimensional subspace $R(P_n)$. We can conclude that P_n is compact (see Theorem 6.11, p. 200).

(2) $P_n x \to x$ as $n \to \infty$ by the definition of a basis (see eqn (8.1), p. 264).

(3) The space X is infinite dimensional. By Theorem 6.11, (p. 200) there exists a bounded set in X,U say, such that U is not sequentially compact. By Theorem 6.3, U is not precompact. But $I(U) = U$, and so I is not compact.

9.6 Let $x \in X$. Then

$$\|STx\| = \|S(Tx)\| \le \|S\| \cdot \|Tx\| \le \|S\| \cdot \|T\| \cdot \|x\| .$$

If $X = Y = Z = \ell_\infty^2$, and S and T are defined, respectively, by the matrices

$$\begin{pmatrix} 0 & 1 \\ 0 & 0 \end{pmatrix} \text{ and } \begin{pmatrix} 1 & 0 \\ 0 & 0 \end{pmatrix} ,$$

then, using Example 5.1 (p. 110) ,

$$\|S\| = 1 , \quad \|T\| = 1 , \quad \|ST\| = 0 .$$

9.7

$$\begin{pmatrix} 0 & 1 \\ 0 & 0 \end{pmatrix} \begin{pmatrix} 0 & 0 \\ 1 & 0 \end{pmatrix} = \begin{pmatrix} 1 & 0 \\ 0 & 0 \end{pmatrix} ,$$

$$\begin{pmatrix} 0 & 0 \\ 1 & 0 \end{pmatrix} \begin{pmatrix} 0 & 1 \\ 0 & 0 \end{pmatrix} = \begin{pmatrix} 0 & 0 \\ 0 & 1 \end{pmatrix} .$$

9.8 Outline of proofs.

(1) T^{-1} exists $\iff ((Tx_1 = Tx_2) \iff (x_1 = x_2))$,

$\iff ((T(x_1 - x_2) = 0) \iff (x_1 - x_2 = 0))$,

$\iff ((Tx = 0) \iff (x = 0))$.

(2) $T[\alpha T^{-1}x_1 + \beta T^{-1}x_2] = \alpha x_1 + \beta x_2$.

(3) Follows from eqns (9.14) and (9.15).

(4) See hint.

(5) Obvious.

(6) Let $x_1, \ldots, x_n$ be a basis for X , and set $y_i = Tx_i$. Then $R(T) = \text{span}(y_1, \ldots, y_n)$.

(7) By (6), $\dim(R(T)) \le \dim(X)$. But $T^{-1}: R(T) \to X$, so that, by (6) again, $\dim(X) \le \dim(R(T))$.

(8) If S^{-1} and T^{-1} exist, then clearly ST is one-to-one and $(ST)^{-1} = T^{-1}S^{-1}$.

Now assume that $(ST)^{-1}$ exists, and use (1). If $Tx = 0 \Rightarrow STx = 0 \Rightarrow x = 0$. Also, if $Sy = 0$ then there exists x such that $y = Tx$; but then $Sy = STx = 0$ so that $x = 0$ and hence $y = Tx = 0$.

9.9 It is a trivial computation to show that $TSy = y$ for $y \in R^1$. T^{-1} does not exist because

$$T\begin{pmatrix}1\\0\end{pmatrix} = T\begin{pmatrix}1\\1\end{pmatrix} = 1 ,$$

so that T is not one-to-one.

9.10 Let $y_n \in R(T)$ satisfy $y_n \to y \in Y$. Let $x_n = T^{-1}y_n$. Then $\{x_n\}$ is a Cauchy sequence in the Banach space X and so $x_n \to x \in X$. Clearly, $Tx = y$. Thus $y \in R(T)$ and so we may conclude that $R(T)$ is closed (as a subset of Y).

9.11 We assume that X and Y are finite dimensional.

As a finite dimensional subspace of a Banach space, $R(T)$ is also a Banach space (Remark 6.7, p. 201).

Since $R(T)$ is closed, $\overline{R(T)} = R(T)$ and so case 9.16(2) cannot occur. In particular, cases 2b and 2c cannot occur.

Let T^{-1} exist. Then T^{-1} is continuous. To see this, let

$$T_1 : x \in X \to Tx \in Y_1 \equiv R(T) .$$

Applying Theorem 8.7 (p. 287) we conclude that T_1^{-1} , which is the same as T^{-1} , is continuous. Thus case 9.17(b), namely T^{-1} exists but is not continuous, cannot occur. In particular, case 3b cannot occur.

9.12 (1a) T is the identity operator.

(1c) Clearly $R(T) = Y$. But $Tx_1 = \underline{0}$ and $x_1 \neq 0$, so T^{-1} does not exist (Theorem 9.1 (1)).

(2b) Let $v = (v_i) = (1,\frac{1}{2},\frac{1}{4},\ldots) = \sum_{k=1}^{\infty} x_k 2^{1-k} \in Y$.

If $u = (u_i) \in X$ and $Tu = v$ then $v_i = 2^{1-i}u_i$ so that $u_i = 1$ for all i . This is impossible since $(1,1,1,\ldots)$ does not belong to X . We conclude that $R(T) \neq Y$.

If $y=(y_i) \in Y$, choose $\varepsilon > 0$. Then, for some n , $\|z_n - y\| \leq \varepsilon$, where $z_n = (y_1,\ldots,y_n,0,0,\ldots)$. Clearly, there exists $u_n \in X$ such that $Tu_n = z_n$. Thus, $R(T)$ is dense in Y .

It follows immediately from the definition of T , that $Tu = 0$ iff $u = 0$, so that T^{-1} exists.

Finally,

$$\|T^{-1}(Tx_k)\| = \|x_k\| = 1 = 2^{k-1}\|Tx\| ,$$

so that T^{-1} is not bounded.

(2c) By the same arguments as in case (2b) above, we conclude that $R(T) \neq Y$ but $\overline{R(T)} = Y$. $Tx_1 = 0$, so that T^{-1} does not exist.

(3a) T is clearly continuous. If $u \in X$ then $\|Tu - x_1\| \geq 1$ so that $\overline{R(T)} \neq Y$.

(3b) See (2b) and (3a).

(3c) Obvious.

9.13 Given $y \in Y$ we have that $Tx = y$ if

$$x(s) = \int_0^s y(t)dt .$$

Thus, $R(T) = Y$. However, if $x(s) = 1$ for $0 < s < 1$ then $x \neq 0$ but $Tx = 0$ and so T^{-1} does not exist.

9.14 Direct computation.

9.15 The solution is

$$A^{+} = \frac{1}{102}\begin{pmatrix} -15 & -18 & 3 & -3 & 18 & 15 \\ 8 & 13 & -5 & 5 & -13 & -8 \\ 7 & 5 & 2 & -2 & -5 & -7 \\ 6 & -3 & 9 & -9 & 3 & -6 \end{pmatrix}$$

9.16 Since $\|A\| < 1$, we know from Theorem 9.2 that $(I-A)^{-1} \in B(X)$. Thus,

$$(I-A)^{-1}BTx = x \ , \quad \text{for} \quad x \in X \ .$$

By Theorem 9.1(4), T^{-1} exists and

$$T^{-1} = (I-A)^{-1}B\Big|_{R(T)} = (I-A)^{-1}\tilde{B} \ .$$

The remaining assertions follow using Theorem 9.2.

9.17 We note that eqn (**) is of the form $BT = I - A$, with appropriate definitions of A and B . It follows from eqn (*) and Problem 9.16 that $T^{-1} \equiv (I-L)^{-1}$

exists and is continuous. Eqn (**) is obtained by substituting the expressions for A and B into Problem 9.16(3).

To prove eqn $({}^{*}_{*}{}^{*})$, one may begin by writing eqn $({}^{**}_{**})$ in the

$$[I+(I-K)^{-1}L](I-L) = I - A . \qquad ({}^{***}_{\;**})$$

Now set $y = (I-L)x$. Applying the operators in eqn $({}^{***}_{\;**})$ to x , we obtain,

$$[I+(I-K)^{-1}L]y = (I-K)^{-1}[I+L-K]y = (I-A)x = (I-A)(I-L)^{-1}y .$$

Rearranging terms we obtain,

$$\begin{aligned}[(I-K)^{-1}-(I-L)^{-1}]y &= -A(I-L)^{-1}y + (I-K)^{-1}(K-L)y , \\ &= A[(I-K)^{-1}-(I-L)^{-1}]y - A(I-K)^{-1}y + \\ &\quad + (I-K)^{-1}(K-L)y .\end{aligned}$$

Thus,

$$\begin{aligned}\|(I-K)^{-1}y-(I-L)^{-1}y\| &\le \Delta \, \|(I-K)^{-1}y-(I-L)^{-1}y\| + \\ &\quad + \Delta \, \|(I-K)^{-1}y\| + \|(I-K)^{-1}\| \, \|(K-L)y\| ,\end{aligned}$$

and eqn $(*^{*}_{*})$ follows.

It may be observed that there is no systematic way of deriving inequalities of this type.

9.18 See the hint.

9.19 Clearly $\|y_{i+1}\| \le q\|y_i\|$, so that $\|y_i\| \le q^i\|y\|$. It follows readily that $\{y_i\}$ and $\{\sum_{k=0}^{i} x_k\}$ are Cauchy sequences converging, respectively, to $\underline{0}$ and x , say. Since $y_{i+1} = y - T(\sum_{k=0}^{i} x_k)$, $Tx = y$.

9.20 Straightforward manipulation.

9.21 (a) Use the identity

$$(I-K_n)(u-u_n) = (K-K_n)u .$$

(b) Starting with eqn(9.25) on page 310, replace u by $(I-K)^{-1}g$ on the right hand side.

(c) Starting with (a), replace u by $[(u-u_n)+u_n]$ on the right hand side.

9.22 Let $\alpha = \frac{2}{5} + \frac{1}{\sqrt{5}}$. The error is

$$e(t) = \alpha - \frac{t}{5} - \frac{1}{1+t} .$$

Interior extremal points t satisfy

$$\dot{e}(t) = -\frac{1}{5} + \frac{1}{(1+t)^2} = 0 ,$$

so that the only interior extremum occurs when $t = \sqrt{5} - 1 = \gamma$, say. Possible extrema are $e(0) = \alpha - 1$; $e(4) = \alpha - \frac{4}{5} - \frac{1}{5} = \alpha - 1$; and

$$e(\gamma) = \alpha - \frac{\gamma}{5} - \frac{1}{1+\gamma} = \frac{2}{5} + \frac{1}{\sqrt{5}} - \frac{1}{\sqrt{5}} + \frac{1}{5} - \frac{1}{\sqrt{5}} = 1 - \alpha .$$

Since these 3 extrema occur with equal magnitude and alternating sign, p(t) is the optimum linear approximation to $1/(1+t)$. (Example 7.2, p.243).

9.23 $u_3(s) \doteq 1.9159 - .2340s^2$.

9.24 We approximate Love's integral equation (9.9) with $d = -1$ by the degenerate equation (9.27).

Noting that

$$k(s,t) \in [-1/5\pi, -1/\pi] \quad \text{for} \quad s,t \in [-1, +1] ,$$

we choose

$$x_1(s) = -\frac{3}{5\pi} , \quad y_1(t) = 1 , \quad k_1(s,t) = x_1(s)y_1(t) ,$$

so that

$$|k(s,t) - k_1(s,t)| \le 2/5\pi \ , \quad \text{for} \quad -1 \le s \ , \ t \le 1 \ .$$

From Example 5.5 (p. 113), and Problem 9.3(part 1),

$$\|K-K_1\|_\infty = \max_{-1\le s\le 1} \int_{-1}^{1} |k(s,t)-k_1(s,t)|dt \le 4/(5\pi) \ ,$$

and

$$\|K\|_\infty = \max_{-1\le s\le 1} \int_{-1}^{+1} |k(s,t)|dt = \frac{1}{2} \ .$$

Then eqns (9.29) became,

$$c_1^{(1)} = \int_{-1}^{+1} y_1(t)g(t)dt = 2 \ ,$$

$$a_{11}^{(1)} = 1 - \int_{-1}^{+1} y_1(t)x_1(t)dt = 1 + 6/(5\pi) \ ,$$

so that

$$w^{(1)} = (A^{(1)})^{-1}c^{(1)} = \frac{10\pi}{5\pi+6} \ ,$$

and

$$\begin{aligned} u_1(s) &= g(s) + w_1^{(1)}x_1(s) \ , \\ &= 5\pi/(5\pi+6) \ . \end{aligned}$$

As an error bound we obtain from eqn (9.25),

$$\begin{aligned} \|u-u_1\| &\le \|(I-K)^{-1}(K-K_1)u_1\| \ , \\ &\le \frac{\|K-K_1\|}{1-\|K\|} \|u_1\| \ , \\ &\le (8/5\pi)\|u_1\| \ , \\ &= \frac{8}{5\pi+6} \ , \end{aligned}$$

which leads to bounds for x:

$$.355 \le \frac{5\pi-8}{5\pi+6} \le u(s) \le \frac{5\pi+8}{5\pi+6} \le 1.093 \ .$$

9.25 Let $x_i(s)$ and $z_j(t)$ be the piecewise linear functions satisfying $x_i(s_k) = \delta_{ik}$, $z_j(t_k) = \delta_{jk}$. Then

$$k_n(s,t) = \sum_{i=1}^{n} x_i(s)y_i(t) \ ,$$

where

$$y_i(t) = \sum_{k=1}^{n} k(s_i,t_k)z_k(t) \ .$$

Also,

$$|k(s,t)-k_n(s,t)| \le h_n[\max |\frac{\partial k}{\partial s}| + \max |\frac{\partial k}{\partial t}|] \ ,$$

$$h_n = \max_{1\le i\le n} \left[\max[|s_{i+1}-s_i|, |t_{i+1}-t_i|]\right].$$

9.26 (1) $\Longleftrightarrow$ (3): Setting $z = 0$, $M = X_n$, and $x = r_n$, we see that $z \in M$. Thus, by Theorem 8.4(2),

$$(r_n-0) \perp M \Longleftrightarrow P_n r_n = 0 \ .$$

(2) $\Longleftrightarrow$ (3): Use eqn (8.10), p. 274.

9.27 Straightforward laborious calculation.

9.28 Apply Hölder's inequality (p. 67) to eqn (*).

9.29 $u_3(s) \doteq 1.9160 - .2857s^2$.

9.30 (1) The matrix $\tilde{S}^{(n)} = I - \tilde{C}^{(n)}\tilde{A}^{(n)}$ is computable, as is $\|\tilde{S}^{(n)}\|_\infty$. If $\|\tilde{S}^{(n)}\|_\infty = r < 1$ then, by Theorem 9.2, $(I-\tilde{S}^{(n)})^{-1} \in B(R^n)$, and $\|(I-\tilde{S}^{(n)})^{-1}\|_\infty \le 1/(1-r)$. Thus, by Problem 9.16,

$(\tilde{A}^{(n)})^{-1}$ exists and satisfies

$$\|(\tilde{A}^{(n)})^{-1}\|_\infty \le \|\tilde{C}^{(n)}\|_\infty/(1-r) = \tilde{\alpha} \text{ , say.}$$

For matrices the Fredholm alternative holds (Theorem 9.5, p. 309), so that $(\tilde{A}^{(n)})^{-1} \in B(R^n)$.

(2) If $\varepsilon < 1/n\tilde{\alpha}$ then

$$\|A^{(n)}-\tilde{A}^{(n)}\|_\infty < \frac{1}{\tilde{\alpha}} \le \frac{1}{\|(\tilde{A}^{(n)})^{-1}\|_\infty} ,$$

so that, by Theorem 9.3, $(A^{(n)})^{-1} \in B(R^n)$ and

$$\|(A^{(n)})^{-1}\|_\infty \le \frac{\tilde{\alpha}}{1-\tilde{\alpha}n\varepsilon} = \alpha \text{ , say.}$$

(3) Using Theorem 9.3 (eqn(9.23)),

$$\begin{aligned}\|\lambda^{(n)}-\hat{\lambda}^{(n)}\|_\infty &\le \|A_n^{-1}-\tilde{A}_n^{-1}\|_\infty\|\tilde{c}^{(n)}\|_\infty + \\ &\quad + \|A_n^{-1}\|_\infty \|\tilde{c}^{(n)}-c^{(n)}\|_\infty + \\ &\quad + \|\tilde{A}_n^{-1}\|_\infty \|\tilde{A}_n\hat{\lambda}^{(n)}-\tilde{c}^{(n)}\|_\infty , \\ &\le \alpha\,\tilde{\alpha}(n\varepsilon)\,\|\tilde{c}^{(n)}\|_\infty + \alpha\varepsilon + \tilde{\alpha}\|\tilde{A}_n\hat{\lambda}^{(n)}-\tilde{c}^{(n)}\|_\infty .\end{aligned}$$

9.31 Let $y = \sum_{j=1}^{n} \lambda_j x_j \in X_n$. If $b \in R^n$, then the conditions $y(s_i) = b_i$ for $1 \le i \le n$ reduce to the system of linear algebraic equations $A\lambda = b$ where $\lambda = (\lambda_j)$ and A is the $n \times n$ matrix $(x_j(s_i))$. It has been assumed that A is non-singular, so that λ , and hence y , is uniquely determined by b .

It follows that, for any $x \in C[a,b]$, eqn (**) uniquely determines $P_n x \in X_n$. Clearly, $P_n(P_n x) = P_n x$, so that P_n is a projection of X onto X_n .

Now consider eqn (*), which we write in the form $(I-K)u = g$. The approximate equations corresponding to the projection P_n are

$$u_n - P_n K u_n = P_n g \ . \qquad ({}^{**}_{**})$$

The elements in X_n on the left and right hand sides of eqn $({}^{**}_{**})$ are uniquely determined by their values at the points s_i; equating these values, and using eqn $(*{}^{*}_{*})$, we obtain eqn (**).

9.32 ________________________________

9.33 Let

$$a_{ij} = \int_a^b \dot{v}_j(s)\dot{v}_i(s)\, ds$$

Then $a_{ij} = 0$ if $|i-j| > 1$.

$$a_{ii} = \int_{t_{i-1}}^{t_{i+1}} [\dot{v}_i(s)]^2 ds = \frac{1}{h^2} \int_{t_{i-1}}^{t_{i+1}} ds = \frac{2}{h} \ .$$

$$a_{i+1,i} = a_{i,i+1} = \int_{t_{i-1}}^{t_{i+1}} \dot{v}_i(s)\dot{v}_{i+1}(s) ds \ ,$$

$$= \frac{-1}{h^2} \int_{t_i}^{t_{i+1}} ds = -\frac{1}{h} \ .$$

9.34 Since

$$\|K\|_\infty = \sup_{\|x\|_\infty = 1} \|Kx\|_\infty \ ,$$

there exists $\tilde{x} \in C[a,b]$ such that $\|\tilde{x}\|_\infty = 1$ and

$$\|K\tilde{x}\|_\infty = \max_{a\le s\le b}\left|\int_a^b k(s,t)\tilde{x}(t)dt\right| ,$$

$$= \left|\int_a^b k(s_0,t)\tilde{x}(t)dt\right| , \text{ say,}$$

$$\ge \|K\|_\infty - \varepsilon/2 .$$

Now, for small h, let $\tilde{x}_h \in C[a,b]$ be constructed so that: (1) $\tilde{x}_h(t) = \tilde{x}(t)$ if $|t-t_j^{(n)}| \ge h$, for $1 \le j \le n$; (2) $\tilde{x}_h(t_j^{(n)}) = 0$ for $1 \le j \le n$; (3) $\tilde{x}_h$ is linear on each of the intervals $[t_j^{(n)}-h, t_j^{(n)}]$ and $[t_j^{(n)}, t_j^{(n)}+h]$. Then $K_n\tilde{x}_h = 0$ and, for sufficiently sufficiently small h, $\|K\tilde{x}_h\|_\infty \ge \|K\|_\infty - \varepsilon$ and $1 + \varepsilon \ge \|\tilde{x}_h\| \ge 1 - \varepsilon$. Set $x = \tilde{x}_h/\|\tilde{x}_h\|$.

9.35 Use Theorem 6.3, p. 191.

9.36 The proof is very similar (but simpler) to that of Lemma 9.9, p. 327.

9.37 $u(-\sqrt{3/5}) = u(+\sqrt{3/5}) \doteq 1.7395$,
$u(0) \doteq 1.9308$.